mathematical programming
with data perturbations

PURE AND APPLIED MATHEMATICS

A Program of Monographs, Textbooks, and Lecture Notes

LECTURE NOTES IN PURE AND APPLIED MATHEMATICS

Additional Volumes in Preparation

mathematical programming with data perturbations

edited by

Anthony V. Fiacco
George Washington University
Washington, D.C.

 **CRC Press**
Taylor & Francis Group
Boca Raton London New York

CRC Press is an imprint of the
Taylor & Francis Group, an **Informa** business

CRC Press
Taylor & Francis Group
6000 Broken Sound Parkway NW, Suite 300
Boca Raton, FL 33487-2742

First issued in hardback 2017

Copyright © 1998 by Taylor & Francis Group, LLC.
CRC Press is an imprint of Taylor & Francis Group, an Informa business

No claim to original U.S. Government works

ISBN 13: 978-1-138-41325-2 (hbk)
ISBN 13: 978-0-8247-0059-1 (pbk)

**Visit the Taylor & Francis Web site at
http://www.taylorandfrancis.com**

**and the CRC Press Web site at
http://www.crcpress.com**

Library of Congress Cataloging-in-Publication Data

Mathematical programming with data perturbations / edited by Anthony V. Fiacco.
    p. cm. — (Lecture notes in pure and applied mathematics ; v. 195)
   Includes bibliographical references and index.
   ISBN 0-8247-0059-7 (pbk.)
   1. Programming (Mathematics) 2. Perturbation (Mathematics) I. Fiacco, Anthony V.
II. Series.
QA402.5.M3558   1997
519.7—dc21

                                         97-35930
                                              CIP

Preface

This volume evolved from a much more modest beginning—an interest in publishing a proceedings of the May 1995 Seventeenth Symposium on Mathematical Programming with Data Perturbation, the 17th conference that I have organized annually at George Washington University. Since I retired from the university on 31 May 1995, after 24 years of service as a professor in the Operations Research Department of the School of Engineering and Applied Science, and since this was to be one of the last such meetings that I would organize, I thought the time for a Symposium-related volume was opportune and submitted a proposal. The publisher then invited me to significantly enlarge the scope of my proposal to include research contributions from authors worldwide. I immediately accepted the more ambitious project, allowing also tutorial expositions, all within the context of mathematical programming with data perturbations. The result is the outstanding collection of papers in this volume, covering a wide spectrum of important topics in the subject area by leading researchers who without exception have conducted cutting-edge research in the respective issues that they address.

I regard this work as my "retirement volume." Much of my professional research effort has been devoted to unifying the incisive and diverse results in sensitivity and stability analysis in mathematical programming, through the annual symposia mentioned and through the publication of books and surveys, particularly edited volumes of multi-authored state-of-the-art contributions. I am grateful to the authors and the publisher for the opportunity to add yet another fine collection of works to the contemporary body of knowledge in this important area.

This book will hopefully contribute to the advancement of the methodology, capturing many of the best known classical results in a modern setting and often as a special case of current results, while introducing several new directions of research. It should serve as a valuable reference for both students and experienced analysts who are looking for authoritative discussions of current results and for new topics of study. Anyone doing serious work in this general area would be well advised to consult this work. Some of the major areas discussed are new characterizations of regularity, the effect of perturbations on the performance of algorithms, the use of approximation techniques to derive optimality and regularity conditions, strong and weak second-order conditions and attendant first- and second-order differential stability results, duality classification using perturbation techniques, embedding and pathfollowing methods, well-posedness and stability, and relations between computational efficiency and data structure and between constraint qualifications and error bounds and regularity. Problem types are finite dimensional and infinite dimensional, including optimal control, linear, nonlinear, and complementarity.

It is appropriate to mention that two other volumes based on the Symposium papers and bearing the same title were published by Marcel Dekker in this same series: Volume 73, 1982, and Volume 85, 1983.

Finally, I wish to thank the Operations Research Department, particularly for the typing assistance of Tessie Abacan, and the School of Engineering and Applied Science of George Washington University for supporting this work. I express my gratitude to Zuhair Nashed and Marcel Dekker, Inc., and especially to the authors for their valuable contributions. A special expression of gratitude is extended to the many highly qualified referees, whose reviews resulted in a significantly enhanced manuscript.

Anthony V. Fiacco

Contents

Contributors

Walter Alt   Institute for Applied Mathematics, University of Jena, Jena, Germany

J. F. Bonnans   INRIA, Rocquencourt, France

Christof Büskens   Institute für Numerische und Instrumentelle Mathematik, Westfälische Wilhelms-Universität Münster, Münster, Germany

S. Dempe   Department of Economics, University of Liepzig, Leipzig, Germany

A. L. Dontchev   Mathematical Reviews, Ann Arbor, Michigan

J. C. Dunn   Mathematics Department, North Carolina State University, Raleigh, North Carolina

Ursula Felgenhauer   Institute of Mathematics, Technical University of Cottbus (BTU), Cottbus, Germany

Sharon Filipowski   The Boeing Company, Seattle, Washington

Jürgen Guddat   Humboldt University, Berlin, Germany

Francisco Guerra   University de las Américas, Puebla, Mexico

Diethard Klatte   Institute für Operations Research, Universität Zürich, Zürich, Switzerland

K. O. Kortanek   Department of Management Sciences, College of Business Administration, The University of Iowa, Iowa City, Iowa

Bernd Kummer   Institut für Mathematik, Humboldt-Universtät zu Berlin, Berlin, Germany

R. Lucchetti   Dipartimento di Matematica, Università di Milano, Milano, Italy

Kazimierz Malanowski   Systems Research Institute, Polish Academy of Sciences, Warszawa, Poland

Helmut Maurer   Institute für Numerische und Instrumentelle Mathematik, Westfälische Wilhelms-Universität Münster, Münster, Germany

Dieter Nowack   Humboldt University, Berlin, Germany

Ryôhei Nozawa   Department of Mathematics, School of Medicine, Sapporo Medical University, Sapporo, Japan

Jean-Paul Penot   Laboratoire de Mathématiques Appliquées, URA, Pau, France

R. T. Rockafellar   Department of Mathematics, University of Washington, Seattle, Washington

Jan-J. Rückmann   Institute for Applied Mathematics, University of Erlangen-Nuremberg, Erlangen, Germany

I. E. Schochetman   Department of Mathematical Sciences, Oakland University, Rochester, Michigan

S. Shiraishi   Faculty of Economics, Toyama University, Toyama, Japan

R. L. Smith   Department of Industrial and Operations Engineering, The University of Michigan, Ann Arbor, Michigan

Klaus Tammer   FB Mathematik, Humboldt-Universität Berlin, Berlin, Germany

S. K. Tsui   Department of Mathematical Sciences, Oakland University, Rochester, Michigan

Doug Ward   Department of Mathematics and Statistics, Miami University, Oxford, Ohio

T. Zolezzi   Dipartimento di Matematica, Università di Genova, Genova, Italy

Discretization and Mesh-Independence of Newton's Method for Generalized Equations

WALTER ALT Institute for Applied Mathematics, University of Jena, D-07740 Jena, Germany

Abstract. This paper investigates local convergence properties of Newton's method for discretized generalized equations. For stable and consistent discretizations it is shown, that the local behaviour of the discretized Newton iterations is asymptotically the same as that for the original iteration and, as a consequence, a mesh-independence principle is derived. The results are applied to the Lagrange-Newton method for optimal control problems.

1 INTRODUCTION

Necessary optimality conditions for nonlinear optimization and optimal control problems define generalized equations, which are the basis for the numerical computation of a local minimizer and an associated Lagrange multiplier by Newton-type methods. We refer for instance to Fiacco [13], Fletcher [14], Robinson [27], Stoer [31], and Kelley/Wright [19]. In these papers the classical implicit-function theorem is the basis to investigate convergence of Newton-type algorithms. Levitin/Polyak [20] extended Newton's method to optimization problems in Hilbert spaces with convex constraints. Machielsen [21] reports numerical results for Newton methods in optimal control.

In [28] Robinson proved an implicit-function theorem, which can be used to show local convergence of Newton type methods for generalized equations. Josephy [17] used this theorem to investigate convergence of Newton's method and quasi-Newton

1

methods for variational inequalities in finite-dimensional spaces (see also Robinson [29]). Applications of Robinson's implicit-function theorem to the Lagrange-
Newton method for infinite optimization problems are given in Alt [2]. Extensions
of the method to nonlinear control problems can be found in Alt/Malanowski [5, 6],
Alt [3], Dontchev et al. [11], and Malanowski [23].

Further generalizations of Newton's methods can be found in Bonnans [7], and
Robinson [30]. Bonnans [7] investigates Newton-type methods for variational inequalities under weaker assumptions than those required by Robinson's implicit-
function theorem. Robinson [30] extends Newton's method to functions having
a so-called point-based approximation. This type of functions includes a class of
nonsmooth functions.

Generalized equations in infinite-dimensional spaces can rarely be solved ana-
lytically. Therefore, one defines suitable dicretizations, and then the discretized
generalized equations are solved in order to obtain approximate solutions of the
original equation. Discretizations of generalized equations, especially equations
arising from Euler's method for optimal control problems, have been investigated
in Dontchev/Hager [9], Dontchev [8], and Malanowski [22] (see also the references
cited in these papers).

For suitable discretizations of generalized equations Newton's method can be
used to solve the discretized equations. One is then interested in the relations
between the infinite process defined by Newton's method applied to the original
equation and the discrete processes defined by Newton's method applied to the
discretized equations. For operator equations such relations have been derived in
Allgower et al. [1]. If a stability and a consistency condition hold, then it is shown
in [1], that the local behaviour of the discretized Newton iterations is asymptotically
the same as that for the original iteration. As a consequence a mesh-independence
principle is derived. The results of [1] are based on the Newton-Kantorovich theorem and a local convergence theorem for Newton's method. Similar results for the
gradient projection method in Hilbert spaces with applications to a special class of
control problems can be found in Kelley/Sachs [18].

The present paper is concerned with discretized Newton methods for generalized equations in infinite-dimensional spaces. In order to extend the results of [1]
to generalized equations we first prove in Section 2 a local convergence theorem
and a semi-local convergence theorem which generalizes the Newton-Kantorovich
theorem. These results are slightly generalized formulations of convergence results
presented in [2], [3], and [5]. In Section 3 we extend the concepts of consistency
and stabiliy of discretizations to generalized equations. For stable and consistent
discretizations we prove existence of solutions to the dicretized equations and we
derive error estimates for these solutions. Further, relations between the infinite
version of the generalized Newton method and the discrete versions are investigated.
Finally, we prove a mesh-independence principle. In Section 4 the results are applied to the Lagrange-Newton method for optimal control problems with convex
control constraints.

Notation: The Fréchet derivative of a map f is denoted by f', the partial Fréchet
derivative with respect to the variable x is denoted by a subscript x, e.g. f_x. By
$B_X(x, r)$ we denote the closed ball with radius r around x in the space X, and 0_X
denotes the zero element of X.   ◇

2 NEWTON'S METHOD FOR GENERALIZED EQUATIONS

Let Z be a Banach spaces, Y a normed space, D an open subset of Z. Further let $F: D \to Y$ be a mapping, and $T: D \to Y$ a multi-valued mapping. Then we consider the problem

$$\text{Find } z \in D \text{ such that } F(z) \in T(z). \tag{2.1}$$

If the mapping F is Fréchet differentiable on D, then Newton's method for operator equations can be extended to the generalized equation (2.1) in the following way.

(GNM): *Choose a starting point* $z_0 \in D$. *Having computed* z_k, *we compute* z_{k+1} *to be the solution of the generalized equation*

$$F(z_k) + F'(z_k)(z - z_k) \in T(z) \,. \tag{2.2}$$

$\diamond$

For this method we prove a local convergence theorem and a semi-local convergence theorem, which generalizes the well-known Newton-Kantorovich theorem. In the next section these convergence results are used to investigate discretizations of (GNM).

As in [27] and [2] we consider the linearized equations (2.2) as perturbations of equation (2.1). Let $w \in Z$. Then we get a family of perturbed generalized equations

$$F(w) + F'(w)(z - w) \in T(z) \tag{2.3}$$

depending on the parameter w. In iteration k of (GNM) we have to solve equation (2.3) for $w = z_k$. For the convergence analysis we use a slightly generalized version of an implicit-function theorem due to Robinson [28]. This theorem requires the concept of strong regularity, which generalizes the assumption of invertibility of $F'(z^*)$ in case of operator equations.

Definition 2.1 Suppose the mapping F is Fréchet differentiable on D. We say that (2.1) is *strongly regular at* $z^* \in D$, if there exist $r_Y(z^*) > 0$ and $c_L(z^*) > 0$, such that for all $y \in B_Y(0_Y, r_Y(z^*))$ the linearized system

$$F(z^*) + F'(z^*)(z - z^*) + y \in T(z) \,, \tag{2.4}$$

has a unique solution $S_L(z^*, y)$, and the mapping $S_L(z^*, \cdot): B_Y(0_Y, r_Y(z^*)) \to Z$ is Lipschitz continuous with modulus $c_L(z^*)$. $\diamond$

Strong regularity of (2.1) at z^* especially requires, that z^* is the unique solution of (2.4) for $y = 0_Y$. Extensions of this concept and applications to optimal control can be found in Hager [15], and Dontchev/Hager [9, 10].

As in case of operator equations the convergence theory of (GNM) requires Lipschitz continuity of F'; more preceisely, we need the following assumption for some point $z^* \in D$.

(A)   The mapping F is Fréchet differentiable on D, and there exist constants $r_1(z^*) > 0$ and $c_F(z^*) > 0$ such that $B_Z(z^*, r_1(z^*)) \subset D$ and

$$\|F'(z_1) - F'(z_2)\|_{Z \to Y} \leq c_F(z^*)\|z_1 - z_2\|_Z$$

for all $z_1, z_2 \in B_Z(z^*, r_1(z^*))$. The multi-valued mapping T has closed graph. $\diamond$

Later on this assumption will be used with $z^* = \tilde{z}$, where $\tilde{z}$ is a solution of (2.1), to prove a local convergence result for (GNM), and with $z^* = z_0$, where z_0 is the starting point for (GNM), in order to generalize the Newton-Kantorovich theorem.

In order to adapt Robinson's implicit-function theorem ([28], Theorem 2.1) for the slightly more general situation considered here, for given $z^* \in D$ we define the function $\ell(z^*; \cdot) \colon D \times D \to Y$ by

$$\ell(z^*; z, w) = F(w) + F'(w)(z - w) - F(z^*) - F'(z^*)(z - z^*). \tag{2.5}$$

The following result will be useful for the convergence analysis of (GNM). For the simple proof we refer to Lemma 3.2 in [2], or Lemma 2.2 in [3].

Lemma 2.2 *Suppose Assumption* (A) *is satisfied. Then the following holds:*

(a) For all $z, w \in B_Z(z^*, r_1(z^*))$,

$$\|\ell(z^*; z, w)\|_Y \le \frac{c_F(z^*)}{2}\|w - z^*\|_Z^2 + c_F(z^*)\|w - z^*\|_Z \|z - z^*\|_Z.$$

(b) Let $r_Y > 0$ *be given and define*

$$r_2(z^*) = \min\left\{ r_1(z^*), \sqrt{\frac{2r_Y}{3c_F(z^*)}} \right\}. \tag{2.6}$$

Then $\|\ell(z^*; z, w)\|_Y \le r_Y$ *for all* $z, w \in B_Z(z^*, r_2(z^*))$.

(c) Let $c^* > 0$ *be given, and define*

$$r_3(z^*) = \min\left\{ r_1(z^*), \frac{2}{3c_F(z^*)c^*} \right\}.$$

Then

$$\|\ell(z^*; z_1, w) - \ell(z^*; z_2, w)\|_Y \le \frac{2}{3c^*}\|z_1 - z_2\|_Z$$

for all $z_1, z_2 \in B_Z(z^*, r_1(z^*))$ *and for all* $w \in B_Z(z^*, r_3(z^*))$. ◇

We can now state a local convergence result for (GNM). Assuming that a solution $\tilde{z}$ of the generalized equation (2.1) exists, we give sufficient conditions such that (GNM) is locally well-defined and that the sequence of iterates z_k, $k = 1, 2, \ldots$, converges quadratically to $\tilde{z}$. We assume that

(A1)   There exists a solution $\tilde{z}$ of (2.1).

(A2)   Assumption (A) is satisfied with $z^* = \tilde{z}$.   ◇

From Lemma 2.2 (a) applied with $z^* = \tilde{z} = z$ we immediately obtain the following result.

Lemma 2.3 *Suppose Assumptions* (A1), (A2) *hold. Then*

$$\|\ell(\tilde{z}; \tilde{z}, w)\|_Y \le \frac{c_F(\tilde{z})}{2}\|w - \tilde{z}\|_Z^2$$

for all $w \in B_Z(\tilde{z}, r_1(\tilde{z}))$. ◇

If Assumption (A2) is satisfied and (2.1) is strongly regular at $\tilde{z}$, then by Lemma 2.2 (b) $\mathcal{S}_L(\tilde{z}, \ell(\tilde{z}; z, w))$ is well-defined for all $z, w \in B_Z(\tilde{z}, r_2(\tilde{z}))$, where $r_2(\tilde{z})$ is defined by (2.6). Hence, for $w \in B_Z(\tilde{z}, r_2(\tilde{z}))$ we can define a mapping

$$\mathcal{S}_w : B_Z(\tilde{z}, r_2(\tilde{z})) \to Z\,, \quad z \mapsto \mathcal{S}_L(\tilde{z}, \ell(\tilde{z}; z, w))\,.$$

Remark 2.4 Since $\ell(\tilde{z}; \tilde{z}, \tilde{z}) = 0_X$, we have $\mathcal{S}_{\tilde{z}}(\tilde{z}) = \mathcal{S}_L(\tilde{z}, 0_Y)$. Therefore, if (2.1) is strongly regular at $\tilde{z}$, then $\tilde{z} = \mathcal{S}_L(\tilde{z}, 0_Y)$, i.e., $\tilde{z}$ is a fixed point of $\mathcal{S}_{\tilde{z}}$. Moreover, z_w is a fixed point of $\mathcal{S}_w$, iff z_w is a solution of (2.3). This fact is used in the proof of the following theorem. For the sequence $\{z_k\}$ computed by (GNM) this implies that z_{k+1} is a fixed point of $\mathcal{S}_{z_k}$.   ◇

Following the argument in the proof of Theorem 2.1 in Robinson [28] we show that for w sufficiently close to $\tilde{z}$ the perturbed generalized equation (2.3) has a unique solution; further we derive an error estimate which can be used to show quadratic convergence of (GNM). A proof of this result for a slightly different situation can be found in Alt [3], Theorem 2.4. For the reader's convenience the proof is included in the appendix.

Theorem 2.5 *Suppose that Assumptions* (A1), (A2) *are satisfied and that* (2.1) *is strongly regular at* $\tilde{z}$*. Let*

$$\rho = \rho(\tilde{z}) = \min\left\{ r_1(\tilde{z}), \sqrt{\frac{2r_Y(\tilde{z})}{3c_F(\tilde{z})}}, \frac{2}{3c_F(\tilde{z})c_L(\tilde{z})} \right\}\,. \tag{2.7}$$

Then there exists a single-valued function $\mathcal{S} \colon B_Z(\tilde{z}, \rho) \to B_Z(\tilde{z}, \rho)$ *such that for each* $w \in B_Z(\tilde{z}, \rho)$*,* $\mathcal{S}(w)$ *is the unique solution in* $B_Z(\tilde{z}, \rho)$ *of* (2.3)*, and*

$$\|\mathcal{S}(w) - \tilde{z}\|_Z \le 3c_L(\tilde{z}) \|\ell(\tilde{z}; \tilde{z}, w)\|_Y \le \frac{3}{2}c_L(\tilde{z})c_F(\tilde{z}) \|w - \tilde{z}\|_Z^2\,. \tag{2.8}$$

◇

Using Theorem 2.5 one can proceed as in case of operator equation to prove the following result on local quadratic convergence of (GNM) (see [2], Theorem 3.3 or [3], Theorem 2.6).

Theorem 2.6 *Suppose that Assumptions* (A1), (A2) *are satisfied and that* (2.1) *is strongly regular at* $\tilde{z}$*. Let* $\rho(\tilde{z})$ *be defined by* (2.7)*. Choose*

$$\tilde{\rho}(\tilde{z}) = \frac{3}{8}c_F(\tilde{z})c_L(\tilde{z})\rho(\tilde{z})^2\,, \quad \delta = \frac{3}{4}c_F(\tilde{z})c_L(\tilde{z})\rho(\tilde{z})\,. \tag{2.9}$$

Then for any starting point $z_0 \in B_Z(\tilde{z}, \tilde{\rho}(\tilde{z}))$ *the generalized Newton method* (GNM) *generates a unique sequence* $\{z_k\}$ *convergent to* $\tilde{z}$*. Moreover,*

$$\|z_{k+1} - \tilde{z}\|_Z \le \frac{3}{2}c_F(\tilde{z})c_L(\tilde{z}) \|z_k - \tilde{z}\|_Z^2 \le \frac{1}{2}\tilde{\rho}(\tilde{z}) \delta^{2^{k+1}-1}\,, \tag{2.10}$$

for $k \ge 1$.   ◇

Remark 2.7 By the definition of $\tilde{\rho}(\tilde{z})$, $\rho(\tilde{z})$, and by (2.10) we obtain

$$\|z_{k+1} - \tilde{z}\|_Z \leq \frac{3}{2}c_L(\tilde{z})c_F(\tilde{z})\tilde{\rho}(\tilde{z})\|z_k - \tilde{z}\|_Z \leq \frac{1}{2}\|z_k - \tilde{z}\|_Z$$

for all $k \in \mathbb{N}$. $\diamond$

In many applications it turns out that the solution $\tilde{z}$ of the generalized equation (2.1) as well as the iterates $\{z_k\}$ have "better smoothness" properties than the elements of Z (compare Section 4 for applications to control problems and Allgower et al. [1] for applications to differential equations). This is a motivation for considering a subset $Z_R \subset Z$ such that

$$\tilde{z} \in Z_R, \quad z_k \in Z_R, \quad z_k - \tilde{z} \in Z_R, \quad z_{k+1} - z_k \in Z_R, \quad k = 0, 1, \dots. \quad (2.11)$$

In order to verify that $\{z_k\}_{k \in \mathbb{N}} \subset Z_R$ if $z_0, \tilde{z} \in Z_R$ we impose the following assumption, which is a strengthened form of strong regularity.

(A3a)   There are closed and convex subsets $\tilde{Z}_R \subset Z_R$ with $\tilde{z} \in \tilde{Z}_R$, $Y_R \subset Y$ with $0_Y \in Y_R$ and constants $r_Y(\tilde{z}) > 0$ and $c_L(\tilde{z}) > 0$, such that for all $y \in Y_R \cap B_Y(0_Y, r_Y(\tilde{z}))$ the linearized system

$$F(\tilde{z}) + F'(\tilde{z})(z - \tilde{z}) + y \in T(z), \quad (2.12)$$

has a unique solution $\mathcal{S}_L(\tilde{z}, y) \in \tilde{Z}_R$, and the mapping $\mathcal{S}_L(\tilde{z}, \cdot): Y_R \cap B_Y(0_Y, r_Y) \to \tilde{Z}_R$ is Lipschitz continuous with modulus $c_L(\tilde{z})$.

(A3b)   There exists $r_R > 0$ such that $\ell(\tilde{z}; z, w) \in Y_R \cap B_Y(0_Y, r_Y(\tilde{z}))$ for all $z, w \in \tilde{Z}_R \cap B_Z(\tilde{z}, r_R)$. $\diamond$

Assumptions of the type (A3) are well-known in stability and sensitivity analysis of optimal control problems (see e.g. Malanowski [24]). From Theorem 2.6 we immediatly obtain

Corollary 2.8 *Suppose that Assumptions (A1)–(A3) are satisfied, and let $\tilde{\rho}(\tilde{z})$ be defined by (2.9). Then for any starting point $z_0 \in \tilde{Z}_R \cap B_Z(\tilde{z}, \tilde{r})$ with $\tilde{r} = \min\{\tilde{\rho}(\tilde{z}), r_R\}$ the generalized Newton method (GNM) generates a unique sequence $\{z_k\} \subset \tilde{Z}_R$ convergent to $\tilde{z} \in \tilde{Z}_R$ satisfying (2.10) for $k \geq 1$.* $\diamond$

Proof.   It can be easily seen from the proof of Theorem 2.5 that (A3) implies $\mathcal{S}(z) \in \tilde{Z}_R$ for $z \in \tilde{Z}_R \cap B_Z(\tilde{z}, \tilde{r})$. Therefore, if $\{z_k\}$ is the sequence defined by (GNM), and $z_0 \in \tilde{Z}_R \cap B_Z(\tilde{z}, \tilde{r})$ then $\{z_k\} \subset \tilde{Z}_R$. $\square$

Remark 2.9 Let the hypothesis of Corollary 2.8 be satisfied, and $Z_R \subset Z$ be any set with the property $Z_R \supset \tilde{Z}_R - \tilde{Z}_R$. Then

$$z_k - \tilde{z} \in Z_R, \quad z_{k+1} - z_k \in Z_R, \quad k = 0, 1, \dots,$$

i.e., (2.11) is satisfied. We will use this fact in Section 4, where the abstract theory is applied to optimal control problems. $\diamond$

Next we derive a result on semi-local convergence of (GNM), which generalizes the well-known Newton-Kantorovich theorem. Similar results have been stated in Levitin/Polyak [20], Theorem 7.1, and in Robinson [29], Theorem 5.1 (see also Josephy [17]). We assume that for the starting point z_0 of (GNM) the following assumptions hold:

(B1) $z_0 \in D$.

(B2)   Assumption (A) is satisfied with $z^* = z_0$.   ◇

Suppose that the generalized equation (2.1) is strongly regular at z_0. Then for all $y \in B_Y(0_Y, r_Y(z_0))$ the linear generalized equation

$$F(z_0) + F'(z_0)(z - z_0) + y \in T(z) \,, \tag{2.13}$$

has a unique solution $\mathcal{S}_L(z_0, y) \in Z$. Especially we denote by $z_1 = \mathcal{S}_L(z_0, 0_Y)$ the unique solution of (2.13) for $y = 0_Y$. Since z_0 is the starting point for (GNM), z_1 is the first iterate. It follows from the definition (2.5) of ℓ that $\ell(z_0; z_1, z_0) = 0_Y$, and hence

$$z_1 = \mathcal{S}_L(z_0, 0_Y) = \mathcal{S}_L(z_0, \ell(z_0; z_1, z_0)) \,. \tag{2.14}$$

By Lemma 2.2 (b), $\mathcal{S}_L(z_0, \ell(z_0; z, w))$ is well-defined for all $z, w \in B_Z(z_0, r_2(z_0))$, where $r_2(z_0)$ is defined by (2.6). Hence, for $w \in B_Z(z_0, r_2(z_0))$ we can define a mapping

$$\mathcal{S}_w \colon B_Z(z_0, r_2(z_0)) \to Z \,, \quad z \mapsto \mathcal{S}_L(z_0, \ell(z_0; z, w)) \,.$$

For $w = z_0$ we obtain from (2.14)

$$\mathcal{S}_{z_0}(z_1) = \mathcal{S}_L(z_0, \ell(z_0; z_1, z_0)) = \mathcal{S}_L(z_0, 0_Y) = z_1 \,,$$

i.e., z_1 is a fixed point of the mapping $\mathcal{S}_{z_0}$. Analogous to Theorem 2.5 one can show that $\mathcal{S}_w$ has a unique fixed point provided that $\|z_1 - z_0\|_Z$ is sufficiently small (see Alt [4], Theorem 2.4).

Theorem 2.10 *Suppose Assumptions* (B1)–(B2) *are satisfied, and let*

$$\rho = \rho(z_0) = \min \left\{ \frac{1}{2} r_2(z_0), \frac{2}{9 c_F(z_0) c_L(z_0)} \right\} \,. \tag{2.15}$$

Assume further that $z_1 \in B_Z(z_0, \rho)$. Then there exists a single-valued function $\mathcal{S} \colon B_Z(z_0, \rho) \to B_Z(z_1, \rho)$ such that for each $w \in B_Z(z_0, \rho)$, $\mathcal{S}(w)$ is the unique fixed point in $B_Z(z_1, \rho)$ of $\mathcal{S}_w$, and

$$\|\mathcal{S}(w) - \mathcal{S}(v)\|_Z \le 3 c_L(z_0) \|\ell(z_0; \mathcal{S}(v), w) - \ell(z_0; \mathcal{S}(v), v)\|_Y \tag{2.16}$$

for all $v, w \in B_Z(z_0, \rho)$.   ◇

Remark 2.11 Let $\{z_k\}$ be the sequence defined by (GNM). It follows from the definition of ℓ that z_{k+1} is the solution of

$$F(z_0) + F'(z_0)(z - z_0) + \ell(z_0; z_{k+1}, z_k) \in T(z) \,.$$

If $\ell(z_0; z_{k+1}, z_k) \in B_Y(0_Y, r_Y(z_0))$ this implies that

$$z_{k+1} = \mathcal{S}_L(z_0, \ell(z_0; z_{k+1}, z_k)) \Leftrightarrow z_{k+1} = \mathcal{S}_{z_k}(z_{k+1}) \,,$$

i.e., z_{k+1} is a fixed point of $\mathcal{S}_{z_k}$, and hence $z_{k+1} = \mathcal{S}(z_k)$. This fact will be used in the proof of Theorem 2.14.   ◇

For the convergence analysis of (GNM) we have to estimate the right hand side of (2.16) in the special case that $w = \mathcal{S}(v)$. To this end, we need the following auxiliary result.

Lemma 2.12 *Suppose Assumptions* (B1)–(B2) *are satisfied. Then*

$$\|\ell(z_0; w, w) - \ell(z_0; w, v)\|_Y \leq \frac{c_F(z_0)}{2}\|w - v\|_Z^2$$

for all $w, v \in B_Z(z_0, r_1(z_0))$.   ◇

Proof. By the definition of ℓ we have

$$\ell(z_0; w, w) - \ell(z_0; w, v) = F(w) - F(v) - F'(v)(w - v)\,.$$

The assertion therefore follows from Assumption (B2).   □

Remark 2.13 Let $\tilde{z} \in B_Z(z_0, \rho(z_0))$. Then by Lemma 2.2 (b)

$$\ell(z_0; \tilde{z}, \tilde{z}) = F(\tilde{z}) - F(z_0) - F'(z_0)(\tilde{z} - z_0) \in B_Y(0_Y, r_Y(z_0))\,.$$

Therefore, $\tilde{z}$ is a fixed point of $\mathcal{S}$, iff

$$\tilde{z} = \mathcal{S}_{\tilde{z}}(\tilde{z}) \Leftrightarrow \tilde{z} = \mathcal{S}_L(z_0, \ell(z_0; \tilde{z}, \tilde{z}))\,.$$

This is equivalent to the fact that $\tilde{z}$ is a solution of

$$F(z_0) + F'(z_0)(z - z_0) + \ell(z_0; \tilde{z}, \tilde{z}) \in T(z)\,,$$

which again is equivalent to

$$F(\tilde{z}) \in T(\tilde{z})\,,$$

i.e., $\tilde{z}$ is a solution of the generalized equation (2.1).   ◇

In order to prove semi-local convergence of (GNM) we now proceed in the same way as in Zeidler [32], Section 5.2 (see also Alt [4], Section 2). We define

$$c_G := 3c_F(z_0)c_L(z_0)\,,$$

and assume that

(B3) $b := \|z_1 - z_0\|_Z \leq \dfrac{3}{2}c_F(z_0)c_L(z_0)\rho(z_0)^2$ ◇

(ball condition) is satisfied. Then $b \leq 2/(9c_G) < 1/(2c_G)$. Therefore, the real quadratic equation

$$h(t) := \frac{c_G}{2}t^2 - t + b = 0$$

has a positive solution

$$t^* = \frac{1 - \sqrt{1 - 2c_Gb}}{c_G}\,.$$

We consider Newton's method for this equation,

$$t_{n+1} = t_n - h'(t_n)^{-1}h(t_n)\,, \quad n = 0, 1, \ldots,$$

with $t_0 = 0$. Then $t_1 = b$,

$$t_0 < t_1 < \ldots < t^* \leq \frac{1}{c_G}, \tag{2.17}$$

and

$$t_{n+1} - t_n = \frac{c_G(t_n - t_{n-1})^2}{2(1 - c_G t_n)}, \quad , n = 1, 2, \ldots. \tag{2.18}$$

As in [32], this method is used as a *majorant method* for (GNM). We define

$$r^* := t^* - b. \tag{2.19}$$

Since $t^* > t_1 = b$, we have $r^* > 0$.

Theorem 2.14 *Suppose that Assumptions* (B1)–(B3) *are satisfied and that* (2.1) *is strongly regular at z_0, where $\rho(z_0)$ is defined by* (2.15). *Let r^* be defined by* (2.19). *Then the generalized Newton method* (GNM) *generates a unique sequence $\{z_k\}$ converging quadratically to a solution $\tilde{z} \in B_Z(z_1, r^*)$ of the generalized equation* (2.1). *This solution is unique on $B_Z(z_0, \rho(z_0))$, and*

$$\|\tilde{z} - z_0\|_Z \leq 2\|z_1 - z_0\|_Z. \tag{2.20}$$

◇

Proof.   Let $\rho = \rho(z_0)$ and define $\delta = \frac{3}{2} c_F(z_0) c_L(z_0) \rho(z_0)$. Then $\delta \leq \frac{1}{3} < 1$, and $b \leq \delta\rho \leq \frac{1}{3}\rho$. We will show by induction on $k = 1, 2, \ldots$, that

$$\|z_{k+1} - z_k\|_Z \leq \rho\, \delta^{2^{k+1}-1}, \quad \|z_{k+1} - z_0\|_Z \leq \tfrac{2}{3}\rho,$$

$$\|z_{k+1} - z_k\|_Z \leq t_{k+1} - t_k, \quad \|z_k - z_1\|_Z \leq r^*, \tag{2.21}$$

for all $k \in \mathbb{N}$. By Assumption (B3), $\|z_1 - z_0\|_Z \leq \delta\rho$. Hence, by Theorem 2.10 and Remark 2.11 $z_2 = \mathcal{S}(z_1) \in B(z_0, \rho(z_0))$ exists and

$$\begin{aligned}
\|z_2 - z_1\|_Z &= \|\mathcal{S}(z_1) - \mathcal{S}(z_0)\|_Z \\
&\leq 3c_L(z_0)\|\ell(z_0; \mathcal{S}(z_0), z_1) - \ell(z_0; \mathcal{S}(z_0), z_0)\|_Y \\
&= 3c_L(z_0)\|\ell(z_0; z_1, z_1) - \ell(z_0; z_1, z_0)\|_Y.
\end{aligned}$$

It then follows from Lemma 2.12 that

$$\|z_2 - z_1\|_Z \leq \frac{3}{2} c_F(z_0) c_L(z_0) \|z_1 - z_0\|_Z^2. \tag{2.22}$$

By the definition of δ this further implies

$$\|z_2 - z_1\|_Z \leq \frac{3}{2} c_F(z_0) c_L(z_0)(\delta\rho)^2 = \rho\delta^3 < \delta\rho \leq \frac{1}{3}\rho,$$

and therefore

$$\|z_2 - z_0\|_Z \leq \|z_2 - z_1\|_Z + \|z_1 - z_0\|_Z \leq \frac{2}{3}\rho.$$

By the definition of c_G we obtain from (2.22)

$$\|z_2 - z_1\|_Z \leq \frac{1}{2} c_G \|z_1 - z_0\|_Z^2 .$$

Since by (2.17) $0 \leq c_G t_n \leq 1$ for all $n \in \mathbb{N}$ we obtain

$$\|z_2 - z_1\|_Z \leq \frac{c_G \|z_1 - z_0\|_Z^2}{2(1 - c_G t_1)} .$$

Since $\|z_1 - z_0\| \leq b = t_1 - t_0$ this implies together with (2.18)

$$\|z_2 - z_1\|_Z \leq \frac{c_G (t_1 - t_0)^2}{2(1 - c_G t_1)} = t_2 - t_1 \leq t^* - t_1 = t^* - b = r^* .$$

This shows that (2.21) holds for $k = 2$. Now suppose (2.21) holds for $k = 1, 2, 3, \ldots, n$. Then we obtain for $k = n$

$$\|z_n - z_1\|_Z \leq \sum_{j=2}^{n} \|z_j - z_{j-1}\|_Z \leq \sum_{j=2}^{n} (t_j - t_{j-1})$$
$$= t_n - t_1 \leq t^* - t_1 = t^* - b = r^* .$$

Since $z_n \in B_Z(z_0, \frac{2}{3}\rho)$, by Theorem 2.10 and Remark 2.11 $z_{n+1} = \mathcal{S}(z_n)$ exists and

$$\begin{aligned}
\|z_{n+1} - z_n\|_Z &= \|\mathcal{S}(z_n) - \mathcal{S}(z_{n-1})\|_Z \\
&\leq 3c_L(z_0)\|\ell(z_0; \mathcal{S}(z_{n-1}), z_n) - \ell(z_0; \mathcal{S}(z_{n-1}), z_{n-1})\|_Y \\
&= 3c_L(z_0)\|\ell(z_0; z_n, z_n) - \ell(z_0; z_n, z_{n-1})\|_Y .
\end{aligned}$$

By Lemma 2.12 it follows that

$$\|z_{n+1} - z_n\|_Z \leq \frac{3}{2} c_F(z_0) c_L(z_0) \|z_n - z_{n-1}\|_Z^2 . \tag{2.23}$$

As for $k = 2$ one then shows that (2.21) holds for $k = n+1$. Next we prove that the sequence $\{z_k\}$ converges to a solution $\tilde{z} \in B_Z(z_1, r^*)$ of the generalized equation (2.1), and that $\|\tilde{z} - z_k\|_Z \leq t^* - t_k$. From (2.21) we get

$$\|z_{k+m} - z_k\|_Z \leq \sum_{j=k+1}^{k+m} \|z_j - z_{j-1}\|_Z \leq t_{k+m} - t_k \leq t^* - t_k . \tag{2.24}$$

Since $t_k \to t^*$ as $k \to \infty$, the sequence $\{z_k\}$ is a Cauchy sequence in $B_Z(z_1, r^*)$ and hence convergent to some $\tilde{z} \in B_Z(z_1, r^*)$. By (2.23) the rate of convergence is quadratic. Since by the definition of (GNM)

$$F(z_k) + F'(z_k)(z_{k+1} - z_k) \in T(z_{k+1}) ,$$

and by Assumption (A) the multi-valued mapping T has closed graph we obtain as $k \to \infty$

$$F(\tilde{z}) \in T(\tilde{z}) ,$$

and by (2.21), $\tilde{z} \in B_Z(z_0, \rho(z_0))$. It follows that $\tilde{z}$ is a solution of the generalized equation (2.1). To show uniqueness of $\tilde{z}$ let $\tilde{y} \in B_Z(z_0, \rho(z_0))$ be a solution of the generalized equation (2.1). We show that $\tilde{y} = \tilde{z}$. By Remark 2.13 we have $\tilde{z} = \mathcal{S}(\tilde{z})$, $\tilde{y} = \mathcal{S}(\tilde{y})$. Therefore by Theorem 2.10

$$
\begin{aligned}
\|\tilde{y} - \tilde{z}\|_Z &= \|\mathcal{S}(\tilde{y}) - \mathcal{S}(\tilde{z})\|_Z \\
&\leq 3c_L(z_0)\|\ell(z_0; \mathcal{S}(\tilde{y}), \tilde{z}) - \ell(z_0; \mathcal{S}(\tilde{y}), \tilde{y})\|_Y \\
&= 3c_L(z_0)\|\ell(z_0; \tilde{y}, \tilde{z}) - \ell(z_0; \tilde{y}, \tilde{y})\|_Y .
\end{aligned}
$$

Since

$$
\|\tilde{y} - \tilde{z}\|_Z \leq 2\rho(z_0) \leq \frac{4}{9c_F(z_0)c_L(z_0)} ,
$$

we obtain together with Lemma 2.12

$$
\|\tilde{y} - \tilde{z}\|_Z \leq \frac{3}{2}c_F(z_0)c_L(z_0)\|\tilde{y} - \tilde{z}\|_Z^2 \leq \frac{2}{3}\|\tilde{y} - \tilde{z}\|_Z ,
$$

and therefore $\tilde{y} = \tilde{z}$. Finally, from

$$
\|\tilde{z} - z_0\|_Z \leq \|\tilde{z} - z_1\|_Z + \|z_1 - z_0\|_Z \leq t^* - b + b = t^* ,
$$

and

$$
t^* = \frac{2c_G b}{c_G(1 - \sqrt{1 - 2c_G b})} \leq 2b ,
$$

we obtain (2.20). $\qquad\square$

3 DISCRETIZATION

Since the formal procedure (GNM) can rarely be executed in infinite-dimensional spaces, (2.1) is replaced in practice by a family of discretized equations

$$
z \in D_N , \quad F_N(z) \in T_N(z) , \tag{3.1}
$$

indexed by $N \in \mathbb{N}$, $N \geq \tilde{N}$ for some $\tilde{N} \in \mathbb{N}$. Here Z_N, Y_N are finite-dimensional spaces, D_N is an open subset of Z_N, $F_N: D_N \to Y_N$ is a mapping, and $T_N: D_N \to Y_N$ a multi-valued mapping. Similar to the discretization methods for operator equations investigated in Allgower et al. [1], the discretization methods to be considered here will be described by

$$
(F_N, T_N, h_N, \Delta_N, \hat{\Delta}_N) , \quad N \geq \tilde{N} , \tag{3.2}
$$

where $\{h_N\}$ is a sequence of mesh sizes with

$$
\lim_{N \to \infty} h_N = 0 , \tag{3.3}
$$

and $\Delta_N: Z \to Z_N$, $\hat{\Delta}_N: Y \to Y_N$ are bounded linear discretization operators.

Applying (GNM) to the discrete generalized equations (3.1) we obtain the following discrete process:

$(\text{GNM})_N$: *Choose a starting point* $z_{0,N} \in D_N$. *Having computed* $z_{k,N}$, *we compute* $z_{k+1,N}$ *to be the solution of the generalized equation*

$$F_N(z_{k,N}) + F_N'(z_{k,N})(z - z_{k,N}) \in T_N(z). \qquad (3.4)$$

$\diamond$

We shall investigate convergence of these discrete processes and their relations to the infinite process (GNM). To this end we use the following assumptions for the discretized generalized equations, where $Z_R \subset Z$, and $Y_R \subset Y$ (compare Assumption (A3)). These assumptions are similar to those used for operator equations in [1]).

(D1)   The mappings F_N are Fréchet differentiable on D_N, the multi-valued mappings T_N have closed graph, and there exists $r_0 > 0$ such that

$$\Delta_N (B_Z(\tilde{z}, r_0) \cap Z_R) \subset D_N, \quad N \geq \tilde{N}.$$

(D2)   The discretization (3.2) is *Lipschitz uniform*, i.e., there exist constants $r_1 > 0$ and $c_F > 0$ such that

$$B_{Z_N}(\Delta_N(\tilde{z}), r_1) \subset D_N, \quad N \geq \tilde{N},$$

and

$$\|F_N'(z_1) - F_N'(z_2)\|_{Z_N \to Y_N} \leq c_F \|z_1 - z_2\|_{Z_N}$$

for all $z_1, z_2 \in B_{Z_N}(\Delta_N(\tilde{z}), r_1)$, and for all $N \geq \tilde{N}$.

(D3)   The discretization (3.2) is *bounded*, i.e., there exists $c_B > 0$ such that

$$\|\Delta_N z\|_{Z_N} \leq c_B \|z\|,,$$

for all $z \in Z_R$, and for all $N \geq \tilde{N}$.

(D4)   The discretization (3.2) is *stable*, i.e., the generalized equations (3.1), $N \geq \tilde{N}$, are *uniformly strongly regular* at $\Delta_N \tilde{z}$, which requires that for $N \geq \tilde{N}$ the following holds with the constants r_Y, c_L from (D1): For each $y \in B_{Y_N}(0, r_Y)$ the linearized system

$$F_N(\Delta_N \tilde{z}) + F_N'(\Delta_N \tilde{z})(z - \Delta_N \tilde{z}) + y \in T_N(z),$$

has a unique solution $\mathcal{S}_{L,N}(\Delta_N \tilde{z}, y)$, and $\mathcal{S}_{L,N}(\Delta_N \tilde{z}, \cdot) : B_{Y_N}(0, r_Y) \to Z_N$ is Lipschitz continuous with modulus c_L.

(D5)   The discretization (3.2) is *consistent* of order p, i.e., there are constants $c_0, c_1 > 0$ such that

$$\|\hat{\Delta}_N F(z) - F_N(\Delta_N z)\|_{Y_N} \leq c_0 h_N^p$$

for all $z \in Z_R \cap B_Z(\tilde{z}, r_1)$, $N \geq \tilde{N}$, and

$$\|\hat{\Delta}_N (F'(u)v) - F_N'(\Delta_N u)(\Delta_N v)\|_{Y_N} \leq c_1 h_N^p \|v\|_Z$$

for all $u \in Z_R \cap B_Z(\tilde{z}, r_1)$, $v \in Z_R$, $N \geq \tilde{N}$. $\diamond$

Remark 3.1 In case of operator equations stability requires that the linearized equation in (D4) has a unique solution for any $y \in Y_N$. In the more general context considered here this property is required only for $y \in B_{Y_N}(0, r_Y)$, where r_Y is independent of N. In some applications however we can choose $r_Y = +\infty$ (see Section 4).

(D5) is the usual definition of consistency for operator equations (see e.g. Allgower et al. [1]). For generalized equations we need the additional assumption (D6) stated below. This assumption is always satisfied for stable and consistent operator equations. $\diamond$

(D6) There exist constants $c_2, c_3 > 0$ such that the following holds: If $y \in Y_R \cap B_Y(0, r_Y)$ and $\bar{z} \in Z_R$ is the solution of the linear generalized equation (2.12), then for each $N \geq \tilde{N}$ there exist $z_N \in Z_N$ and $y_N \in Y_N$ such that

$$\|z_N - \Delta_N \bar{z}\|_{Z_N} \leq c_2 h_N^p, \quad \|y_N - \hat{\Delta}_N y\|_{Y_N} \leq c_3 h_N^p,$$

and z_N is the solution of the linear generalized equation

$$F_N(\Delta_N \tilde{z}) + F_N'(\Delta_N \tilde{z})(z - \Delta_N \tilde{z}) + y_N \in T_N(z). \tag{3.5}$$

$\diamond$

Using Theorem 2.14 on semi-local convergence of (GNM) we first prove existence of solutions of the discretized generalized equations (3.1) and error estimates for stable and consistent discretizations.

Theorem 3.2 *Let $\tilde{z} \in Z_R \subset Z$ be a solution of (2.1), and let a discretization be defined by (3.2) which satisfies Assumptions* (D1), (D2), (D4), *and* (D6) *with $Y_R = \{0_Y\}$. Then there exists $N_1 \geq \tilde{N}$, such that for $N \geq N_1$ the generalized equation (3.1) has a locally unique solution $\tilde{z}_N$, and*

$$\|\tilde{z}_N - \Delta_N \tilde{z}\|_{Z_N} \leq 2(c_2 + c_L c_3)h_N^p. \tag{3.6}$$

$\diamond$

Proof. Let $N \geq \tilde{N}$. We apply Theorem 2.14 to (3.1) with the starting point $z_{0,N} = \Delta_N \tilde{z}$. The seqence generated by (GNM) is denoted by $\{z_{k,N}\}$. By (D1) and (D2) Assumptions (B1) and (B2) are satisfied. By Assumption (D4), (3.1) is strongly regular at $z_{0,N}$. Since $\tilde{z}$ is the solution of the linear generalized equation

$$F(\tilde{z}) + F'(\tilde{z})(z - \tilde{z}) + 0_Y \in T(z),$$

and $\hat{\Delta}_N 0 = 0$, (D6) with $Y_R = \{0_Y\}$ implies that there exist $z_N \in Z_N$ and $y_N \in Y_N$ such that

$$\|z_N - \Delta_N z\|_{Z_N} \leq c_2 h_N^p, \quad \|y_N\|_{Y_N} \leq c_3 h_N^p, \tag{3.7}$$

and z_N is the solution of the linear generalized equation

$$F_N(\Delta_N \tilde{z}) + F_N'(\Delta_N \tilde{z})(z - \Delta_N \tilde{z}) + y_N \in T_N(z).$$

Since by the definition of (GNM), $z_{1,N}$ is the solution of

$$F_N(\Delta_N \tilde{z}) + F_N'(\Delta_N \tilde{z})(z - \Delta_N \tilde{z}) + 0 \in T_N(z).$$

it follows by (D4) and (3.7) that if $c_3 h_N^p \leq r_Y$ then

$$\|z_{1,N} - z_N\|_{Z_N} = \|\mathcal{S}_{L,N}(\Delta_N \tilde{z}; 0) - \mathcal{S}_{L,N}(\Delta_N \tilde{z}; y_N)\|_{Z_N} \leq c_L \|y\|_{Y_N} \leq c_L c_3 h_N^p .$$

By (3.7) we further obtain

$$\|z_{1,N} - \Delta_N \tilde{z}\|_{Z_N} \leq \|z_{1,N} - z_N\|_{Z_N} + \|z_N - \Delta_N \tilde{z}\|_{Z_N} \leq (c_2 + c_L c_3) h_N^p . \quad (3.8)$$

In order to satisfy Assumption (B3) we must have

$$b_N = \|z_{1,N} - \Delta_N \tilde{z}\|_{Z_N} \leq \frac{3}{2} c_F c_L \rho^2 ,$$

where

$$\rho = \min \left\{ \frac{1}{2} r_1, \frac{1}{2} \sqrt{\frac{2 r_Y}{3 c_F}}, \frac{2}{9 c_F c_L} \right\} .$$

Hence, by (3.8), (B3) is satisfied, if

$$h_N^p \leq \frac{3 c_F c_L \rho^2}{2(c_2 + c_L c_3)} , \quad c_3 h_N^p \leq r_Y . \quad (3.9)$$

By (3.3) there exists $N_1 \geq \tilde{N}$ such that for $N \geq N_1$, (3.9) holds. Therefore, if $N \geq N_1$, Theorem 2.14 implies the existence of a solution $\tilde{z}_N$ of (3.1). This solution is unique on $B_{Z_N}(\Delta_N \tilde{z}, \rho)$, and

$$\|\tilde{z}_N - \Delta_N \tilde{z}\|_{Z_N} \leq 2 \|z_{1,N} - \Delta_N \tilde{z}\|_{Z_N} .$$

Together with (3.8) this implies (3.6). $\qquad \square$

Based on the local convergence result for (GNM) stated in Theorem 2.6 we now investigate relations between the infinite process (GNM) and the discrete processes $(\text{GNM})_N$. For stable and consistent discretizations we show, that the local behaviour of the discrete Newton iterations is asymptotically the same as that for the original iteration.

Theorem 3.3 *Let the hypothesis of Theorem 2.6 be satisfied, and let $Z_R \subset Z$ be such that (2.11) is satisfied for $z_0 \in B_Z(\tilde{z}, \tilde{\rho}(\tilde{z}))$, and $Y_R \subset Y$ such that (A3) is satisfied. Further let a discretization be defined by (3.2) which satisfies Assumptions (D1)–(D6). Then there exists $N_2 \geq N_1$, $\rho_2 > 0$ such that the sequence $\{z_{N,k}\}_{k \in \mathbb{N}}$ generated by $(\text{GNM})_N$ with starting point $\Delta_N z_0$ converges to $\tilde{z}_N$, and that*

$$\|z_{k,N} - \Delta_N z_k\|_{Z_N} \leq \tilde{c}_1 h_N^p , \quad k = 0, 1, \dots , \quad (3.10)$$

for all $N \geq N_2$ and all $z_0 \in Z_R \cap B_Z(\tilde{z}, \rho_2)$. $\qquad \diamond$

Proof. Without loss of generality we may assume $r_1(\tilde{z}) = r_1$, $c_F(\tilde{z}) = c_F$, $c_L(\tilde{z}) = c_L$, and $r_Y(\tilde{z}) = r_Y$. Let $N \geq N_1$. By Theorem 3.2 the generalized equation (3.1) has a locally unique solution $\tilde{z}_N$. We define

$$\rho = \min \left\{ r_1, \sqrt{\frac{2 r_Y}{3 c_F}}, \frac{2}{3 c_F c_L} \right\} , \quad \tilde{\rho} = \frac{3}{8} c_F c_L \rho^2 , \quad \rho_1 = \min \left\{ \frac{\tilde{\rho}}{2}, \frac{\tilde{\rho}}{2 c_B} \right\} .$$

By Theorem 2.6 the sequence $\{z_{N,k}\}_{k\in\mathbb{N}}$ generated by $(GNM)_N$ with starting point $\Delta_N z_0$ converges to $\tilde{z}_N$, if

$$\|\Delta_N z_0 - \tilde{z}_N\|_Z \leq \tilde{\rho}. \tag{3.11}$$

By (3.3) there exists $M_1 \geq N_1$, such that

$$2(c_2 + c_L c_3)h_N^p \leq \rho_1 \quad \forall N \geq M_1.$$

Then by (D3) and Theorem 3.2 we obtain

$$\|\Delta_N z_0 - \tilde{z}_N\|_Z \leq \|\Delta_N z_0 - \Delta_N \tilde{z}\|_Z + \|\Delta_N \tilde{z} - \tilde{z}_N\|_Z$$
$$\leq c_B \|z_0 - \tilde{z}\|_Z + 2(c_2 + c_L c_3)h_N^p \leq \tilde{\rho}.$$

Therefore, (3.11) is satisfied, if $z_0 \in B_Z(\tilde{z}, \rho_1)$, and if $N \geq M_1$. In order to prove (3.10), we define

$$\rho_2 = \min\{\,\rho_1, \frac{1}{18 c_F c_L c_B}\,\}, \tag{3.12}$$

and

$$\tilde{c}_1 = 24 c_L(c_0 + c_1 \rho_2) + 4(c_2 + c_L c_3). \tag{3.13}$$

Further, we choose $N_2 \geq M_1$ such that

$$c_F \tilde{c}_1 h_N^p \leq \frac{1}{6 c_L} \quad \text{for } N \geq N_2. \tag{3.14}$$

Now let $N \geq N_2$ and $z_0 \in B_Z(\tilde{z}, \rho_2)$ be given. For $k = 0$ we have $z_{0,N} = \Delta_N(z_0)$. Hence (3.10) is satisfied. Suppose that (3.10) holds for $k = 0, 1, \ldots, n$. By the definition of $(GNM)_N$, $z_{n+1,N}$ is the solution of

$$F_N(\Delta_N \tilde{z}) + F_N'(\Delta_N \tilde{z})(z - \Delta_N \tilde{z}) + v_N \in T_N(z), \tag{3.15}$$

where

$$v_N = F_N(z_{n,N}) + F_N'(z_{n,N})(z_{n+1,N} - z_{n,N})$$
$$- F_N(\Delta_N \tilde{z}) - F_N'(\Delta_N \tilde{z})(z_{n+1,N} - \Delta_N \tilde{z}).$$

By the definition of (GNM), z_{n+1} is the solution of

$$F(\tilde{z}) + F'(\tilde{z})(z - \tilde{z}) + y \in T(z),$$

where

$$y = F(z_n) + F'(z_n)(z_{n+1} - z_n) - F(\tilde{z}) - F'(\tilde{z})(z_{n+1} - \tilde{z}) = \ell(\tilde{z}; z_{n+1}, z_n) \in Y_R.$$

By Assumption (D6) there exist $w_N \in Z_N$ and $y_N \in Y_N$ such that

$$\|w_N - \Delta_N z_{n+1}\|_{Z_N} \leq c_2 h_N^p, \quad \|y_N - \hat{\Delta}_N y\|_{Y_N} \leq c_3 h_N^p,$$

and w_N is the solution of

$$F_N(\Delta_N \tilde{z}) + F_N'(\Delta_N \tilde{z})(z - \Delta_N \tilde{z}) + y_N \in T_N(z). \tag{3.16}$$

Thus, we obtain

$$\|z_{n+1,N} - \Delta_N z_{n+1}\|_{Z_N} \leq \|z_{n+1,N} - w_N\|_{Z_N} + \|w_N - \Delta_N z_{n+1}\|_{Z_N}$$
$$\leq \|z_{n+1,N} - w_N\|_{Z_N} + c_2 h_N^p \,.$$

Since $z_{n+1,N}$ is the solution of (3.15) and w_N is the solution of (3.16), it follows from Assumption (D4) that

$$\|z_{n+1,N} - w_N\|_{Z_N} \leq c_L \|v_N - y_N\|_{Y_N}$$
$$\leq c_L \|v_N - \hat{\Delta}_N y\|_{Y_N} + c_L \|\hat{\Delta}_N y - y_N\|_{Y_N}$$
$$\leq c_L \|v_N - \hat{\Delta}_N y\|_{Y_N} + c_L c_3 h_N^p \,.$$

Therefore, we have

$$\|z_{n+1,N} - \Delta_N z_{n+1}\|_{Z_N} \leq c_L \|v_N - \hat{\Delta}_N y\|_{Y_N} + (c_2 + c_L c_3) h_N^p \,. \tag{3.17}$$

A simple calculation shows that

$$v_N - \hat{\Delta}_N y = E_1 + E_2 + E_3 + E_4 \,, \tag{3.18}$$

where

$$E_1 = F_N(z_{n,N}) + F_N'(z_{n,N})(\Delta_N z_n - z_{n,N}) - F_N(\Delta_N z_n) \,,$$
$$E_2 = F_N(\Delta_N z_n) - \hat{\Delta}_N F(z_n) - F_N(\Delta_N \tilde{z}) + \hat{\Delta}_N F(\tilde{z})$$
$$+ F_N'(\Delta_N z_n)(\Delta_N z_{n+1} - \Delta_N z_n) - \hat{\Delta}_N F'(z_n)(z_{n+1} - z_n)$$
$$- F_N'(\Delta_N \tilde{z})(\Delta_N z_{n+1} - \Delta_N \tilde{z}) + \hat{\Delta}_N F'(\tilde{z})(z_{n+1} - \tilde{z}) \,,$$
$$E_3 = [F_N'(z_{n,N}) - F_N'(\Delta_N z_n)](\Delta_N z_{n+1} - \Delta_N z_n) \,,$$
$$E_4 = [F_N'(z_{n,N}) - F_N'(\Delta_N \tilde{z})](z_{n+1,N} - \Delta_N z_{n+1}) \,.$$

From this and Assumptions (D1), (D2) we obtain the estimate

$$\|E_1\|_{Y_N} \leq \frac{1}{2} c_F \|\Delta_N z_n - z_{n,N}\|_{Z_N}^2$$

By the induction assumption and (3.14) it follows then that

$$\|E_1\|_{Y_N} \leq \frac{c_F}{2} \left(\tilde{c}_1 h_N^p \right)^2 \leq \frac{1}{12 c_L} \tilde{c}_1 h_N^p \,.$$

By Assumption (D5) we have

$$\|E_2\|_{Y_N} \leq 2 c_0 h_N^p + c_1 h_N^p \|z_{n+1} - z_n\|_Z + c_1 h_N^p \|z_{n+1} - \tilde{z}\|_Z \,.$$

Since by Remark 2.7,

$$\|z_{n+1} - \tilde{z}\|_Z \leq \frac{1}{2} \|z_0 - \tilde{z}\|_Z \leq \frac{1}{2} \rho_2 \,,$$

and

$$\|z_{n+1} - z_n\|_Z \leq \|z_{n+1} - \tilde{z}\|_Z + \|z_n - \tilde{z}\|_Z \leq \frac{3}{2} \|z_0 - \tilde{z}\|_Z \leq \frac{3}{2} \rho_2 \,, \tag{3.19}$$

we obtain

$$\|E_2\|_{Y_N} \le 2(c_0 + c_1\rho_2)h_N^p \,.$$

By (3.13) this implies

$$\|E_2\|_{Y_N} \le \frac{1}{12c_L}\tilde{c}_1 h_N^p \,.$$

By Assumptions (D1), (D2) we have

$$\|E_3\|_{Y_N} \le c_F\|z_{n,N} - \Delta_N z_n\|_{Z_N}\|\Delta_N z_{n+1} - \Delta_N z_n\|_{Z_N} \,.$$

Using the induction assumption and (D3) we obtain

$$\|E_3\|_{Y_N} \le c_F\tilde{c}_1 h_N^p c_B\|z_{n+1} - z_n\|_Z \,.$$

By (3.19) and (3.12) this implies

$$\|E_3\|_{Y_N} \le \frac{3}{2}c_F\tilde{c}_1 h_N^p c_B\rho_2 \le \frac{1}{12c_L}\tilde{c}_1 h_N^p \,.$$

By Assumptions (D1), (D2) we have

$$\|E_4\|_{Y_N} \le c_F\|z_{n,N} - \Delta_N\tilde{z}\|_{Z_N}\|z_{n+1,N} - \Delta_N z_{n+1}\|_{Z_N} \,.$$

Using the induction assumption, (D3), and Remark 2.7 (compare (3.19)) we obtain

$$\begin{aligned}
\|z_{n,N} - \Delta_N\tilde{z}\|_{Z_N} &\le \|z_{n,N} - \Delta_N z_n\|_{Z_N} + \|\Delta_N z_n - \Delta_N\tilde{z}\|_{Z_N} \\
&\le \tilde{c}_1 h_N^p + c_B\|z_n - \tilde{z}\|_Z \le \tilde{c}_1 h_N^p + c_B\rho_2 \,.
\end{aligned} \tag{3.20}$$

Hence, E_4 can be estimated by

$$\|E_4\|_{Y_N} \le c_F\left(\tilde{c}_1 h_N^p + c_B\rho_2\right)\|z_{n+1,N} - \Delta_N z_{n+1}\|_{Z_N} \,.$$

By (3.12) and (3.14) this implies

$$\begin{aligned}
\|E_4\|_{Y_N} &\le \left(\frac{1}{6c_L} + \frac{1}{18c_L}\right)\|z_{n+1,N} - \Delta_N z_{n+1}\|_{Z_N} \\
&\le \frac{1}{2c_L}\|z_{n+1,N} - \Delta_N z_{n+1}\|_{Z_N} \,,
\end{aligned}$$

From (3.17), (3.18) and from the estimates for E_i, $i = 1,\ldots,4$, we finally obtain

$$\begin{aligned}
\|z_{n+1,N} &- \Delta_N z_{n+1}\|_{Z_N} \\
&\le c_L\left(\|E_1\| + \|E_2\| + \|E_3\| + \|E_4\|\right) + (c_2 + c_L c_3)h_N^p \\
&\le \frac{1}{4}\tilde{c}_1 h_N^p + \frac{1}{2}\|z_{n+1,N} - \Delta_N z_{n+1}\|_{Z_N} + (c_2 + c_L c_3)h_N^p \,.
\end{aligned}$$

which implies

$$\|z_{n+1,N} - \Delta_N z_{n+1}\|_{Z_N} \le \frac{1}{2}\tilde{c}_1 h_N^p + 2(c_2 + c_L c_3)h_N^p \,.$$

This completes the induction. □

In view of the mesh-independence principle we need some additional estimates which are easily obtained from Theorem 3.3.

Corollary 3.4 *Let the hypothesis of Theorem 3.3 be satisfied. Then there are constants $\tilde{c}_2$, $\tilde{c}_3$ such that*

$$\|F_N(z_{k,N}) - \hat{\Delta}_N F(z_k)\|_{Y_N} \leq \tilde{c}_2 h_N^p, \quad k = 0, 1, \dots, \tag{3.21}$$

and

$$\|z_{k,N} - \tilde{z}_N - \Delta_N(z_k - \tilde{z})\|_{Z_N} \leq \tilde{c}_3 h_N^p, \quad k = 0, 1, \dots, \tag{3.22}$$

for all $N \geq N_2$ and all $z_0 \in Z_R \cap B_Z(\tilde{z}, \rho_2)$. $\qquad\qquad\qquad\diamond$

Proof. By Assumptions (D2) there exists $\tilde{c}_F$ such that

$$\|F'(z)\|_{Z_n \to Y_N} \leq \tilde{c}_F \quad \text{for all } z \in B_{Z_N}(\Delta_N \tilde{z}, r_1).$$

Together with (D5) we therefore obtain

$$\|F_N(z_{k,N}) - \hat{\Delta}_N F(z_k)\|_{Y_N}$$
$$\leq \|F_N(z_{k,N}) - F_N(\Delta_N z_k)\|_{Y_N} + \|F_N(\Delta_N z_k) - \hat{\Delta}_N F(z_k)\|_{Y_N}$$
$$\leq \tilde{c}_F \|z_{k,N} - \Delta_N z_k\|_{Z_N} + c_0 h_N^p.$$

By Theorem 3.3 this implies (3.21). Since

$$\|z_{k,N} - \tilde{z}_N - \Delta_N(z_k - \tilde{z})\|_{Z_N} \leq \|z_{k,N} - \Delta_N z_k\|_{Z_N} + \|\tilde{z}_N - \Delta_N \tilde{z}\|_{Z_N},$$

inequality (3.22) follows from Theorem 3.3 and Theorem 3.2. $\qquad\qquad\qquad\square$

As a consequence of the preceeding results we can now prove a mesh-independence principle, which states that for sufficiently large N there is at most a difference of one between the number of iteration steps required by the two processes (GNM) and $(\text{GNM})_N$ to converge to within a given tolerance $\varepsilon > 0$. The proof is a slight modification of the proof of Corollary 1 in Allgower et al. [1].

Theorem 3.5 *Suppose that the hypothesis of Theorem 3.3 hold and that there is a constant $c_D > 0$ for which*

$$\liminf_{N \geq N_1} \|\Delta_N z\|_{Z_N} \geq 2c_D \|z\|_Z \quad \text{for each } z \in Z_R. \tag{3.23}$$

Then for some $\rho_3 \in]0, \rho_2]$, and for any fixed $\varepsilon > 0$ and $z_0 \in B_Z(\tilde{z}, \rho_3)$ there is a $N_3 = N_3(z_0, \varepsilon)$ such that

$$\left| \min\{k \geq 0, \|z_k - \tilde{z}\|_Z < \varepsilon\} - \min\{k \geq 0, \|z_{k,N} - \tilde{z}_N\|_{Z_N} < \varepsilon\} \right| \leq 1 \tag{3.24}$$

for all $N \geq N_3$. $\qquad\qquad\qquad\diamond$

Proof. Let i be the unique integer defined by

$$\|z_{i+1} - \tilde{z}\|_Z < \varepsilon \leq \|z_i - \tilde{z}\|_Z \tag{3.25}$$

(compare Remark 2.7). By (3.23) there exists $M \geq N_2$ such that

$$\|\Delta_N(z_i - \tilde{z})\|_{Z_N} \geq c_D \|z_i - \tilde{z}\|_Z \qquad (3.26)$$

for $N \geq M$. We choose $N_3 \geq M$ such that

$$\max\{\tilde{c}_2, 2\tilde{c}_3\} h_N^p \leq c_B \varepsilon, \quad \max\left\{3c_F c_L c_B \tilde{c}_1, \frac{3c_F c_L \tilde{c}_1}{2c_D}\right\} h_N^p \leq \frac{1}{4} \qquad (3.27)$$

for $N \geq N_3$, and $0 < \rho_2 \leq \rho_2$ such that

$$\max\left\{3c_F c_L c_B^2, \frac{3c_F c_L c_B}{2c_D}\right\} \rho_3 \leq \frac{1}{4}. \qquad (3.28)$$

By (3.22), Assumption (D3), and (3.27)

$$\|z_{i+1,N} - \tilde{z}_N\| \lesssim \|\Delta_N(z_{i+1} - \tilde{z})\|_{Z_N} + \tilde{c}_3 h_N^p \leq c_B \varepsilon + \tilde{c}_3 h_N^p \leq 2c_B \varepsilon.$$

By Theorem 2.6 and (3.20) it then follows that

$$\|z_{i+2,N} - \tilde{z}_N\|_{Z_N} \leq \frac{3}{2} c_F c_L \|z_{i+1,N} - \tilde{z}_N\|_{Z_N}^2 \leq \frac{3}{2} c_F c_L (\tilde{c}_1 h_N^p + c_B \rho_2) 2 c_B \varepsilon,$$

and from (3.27), (3.28) we obtain

$$\|z_{i+2,N} - \tilde{z}_N\|_{Z_N} \leq \frac{1}{2}\varepsilon < \varepsilon. \qquad (3.29)$$

Because of (3.25), (3.25), and (3.22) we have

$$\varepsilon \leq \|z_i - \tilde{z}\|_Z \leq \frac{1}{c_D}\|\Delta_N(z_i - \tilde{z})\|_{Z_N} \leq \frac{1}{c_D}\|z_{i,N} - \tilde{z}_N\|_{Z_N} + \tilde{c}_3 h_N^p,$$

or, using (3.27),

$$\|z_{i,N} - \tilde{z}_N\|_{Z_N} \geq c_D \varepsilon - \tilde{c}_3 h_N^p \geq c_D \varepsilon - \frac{c_D}{2}\varepsilon = \frac{c_D}{2}\varepsilon. \qquad (3.30)$$

If $\|z_{i-1,N} - \tilde{z}_N\|_{Z_N} < \varepsilon$ then by (3.27) and (3.28) we obtain analogously to (3.29)

$$\|z_{i,N} - \tilde{z}_N\|_{Z_N} \leq \frac{c_D}{2}\varepsilon,$$

which contradicts (3.30). Therefore, we must have

$$\|z_{i-1,N} - \tilde{z}_N\|_{Z_N} \geq \varepsilon, \qquad (3.31)$$

and it is easily seen that (3.25), (3.29) and (3.31) imply (3.24). $\qquad \square$

As in case of operator equations, condition (3.23) is an immediate consequence of the convergence condition

$$\lim_{N \to \infty} \|\Delta_N z\|_{Z_N} = \|z\|_Z \quad \text{for each } z \in Z_R.$$

Moreover, for some discretizations we have

$$\lim_{N \to \infty} \|\Delta_N z\|_{Z_N} = \|z\|_Z \quad \text{uniformly for } z \in Z_R. \qquad (3.32)$$

(see the application in the next section). In such cases the following stronger formulation of the mesh-independence principle applies, where N_3 is independent of the starting point (compare Allgower et al. [1], Corollary 2).

Corollary 3.6 *Suppose that the hypothesis of Theorem 3.5 is satisfied and that (3.32) holds. Then there exists a constant $\rho_3 \in\,]0, \rho_2]$, and, for any fixed $\varepsilon > 0$ there exists some $N_3 = N_3(\varepsilon)$, such that (3.24) holds for all $N \geq N_3$ and all starting points $z_0 \in B_Z(\tilde{z}, \rho_3)$.* $\diamond$

4 APPLICATION TO OPTIMAL CONTROL PROBLEMS

In this section we give an application of the results of the previous section to a class of nonlinear optimal control problems and their discretizations by Euler's method. Most of the results needed to verify Assumptions (A1)–(A3) and (D1)–(D6) can be found in Hager [15] and Dontchev/Hager [9].

We consider the following optimal control problem:

$$(O) \qquad \text{Min}_{(x,u)} \quad J(x,u) = \int_0^T g(x(t), u(t))\, dt$$

subject to

$$\dot{x}(t) = f(x(t), u(t)) \quad \forall' t \in [0, T]\,,$$
$$x(0) = a\,,$$
$$u(t) \in U \qquad\qquad \forall' t \in [0, T]\,,$$
$$x \in W^{1,\infty}(0, T; \mathbb{R}^n)\,, \quad u \in L^\infty(0, T; \mathbb{R}^m)\,,$$

where $g: \mathbb{R}^n \times \mathbb{R} \to \mathbb{R}$, $f: \mathbb{R}^n \times \mathbb{R}^m \to \mathbb{R}$, $U \subset \mathbb{R}^m$ is nonempty, closed, and convex, and a is the given initial state.

Let H denote the Hamiltonian defined by

$$H(x, u, \lambda) = g(x, u) + \langle \lambda, f(x, u) \rangle\,,$$

By the minimum principle (see e.g. [16]) a solution $(\tilde{x}, \tilde{u})$ of (O) satisfies

$$H_u(\tilde{x}(t), \tilde{u}(t), \tilde{\lambda}(t))\, (u - \tilde{u}(t)) \geq 0 \quad \text{a.e. } t \in [0, T] \text{ and for every } u \in U,$$

where $\tilde{\lambda}$ is the solution of the adjoint equation

$$\dot{\lambda}(t) = -H_x(x(t), u(t), \lambda(t)) \quad \text{a.e. } t \in [0, T], \lambda(T) = 0.$$

Therefore, $(\tilde{x}, \tilde{u}, \tilde{\lambda})$ is a solution of the generalized equation

$$(x, u, \lambda) \in C\,, \quad \begin{bmatrix} H_x(x(\cdot), u(\cdot), \lambda(\cdot)) + \dot{\lambda} \\ H_u(x(\cdot), u(\cdot), \lambda(\cdot)) \\ f(x(\cdot), u(\cdot)) - \dot{x} \end{bmatrix} \in \begin{bmatrix} 0 \\ \partial U\,(u(\cdot)) \\ 0 \end{bmatrix}\,, \qquad (4.1)$$

where

$$C = \{(x, u, \lambda) \in Z \mid x(0) = a,\ \lambda(T) = 0\}\,,$$

and $Z = Z_1 \times Z_2 \times Z_3$ is defined by

$$Z_1 = Z_3 = W^{1,\infty}(0, T; \mathbb{R}^n)\,, \quad Z_2 = L^\infty(0, T; \mathbb{R}^m)\,.$$

In order to put equation (4.1) in the general framework of the previous sections we further define $Y = Y_1 \times Y_2 \times Y_3$ by

$$Y_1 = Y_3 = L^\infty(0, T; \mathbb{R}^n)\,, \quad Y_2 = L^\infty(0, T; \mathbb{R}^m)\,,$$

$F: Z \to Y$ by

$$F(x, u, \lambda) = \begin{bmatrix} H_x(x(\cdot), u(\cdot), \lambda(\cdot)) + \dot{\lambda} \\ H_u(x(\cdot), u(\cdot), \lambda(\cdot)) \\ f(x(\cdot), u(\cdot)) - \dot{x} \end{bmatrix},$$

and $T: Z \to Y$ by

$$T(x, u, \lambda) = \begin{bmatrix} 0 \\ \partial U(u(\cdot)) \\ 0 \end{bmatrix},$$

if $(x, u, \lambda) \in C$, and $T(x, u, \lambda) = \emptyset$ if $(x, u, \lambda) \notin C$. We assume that

(C1)   There exists a (local) solution $(\tilde{x}, \tilde{u})$ of (O).

(C2)   The mappings f, and g are two times Fréchet differentiable in all arguments, and the respective derivatives are locally Lipschitz continuous.   ◇

We use the notation $\tilde{z} = (\tilde{x}, \tilde{u}, \tilde{\lambda})$, where $\tilde{\lambda}$ is the solution of the adjoint equation. Then Assumptions (A1), (A2) are satisfied.

Next we want to verify Assumption (A3). Sufficient conditions for strong regularity of (4.1) have been given in Hager [15]. According to Assumption (8) in this paper and (64) in Dontchev/Hager [9], we assume that the following coercitivity condition holds with $\tilde{H}(t) = H(\tilde{z}(t))$, $\tilde{f}(t) = f(\tilde{x}(t), \tilde{u}(t))$.

(C3)   There exists $\alpha > 0$ such that

$$\int_0^T [x^T(t), x^T(t)] \begin{bmatrix} \tilde{H}_{xx}(t) & \tilde{H}_{xu}(t) \\ \tilde{H}_{ux}(t) & \tilde{H}_{uu}(t) \end{bmatrix} \begin{bmatrix} x(t) \\ u(t) \end{bmatrix} dt \ge \alpha \int_0^T (|u|^2 + |x|^2) \, dt,$$

for all pairs $(x, u) \in Z_1 \times Z_2$ satisfying $x(0) = 0$,

$$\dot{x}(t) = \tilde{f}_x(t)x(t) + \tilde{f}_u(t)u(t) \quad \forall' t \in [0, T],$$

and $u = u_1 - u_2$ for some $u_1, u_2 \in Z_2$ with $u_1(t), u_2(t) \in U$ for almost every $t \in [0, T]$.   ◇

It can be shown, that (C3) implies the following pointwise coercivity condition (see [12], [9]).

(C4)   There exists $\alpha_2 > 0$ independent of t such that

$$\langle u, \tilde{H}_{uu}(t)u \rangle \ge \alpha_2 |u|^2$$

for all $u = u_1 - u_2$ with $u_1, u_2 \in U$, and for almost every $t \in [0, T]$.   ◇

By Lemma 3 of Hager [15] (see also Theorem 5 in Dontchev/Hager [9]), Assumptions (C1)–C4) imply that the generalized equation (4.1) is strongly regular at $\tilde{z} = (\tilde{x}, \tilde{u}, \tilde{\lambda})$ with $r_Y = +\infty$. Therefore, by Theorem 2.6 the generalized Newton method applied to (4.1) converges if the starting point (x_0, u_0, λ_0) is sufficiently close to $(\tilde{x}, \tilde{u}, \tilde{\lambda})$, i.e., if

$$\|x_0 - \tilde{x}\|_{1,\infty}, \; \|u_0 - \tilde{u}\|_\infty, \; \|\lambda_0 - \tilde{\lambda}\|_{1,\infty} \le \rho(\tilde{z}).$$

In iteration k of (GNM) we have to solve the linear generalized equation

$$z \in C, \quad \begin{bmatrix} H_x(\bar{z}) + H_x'(\bar{z}(\cdot))(z - \bar{z}) + \dot{\lambda} \\ H_u(\bar{z}) + H_u'(\bar{z}(\cdot))(z - \bar{z}) \\ f_x(\bar{z})(x - \bar{x}) + f_u(\bar{z})(u - \bar{u}) - \dot{x} \end{bmatrix} \in \begin{bmatrix} 0 \\ \partial U\left(u(\cdot)\right) \\ 0 \end{bmatrix} \quad (4.2)$$

for $z = (x, u, \lambda)$, $\bar{z} = (x_k, u_k, \lambda_k)$. Here $H_v'(x, u, \lambda)$ denotes the Fréchet derivative of H_v, i.e.,

$$H_v'(\bar{x}, \bar{u}, \bar{\lambda})(x, u, \lambda) = H_{vx}(\bar{x}, \bar{u}, \bar{\lambda})x + H_{vu}(\bar{x}, \bar{u}, \bar{\lambda})u + H_{v\lambda}(\bar{x}, \bar{u}, \bar{\lambda})\lambda.$$

It is well-known that (4.2) defines the necessary optimality conditions for the following quadratic control problem.

(QO) $\quad \mathrm{Min}_{(x,u) \in Z_1 \times Z_2} \quad \displaystyle\int_0^T G(x(t), u(t)) \, dt$

subject to

$$\dot{x}(t) = f_x(\bar{x}(t), \bar{u}(t))x(t) + f_u(\bar{x}(t), \bar{u}(t))u(t) \qquad \forall' t \in [0, T],$$

$$x(0) = a,$$

$$u(t) \in U \qquad\qquad\qquad\qquad\qquad\qquad\qquad \forall' t \in [0, T],$$

where

$$G(x(t), u(t), w(t)) = \langle g_x(\bar{x}(t), \bar{u}(t)), x(t) - \bar{x}(t) \rangle + \langle g_u(\bar{x}(t), \bar{u}(t)), u(t) - \bar{u}(t) \rangle$$

$$+ \frac{1}{2} \begin{bmatrix} x(t) - \bar{x}(t) \\ u(t) - \bar{u}(t) \end{bmatrix}^T \begin{bmatrix} H_{xx}(\bar{z}(t)) & H_{xu}(\bar{z}(t)) \\ H_{ux}(\bar{z}(t)) & H_{uu}(\bar{z}(t)) \end{bmatrix} \begin{bmatrix} x(t) - \bar{x}(t) \\ u(t) - \bar{u}(t) \end{bmatrix}.$$

It follows from the results of Hager [15], Section 2, that the generalized equation (4.2) and Problem (QO) have the same solutions. Therefore, (GNM) can equivalently be formulated as a sequential quadratic programming method (see also Alt [2], Alt/Malanowski [5]).

Since the generalized equation (4.1) is strongly regular at $\tilde{z} = (\tilde{x}, \tilde{u}, \tilde{\lambda})$ with $r_Y = +\infty$, there exists $c_L(\tilde{z}) > 0$, such that for all $y = (y_1, y_2, y_3) \in Y$ the linearized system (2.12), which is equivalent to

$$z \in C, \quad \begin{bmatrix} H_x(\tilde{z}) + H_x'(\tilde{z})(z - \tilde{z}) + \dot{\lambda} + y_1 \\ H_u(\tilde{z}) + H_u'(\tilde{z})(z - \tilde{z}) + y_2 \\ f(\tilde{x}, \tilde{u}) + f'(\tilde{x}, \tilde{u})((x, u) - (\tilde{x}, \tilde{u})) - \dot{x} + y_3 \end{bmatrix} \in \begin{bmatrix} 0 \\ \partial U\left(u(\cdot)\right) \\ 0 \end{bmatrix}, \quad (4.3)$$

has a unique solution $\mathcal{S}_L(\tilde{z}, y) \in Z$, and the mapping $\mathcal{S}_L(\tilde{z}, \cdot): Y \to Z$ is Lipschitz continuous with modulus $c_L(\tilde{z})$.

Remark: If (C1)–(C4) hold then all Assumptions of Theorem 2.6 are satisfied, i.e., (GNM) applied to the control problem (O) generates a unique sequence $z_k = (x_k, u_k, \lambda_k)$ converging locally quadratically to $\tilde{z} = (\tilde{x}, \tilde{u}, \tilde{\lambda})$. $\qquad\qquad \diamond$

In order to verify Assumption (A3) and to get uniform convergence of discretizations of (GNM), we assume that

(C5) $\tilde{z}$ is Lipschitz with modulus $\tilde{L}$.   ◇

Let $L_Y > 0$ be an arbitrary constant. Then we define the set Y_R by

$$Y_R = \{\, y \in B_Y(0_Y, 1) \mid y \text{ is Lipschitz with modulus } L_Y \,\} \,.$$

If $y \in Y_R$, then for the unique solution $z = (x, u, \lambda) = \mathcal{S}_L(\tilde{z}, y) \in Z$ of (4.3) we have

$$\|z - \tilde{z}\|_Z = \|\mathcal{S}_L(\tilde{z}, y) - \mathcal{S}_L(\tilde{z}, 0_Y)\|_Z \le c_L(\tilde{z}) \|y\|_Y \le c_L \,.$$

This especially implies

$$\|\dot{x} - \dot{\tilde{x}}\|_\infty \,, \ \|\dot{\lambda} - \dot{\tilde{\lambda}}\|_\infty \le c_L \,.$$

Hence, it follows by Assumption (C5) that x and λ are Lipschitz with modulus $\tilde{L} + c_L$. Furthermore, it follows from (4.3) that for a.e. $t \in [0, T]$, $u(t)$ is the solution of

$$\langle \widetilde{H}_{uu}(t) u(t) + q(t) + y_2(t), v - u(t) \rangle \ge 0 \quad \forall v \in U \,, \tag{4.4}$$

where

$$q(t) = g_u(\tilde{x}(t), \tilde{u}(t)) + \lambda(t)^T f_u(\tilde{x}(t), \tilde{u}(t)) - \widetilde{H}_{uu} \tilde{u}(t) + \widetilde{H}_{ux}(t)(x(t) - \tilde{x}(t)) \,.$$

Choosing in (4.4) $t = t_1$, and $v = u(t_2)$ followed by $t = t_2$, and $v = u(t_1)$ and adding the resulting relations yields after simple rearrangements

$$\langle \widetilde{H}_{uu}(t_2)(u(t_2) - u(t_1)), u(t_2) - u(t_1) \rangle$$
$$\le \langle q(t_1) - q(t_2) - (\widetilde{H}_{uu}(t_2) - \widetilde{H}_{uu}(t_1)) u(t_1), u(t_2) - u(t_1) \rangle \,.$$

By Assumptions (C1), (C2), (C4), and (C5), and the Lipschitz continuity of x, λ, and y_2 this implies

$$|u(t_1) - u(t_2)| \le \tilde{c}_1 |t_1 - t_2|$$

with some constant $\tilde{c}_1$ independent of $y \in Y_R$. Finally, it follows from the state equation and the adjoint equation that

$$|\dot{x}(t_1) - \dot{x}(t_2)| \,, \ |\dot{\lambda}(t_1) - \dot{\lambda}(t_2)| \le \tilde{c}_2 |t_1 - t_2|$$

with some constant $\tilde{c}_2$ independent of $y \in Y_R$. Therefore, a suitable choice for Z_R will be

$$Z_R = \{\, z = (x, u, \lambda) \in Z \mid x, \dot{x}, u, \lambda, \dot{\lambda} \text{ are Lipschitz with modulus } 2L_Z \,\} \,,$$

where $L_Z = \max\{\tilde{c}_1, \tilde{c}_2, \tilde{L} + c_L\}$. We further define

$$\widetilde{Z}_R = \{\, z = (x, u, \lambda) \in Z \mid x, \dot{x}, u, \lambda, \dot{\lambda} \text{ are Lipschitz with modulus } L_Z \,\} \,.$$

Next we show that there exists $r_R > 0$ such that $\ell(\tilde{z}; z, w) \in Y_R$ for all $z, w \in \widetilde{Z}_R \cap B_Z(\tilde{z}, r_R)$. By Lemma 2.2(b) there exists $r_2(\tilde{z}) > 0$ such that $\|\ell(\tilde{z}; z, w)\|_Y \le 1$ for all $z, w \in B_Z(\tilde{z}, r_2(\tilde{z}))$. Let $\overline{H}(t) = H(w(t))$. Then we have $\ell(\tilde{z}; z, w)(t) = (y_1(t), y_2(t), y_3(t))^T$, where

$$\begin{pmatrix} y_1(t) \\ y_2(t) \\ y_3(t) \end{pmatrix} = \begin{pmatrix} \overline{H}_x(t) + \overline{H}'_x(t)(z(t) - w(t)) \\ \overline{H}_u(t) + \overline{H}'_u(t)(z(t) - w(t)) \\ \overline{H}_\lambda(t) + \overline{H}'_\lambda(t)(z(t) - w(t)) \end{pmatrix} - \begin{pmatrix} \widetilde{H}_x(t) + \widetilde{H}'_x(t)(z(t) - \tilde{z}(t)) \\ \widetilde{H}_u(t) + \widetilde{H}'_u(t)(z(t) - \tilde{z}(t)) \\ \widetilde{H}_\lambda(t) + \widetilde{H}'_\lambda(t)(z(t) - \tilde{z}(t)) \end{pmatrix} \,.$$

Thus we have

$$y_1(t) = \int_0^1 \left[H_x'(s\tilde{z}(t) + (1-s)w(t)) - H_x'(w(t)) \right] (w(t) - \tilde{z}(t)) \, ds$$
$$+ \left[H_x'(w(t)) - H_x'(\tilde{z}(t)) \right] (z(t) - \tilde{z}(t)) .$$

For $t_1, t_2 \in [0, T]$ we therefore obtain

$$y_1(t_1) - y_1(t_2) =$$
$$\int_0^1 \left[H_x'(s\tilde{z}(t_1) + (1-s)w(t_1)) - H_x'(w(t_1)) \right] (w(t_1) - w(t_2)) \, ds$$
$$+ \int_0^1 \left[H_x'(s\tilde{z}(t_1) + (1-s)w(t_1)) - H_x'(w(t_1)) \right.$$
$$\left. - H_x'(s\tilde{z}(t_2) + (1-s)w(t_2)) + H_x'(w(t_2)) \right] (w(t_2) - \tilde{z}(t_2)) \, ds$$
$$- \int_0^1 \left[H_x'(s\tilde{z}(t_1) + (1-s)w(t_1)) - H_x'(w(t_1)) \right] (\tilde{z}(t_1) - \tilde{z}(t_2)) \, ds$$
$$+ \left[H_x'(w(t_1)) - H_x'(\tilde{z}(t_1)) \right] (z(t_1) - z(t_2))$$
$$+ \left[H_x'(w(t_1)) - H_x'(\tilde{z}(t_1)) - H_x'(w(t_2)) + H_x'(\tilde{z}(t_2)) \right] (z(t_2) - \tilde{z}(t_2))$$
$$+ \left[H_x'(w(t_1)) - H_x'(\tilde{z}(t_1)) \right] (\tilde{z}(t_2) - \tilde{z}(t_1)) .$$

By Assumptions (C1), C(2), (C5) we obtain for $z, w \in \widetilde{Z}_R \cap B_Z(\tilde{z}, r_R)$ with some $0 < r_R \le r_2(\tilde{z})$

$$|y_1(t_1) - y_1(t_2)| \le \tilde{c}_R r_R |t_1 - t_2| ,$$

where the constant $\tilde{c}_R$ is independent of z, w. In the same way it follows that

$$|y_2(t_1) - y_2(t_2)| , \ |y_3(t_1) - y_3(t_2)| \le \tilde{c}_R r_R |t_1 - t_2| .$$

If we choose r_R sufficiently small, this implies $\ell(\tilde{z}; z, w) \in Y_R$. Thus, (A3) is satisfied. It therefore follows from Corollary 2.8 and Remark 2.9 that (2.11) holds.

As an example of a discretization of Problem (O) we consider Euler's method. This method has also been investigated by Dontchev/Hager [9]. Given a natural number N, let $h_N = 1/N$ be the mesh size. We approximate Z_2, Y_1, Y_2, Y_3 by piecewise constant functions and Z_1, Z_3 by continuous, piecewiese linear functions, and we represent the approximate functions by their values at the breakpoints ih_N, $i = 0, 1, \dots, N$. Let $\| \cdot \|_{Z_N} = \| \cdot \|_Z$, and $\| \cdot \|_{Y_N} = \| \cdot \|_Y$ be the norms induced by the spaces Z, respectively, Y. Then the Euler discretization of (O) is given by

$(O)_N$ $\text{Min}_{(x,u) \in Z_{1,N} \times Z_{2,N}}$ $J_N(x, u) = \displaystyle\sum_{i=0}^{N-1} h_N g(x_i, u_i)$

       subject to

$$x_{i+1} = x_i + h_N f(x_i, u_i), \quad i = 0, 1, \dots, N-1 ,$$
$$x_0 = a ,$$
$$u_i \in U \qquad\qquad\qquad i = 0, 1, \dots, N-1 .$$

Let the set C_N be defined by

$$C_N = \{(x, u, \lambda) \in Z_N \mid x_0 = a, \ \lambda_N = 0\}.$$

Then component i, $0 \leq i \leq N - 1$, of the mapping F_N is defined by

$$F_{N,i}(x, u, \lambda) = \begin{bmatrix} H_x(x_i, u_i, \lambda_{i+1}) + (\lambda_{i+1} - \lambda_i)/h_N \\ H_u(x_i, u_i, \lambda_{i+1}) \\ f(x_i, u_i) - (x_{i+1} - x_i)/h_N \end{bmatrix},$$

and component i, $0 \leq i \leq N - 1$, of the mapping T_N is defined by

$$T_{N,i}(x, u, \lambda) = \begin{bmatrix} 0 \\ \partial U(u_i) \\ 0 \end{bmatrix}$$

if $(x, u, \lambda) \in C_N$, and $T_{N,i}(x, u, \lambda) = \emptyset$ if $(x, u, \lambda) \notin C_N$ (compare [9], proof of Theorem 6). Then Assumptions (D1), (D2) are satisfied, and Assumption (D3) is satisfied with $c_B = 1$. Further, it follows from Assumption (C5), Lemma 11 of [9] and the proof of Theorem 6 in [9] that Assumption (D4) is satified with $r_Y = +\infty$. Moreover, as in the infinite case, the linear generalized equations (3.4) can be equivalently formulated as finite-dimensional, quadratic optimization problems (see [9]).

Let $z = (x, u, \lambda) \in Z_R$. For $t_i = ih_N$, $0 \leq i \leq N - 1$, we define

$$z_i := z(t_i) = (x(t_i), u(t_i), \lambda(t_i)), \quad z_i^+ := (x(t_i), u(t_i), \lambda(t_{i+1})).$$

Then we have

$$\left[\hat{\Delta}_N F(z) - F_N(\Delta_N z)\right](t_i)$$

$$= \begin{bmatrix} H_x(z_i) + \dot{\lambda}(t_i) \\ H_u(z_i) \\ f(x(t_i), u(t_i)) - \dot{x}(t_i) \end{bmatrix} - \begin{bmatrix} H_x(z_i^+) + (\lambda(t_{i+1}) - \lambda(t_i))/h_N \\ H_u(z_i^+) \\ f(x(t_i), u(t_i)) - (x(t_{i+1}) - x(t_i))/h_N \end{bmatrix}.$$

Using the definition of Z_R and Assumption (C2) one easily obtains the first inequality of the consistency requirement (D5) with $p = 1$. Now let $z = (x, u, \lambda), \bar{z} = (\bar{x}, \bar{u}, \bar{\lambda}) \in Z_R$ be given. Using the abbreviations

$$\bar{z}_i := \bar{z}(t_i) = (\bar{x}(t_i), \bar{u}(t_i), \bar{\lambda}(t_i)), \quad \bar{z}_i^+ := (\bar{x}(t_i), \bar{u}(t_i), \bar{\lambda}(t_{i+1})),$$

for $t_i = ih_N$, $0 \leq i \leq N - 1$, we have

$$\hat{\Delta}_N(F'(\bar{z})z) - F_N'(\Delta_N \bar{z})(\Delta_N z)$$

$$= \begin{bmatrix} H_{xx}(\bar{z}_i)x(t_i) + H_{xu}(\bar{z}_i)u(t_i) + H_{x\lambda}(\bar{z}_i)\lambda(t_i) + \dot{\lambda}(t_i) \\ H_{ux}(\bar{z}_i)x(t_i) + H_{uu}(\bar{z}_i)u(t_i) + H_{u\lambda}(\bar{z}_i)\lambda(t_i) \\ f_x(x_i, u_i)x_i + f_u(x_i, u_i)x_i - \dot{x}(t_i) \end{bmatrix}$$

$$- \begin{bmatrix} H_{xx}(\bar{z}_i^+)x(t_i) + H_{xu}(\bar{z}_i^+)u(t_i) + H_{x\lambda}(\bar{z}_i^+)\lambda(t_{i+1}) + (\lambda(t_{i+1}) - \lambda(t_i))/h_N \\ H_{ux}(\bar{z}_i^+)x(t_i) + H_{uu}(\bar{z}_i^+)u(t_i) + H_{u\lambda}(\bar{z}_i^+)\lambda(t_{i+1}) \\ f_x(x_i, u_i)x_i + f_u(x_i, u_i)x_i - (x(t_{i+1}) - x(t_i))/h_N \end{bmatrix}.$$

Using the definition of Z_R and Assumption (C2) one easily obtains the second inequality of the consistency requirement (D5) with $p = 1$.

Let $y = (y_1, y_2, y_3) \in Y_R$ be given and let $z \in Z_R$ be the unique solution of the linearized system (4.3). Then

$$
\begin{bmatrix}
H_x(\tilde{z}(t)) + H'_x(\tilde{z}(t))(z(t) - \tilde{z}(t)) + \dot{\lambda}(t) \\
H_u(\tilde{z}(t)) + H'_u(\tilde{z}(t))(z(t) - \tilde{z}(t)) \\
f_x(\tilde{x}(t), \tilde{u}(t))(x(t) - \tilde{x}(t)) + f_u(\tilde{x}(t), \tilde{u}(t))(u(t) - \tilde{u}(t)) - \dot{x}(t)
\end{bmatrix}
$$
$$
+ \begin{bmatrix} y_1(t) \\ y_2(t) \\ y_3(t) \end{bmatrix} \in \begin{bmatrix} 0 \\ \partial U\,(u(t)) \\ 0 \end{bmatrix} \tag{4.5}
$$

for a.e. $t \in [0, T]$. We define $z_N = \Delta_n z$,

$$
\tilde{z}_i^+ := (\tilde{x}(t_i), \tilde{u}(t_i), \tilde{\lambda}(t_{i+1})),
$$

and y_N by

$$
\begin{aligned}
y_{N,1}(t_i) &= y_1(t_i) + H_x(\tilde{z}(t_i)) - H_x(\tilde{z}_i^+) + H'_x(\tilde{z}(t_i))(z(t_i) - \tilde{z}(t_i)) \\
&\quad - H'_x(\tilde{z}_i^+)(z(t_i) - \tilde{z}(t_i)) + \dot{\lambda}(t_i) - (\lambda(t_{i+1} - \lambda(t_i))/h_N\,, \\
y_{N,2}(t_i) &= y_2(t_i) + H_u(\tilde{z}(t_i)) - H_u(\tilde{z}_i^+) + H'_u(\tilde{z}(t_i))(z(t_i) - \tilde{z}(t_i)) \\
&\quad - H'_u(\tilde{z}_i^+)(z(t_i) - \tilde{z}(t_i)) \\
y_{N,3}(t_i) &= y_3(t_i) - \dot{x}(t_i) + (x(t_{i+1} - x(t_i))/h_N\,,
\end{aligned}
$$

for $0 \le i \le N - 1$. Then (4.5) for $t = t_i$ constitutes equation (3.5) for $(O)_N$. Using the definition of Z_R and Assumption (C2) one easily verifies that $\|y_N - \hat{\Delta}_N y\|_{Y_N} \le c_3 h_N$, so that Assumption (D6) is satified.

Finally, by the definition of Z_R, (3.32) holds. Thus, all assumptions of Corollary 3.6 are satisfied, and the mesh-independence principle applies.

Example 4.1 In order to illustrate the theoretical results by a numerical example we consider the following control problem:

$$
\text{(O1)} \quad \min_{(u,x)} \quad \frac{1}{2}\int_0^1 \left(x(t)^3 + u(t)^2\right)\,dt
$$
$$
\text{subject to}
$$
$$
\dot{x}(t) = u(t) \qquad \forall' t \in [0, 1]\,,
$$
$$
x(0) = 4\,,
$$
$$
-5 \le u(t) \le -1 \qquad \forall' t \in [0, 1]\,.
$$

The same control problem with an end condition instead of a pointwise control constraints has been analyzed by Maurer/Pesch [25, 26], who investigated solution differentiability for nonlinear parametric control problems. Variants of this problem have also been used in Alt/Malanowski [5, 6] for numerical illustrations of the convergence of the Lagrange-Newton method.

It can be easily seen, that Assumptions (C1)–(C4) are satisfied for Problem (O1). By the numerical solution for $N = 100$ shown in Figure 4.1 we can expect that

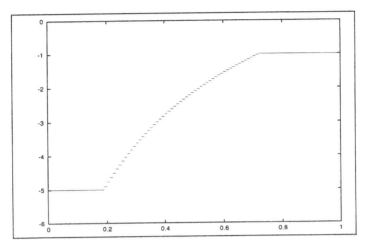

Figure 4.1: Optimal control for Problem (O1), $N = 100$

(C5) is also satisfied. For the solution of the finite-dimensional, quadratic optimization problems equivalent to the linear generalized equations (3.4) we used subroutine E04NCF of the NAG library in double precision. With the starting functions $x_0 \equiv 4$, $u_0 \equiv -5$, $\lambda_0 \equiv 0$, and the tolerance $\varepsilon = 10^{-10}$ the number of iterations of $(\text{GNM})_N$ for $N = 20, 50, 75, 100, 150, 200$, is 6, independent of N. Table 4.1 shows some numerical results. The numbers in the last column are bounded

Table 4.1: Numerical results for Problem (O1)

N	$J_N(\tilde{x}_N, \tilde{u}_N)$	$\|\tilde{u}_N - \tilde{u}_{200}\|_\infty$	$\|\tilde{u}_N - \tilde{u}_{200}\|_\infty / h_N$
20	4.646721319195805	0.831758975982666	16.635179519653300
50	4.626057018967789	0.373470008373261	18.673500418663000
75	4.619794136568125	0.220128998160362	16.509674862027200
100	4.617278015172680	0.170686006546021	17.068600654602100
150	4.614495469361319	0.104313999414444	15.647099912166600
200	4.613239875794786		

which indicates linear convergence of the optimal solutions in terms of the mesh size. This is consistent with the error estimate in Theorem 3.2. ◇

5 APPENDIX

Proof of Theorem 2.5: By Remark 2.4 the first part of the assertion is proved, if we show that for each $w \in B_Z(\tilde{z}, \rho)$ there exists a unique fixed point $\mathcal{S}(w)$ in $B_Z(\tilde{z}, \rho)$ of $\mathcal{S}_w$. Choose any $w \in B_Z(\tilde{z}, \rho)$. By Lemma 2.3 and the definition of ρ we have

$$c_L(\tilde{z}) \, \|\ell(\tilde{z}; \tilde{z}, w)\|_Y \leq \tfrac{1}{2} c_L(\tilde{z}) c_F(\tilde{z}) \rho^2 \leq \tfrac{1}{3}\rho. \tag{5.1}$$

Let $z_1, z_2 \in B_Z(\tilde{z}, \rho)$ be given. By Lemma 2.2 (a), strong regularity of (2.1) at $\tilde{z}$, and Lemma 2.2 (c) (with $\tilde{c} = c(\tilde{z})$) it follows that

$$\|\mathcal{S}_w(z_1) - \mathcal{S}_w(z_2)\|_Z = \|\mathcal{S}_L(\tilde{z}, \ell(\tilde{z}; z_1, w)) - \mathcal{S}_L(\tilde{z}, \ell(\tilde{z}; z_2, w))\|_Z$$
$$\leq c_L(\tilde{z}) \|\ell(\tilde{z}; z_1, w) - \ell(\tilde{z}; z_2, w)\|_Y \leq \tfrac{2}{3} \|z_1 - z_2\|_Z . \quad (5.2)$$

Hence, $\mathcal{S}_w$ is strongly contractive on $B_Z(\tilde{z}, \rho)$. Since $\mathcal{S}_w(\tilde{z}) = \mathcal{S}_L(\tilde{z}, \ell(\tilde{z}; \tilde{z}, w))$ and $\tilde{z} = \mathcal{S}_L(\tilde{z}, \ell(\tilde{z}; \tilde{z}, \tilde{z})) = \mathcal{S}_L(\tilde{z}, 0_Y)$ we have by (5.1)

$$\|\mathcal{S}_w(\tilde{z}) - \tilde{z}\|_Z \leq c_L(\tilde{z}) \|\ell(\tilde{z}; \tilde{z}, w)\|_Y \leq \tfrac{1}{3}\rho ,$$

and therefore for any $z \in B_Z(\tilde{z}, \rho)$ by (5.2)

$$\|\mathcal{S}_w(z) - \tilde{w}\|_Z \leq \|\mathcal{S}_w(z) - \mathcal{S}_w(\tilde{z})\|_Z + \|\mathcal{S}_w(\tilde{z}) - \tilde{w}\|_Z \leq \tfrac{2}{3} \|z - \tilde{z}\|_Z + \tfrac{1}{3}\rho \leq \rho ,$$

so that $\mathcal{S}_w$ is a self-map on $B_Z(\tilde{z}, \rho)$. By the contraction principle, $\mathcal{S}_w$ has a unique fixed point $\mathcal{S}(w) \in B_Z(\tilde{z}, \rho)$. Thus, we have established the existence of the function $\mathcal{S}: B_Z(\tilde{z}, \rho) \to B_Z(\tilde{z}, \rho)$. In addition, by the contraction principle for each $z \in B_Z(\tilde{z}, \rho)$ one has the bound

$$\|\mathcal{S}(w) - z\|_Z \leq (1 - \tfrac{2}{3})^{-1} \|\mathcal{S}_w(z) - z\|_Z = 3 \|\mathcal{S}_w(z) - z\|_Z . \quad (5.3)$$

To obtain the bound (2.8), we take any $w \in B_Z(\tilde{z}, \rho)$, and apply (5.3) with $z = \tilde{z} = \mathcal{S}(\tilde{z})$ to get

$$\|\mathcal{S}(w) - \mathcal{S}(\tilde{z})\|_Z \leq 3 \|\mathcal{S}_w(\tilde{z}) - \mathcal{S}(\tilde{z})\|_Z .$$

If we now recall that $\ell(\tilde{z}; \tilde{z}, \tilde{z}) = 0_Y$, $\mathcal{S}(\tilde{z}) = \mathcal{S}_{\tilde{z}}(\mathcal{S}(\tilde{z}))$ and employ the bound

$$\|\mathcal{S}_w(\tilde{z}) - \mathcal{S}_{\tilde{z}}(\tilde{z})\|_Z = \|\mathcal{S}_L(\tilde{z}, \ell(\tilde{z}; \tilde{z}, w)) - \mathcal{S}_L(\tilde{z}, \ell(\tilde{z}; \tilde{z}, \tilde{z}))\|_Z$$
$$\leq c_L(\tilde{z}) \|\ell(\tilde{z}; \tilde{z}, w) - \ell(\tilde{z}; \tilde{z}, \tilde{z})\|_Z ,$$

we immediately obtain the first inequality of (2.8). The second inequality follows from Lemma 2.3.   □

Acknowledgment: The author thanks the referees for their careful reading of this paper and their valuable suggestions for improving the paper.

REFERENCES

[1] E. L. Allgower, K. Böhmer, F. A. Potra, and W. C. Rheinboldt. A mesh-independence principle for operator equations and their discretizations. *SIAM Journal Numerical Analysis*, 23:160–169, 1986.

[2] W. Alt. The Lagrange-Newton method for infinite-dimensional optimization problems. *Numerical Functional Analysis and Optimization*, 11:201–224, 1990.

[3] W. Alt. Local convergence of the Lagrange-Newton method with applications to optimal control. *Control and Cybernetics*, 23:87–106, 1994.

[4] W. Alt. Semi-local convergence of the Lagrange-Newton method with applications to optimal control. In R. Durier and C. Michelot, editors, *Recent Developments in Optimization*, volume 429 of *Lecture Notes in Economics and Mathematical Systems*, pages 1–16. Springer Verlag, 1995.

[5] W. Alt and K. Malanowski. The Lagrange-Newton method for nonlinear optimal control problems. *Computational Optimization and Applications*, 2:77–100, 1993.

[6] W. Alt and K. Malanowski. The Lagrange-Newton method for state constrained optimal control problems. *Computational Optimization and Applications*, 4:217–239, 1995.

[7] J. F. Bonnans. Local analysis of Newton-type methods for variational inequalities and nonlinear programming. *Applied Mathematics and Optimization*, 29:161–186, 1994.

[8] A. L. Dontchev. Discrete approximations in optimal control. to appear, 1995.

[9] A. L. Dontchev and W. W. Hager. Lipschitz stability in nonlinear control and optimization. *SIAM Journal Control and Optimization*, 31:569–603, 1993.

[10] A. L. Dontchev and W. W. Hager. Implicit functions, Lipschitz maps, and stability in optimization. *Mathematics of Operations Research*, 19:753–768, 1994.

[11] A. L. Dontchev, W. W. Hager, A. B. Poore, and B. Yang. Optimality, stability, and convergence in nonlinear control. *Applied Mathematics and Optimization*, 31:297–326, 1995.

[12] J. C. Dunn and T. Tian. Variants of the Kuhn-Tucker sufficient conditions in cones of nonnegative functions. *SIAM Journal Control and Optimization*, 30:1361–1384, 1992.

[13] A. V. Fiacco. *Introduction to Sensitivity and Stability Analysis in Nonlinear Programming*. Academic Press, New York, 1983.

[14] R. Fletcher. *Practical Methods of Optimization*. John Wiley & Sons, New York, second edition, 1987.

[15] W. W. Hager. Multiplier methods for nonlinear optimal control. *SIAM Journal Numerical Analysis*, 17:1061–1080, 1990.

[16] A. D. Ioffe and V. M. Tihomirov. *Theory of Extremal Problems*. North Holland, Amsterdam, 1979.

[17] N. H. Josephy. Newton's method for generalized equations. Technical Summary Report No. 1965, Mathematics Research Center, University of Wisconsin–Madison, 1979.

[18] C. T. Kelley and E. W. Sachs. Mesh independence of the gradient projection method for optimal control problems. *SIAM Journal Control and Optimization*, 30:477–483, 1992.

[19] C. T. Kelley and S. J. Wright. Sequential quadratic programming for certain parameter identification problems. Preprint, 1990.

[20] E. S. Levitin and B. T. Polyak. Constrained minimization methods. *USSR Journal on Computational Mathematics and Mathematical Physics*, 6(5):1–50, 1966.

[21] K. C. P. Machielsen. Numerical solution of optimal control problems with state constraints by sequential quadratic programming in function space. *CWI Tract*, 53, 1987.

[22] K. Malanowski. Convergence of approximations to nonlinear optimal control problems. Working Paper, Systems Research Institute, Polish Academy of Sciences, 1995.

[23] K. Malanowski. Lagrange-Newton method for control state and pure state constrained optimal control problems. Working Paper, Systems Research Institute, Polish Academy of Sciences, 1995.

[24] K. Malanowski. Stability and sensitivity analysis of solutions to nonlinear optimal control problems. *Applied Mathematics and Optimization*, 32:111–141, 1995.

[25] H. Maurer and H. J. Pesch. Solution differentiability for nonlinear parametric control problems. *SIAM Journal Control and Optimization*, 32:1542–1554, 1994.

[26] H. Maurer and H. J. Pesch. Solution differentiability for parametric nonlinear control problems with control-state-constraints. *Control and Cybernetics*, 1994. to appear.

[27] S. M. Robinson. Perturbed Kuhn-Tucker points and rates of convergence for a class of nonlinear-programming algorithms. *Mathematical Programming*, 7:1–16, 1974.

[28] S. M. Robinson. Strongly regular generalized equations. *Mathematics of Operations Research*, 5:43–62, 1980.

[29] S. M. Robinson. Generalized equations. In A. Bachem, M. Grötschel, and B. Korte, editors, *Mathematical Programming: The State of the Art*, pages 165–207, Berlin, 1983. Springer-Verlag.

[30] S. M. Robinson. Newton's method for a class of nonsmooth functions. *Set-Valued Analysis*, 2:291–305, 1994.

[31] J. Stoer. Principles of sequential quadratic programming methods for solving nonlinear programs. In K. Schittkowski, editor, *Computational Mathematical Programming*, volume F15, pages 165–207. Nato ASI Series, 1985.

[32] E. Zeidler. *Nonlinear Functional Analysis and its Applications. Part I: Fixed-Point Theorems*. Springer Verlag, New York, 1991.

Extended Quadratic Tangent Optimization Problems

J.F. BONNANS   INRIA, B.P. 105, 78153 Rocquencourt, France.

Abstract   We associate an *extended quadratic tangent* (EQT) optimization problem with any feasible point of a nonlinear programming problem. The EQT problem has linear homogeneous constraints and its cost is the sum of a linear term and of a finite supremum of quadratic terms. The second-order necessary or sufficient conditions have a natural formulation in terms of the EQT problem. The strong regularity condition may also be formulated using EQT problems to simple perturbations of the original optimization problem. Local properties of sequential quadratic programming algorithms are best understood using this concept. We give a partial extension of the theory to nonisolated solutions. Finally, we discuss the extension of the theory to problems with abstract constraints.

1   INTRODUCTION

This paper gives a new presentation of the second order theory for nonlinear programming problems (a finite dimensional optimization problem with finitely many constraints) and discusses some extensions. The theory of second-order necessary or sufficient conditions is due to Levitin, Miljutin and Osmolovski (1974) (see also Ioffe (1979), Ben-Tal (1980)). The results are expressed in terms of the curvature of the Hessian of the Lagrangian along critical directions. The second order necessary condition tells that, if some qualification condition is satisfied, then this curvature is, for a certain Lagrange multiplier depending on the critical direction, nonnegative. That this curvature is positive is a sufficient condition for optimality. For non-qualified problems, similar results can be expressed in terms of a generalized Lagrangian.

Our presentation of the second-order theory is based on the notion of *extended quadratic tangent optimization problem*. Consider an unconstrained minimization problem

$$\operatorname*{Min}_{x} f(x): \quad x \in \mathbb{R}^n.$$

31

With $x \in \mathbb{R}^n$ we associate the *tangent quadratic problem*

$$(Q_x) \qquad\qquad \underset{d}{\text{Min}} \, f'(x)d + \frac{1}{2}f''(x)dd; \quad d \in \mathbb{R}^n.$$

The second-order necessary optimality condition may be expressed as follows: 0 is a solution of (Q_x), while the second-order sufficient optimality condition is that 0 is the unique solution of (Q_x). Problem (Q_x) is also a central object in many optimization algorithms.

For constrained problems the situation is not so simple. We nevertheless show that under a certain qualification condition, one may state an extended quadratic problem (in the sense of section 2) that plays the same role. This extended quadratic tangent (EQT) problem has linear homogeneous constraints and its cost is the sum of a linear term and of a finite supremum of quadratic terms. We reformulate in §3 the second-order analysis as follows: a local solution of a nonlinear programming problem (P) is, assuming the constraints to be qualified, such that 0 is a local minimum of the EQT problem; if 0 is an strict local minimum of the EQT problem, then the point is a local minimum of (P). The case of local solutions with unqualified constraints is discussed in §4. In §5 we show that the strong regularity condition of Robinson (1980)) may be formulated by reference to the EQT problem of a simple perturbations of the original optimization problem. We discuss also the directional second-order condition of Shapiro (1988). Then in §6 we express a necessary and sufficient condition for sequential quadratic programming algorithms to converge in the vicinity of a local solution. This condition, whose formulation is based on the quadratic tangent problem at the solution, implies the quadratic convergence of the sequence. In §7 we present an extension of EQT optimization problems to sets of nonisolated solutions. Taking advantage of results in Bonnans and Ioffe (1995,1996), we give a characterization of quadratic growth for convex programs in terms of the EQT optimization problem. Finally, in §8 we discuss an extension of EQT problems to optimization problems with abstract constraints.

2   EXTENDED QUADRATIC PROGRAMMING

We define an *extended quadratic optimization problem* as a problem of the form

$$(EQP) \qquad\qquad \underset{d \in \mathbb{R}^n}{\text{Min}} \, c^t d + Q(d); \, Ad \leq 0,$$

where $c \in \mathbb{R}^n$, A is a $p \times n$ matrix, the inequality $Ad \leq 0$ is taken componentwise, and

$$Q(d) := \sup\{d^t H d \; ; \; H \in \mathcal{H}\},$$

where $\mathcal{H}$ is a nonempty bounded set of $n \times n$ symmetric matrices. When $\mathcal{H}$ is a singleton, problem (EQP) is quadratic (i.e. it has a quadratic cost and linear constraints). Since $\mathcal{H}$ is bounded, $Q(d) = O(\|d\|^2)$, and an obvious necessary condition for $d = 0$ to be a local minimum of (EQP) is that

$$c^t d \geq 0, \quad \forall d \in \mathbb{R}; \quad Ad \leq 0. \qquad\qquad (1)$$

This is actually equivalent to existence of some Lagrange multipliers associated with the linear problem $\text{Min}\{c^t d; Ad \leq 0\}$, but we do not need to introduce these multipliers in the subsequent analysis. The cone of critical directions is defined as

$$C := \{d \in \mathbb{R}^n; \quad c^t d \leq 0; \quad Ad \leq 0\}.$$

The theorem below is the cornerstone of the paper. In the case of a quadratic programming problem it reduces to Majthay's result (Majthay (1971)). The statement uses the following concept: A local solution $\bar{x}$ of the optimization problem $\text{Min}\{f(x); x \in X\}$, where X is a subset of a normed space, is said to be a *strict local solution* if

$$\exists \epsilon > 0; \quad f(x) > f(\bar{x}) \quad \text{whenever} \quad x \in X, \ \|x - \bar{x}\| \leq \epsilon,$$

and to satisfy the *quadratic growth condition* if

$$\exists \alpha > 0; \quad f(x) \geq f(\bar{x}) + \alpha \|x - \bar{x}\|^2 + o(\|x - \bar{x}\|^2) \quad \text{whenever} \ x \in X.$$

The feasible set, set of solutions and optimal value function of an optimization problem (P) are denoted $F(P)$, $S(P)$ and $v(P)$.

THEOREM 1 *(i) Problem (EQP) has a local solution at 0 iff (1) holds and $Q(d) \geq 0$ for all critical direction d.*

(ii) The point 0 is a strict local solution of (EQP) satisfying the quadratic growth condition iff (1) holds and $Q(d) > 0$ for all non zero critical direction d.

Proof. (i) **"Only if" part.** If 0 is a local solution of (EQP) and d is a feasible direction, then for all $\sigma > 0$

$$0 \leq c^t(\sigma d) + Q(\sigma d) = \sigma c^t d + \sigma^2 Q(d).$$

Dividing by $\sigma \downarrow 0$, we get $c^t d \geq 0$, whence (1) holds. If $c^t d = 0$ then d is a critical direction, and dividing by σ^2 the above relation, we get $Q(d) \geq 0$, as had to be proved.

"If part" Let us proceed by contradiction. If 0 is not a local minimum of (EQP), then there exists sequences $d^k \in \mathbb{R}^n$, $\|d^k\| = 1$, $\sigma_k \downarrow 0$, such that

$$c^t(\sigma_k d^k) + Q(\sigma_k d^k) < 0.$$

Note that $c^t d^k \geq 0$, by (1) and feasibility of direction d^k. We have then

$$Q(d^k) < -c^t d^k / \sigma_k \leq 0. \tag{2}$$

By Hoffmann's lemma (Hoffman (1952)) there exists a critical direction $\hat{d}^k$ such that

$$\hat{d}^k = d^k + O(c^t d^k).$$

As $\mathcal{H}$ is bounded, the function Q is Lipschitz continuous. Using $Q(\hat{d}^k) \geq 0$ and the above equality, we get

$$Q(d^k) = Q(\hat{d}^k) + O(d^k - \hat{d}^k) \geq O(d^k - \hat{d}^k) = O(c^t d^k).$$

Combining with (2), we deduce that $c^t d^k / \sigma_k \leq O(c^t d^k)$, a contradiction.

(ii) **"Only if"** **part** Let 0 be a strict local solution of (EQP). By (i) we know that (1) holds. Let d be a critical direction. Then $d \in F(EQP)$ and $c^t d = 0$, therefore $Q(d) > 0$ as was to be proved.

"If part" Set $\alpha := \inf\{Q(d); d \in C, \|d\| = 1\}$. The continuous mapping Q attains its minimum on the compact set $\{d \in C, \|d\| = 1\}$; therefore $\alpha > 0$. We claim that $\bar{x}$ satisfies the quadratic groth condition with this value of the parameter α: it is enough to notice that $Q_\alpha(d) := Q(d) - \alpha\|d\|^2$ has the same set of critical directions C than Q and satisfies $Q_\alpha(d) \geq 0$, for all $d \in C$. Therefore, by point (i), $\bar{x}$ is a local minimum of $c^t d + Q_\alpha(d)$, which is equivalent to the desired conclusion. $\square$

3 APPLICATION TO NONLINEAR PROGRAMMING

We consider the nonlinear programming problem

$$(P) \qquad\qquad \text{Min } f(x); \quad g(x) \prec 0$$

where f and g are smooth mappings from $\mathbb{R}^n$ to $\mathbb{R}$ and $\mathbb{R}^p$, respectively, and $\prec$ stands for a finite number of equalities and inequalities, i.e. for $z \in \mathbb{R}^p$, $z \prec 0$ iff

$$z_i \leq 0 \text{ , for all } i \in I := \{1, \cdots, q\}, \text{ and } z_j = 0 \text{ for all } j \in J := \{q+1, \cdots, p\}.$$

The set of *active* inequality constraints is denoted

$$I(x) := \{i \in I; g_i(x) = 0\}.$$

For future references, let us denote the *critical cone* as

$$C(x) := \{d \in \mathbb{R}^n; \ f'(x)d \leq 0; \ g_i'(x)d \leq 0. i \in I(x); \ g_J'(x)d = 0\}.$$

We start with the case of linear constraints. i.e. $g(x) = Ax + b$, with A being a $p \times n$ matrix and $b \in \mathbb{R}^p$. Define the *tangent quadratic problem* to the linearly constrained problem (P) at x as

$$(Q_x^a) \qquad \underset{d}{\text{Min}} f'(x)d + \frac{1}{2}f''(x)dd \ ; \ g_i'(x)d \leq 0, \ i \in I(x) \ ; \ g_j'(x)d = 0, \ j \in J.$$

THEOREM 2 *Let x be a feasible point of the linearly constrained problem (P). Then:*

(i) If x is a local solution of (P). then $d = 0$ is a local solution of (Q_x^a).

(ii) The point 0 is a strict local solution of (Q_x^a) iff x is a local solution of (P) satisfying the quadratic growth condition.

Proof. (i) Let x be a local solution of (P). and let $d \in F(Q_x^a)$. Then $x(\sigma) := x + \sigma d$ is feasible for small enough σ. Since

$$f(x(\sigma)) = f(x) + \sigma f'(x)d + \frac{\sigma}{2}f''(x)dd + o(\sigma^2),$$

we deduce that $f'(x)d \geq 0$, and $f''(x)dd \geq 0$ whenever $f'(x)d = 0$. Therefore, by theorem 1(i), 0 is a local solution of (Q_x^a).

(*ii*) The constraints of (P) being linear, problems (P) and (Q_x^a) have (up to a translation) the same feasible set and second-order expansion of the cost function. Therefore, the quadratic growth condition holds for one of them iff it holds for the other. Consequently, point (*ii*) is a consequence of theorem 1(*ii*). □

We now turn to nonlinearly constrained problems. Let x be a feasible point of (P). The set of *Lagrange multipliers* associated with x is defined as:

$$\Lambda(x) := \left\{ \lambda \in \mathbb{R}^p; \nabla f(x) + g'(x)^t \lambda = 0; \lambda_i \geq 0; g_i(x) \leq 0; \lambda_i g_i(x) = 0, \forall i \in I \right\}.$$

We say that the Mangasarian-Fromovitz (1967) qualification condition holds at x if

$$(MF) \quad \begin{cases} \text{(i)} & \{\nabla_x g_i(x)\}, \; i \in J \text{ are linearly independent}, \\ \text{(ii)} & \exists d \in \mathbb{R}^n \; ; \; g_J'(x)d = 0 \; ; \; g_i'(x)d < 0, \; i \in I(x). \end{cases}$$

It is known that if $\bar{x}$ is a local minimum of (P), then $\Lambda(\bar{x})$ is non empty and bounded iff (MF) holds at $\bar{x}$ (Gauvin (1979)). The *Lagrangian function* associated with (P) is

$$\mathcal{L}(x, \lambda) := f(x) + \lambda^t g(x).$$

With x we associate the *extended quadratic tangent* (EQT) problem below:

$$(Q_x) \quad \underset{d}{\text{Min}} \; f'(x)d + \frac{1}{2} \max_{\lambda \in \Lambda(x)} \mathcal{L}''_{x^2}(x, \lambda)dd \; ; \quad g_i'(x)d \leq 0, \; i \in I(x); \quad g_J'(x)d = 0.$$

We use the convention that the maximum over an empty set is $-\infty$. Consequently, (Q_x) is defined at every feasible point of (P), and has value $-\infty$ if $\Lambda(x)$ is empty. Note that, if $\Lambda(x) \neq \emptyset$ and the constraints are linear, then (Q_x) is identical to (Q_x^a).

THEOREM 3 *let x be a feasible point of (P) satisfying (MF). Then:*
  (*i*) *If x is a local solution of (P), then $d = 0$ is a local solution of (Q_x).*
  (*ii*) *The point $d = 0$ is a strict local solution of (Q_x) if and only if x is local solution of (P) satisfying the quadratic growth condition.*

  Proof. Point (*i*) is a consequence of the second-order necessary condition (e.g. [2]) and theorem 1(*i*). The proof of (*ii*) is similar to the one of the second-order sufficient condition (e.g. [2]). □

Note that under the (MF) hypothesis, if no inequality constraint is active, then (Q_x) is an equality constrained quadratic problem, and local optimality for (Q_x) is equivalent to global optimality.

Let (P_1) and (P_2) be two nonlinear programming optimization problems over $\mathbb{R}^n$. We say that (P_1) and (P_2) are *tangent* at $x \in F(P_1) \cap F(P_2)$ if (P_1) and (P_2) have the same EQT problem at x. If x satisfies (MF) for both (P_1) and (P_2), then if follows from the above theorems that x satisfies the second order necessary (resp. sufficient) second order optimality condition for (P_1) iff it satisfies this condition for (P_2).

4 SINGULAR POINTS

A classical idea for dealing with points that do not necessarily satisfy qualification conditions is the following: with a feasible point $\bar{x}$ for (P), we associate the optimization problem (where $x \in \mathbb{R}^n$ and $z \in \mathbb{R}$):

$$(P_{\bar{x}}^s) \qquad \underset{x,z}{\text{Min}}\, z; \quad f(x) - f(\bar{x}) - z \leq 0; \quad g_i(x) - z \leq 0, i \in I, \quad g_J(x) = 0.$$

It is easily checked that if $\bar{x}$ is a local solution of (P), then $(\bar{x}, 0)$ is a local solution of $(P_{\bar{x}}^s)$. The displacement $(0_{\mathbb{R}^n}, 1)$ belongs to the kernel of equality constraints and satisfies strictly the linearized active inequality constraints. It follows that $(\bar{x}, 0)$ satisfies (MF) for $(P_{\bar{x}}^s)$ iff $g_J'(x)$ is onto.

The *singular Lagrangian function* associated with (P) is defined as

$$\mathcal{L}^s(x, \lambda) := \lambda_0 f(x) + \sum_{i \in I \cup J} \lambda_i g_i(x),$$

with $\lambda = (\lambda_0, \cdots, \lambda_p)$. The EQT problem for problem $(P_{\bar{x}}^s)$, associated with $(\bar{x}, 0)$, may be written as

$$(Q_{\bar{x}}^s) \qquad \underset{d^x, d^z}{\text{Min}}\, d^z + \frac{1}{2} \max_{\lambda \in \Lambda^g(x)} (\mathcal{L}^s)_{x^2}''(\bar{x}, \lambda) d^x d^x; f'(\bar{x}) d^x \leq d^z;$$
$$g_i'(\bar{x}) d^x \leq d^z, i \in I(\bar{x}); g_J'(\bar{x}) d = 0,$$

where $\Lambda^g(x)$ is the set of normalized *generalized Lagrange multipliers* associated with $(\bar{x}, 0)$ (John (1948)):

$$\Lambda^g(\bar{x}) := \{\lambda = (\lambda_0, \cdots, \lambda_p); \ \lambda_0 + \sum_{i \in I(x)} \lambda_i = 1; \ \lambda_0 \geq 0;$$
$$\lambda_i \geq 0, \ \lambda_i g_i(\bar{x}) = 0, \ i \in I \ ; \ \lambda_0 f'(\bar{x}) + \sum_{i=1}^{p} \lambda_i g'(\bar{x}) = 0\}.$$

Note that an equivalent formulation of $(Q_{\bar{x}}^s)$ is:

$$\underset{d \in \mathbb{R}^n}{\text{Min}}\, \max[f'(\bar{x}) d, g_i'(\bar{x}) d, i \in I(\bar{x})] + \frac{1}{2} \max_{\lambda \in \Lambda^g(\bar{x})} (\mathcal{L}^s)_{x^2}''(\bar{x}, \lambda) dd; \ g_J(\bar{x}) d = 0.$$

THEOREM 4 *Let $\bar{x}$ be a feasible point of (P) such that $g_J'(\bar{x})$ is onto. Then:*

(i) If $\bar{x}$ is a local solution of (P), then $(\bar{x}, 0)$ is a local solution of $(P_{\bar{x}}^s)$, and 0 is a local solution of $(Q_{\bar{x}}^s)$.

(ii) The point 0 is a strict local solution of $(Q_{\bar{x}}^s)$ if and only if $(\bar{x}, 0)$ is a local solution of $(P_{\bar{x}}^s)$ satisfying the quadratic growth condition. If in addition the (MF) hypothesis holds, this is equivalent to the fact that $\bar{x}$ is a local solution of (P) satisfing the quadratic growth condition.

Proof.   Point (i) is a consequence of theorem $3(i)$, while point (ii) is a consequence of theorem $3(ii)$ and of the fact that, under the (MF) condition, $\bar{x}$ satisfies the quadratic growth condition for (P) iff it satisfies the quadratic growth condition for $(P_{\bar{x}}^s)$: see e.g. Bonnans and Ioffe (1995), section 5. $\square$

5 STRONG SECOND-ORDER CONDITIONS

Some strong forms of the second-order sufficient conditions are used in the stability analysis of solutions of perturbed nonlinear programs, see Fiacco (1983), Levitin (1994) and Bonnans and Shapiro (1996). We show that two of them, that maybe are the most important, may be expressed using the concepts presented here.

Robinson's *strong stability* condition (Robinson (1980)) is in fact a general stability condition for the sum of a smooth mapping and of a multivalued operator. It says that the *linearized operator*, that is the sum of the the multivalued operator and of the linearization of the smooth mapping, is locally the inverse of a Lipschitz mapping. This concept is useful in the study of numerical algorithms and for conducting a perturbation analysis. When applied to the first-order optimality system of a nonlinear program, strong stability is known to have a simple characterization (see e.g. Bonnans, Sulem (1995)) for a simple proof and references therein). Let (x, λ) be a solution of the first order optimality system of (P) (i.e., $\lambda \in \Lambda(x)$). Define

$$I_+(x) := \{i \in I; g_i(x) = 0 \text{ and } \lambda_i > 0\}.$$

A characterization of strong regularity is

$$(SR) \quad \begin{cases} (i) & \{\nabla_x g_i(x)\}, \ i \in J \cup I(x) \text{ are linearly independent,} \\ (ii) & \mathcal{L}''_{x^2}(x, \lambda)dd > 0, \ \forall d \in \mathbb{R}^n \backslash \{0\}, \ g'_i(x)d = 0, \ i \in J \cup I_+(x). \end{cases}$$

Point (i) is a constraint qualification hypothesis, that implies uniqueness of the Lagrange multiplier, while (ii) is a strengthened form of the second-order sufficient condition (since the set of directions d in (ii) contains the critical cone).

As (x, λ) satisfies the first order optimality system of (P), a statement equivalent to (ii) is: 0 is the unique solution of the equality constrained quadratic problem

$$\underset{d}{\text{Min}} \ f'(x)d + \frac{1}{2}\mathcal{L}''_{x^2}(x, \lambda)dd; g'_i(x)d = 0, \ i \in J \cup I_+(x).$$

Taking point $(SR)(i)$ into account, we see that (SR) is equivalent to the fact that 0 is the unique solution, associated with a unique multiplier, of the equality constrained quadratic problems

$$\underset{d}{\text{Min}} \ f'(x)d + \frac{1}{2}\mathcal{L}''_{x^2}(x, \lambda)dd; g'_i(x)d = 0, \ i \in K,$$

for any K such that

$$J \cup I_+(x) \subset K \subset J \cup I(x).$$

We now turn to the discussion of a condition that proved to be useful for the study of optimization problems of the form

$$\text{Min} \ f(x, u); \quad g(x, u) \prec 0$$

where for simplicity we assume that $u \in \mathbb{R}_+$. An associated linearized problem, expressed at the point $(\bar{x}, u = 0)$ is

$$(L^*) \quad \underset{d}{\text{Min}} \ f'(\bar{x}, 0)(d, 1); \quad g'_i(\bar{x}, u)(d, 1) \leq 0, i \in I(\bar{x}); \ g'_j(\bar{x}, 0)(d, 1) = 0, j \in J.$$

Its dual problem is

$$(D^*) \qquad\qquad \operatorname*{Max}_\lambda \mathcal{L}'_u(\bar{x}, \lambda, 0); \quad \lambda \in \Lambda(\bar{x}).$$

The set of solutions of the dual problem is therefore a subset of the set of Lagrange multipliers. It may be characterized through the complementarity conditions, as follows:

$$S(D^*) = \{\lambda \in \Lambda(\bar{x}); \ \lambda_i = 0, \forall i \in I^*(\bar{x})\},$$

where

$$I^*(\bar{x}) = \{i \in I(\bar{x}); \ \exists d \in S(L); d_i > 0\}.$$

The statement of Shapiro's condition (Shapiro 1988) is

$$\forall d \in C(\bar{x}); \ \exists \lambda \in S(D^*); \ \mathcal{L}''_{x^2}(\bar{x}, \lambda) dd > 0.$$

Used in connection with a certain directional qualification hypothesis, this condition allows to check that the variation of the solution of the optimization problem is of the order of the perturbation in the data (Shapiro (1988), Auslender-Cominetti (1990), Bonnans-Ioffe-Shapiro (1992)).

Using theorem 1, we may restate Shapiro's condition as follows: 0 is a strict local solution of the extended quadratic optimization problem

$$\operatorname*{Min}_d f'(\bar{x})d + \frac{1}{2} \operatorname*{max}_{\lambda \in S(D^*)} \mathcal{L}''_{x^2}(\bar{x}, \lambda) dd; g_i'(\bar{x})d \leq 0, i \in I(\bar{x}); \ g_J'(\bar{x})d = 0.$$

It is not clear, however, if the above problem may be interpretated as an EQT problem.

6   RELATION WITH NEWTON'S METHOD FOR CONSTRAINED OPTIMIZATION

A natural extension of Newton's methods (for solving systems of non linear equations) to constrained optimization consists in linearizing the first order optimality system at a candidate point (x, λ) in order to compute a convenient displacement (d_0, μ_0). The linearization may be done in such a way that d_0 and $\lambda + \mu_0$ are solution of the optimality system of the quadratic problem

$$(Q(x, \lambda)) \ \operatorname*{Min}_d f(x)'d + \frac{1}{2}\mathcal{L}''_{x^2}(x, \lambda)dd; \quad g_I(x) + g_I'(x)d \leq 0; \quad g_J(x) + g_J'(x)d = 0.$$

Note that when $(x, \lambda) = (\bar{x}, \bar{\lambda})$, with $\bar{x}$ a local solution and $\bar{\lambda}$ the unique associated Lagrange multiplier, then $(Q(x, \lambda))$ coincides, up to the non active inequality constraints, with the quadratic tangent problem. The basic algorithm (without linesearches) is as follows:

Algorithm
Choose $x^0 \in \mathbb{R}^n$ and $\lambda_0 \in \mathbb{R}^p$. such that $\lambda_I \geq 0; k - 0.$
1) Compute (d^k, λ^{k+1}). solution of the optimality system of $Q(x^k, \lambda^k)$.
2) $x^{k+1} := x^k + d^k. k := k + 1.$ Go to 1.

We discuss the convergence of (x^k, λ^k). We say that an algorithm is *locally convergent* if, given a starting point close enough to the solution, convergence to the solution always occurs. Because the above algorithm is based on Taylor expansion of data, and reduces to Newton's method applied to the gradient of the value function for unconstrained problems, we may hope no more than local convergence. Now let us assume that the algorithm starts from $(\bar{x}, \bar{\lambda})$, a local solution of (P) and an associated Lagrange multiplier. If $\bar{\lambda}$ is not the unique Lagrange multiplier associated with $\bar{x}$, then local convergence does not occur (take $x^k = \bar{x}$ and λ^k a non constant sequence of Lagrange multiplier associated with $\bar{x}$). Therefore, uniqueness of the Lagrange multiplier is a *necessary* condition for local convergence.

A desirable property, that is independent from local convergence, is that if the algorithm starts from $(\bar{x}, \bar{\lambda})$, then $(x^k, \lambda^k) = (\bar{x}, \bar{\lambda})$ for all $k > 0$. A necessary condition for this is that 0 is a strict local minimum of $(Q_{\bar{x}})$. (Otherwise, there exists a nonzero critical direction d such that $\mathbb{R}_+ d$ is solution of $(Q_{\bar{x}})$). Therefore, we see that uniqueness of the Lagrange multiplier and condition that 0 is a strict local minimum of $(Q_{\bar{x}})$ are necessary conditions for well posedness of the algorithm. The theorem below shows that they are also sufficient, provided the algorithm computes a displacement of "small" norm. This is a natural restriction, as otherwise the sequence might not converge, even if the starting point satisfies the hypotheses of the theorem. So, for simplicity, we assume the displacement of primal variables to be of minimum norm. It is remarkable that these weak hypotheses imply quadratic convergence.

THEOREM 5 *Let $\bar{x}$ be a local solution of (P) and $\bar{\lambda}$ be its unique associated Lagrange multiplier, such that $d = 0$ is an isolated local solution of $Q(\bar{x}, \bar{\lambda})$. Assume that (x^0, λ^0) is close enough to $(\bar{x}, \bar{\lambda})$ and (d^k, λ^k) is such that $\|d^k\|$ is of minimum norm among all solutions of the optimality system of $Q(x^k, \lambda^k)$. Then $(x^k, \lambda^k) \rightarrow (\bar{x}, \bar{\lambda})$ quadratically.*

Proof.   This is just a reformulation of theorem 6.1 of Bonnans (1994), that is stated in terms of the second-order sufficient condition. Therefore, the equivalence between the two theorems is a consequence of theorem 2. □

7   NONISOLATED SOLUTIONS

The standard second-order sufficient condition implies that the considered point is a strict local solution of problem (P). However, there are important classes of problems that have nonisolated solutions. In particular, convex problems, in the case the solution is not unique, do not have isolated solution. We briefly review some recent results concerning the links between an extended notion of quadratic growth and some recent second-order conditions. Then we extend the notion of EQT problem to nonisolated solutions.

Let S be a closed set of feasible points of (P) such that f has over S a constant value denoted $f(S)$. The distance of x to S is defined as

$$d_S(x) := \min\{\|x - y\|, y \in S\}.$$

A projection of x onto S, denoted $P_S x$, is a point of S where the minimum is attained. We say that S satisfies the *quadratic growth condition* if

$\exists \epsilon > 0; \, \exists \alpha > 0$ such that $f(x) \geq f(S) + \alpha d_S(x)^2$ if $x \in F(P), d_S(x) \leq \epsilon$.

Note that this is an extension of the definition given in section 2. Let us discuss some material borrowed from Bonnans-Ioffe (1995,1996). The *contingent cone* to S at x is defined as

$$T_S(x) := \{d \in \mathbb{R}^n; \, \exists \{x^k\} \in S, \, t_k \downarrow 0, (t_k)^{-1}(x^k - x) \to d\}.$$

The *normal cone* to S at x is the polar cone of $T_s(x)$, i.e.

$$N_S(x) := \{d \in \mathbb{R}^n; \, d^t y \leq 0, \, \forall y \in T_s(x)\}.$$

Given $\epsilon > 0$, the *approximate critical cone* at x is

$$C^\epsilon(x) := \{d \in \mathbb{R}^n; \, \text{dist}(d, C(x)) \leq \epsilon \|d\|\}.$$

We say that (P) is a *stable convex problem* if the set of Lagrange multipliers associated with a solution (that is the same at every solution) is non empty and bounded.

THEOREM 6 *Let S be a compact set of points over which f has a constant value. Assume (MF) to hold at each $x \in S$. Then*
  (i) A necessary condition for S to satisfy the quadratic growth condition is

$$\exists \alpha > 0, \epsilon > 0; \, \forall x \in S \text{ and } d \in C^\epsilon(x) \cap N_S(x): \max_{\lambda \in \Lambda(x)} \mathcal{L}''_{x^2}(x, \lambda) dd \geq \alpha \|d\|^2.$$

  (ii) Assume that the following estimate of distance to critical cones holds:

$$\exists \gamma > 0; \quad \text{dist}(d, C(x)) \leq \gamma \nabla f(x)^t d, \, \forall (x, d) \in S \times \mathbb{R}^n;$$
$$g'_i(x)d \leq 0, i \in I(x); \, g'_J(x)d = 0.$$

Then the condition of point (i) is necessary and sufficient for quadratic growth.
  (iii) Assume that (P) is a stable convex problem. Then the estimate of distance to critical cones of point (ii) holds and, consequently, the condition of point (i) is necessary and sufficient for quadratic growth.

  Proof.   Point (i) is proved in Bonnans-Ioffe (1995), theorem 3, point (ii) in Bonnans-Ioffe (1995), theorem 1 and proposition 2, and point (iii) in Bonnans-Ioffe (1996), theorems 2.3 and 4.1.   □

  By *necessary condition for quadratic growth* and *uniform estimate of distance to critical cones* we will refer to the conditions in points (i) and (ii), respectively, of the theorem. Note that in the case of an isolated solution, the estimate of distance to critical cones holds, being a consequence of Hoffmann's lemma (Hoffman (1952)). A simple example of a nonconvex problem where this does not hold is

$$\text{Min } x_1 x_2; \, x \in \mathbb{R}^2, \quad 0 \leq x_1 \leq 1, \quad 0 \leq x_2 \leq 1.$$

Several other sufficient second-order conditions for quadratic growth, in the framework of nonconvex programming, may be found in Bonnans-Ioffe (1995). An early reference on this subject, where the set of active constraints is assumed to be constant over S, is Shapiro (1988a).

We now define an EQT problem associated with a set of possible solutions S, as follows. With $x \in S$ we associate the optimal value function of problem (Q_x):

$$q_x(d) := f'(x)d + \frac{1}{2} \max_{\lambda \in \Lambda(x)} \mathcal{L}''_{x^2}(x, \lambda)dd \text{ if } d \in F(Q_x), +\infty \text{ otherwise.}$$

Then we define

$$\chi(x) := \inf\{q_{\bar{x}}(x - \bar{x}); \|x - \bar{x}\| = \text{dist}(x, S)\}.$$

In the case $S = \{\bar{x}\}$, we have of course $\chi(x) = q_{\bar{x}}(x)$.

Note that if (P) is a stable convex problem, then S is convex and $\chi(x) := q_{\bar{x}}(x - \bar{x})$, where $\bar{x}$ is the unique projection of x onto S. In addition, as the set of Lagrange multipliers is bounded and constant over S, and $F(Q_x)$ contains $F(P) - x$, $\chi(x)$ is a continuous function over $F(P)$.

THEOREM 7 *Let S be a compact set of points over which f has a constant value. Assume (MF) to hold at each $x \in S$, and the uniform estimate of distance to critical cones to hold. Then S satisfies the quadratic growth for (P) iff the problem*

$$(Q_S) \qquad\qquad \text{Min } \chi(x); \text{ dist}(x, S) \leq \epsilon$$

has solution S, and S satisfies the quadratic growth condition for (Q_S).

Note that, by theorem 6, the hypotheses on problem (P) are satisfied if (P) is a stable convex problem with a bounded set of solutions.

Proof. (a) Assume that S satisfies the quadratic growth condition for (P). If S does not satisfy the quadratic growth condition for (Q_S), then there exist $\{x^k\} \subset F(Q_S)$ such that $\text{dist}(x^k, S) \to 0$ and $\chi(x^k) \leq \text{dist}(x^k, S)^2/k$ (as the value of χ over S is 0). Let $\bar{x}^k = P_S x^k$ be such that $\chi(x^k) = q_{\bar{x}^k}(x^k - \bar{x}^k)$. Denote

$$\sigma_k := \|x^k - \bar{x}^k\|, \quad d^k := (x^k - \bar{x}^k)/\sigma_k.$$

Set $\epsilon > 0$. As

$$0 \geq \limsup \chi(x^k)/\sigma_k = q_{x^k}(d^k)/\sigma_k = \nabla f(\bar{x}^k)^t d^k,$$

we have by the uniform estimate of distance to critical cones that $d^k \in C^\epsilon(\bar{x}^k)$ for large enough k. Note that $f'(x^k)d^k \geq 0$ as $d^k \in F(Q_{\bar{x}^k})$ and (MF) condition holds over S. By theorem 6, the second-order necessary condition for quadratic growth implies

$$\chi(x^k) = q_{x^k}(d^k) \geq \frac{1}{2} \max_{\lambda \in \Lambda(x)} \mathcal{L}''_{x^2}(x, \lambda)dd \geq \alpha\|d\|^2,$$

for some $\alpha > 0$ not depending on k, which gives the desired contradiction.

(b) Let S satisfies the quadratic growth condition for (Q_S). Then in particular it satisfies the second-order necessary condition for quadratic growth. By theorem 6, S satisfies the quadratic growth condition for (P). $\qquad\square$

8   EXTENSION TO ABSTRACT CONSTRAINTS

We now turn our attention to a general optimization problem in Banach spaces (e.g. Bonnans-Cominetti (1996))

$$(P) \qquad\qquad \min_x f(x) : G(x) \in K,$$

where f and G are C^2 mappings from X to $\mathbb{R}$ and Y respectively, X and Y are Banach spaces, and K is a closed convex subset of Y. In order to state a natural extension of the EQT problem, we recall the definition of the first and second order tangent sets to K

$$T_K(y) \quad := \quad \{h \in Y : \text{ there exists } o(t) \text{ such that } y + th + o(t) \in K\},$$

$$T_K^2(y, h) \quad := \quad \{k \in Y : \text{ there exists } o(t^2) \text{ such that } y + th + \frac{1}{2}t^2 k + o(t^2) \in K\}.$$

A natural first-order approximation of the optimization problem at a point $\bar{x}$ is

$$(L) \qquad\qquad \min_d f'(\bar{x})d; \quad G'(\bar{x})d \in T_K(G(\bar{x})).$$

We assume that $\bar{x}$ satisfies an extension of (MF) condition, due to Robinson (1976)

$$(EMF) \qquad\qquad 0 \in \text{int } [G(x_0, 0) + G'_x(x_0, 0)X - K].$$

Under this condition, if $\bar{x}$ is a local solution of (P), then $v(L) = v(D) = 0$ where (D) is the dual of (L), i.e.

$$(D) \qquad\qquad \max_\lambda 0; \quad \mathcal{L}'_x(\bar{x}, \lambda) = 0; \quad \lambda \in N_K(G(\bar{x})),$$

and $S(D)$ is the set of Lagrange multipliers associated with $\bar{x}$. The analysis of critical directions leads to the problem (Cominetti (1990))

$$(L_d) \qquad \min_w f'(\bar{x})w + f''(\bar{x})dd; \quad G'(\bar{x})w + G''(\bar{x})dd \in T_K^2(G(\bar{x}), G'(\bar{x})d).$$

Assuming (EMF), a *second-order necessary condition* is that $v(L_d) \geq 0$ for all critical direction d (Cominetti (1990)). Another consequence of (EMF) is that the value of (L_d) is equal to that of its dual

$$(D_d) \qquad\qquad \max_\lambda \mathcal{L}''_{x^2}(\bar{x}, \lambda)dd - \sigma(\lambda, T_K^2(G(\bar{x}), G'(\bar{x})d)); \quad \lambda \in S(D).$$

Here σ stands for the support function

$$\sigma(y, Z) := \sup\{\langle y, z\rangle; z \in Z\}.$$

That $v(D_d) \geq 0$ appears as a natural extension of the second-order necessary condition in nonlinear programming (as this σ term is zero in that case). Therefore, a natural extension of the EQT problem is

$$(Q_x) \qquad \min_d f'(\bar{x})d + \frac{1}{2}v(D_d) \quad \text{if} \quad G'(\bar{x})d \in T_K(G(\bar{x})), \quad +\infty \text{ if not,}$$

and a natural question is whether the second-order necessary condition is equivalent to the fact that 0 is a local solution of (Q_x).

We say that the cone of critical directions $S(L)$ has the *approximation property* if, given $d \in F(L)$, there exists $\bar{d} \in S(L)$ with $\|d - \bar{d}\| = O(f'(\bar{x})d)$. By Hoffman (1952), the approximation property holds for a qualified problem with a finite number of equality and inequality constraints.

THEOREM 8 *Assume that the* (EMF) *condition is satisfied, that the cone of critical directions* $S(L)$ *has the approximation property, and that the cost function of* (Q_x) *is Lipschitz continuous over* $F(Q_x)$. *Then the second-order necessary condition holds iff 0 is a local solution of* (Q_x).

Proof. The proof is similar to the one of theorem $1(i)$. Indeed, set $Q(q) := \frac{1}{2}v(D_d)$. Then Q is positively homogeneous of order 2 by definition of second order tangent sets, and Lipschitz continuous over $F(Q_x)$ by hypothesis. These are the two properties used in the proof of theorem $1(i)$. $\square$

We note that the hypothesis that the cost function of (Q_x) is Lipschitz continuous over $F(Q_x)$ is always satisfied for positive definite optimization problems, as follows from section 4 in Shapiro (to appear).

Another approach to tangent quadratic problems is presented in Bonnans (1996), in the context of optimal control problems with polyhedric feasible sets.

REFERENCES

1 A. AUSLENDER AND R. COMINETTI, *First and second order sensitivity analysis of nonlinear programs under directional constraint qualification conditions*, Optimization 21(1990), pp. 351-363.

2 A. BEN-TAL, *Second-order and related extremality conditions in nonlinear programming*, J. Optimization Theory Applications 21(1980), pp. 143–165.

3 J.F. BONNANS, *Local analysis of Newton-type methods for variational inequalities and nonlinear programming*, Applied Mathematics Optimization 29(1994), pp. 161–186.

4 J.F. BONNANS, *Second order analysis for control constrained optimal control problems of semilinear elliptic systems.* Rapport de Recherche INRIA 3014, 1996.

5 J.F. BONNANS AND R. COMINETTI, *Perturbed optimization in Banach spaces I: a general theory based on a weak directional constraint qualification.* SIAM J. Control Optimization 34 (1996), 1151–1171.

6 J.F. BONNANS, R. COMINETTI AND A. SHAPIRO, *Second order necessary and sufficient optimality conditions under abstract constraints*, Rapport de Recherche INRIA 2952 (1996).

7 J.F. BONNANS AND A.D. IOFFE, *Quadratic growth and stability in convex programming problems with multiple solutions.* J. Convex Analysis 2(1995) (Special issue dedicated to R.T. Rockafellar), pp. 41–57.

8 J.F. BONNANS AND A.D. IOFFE, *Second-order sufficiency and quadratic growth for non isolated minima.* Mathematics of Operations Research 20 (1996), 801–817.

9 J.F. BONNANS, A.D. IOFFE AND A. SHAPIRO, *Développement de solutions exactes et approchées en programmation non linéaire.* Comptes Rendus de l'Académie des Sciences de Paris, t. 315, Série I. p. 119–123.

10 J.F. BONNANS AND A. SHAPIRO, *Optimization Problems with perturbations, A guided tour.* Rapport de Recherche INRIA 2872, 1996.

11 J.F. BONNANS AND A. SULEM, *Pseudopower expansion of solutions of generalized equations and constrained optimization problems,* Mathematical Programming 70(1995), pp. 123–148.

12 R. COMINETTI, *Metric regularity, tangent sets and second order optimality conditions,* Applied Mathematics and Optimization 21(1990), pp. 265-287.

13 A.V. FIACCO, *Introduction to sensitivity and stability analysis in nonlinear programming,* Academic Press, New York, 1983.

14 J. GAUVIN, *A necessary and sufficient regularity condition to have bounded multipliers in nonconvex programming,* Mathematical Programming 12(1977), pp. 136–138.

15 A. HOFFMAN, *On approximate solutions of systems of inequalities,* J. Research National Bureau of Standards, Sect. B49(1952), pp. 629-649.

16 A.D. IOFFE, *Necessary and sufficient conditions for a minimum,* SIAM J. Control and Optimization (1979), pp. 245-288.

17 F. JOHN, *Extremum problems with inequalities as subsidiary conditions,* in Studies and Essays, R. Courant anniversary volume, Interscience, New York, 1948, pp. 187-204.

18 E.S. LEVITIN, *Perturbation theory in mathematical programming and its applications.* J. Wiley, Chichester, 1994.

19 E.S. LEVITIN, A.A. MILJUTIN AND N.P. OSMOLOVSKI, *On conditions for a local minimum in a problem with constraints.* In "Mathematical economics and functional analysis, B.S. Mitjagin ed., Nauka, Moscow, 1974, pp. 139-202 (In Russian).

20 A. MAJTHAY, *Optimality conditions for quadratic programming,* Mathematical Programming 1(1971), pp. 359–365.

21 O.L. MANGASARIAN AND S. FROMOVITZ, *The Fritz John necessary optimality conditions in the presence of equality and inequality constraints,* J. of Mathematical Analysis and Applications 17(1967), pp. 37–47.

22 S.M. ROBINSON, *Stability theory for systems of inequalities, part II: differentiable nonlinear systems.* SIAM J. Numerical Analysis 13(1976). pp. 497–513.

23 S.M. ROBINSON, *Strongly regular generalized equations*, Mathematics of Operations Research 5(1980), pp. 43–62.

24 A. SHAPIRO, *Sensitivity analysis of nonlinear programs and differentiability properties of metric projections*, SIAM J. Control and Opt. 26(1988), pp. 628–645.

25 A. SHAPIRO, *Perturbation theory of nonlinear programs when the set of optimal solutions is not a singleton*, Applied Mathematics and Optimization 18(1988a), pp. 215-229.

26 A. SHAPIRO, *First and second order analysis of nonlinear semidefinite programs*, Mathematical Programming. Series B, to appear.

On Generalized Differentiability of
Optimal Solutions in Nonlinear Parametric Optimization

S. DEMPE, University of Leipzig, Department of Economics, 04109 Leipzig, Germany

Abstract: In this paper a smooth parametric optimization problem is considered under conditions guaranteeing strong stability of a (local) optimal solution x^0 of this problem for some fixed parameter y^0. It is well known that the function $x(\cdot)$ locally uniquely describing (local) optimal solutions near x^0 for perturbed problems, is then directionally differentiable, too. If some additional assumption is also valid, $x(\cdot)$ admits also a generalized Jacobian in the sense of Clarke and is quasidifferentiable in the sense of Dem'yanov and Rubinov. We give ideas for computing the generalized Jacobian and some representation of the quasidifferential and show that the additional assumption used is essential.

1 Introduction

The investigation of the local behavior of an optimal solution for a parametric optimization problem is one of the central topics in sensitivity and stability analysis [11]. Especially quantitative results on the dependency of a solution on the parameters (input data) of the problem are used and needed in many cases: They can be applied directly to give a local approximation of the set of efficient points in multicriterial optimization or to construct algorithms computing local or global optimal solutions (e.g. by means of embedding techniques) for optimization problems [14]. One of the main approaches in bilevel and multilevel programming uses these results [6, 8]. Last but not least we will mention that the verification of the quality of a computed solution for an optimization problem with respect to the initial situation, modelled by this problem, is supported by quantitative stability of the solution [1].

First results on this topic can be found in [12]. There it has been shown by the help of the classical implicit function theorem that the optimal solution of a parametric optimization problem

$$\min_x\{f(x,y) : h(x,y) = 0\} \tag{1}$$

$(f \in C^2(\mathbb{R}^n \times \mathbb{R}^m, \mathbb{R})$, $h \in C^2(\mathbb{R}^n \times \mathbb{R}^m, \mathbb{R}^q))$ is locally uniquely determined by a continuously differentiable function $x : U(y^0) \to V(x^0)$, mapping some open neighborhood $U(y^0)$ into an open neighborhood $V(x^0)$, provided that the linear independence constraint qualification (LICQ) together with a sufficient optimality condition of second order (SOC) are satisfied at the (local) optimal solution $x^0 = x(y^0)$ for problem (1) at $y = y^0$. Later on, inequality constraints have been introduced to problem (1):

$$\min_x\{f(x,y) : g(x,y) \leq 0, \; h(x,y) = 0\}, \tag{2}$$

where g is also assumed to be at least twice continuously differentiable ($g \in C^2(\mathbb{R}^n \times \mathbb{R}^m, \mathbb{R}^p)$). Then, the same result can be obtained if the strict complementarity slackness condition (SC) is added to (LICQ) and (SOC) [11]. Under these assumptions a local reduction to a problem with equality constraints only is possible. This transformation can not longer be used if the assumption (SC) is dropped. Moreover, it is even not true that the perturbed problems have unique local optimal solutions in $V(x^0)$ without strengthening the sufficient optimality condition of second order [2]. There it has been shown that perturbed problems can have only a finite number of (local) optimal solutions in $V(x^0)$ which define smooth functions of (local) optimal solutions provided that a so-called semistrong sufficient optimality condition of second order is satisfied. If we suppose that a strong sufficient optimality condition of second order (SSOC) is valid at (x^0, y^0) then local uniqueness of a (local) optimal solution $x(y)$ is retained [20]. But, in general, the function $x : U(y^0) \to V(x^0)$ can have kinks, it is in general not continuously differentiable at y^0. If a direction r is fixed, then the function $\xi(t) := x(y^0 + tr)$ depending on the scalar parameter t has been shown to permit a pseudopower expansion for $t \geq 0$ [3]. This is in general not true for general perturbations. The function $x(\cdot)$ is directionally differentiable in the sense that the limit

$$x'(y^0; r) := \lim_{t \to +0} t^{-1}[x(y^0 + tr) - x^0]$$

exists for all $r \in \mathbb{R}^m$ [16]. This function proves also to be locally Lipschitz continuous, i.e. there exists a constant $L \geq 0$ such that

$$\|x(y) - x(y')\| \leq L\|y - y'\|$$

for all y, y' in an open neighborhood which can be supposed to coincide with $U(y^0)$ [15, 16]. Additionally, the function $x(\cdot)$ is a PC^1-function, i.e. it is locally composed by a finite number of continuously differentiable functions [2]. As a locally Lipschitzian function, $x(\cdot)$ admits a generalized Jacobian in the sense of Clarke [4]. Formulae for the computation of at least one element of this generalized Jacobian can be found in [25, 26].

  In this paper we will investigate the quantitative behavior of a local optimal solution of the problem (2) under a weaker regularity assumption, namely the Mangasarian-Fromowitz constraint qualification instead of the (LICQ). By adding further assumptions we will be able to maintain all of the above results: Lipschitz continuity, directional differentiability, the possibility to compute elements of the generalized Jacobian. Additionally we give a method to compute the quasi-differential in the sense of Dem'yanov and Rubinov [9]. It should be mentioned that our regularity assumptions are substantially weaker then (LICQ) in the sense that they permit the existance of nonuniquely determined Lagrange multipliers of the unperturbed problem.

It should be noted that we consider only sufficiently smooth problems, i.e. problems with defining functions which are at least twice continuously differentiable. For problems with e.g. $C^{1,1}$ data, the reader is referred to [17, 19] and the references therein. We will also consider only problems in which the local optimal solution of the unperturbed problem is locally uniquely determined. For papers which do not use this assumption see e.g. [18, 31].

2   Directional differentiability

We consider the parametric nonlinear programming problem (2)

$$\min_x \{f(x, y) : g(x, y) \le 0, \ h(x, y) = 0\},$$

which depends on a given parameter $y \in R^m$, where $f : R^n \times R^m \to R$, $g : R^n \times R^m \to R^p$, and $h : R^n \times R^m \to R^q$. Let the following assumptions be satisfied:

(C) f, g and h are at least twice continuously differentiable near $(x^0, y^0) \in R^n \times R^m$, where x^0 is a local solution of the problem (2) for the fixed value $y = y^0$ of the parameter.

(MFCQ)   (Mangasarian-Fromowitz constraint qualification) The set of vectors $\{\nabla_x h_j(x^0, y^0) : j = 1, \ldots, q\}$ is linearly independent, and there exists $d \in R^n$ such that $\nabla_x g_i(x^0, y^0)d < 0$ for all $i : g_i(x^0, y^0) = 0$, and $\nabla_x h(x^0, y^0)d = 0$.

Let $L(x, y, \lambda, \mu) = f(x, y) + \lambda^\top g(x, y) + \mu^\top h(x, y)$ denote the Lagrangian of problem (2). The assumption (MFCQ) guarantees that the set of Lagrange multipliers

$$\Lambda(x^0, y^0) = \{(\lambda, \mu) : \nabla_x L(x^0, y^0, \lambda, \mu) = 0, \ \lambda \ge 0, \ \lambda^\top g(x^0, y^0) = 0\}$$

is a nonempty and compact polyhedron if x^0 is a local optimal solution of the problem (2) at $y = y^0$. Let, for $(\lambda, \mu) \in \Lambda(x^0, y^0)$, the set

$$K_0(\lambda) = \{d \in R^n \setminus \{0\} : \nabla_x g_i(x^0, y^0)d = 0 \text{ for } \lambda_i > 0, \ \nabla_x h(x^0, y^0)d = 0\}$$

be defined.

(SSOC)   (strong sufficient optimality condition of second order) For each $(\lambda, \mu) \in \Lambda(x^0, y^0)$ and each $d \in K_0(\lambda)$,

$$d^T \nabla_{xx}^2 L(x^0, y^0, \lambda, \mu)d > 0.$$

These assumptions together imply that the solution x^0 is strongly stable in the sense of Kojima:

Theorem 2.1 *[20] Under assumptions (C), (MFCQ) and (SSOC), there are open neighborhoods U of y^0 and V of x^0, and a uniquely determined function $x(\cdot) : U \to V$ satisfying: $x(\cdot)$ is continuous and, for each $y \in U$, $x(y)$ is the unique local solution of problem (2) in V.*

To get even stronger results about the function $x(\cdot)$ we need one additional constraint qualification:

(CRCQ) (constant rank constraint qualification) There exists an open neighborhood W of (x^0, y^0) such that for any subsets $I \subseteq I^0 = \{i : g_i(x^0, y^0) = 0\}$ and $J \subseteq \{1, \dots, q\}$, the family of gradient vectors $\{\nabla_x g_i(x, y) : i \in I\} \cup \{\nabla_x h_j(x, y) : j \in J\}$ has the same rank (depending on I and J) for all vectors $(x, y) \in W$.

The function $x(\cdot)$ given in Theorem 2.1 is called a PC^1-function near y^0 provided that it is continuous and there exists a finite number of continuously differentiable functions $x^1(\cdot), \dots, x^k(\cdot)$ mapping some open neighborhood U of y^0 to some open neighborhood V of x^0 with

$$x(y) \in \{x^1(y), \dots, x^k(y)\}$$

for each $y \in U$.

Theorem 2.2 *[27] Let the assumptions (C), (MFCQ), (SSOC) and (CRCQ) be satisfied at some local solution x^0 for the problem (2) with $y = y^0$. Then, the function $x(\cdot)$ defined in Theorem 2.1 is a PC^1-function near y^0.*

In the proof of this theorem it is shown that, in place of the functions $x^i(\cdot)$, we can use functions given by local optimal solutions of the perturbed problems

$$\min_x \{f(x, y) : g_i(x, y) = 0, \; i \in I, \; h(x, y) = 0\}, \tag{3}$$

where some of the inequality constraints have been replaced by equations and the others have been dropped. Here, the set I can be chosen such that it satisfies the following two conditions:

(D1) $\{j : \lambda_j > 0\} \subseteq I \subseteq I^0$ for some Lagrange multiplier vector $(\lambda, \mu) \in \Lambda(x^0, y^0)$, and

(D2) $\{\nabla_x g_i(x^0, y^0) : i \in I\} \cup \{\nabla_x h_j(x^0, y^0) : j = 1, \dots, q\}$ are linearly independent.

Corollary 2.3 *Under the assumptions of Theorem 2.2, the function $x : U \to V$ defined in Theorem 2.1 is*

1. *locally Lipschitz continuous at y^0, i.e. there exists a constant c such that $\|x(y) - x(y')\| \le c\|y - y'\| \; \forall \; y, y' \in V$ [15],*

2. *directionally differentiable at y^0 [5, 32].*

3. *As a function of the direction only, the directional derivative of the function x is also Lipschitz continuous at y^0 with the same constant c [9].*

It should be noticed that the third condition of this Corollary follows from the first one. Also, directional differentiability of the function x has been shown in [5] without the assumption (CRCQ) and can be obtained as a direct consequence of the results in [32] under even weaker conditions than (MFCQ) and (SSOC). The following example (which is borrowed from [32]) shows that the above assumptions are necessary for obtaining local Lipschitz continuity of the function x:

Example 2.4 Consider the problem

$$\tfrac{1}{2}(x_1 - 1)^2 + \tfrac{1}{2}x_2^2 \ \rightarrow \min$$

$$x_1 \leq 0$$

$$x_1 + y_1 x_2 + y_2 \leq 0$$

at $y^0 = (0,0)^\top$. This is a convex parametric optimization problem satisfying the assumption (C). At the unique optimal solution $x^0 = (0,0)^\top$, the assumptions (MFCQ) and (SSOC) are also valid. Hence, the solution x^0 is strongly stable. It can easily be seen that the unique optimal solutions $x(y)$ are given by

$$x(y) = \begin{cases} (0,0)^\top & \text{if } y_2 \leq 0, \\ \left(0, -\frac{y_2}{y_1}\right)^\top & \text{if } 0 < y_2 \leq y_1^2, \\ \left(\frac{y_1^2 - y_2}{1 + y_1^2}, -\frac{y_1 + y_1 y_2}{1 + y_1^2}\right)^\top, & \text{if } y_1^2 \leq y_2. \end{cases}$$

But, the function $x(\cdot)$ is not locally Lipschitz continuous at y^0. This can be seen by the following: Take $y_2 = y_1^2 - y_1^3 > 0$. Then

$$\frac{\|x(y_1, y_1^2 - y_1^3) - x(y_1, 0)\|}{\|(y_1, y_1^2 - y_1^3) - (y_1, 0)\|} = \frac{y_1 - y_1^2}{y_1^2 - y_1^3} \rightarrow \infty$$

for $y_1 \rightarrow 0$ from above. Hence, the function $x(\cdot)$ cannot be locally Lipschitz continuous. $\qquad\Box$

Another example in [24] shows that it also not possible to weaken the smoothness assumptions related to the functions describing the problem (2) while maintaining Lipschitz continuity of the solution function.

For the computation of the directional derivative of the function x an practicable way has been found in [27]. This method says that it is sufficient to solve first a linear optimization problem depending on the direction r and then to compute $x'(y^0; r)$ by solving a quadratic optimization problem depending on the direction r <u>and</u> an arbitrary optimal solution of the linear problem: Let (λ^0, μ^0) be an arbitrary optimal solution of the linear problem

$$\max_{(\lambda, \mu)} \{\nabla_y L(x^0, y^0, \lambda, \mu) r : (\lambda, \mu) \in \Lambda(x^0, y^0)\}. \tag{4}$$

Then, $x'(y^0; r)$ is equal to the unique optimal solution of the following quadratic optimization problem $(QP(r))$:

$$\frac{1}{2} d^\top \nabla_{xx}^2 L(x^0, y^0, \lambda^0, \mu^0) d + d^\top \nabla_{xy}^2 L(x^0, y^0, \lambda^0, \mu^0) r \rightarrow \min_d$$

$$\nabla_x g_i(x^0, y^0) d + \nabla_y g_i(x^0, y^0) r \begin{cases} = 0, & \lambda_i^0 > 0 \\ \leq 0, & \lambda_i^0 = g_i(x^0, y^0) = 0 \end{cases}$$

$$\nabla_x h_j(x^0, y^0) d + \nabla_y h_j(x^0, y^0) r = 0, \ \forall \ j = 1, \ldots, q.$$

If (LICQ) is satisfied then the set of Lagrange multiplier vectors $\Lambda(x^0, y^0)$ reduces to a singleton and it is not longer necessary to solve the problem (4) in order to select a suitable Lagrange multiplier. In this case, our result reduces to the one by Jittorntrum [16]. If the assumption (CRCQ) is not satisfied but (MFCQ) together with (SSOC), then the directional derivative of the function $x(\cdot)$ exists and can be computed by solving the problem $(QP(r))$ for some element of the set $\Lambda(x^0, y^0)$. In this case it is difficult to determine which Lagrange multiplier is to be selected [5]. It has been shown in [32], that under this (and even slightly weaker) assumptions the computation of the directional derivative of $x(\cdot)$ is possible by solving a certain minimax problem. Under (SSOC) and (MFCQ) the inner maximization in this problem is not longer necessary since the optimal solution of this inner problem is uniquely determined by strong convexity [30]. The following example taken from [27] can be used to verify that it is really not possible to suppress the special selection of the Lagrange multiplier in the quadratic problem:

Example 2.5 The problem below, which satisfies all assumptions, shows that $x'(y; r)$ is not necessarily determined by $(QP(r))$ for any multiplier $\lambda \in \Lambda(x^0, y^0)$:

$$\min_x \{(x-1)^2 : x + y \le 0, \ x - y \le 0\}.$$

Let $y^0 = 0$. Then, $x^0 = 0$ and $\Lambda(x^0, y^0) = $ conv $\{(2, 0), (0, 2)\}$, where conv A denotes the convex hull of the set A. Taking $\lambda = (2, 0)$ and $r = -1$ yields an empty set of feasible points for the problem $(QP(r))$. The optimal solution of the problem (4) for this direction is $\lambda = (0, 2)$. With this solution problem $(QP(-1))$ has the solution $d = -1$ which is correct since $x(y) = -|y|$. $\qquad \square$

3   The generalized Jacobian of the optimal solution

Since the function $x(\cdot)$ is locally Lipschitz continuous at y^0 it admits a generalized Jacobian $\partial x(y^0)$ in the sense of Clarke [4]:

$$\partial x(y^0) := \text{conv } \{ \ H \in \mathbf{R}^n \times \mathbf{R}^m : \exists \{y^k\}_{k=1}^{\infty} \text{ converging to } y^0 \text{ such that} \\ \nabla x(y^k) \text{ exists for all } k \text{ and } \lim_{k \to \infty} \nabla x(y^k) = H \}.$$

By the local representation of $x(\cdot)$ as a PC^1-function and by the choice of the selection functions in the previous section, the Jacobians of the latter functions should give elements of the generalized Jacobian of the function $x(\cdot)$. These Jacobians can be computed as follows: Denote the optimal solution and the unique Lagrange multiplier vector of problem (3) for a fixed set I by $x^I(\cdot)$ and $(\lambda^I(\cdot), \mu^I(\cdot))$, resp. Then, applying the implicit function theorem (cf. [12]), we get

$$\begin{pmatrix} \nabla x^I(y^0) \\ \nabla \lambda^I(y^0) \\ \nabla \mu^I(y^0) \end{pmatrix} = B_I^{-1} N_I, \tag{5}$$

where

$$B_I = \begin{pmatrix} \nabla_{xx}^2 L_I(x^0, y^0, \lambda^I(y^0), \mu^I(y^0)) & \nabla_x^\top g_I(x^0, y^0) & \nabla_x^\top h(x^0, y^0) \\ \nabla_x g_I(x^0, y^0) & 0 & 0 \\ \nabla_x h(x^0, y^0) & 0 & 0 \end{pmatrix},$$

$$N_I = \begin{pmatrix} -\nabla^2_{xy} L_I(x^0, y^0, \lambda^I(y^0), \mu^I(y^0)) \\ -\nabla_y g_I(x^0, y^0) \\ -\nabla_x h(x^0, y^0) \end{pmatrix},$$

and notations $L_I(x, y, \lambda^I, \mu^I) = f(x, y) + {\lambda^I}^\top g_I(x, y) + {\mu^I}^\top h(x, y)$, $g_I(x, y) = (g_i(x, y))_{i \in I}$ are used. This shows that the main problem for computing elements of the generalized Jacobian $\partial x(y^0)$ is the choice of a suitable set I (which has to satisfy the conditions (D1) and (D2)). Define the sets

$$Y^I := \{y : x(y) = x^I(y)\}.$$

Then we have

Theorem 3.1 *[21, 22, 29] Let (C), (MFCQ), (SSOC), and (CRCQ) be satisfied for problem (2) at (x^0, y^0). Then,*

$$\partial x(y^0) = \text{conv} \{\nabla x^I(y^0) : y^0 \in \text{cl (int } Y^I)\}.$$

Having only full information about the problems (2) at the point $y = y^0$ and $(QP(r))$ for each r, the verification if int $Y^I \neq \emptyset$ is not an easy task. An example showing this can be found in [6]. On the other hand it is not very difficult to see that the properties of the problem $(QP(r))$ can be used to get some information of whether a certain Jacobian matrix $\nabla x^I(y^0)$ will indeed contribute to the generalized Jacobian of $x(\cdot)$ at $y = y^0$.

Theorem 3.2 *[6] Let the assumptions (C), (MFCQ), (SSOC), and (CRCQ) be satisfied for problem (2). Take a set I satisfying (D1), (D2) for some $(\lambda^0, \mu^0) \in \Lambda(x^0, y^0)$ and let the system*

$$\nabla_x g_i(x^0, y^0)d + \nabla_y g_i(x^0, y^0)r \begin{cases} = 0, & i \in I, \\ < 0, & i \in I^0 \setminus I, \end{cases}$$
$$\nabla_x h(x^0, y^0)d + \nabla_y h(x^0, y^0)r = 0$$

have a feasible solution (d^0, r^0). Then, if strict complementary slackness is satisfied for the optimal solution $x'(y^0; r^0)$ of the problem $(QP(r^0))$, i.e. if the following equality holds:

$$I = \{i : \nabla_x g_i(x^0, y^0)x'(y^0; r^0) + \nabla_y g_i(x^0, y^0)r^0 = 0\} = \{i : \nu_i > 0\} \cup \{j : \lambda_j^0 > 0\}$$

then $\nabla x^I(y^0) \in \partial x(y^0)$ for the corresponding function $x^I(\cdot)$.

Here, ν_i are the Lagrange multipliers corresponding to the first set of constraints in problem $(QP(r))$. This Theorem enables us to compute at least one element of the generalized Jacobian. For computing the whole generalized Jacobian, we need a more restrictive assumption.

(E) For each vertex (λ^0, μ^0) of $\Lambda(x^0, y^0)$, the matrix M given as

$$M = \begin{pmatrix} \nabla^2_{xx} L^0 & \nabla^\top_x g^0_{J(\lambda^0)} & \nabla^\top_x h^0 & \nabla^2_{xy} L^0 \\ \nabla_x g^0_{I^0} & 0 & 0 & \nabla_y g^0_{I^0} \\ \nabla_x h^0 & 0 & 0 & \nabla_y h^0 \end{pmatrix}$$

has full row rank $n + |I^0| + q$, where $L^0 := L(x^0, y^0, \lambda^0, \mu^0)$, $g^0_I := g_I(x^0, y^0)$ and $h^0 := h(x^0, y^0)$, $I^0 := \{j : g_j(x^0, y^0) = 0\}$ and $J(\lambda^0) := \{j : \lambda_j^0 > 0\}$.

By [20], the rank of the matrix M is not less than $n + |J(\lambda^0)| + q$, where (λ^0, μ^0) is a vertex of $\Lambda(x^0, y^0)$ (since $B_{J(\lambda^0)}$ is regular). Hence the addition of $|I^0 \setminus J(\lambda^0)|$ rows and m columns to the matrix $B_{J(\lambda^0)}$ must increase the rank of this matrix by $|I^0 \setminus J(\lambda^0)|$ if assumption (E) is to be satisfied.

Theorem 3.3 *Consider the problem (2) at a point (x^0, y^0) and let the assumptions (C), (MFCQ), (SSOC), (CRCQ), and (E) be satisfied there. Take any set I satisfying the conditions (D1) and (D2). Then, $\nabla x^I(y^0) \in \partial x(y^0)$.*

Proof: By the assumptions (D2) and (SSOC), the optimal solution $x^I(\cdot)$ of the problem (3) is continuously differentiable [12]. By assumption (D1), $x^I(y^0) = x(y^0)$. Let $(\lambda^0, \mu^0) \in \Lambda(x^0, y^0)$ be taken accordingly to (D1). Consider the necessary and sufficient optimality conditions of first order for problem $(QP(r))$:

$$\nabla^2_{xx} L(x^0, y^0, \lambda^0, \mu^0)d + \nabla^2_{xy} L(x^0, y^0, \lambda^0, \mu^0)r +$$
$$\nabla^T_x g_{I^0}(x^0, y^0)\nu + \nabla^T_x h(x^0, y^0)\eta = 0,$$
$$\nabla_x g_i(x^0, y^0)d + \nabla_y g_i(x^0, y^0)r = 0, \; i \in J(\lambda^0),$$
$$\nabla_x g_i(x^0, y^0)d + \nabla_y g_i(x^0, y^0)r \le 0, \; i \in I^0 \setminus J(\lambda^0),$$
$$\nabla_x h_j(x^0, y^0)d + \nabla_y h_j(x^0, y^0)r = 0, \; j = 1, \ldots, q,$$
$$\nu_i(\nabla_x g_i(x^0, y^0)d + \nabla_y g_i(x^0, y^0)r) = 0, \; i \in I^0 \setminus J(\lambda^0),$$
$$\nu_i \ge 0, i \in I^0 \setminus J(\lambda^0).$$

Let, without loss of generality,

$$J(\lambda^0) = \{1, \ldots, s\}, \; I = \{1, \ldots, u\}, \; I^0 = \{1, \ldots, v\}$$

for $s \le u \le v$. Then, by (E), the matrix

$$M^0 := \begin{pmatrix} \nabla^2_{xx} L^0 & \nabla^T_x g^0_{I^0} & \nabla^T h^0 & \nabla^2_{xy} L^0 \\ \nabla_x g^0_{I^0} & 0 & 0 & \nabla_y g^0_{I^0} \\ \nabla_x h^0 & 0 & 0 & \nabla_y h^0 \\ 0 & e^T_{s+1} & 0 & 0 \\ 0 & \vdots & 0 & 0 \\ 0 & e^T_v & 0 & 0 \end{pmatrix}$$

has rank $n + |I^0| + q + |I^0 \setminus J(\lambda^0)|$ (note that we have added $|I^0 \setminus J(\lambda^0)|$ columns and the same number of rows which contain a unit matrix of full dimension). Here e_i denotes the i-th unit vector. Hence, the system of linear equations $M^0(d, \nu, \eta, r)^T = a$ has a solution for arbitrary right-hand side a. Take a right-hand side vector a which has the value $-\varepsilon < 0$ for each component corresponding to a left-hand side

$$\nabla_x g_i(x^0, y^0)d + \nabla_y g_i(x^0, y^0)r, \; i \in I^0 \setminus I,$$

the value ε in each component for left-hand side ν_i, $i \in I \setminus J(\lambda^0)$, and vanishes in all other components. Let $(d^0, \nu^0, \eta^0, r^0)^T$ be a solution of the resulting linear system. Then, $(d^0, \nu^0, \eta^0)^T$ satisfies the Karush-Kuhn-Tucker conditions for the problem $(QP(r^0))$. Moreover, strict complementarity slackness is satisfied for this system. This implies the assumptions of Theorem 3.2 and proves the theorem. $\square$

Remark 3.4 *Under assumptions which are much more restrictive than ours, Malanowski [25] has obtained a result which is similar to Theorem 3.3.*

Corollary 3.5 *Under the assumptions of Theorem 3.3 we have*

$$\partial x(y^0) = \operatorname{conv}\left\{\bigcup_{(\lambda,\mu)\in E\Lambda(x^0,y^0)}\bigcup_{I\in\mathcal{M}(\lambda,\mu)}\{\nabla x^I(y^0)\}\right\},$$

where $E\Lambda(x^0,y^0)$ denotes the vertex set of $\Lambda(x^0,y^0)$ and $\mathcal{M}(\lambda,\mu)$ is the family of all sets I satisfying conditions (D1) and (D2) for a fixed pair (λ,μ).

Remark 3.6 *For computing the generalized Jacobian of $x(\cdot)$ we have to proceed as follows: First compute the set of all vertices of the bounded polyhedron $\Lambda(x^0,y^0)$ by means of appropriate algorithms (cf.e.g. [10]). Then, for each vertex (λ^0,μ^0) determine the family of all sets $I\in\mathcal{M}(\lambda^0,\mu^0)$ and compute $\nabla x^I(y^0)$ by means of equation (5) or by solving an equivalent system of linear equations.*

The following example can be used to verify that the assumptions of the Theorems 3.2 and 3.3 are indeed necessary.

Example 3.7 Consider the simple problem

$$\min_x\{(x_1 - y)^2 + (x_2 - 1)^2 : x_1 + x_2 \le 1,\ -x_1 + x_2 \le 1\}$$

at the point $y^0 = 0$. Then, $x = (0,1)^\top$ is the unique optimal solution and the assumptions (C), (MFCQ), (SSOC), and (CRCQ) are satisfied there. It is easy to see that

$$x(y) \in \{(y/2, 1 - y/2)^\top, (y/2, 1 + y/2)^\top\}$$

for y near y^0, and $\Lambda(x^0,y^0) = \{(0,0)^\top\}$. The sets $I \in \{\emptyset, \{1\}, \{2\}, \{1,2\}\}$ are to be considered in order to compute the generalized Jacobian of the function $x(\cdot)$. The assumption (E) is not satisfied for this example. The assumptions of the Theorem 3.2 are satisfied for the sets $I = \{1\}$ and $I = \{2\}$ but not for $I = \emptyset$ and $I = \{1,2\}$. For $I = \{1\}$ we compute $\nabla x^I(y^0) = (0.5, -0.5)^\top$. For $I = \{2\}$ we get $\nabla x^I(0) = (0.5, 0.5)^\top$. Hence, $\nabla x^I(0) \in \partial x(0)$ for both sets $I = \{1\}$ and $I = \{2\}$. But, for $I = \{1,2\}$ we compute $x^I(0) = (0,0)^\top \notin \partial x(0)$. Here, strict complementary slackness cannot be satisfied for the problem $(QP(r))$ since we get $\nu = (1,-1)^\top$. Last but not least, we see that $\nabla x^I(0) = (1,0)^\top \notin \partial x(0)$ for $I = \emptyset$. Clearly, there is also no solution (d,ν) of the problem $(QP(r))$ with $d = (1,0)^\top$. The strict complementary slackness assumption for the problem $(QP(r))$ is also not satisfied with $I = \emptyset$. $\qquad\Box$.

4 The quasidifferential of the solution function

In the last part of this paper we will show that the function $x(\cdot)$ is also quasi-differentiable in the sense of Dem'yanov and Rubinov [9] provided that the above assumptions are satisfied. A function $z : \mathbb{R}^l \to \mathbb{R}$ is called <u>quasidifferentiable</u> at $w^0 \in \mathbb{R}^l$ if it is directionally differentiable at w^0 and there exist two convex compact sets $V, W \subseteq \mathbb{R}^l$ such that the directional derivative $z'(w^0; r)$ has the representation

$$z'(w^0; r) = \max_{v\in V}\langle v, r\rangle + \min_{w\in W}\langle w, r\rangle,\ \forall r \in \mathbb{R}^l.$$

The pair of sets (V, W) is called quasidifferential of z at w^0. The quasidifferential of a function is not uniquely determined, there is rather an equivalence relation in the product space $\mathcal{C} \times \mathcal{C}$, where $\mathcal{C}$ denotes the family of all compact convex sets in $\mathbb{R}^l$, such that the quasidifferential is an equivalence class in that space. Our aim is it to derive formulae for at least one (presumably) not too large representation of the quasidifferential for the solution function of the problem (2). Since the function $x(\cdot)$ is vector-valued, we have to understand the sets V, W as being families of matrices and the max and min operators in the definition of the notion quasidifferentiable as being taken component-wise. Then, the rows of the matrices V, W are the quasi-differentials of the component functions $x_i(\cdot)$ of the function $x(\cdot)$. Hence, we can reduce our considerations to these component functions. These functions are also PC^1-functions, the directional derivatives of them are continuous selections of linear functions, they are positively homogeneous and piecewise linear. Let

$$x_i'(y^0; r) \in \{\langle a^1, r \rangle, \ldots, \langle a^l, r \rangle\} \ \forall r \in \mathbb{R}^m \tag{6}$$

be a corresponding representation of the directional derivative of one component function. Here, the generalized gradient of the function $x_i(\cdot)$ is

$$\partial x_i(y^0) = \text{conv} \ \{a^1, \ldots, a^l\}.$$

One possibility to show quasidifferentiability of a directionally differentiable function is to give formulae for its quasidifferential. This can be done for PC^1-functions by use of a minmax-formula for the directional derivative. This formula is valid for arbitrary selections of linear functions as (6). For $x'(y^0; \cdot)$ given by (6) there exist sets $M_i \subseteq \{1, \ldots, l\}$, $i = 1, \ldots, p$, such that

$$x_i'(y^0; r) = \min_{1 \leq i \leq p} \max_{j \in M_i} \langle a^j, r \rangle$$

[13]. This can be transformed into

$$x_i'(y^0; r) = \sum_{i=1}^{p} \max_{j \in M_i} \langle a^j, r \rangle - \max_{1 \leq i \leq p} \sum_{\substack{k=1 \\ k \neq i}}^{p} \max_{j \in M_k} \langle a^j, r \rangle \tag{7}$$

(cf. e.g. [9, formula (6.9)], [23]). Formula (7) can be used to derive a description for the quasidifferential of the function $x_i(\cdot)$ at $y = y^0$ [9]. Hence, it shows quasidifferentiability of the function $x(\cdot)$. The drawback of this approach is due to the difficulties arising when the sets M_i are to be determined.

The following lemma describes a more practical approach for deriving formulae for the quasidifferential. It is a direct consequence of the mixing lemma in [33]. We will add also its proof since it gives a constructive tool for our later investigations. Note that our proof uses a sum with less elements than that in [33]. Note that sublinearity is equivalent to convexity for positively homogeneous functions [9].

Lemma 4.1 *Let $z : D \to \mathbb{R}$ be a positively homogeneous, piecewise linear and continuous function and $0 \in \text{int } D$, $D \subseteq \mathbb{R}^k$. Then, z can be represented as difference of two positively homogeneous, sublinear functions.*

Proof: By the properties of the function z there are convex cones T^i, $i \in \mathcal{M}$, satisfying

$$\bigcup_{i \in \mathcal{M}} T^i = \mathbb{R}^k, \ \text{int } T^i \neq \emptyset, \ \text{int } T^i \cap \text{int } T^j = \emptyset, \ i \neq j, \ i, j \in \mathcal{M}$$

and linear functions $z^i : \mathbb{R}^k \to \mathbb{R}$, $i \in \mathcal{M}$, such that

$$z(r) = z^i(r) \ \forall r \in D \cap T^i, \ i \in \mathcal{M}.$$

Let $\mathcal{N} := \{(i, j) \in \mathcal{M} \times \mathcal{M} : i < j, \ \text{cl } T^i \cap \text{cl } T^j \neq \emptyset\}$. Define the function

$$u(r) = \sum_{(i,j) \in \mathcal{N}} |z^i(r) - z^j(r)|.$$

The function $u(\cdot)$ is also positively homogeneous, piecewise linear and continuous. Moreover, it is convex as the sum of convex functions and, hence, sublinear. Without loss of generality, we can assume that the function $u(\cdot)$ is linear on each cone T^i (else use a finer partition of $\mathbb{R}^k$ than $\{T^i\}_{i \in \mathcal{M}}$ in order to guarantee this assumption). Since $z(r) = z(r) + u(r) - u(r)$, the theorem follows from the sublinearity of the function $w(r) = z(r) + u(r)$ which will be shown in what follows.

Let $u, v \in D$, $\alpha \in (0, 1)$ be arbitrarily chosen. Then, we have to show

$$w(\alpha u + (1 - \alpha)v) \leq \alpha w(u) + (1 - \alpha)w(v). \tag{8}$$

For doing so, we consider the three possible cases:

1. If $u, v \in T^i$ for some $i \in \mathcal{M}$, (8) is implied by linearity of $w(\cdot)$ on T^i.

2. Let $u \in T^i$, $v \in T^j$ and let there exist $a \in \text{cl } T^i \cap \text{cl } T^j$ such that

$$a = \hat{\alpha} u + (1 - \hat{\alpha})v \ \text{for some } \hat{\alpha} \in (0, 1).$$

By linearity of $w(\cdot)$ on T^i, T^j we have (8) if and only if

$$w(a) \leq \hat{\alpha} w(u) + (1 - \hat{\alpha})w(v).$$

By our assumptions,

$$w(a) = z(a) + u(a) = z^i(a) + u(a) =$$
$$z^i(a) + |z^i(a) - z^j(a)| + \sum_{\substack{(k,l) \in \mathcal{N} \\ \{i,j\} \neq \{k,l\}}} |z^k(a) - z^l(a)|. \tag{9}$$

(a) Take in the first case $z^i(u) > z^j(u)$. Then, by linearity and due to $z^i(a) = z^j(a)$, we have $z^i(v) < z^j(v)$. This implies,

$$\begin{aligned}
0 &= |z^i(a) - z^j(a)| \\
&= \hat{\alpha} z^i(u) - \hat{\alpha} z^j(u) + (1 - \hat{\alpha})z^i(v) - (1 - \hat{\alpha})z^j(v) \\
&\leq \hat{\alpha}(z^i(u) - z^j(u)) + (1 - \hat{\alpha})(z^i(v) - z^j(v)) \\
&+ (1 - \hat{\alpha})(z^j(v) - z^i(v)).
\end{aligned} \tag{10}$$

Inserting this into inequality (9) we derive

$$w(a) \leq$$
$$\hat{\alpha} z^i(u) + (1 - \hat{\alpha}) z^i(v) + (1 - \hat{\alpha})(z^j(v) - z^i(v)) +$$
$$+ \hat{\alpha}|z^i(u) - z^j(u)| + (1 - \hat{\alpha})|z^i(v) - z^j(v)| +$$
$$\hat{\alpha} \sum_{\substack{(k,l) \in \mathcal{N} \\ \{i,j\} \neq \{k,l\}}} |z^k(u) - z^l(u)| + (1 - \hat{\alpha}) \sum_{\substack{(k,l) \in \mathcal{N} \\ \{i,j\} \neq \{k,l\}}} |z^k(v) - z^l(v)|$$
$$= \hat{\alpha} w(u) + (1 - \hat{\alpha}) w(v).$$

(b) If in the other case the opposite inequality $z^i(u) \leq z^j(u)$ holds, then $z^i(v) \geq z^j(v)$, and inequality (10) has to be replaced by

$$0 = |z^i(a) - z^j(a)| =$$
$$(1 - \hat{\alpha})(z^i(v) - z^j(v)) - (1 - \hat{\alpha})(z^i(v) - z^j(v)) \leq$$
$$\hat{\alpha}(z^j(u) - z^i(u)) + (1 - \hat{\alpha})(z^i(v) - z^j(v)) +$$
$$+ (1 - \hat{\alpha})(z^j(v) - z^i(v)) \tag{11}$$

Inserting (11) into (9), we get the desired inequality also in this case.

3. Now, let there exist a sequence $\{T^{i_j}\}_{j=1}^s$ of convex cones and a sequence $\{\alpha_{i_j}\}_{j=1}^s \subseteq (0,1)$ with $\alpha_{i_j} < \alpha_{i_{j+1}}$, $j = 1, \dots, s-1$, such that

$$a_j := \alpha_{i_j} u + (1 - \alpha_{i_j}) v \in \mathrm{cl}\, T^{i_j} \cap T^{i_{j+1}}, \quad j = 1, \dots, s-1.$$

Let $\alpha \in (\alpha_{i_k}, \alpha_{i_{k+1}})$, for some $k > 1$, $a_0 := u$, $a_{s+1} := v$. Then, by applying the result just obtained, we get $w(a) \leq \hat{\alpha} w(a_k) + (1 - \hat{\alpha}) w(a_{k+1})$ for $\hat{\alpha} \in (0,1)$ such that $a = \hat{\alpha} a_k + (1 - \hat{\alpha}) a_{k+1}$. Similarly, $w(a_k) \leq \alpha^0 w(a_{k-1}) + (1 - \alpha^0) w(a_{k+1})$ for that value of α^0 with $\alpha^0 \in (0,1)$ satisfying $a_k = \alpha^0 a_k + (1 - \alpha^0) a_{k+1}$. This implies

$$w(a) \leq \hat{\alpha} \alpha^0 w(a_{k-1}) + (1 - \hat{\alpha} \alpha^0) w(a_{k+1}).$$

Proceeding in this way we easily derive the conclusion of the theorem. □

Since, as a function of the direction only, $x'(y^0; \cdot)$ satisfies the conditions of Lemma 4.1, this result can now be applied to the directional derivative of the components of the function $x(\cdot)$. Then, we derive the formula

$$x_i'(y^0; r) = x_i'(y^0; r) + \sum_{(k,l) \in \mathcal{N}} |\nabla x_i^k(y^0) r - \nabla x_i^l(y^0) r| -$$
$$- \sum_{(k,l) \in \mathcal{N}} |\nabla x_i^k(y^0) r - \nabla x_i^l(y^0) r|,$$

where, of course, the set $\mathcal{N}$ is an index set related to the family of all sets I for which the conditions (D1) and (D2) hold. It will be convenient for the subsequent considerations to use the natural ordering with respect to inclusions of the sets

I. This ordering is closely related to the polytope graph of the set of Lagrange multipliers of the problem (2). This graph $\mathcal{G} = (\mathcal{V}, \mathcal{E})$ has the vertex set

$$\mathcal{V} = \{I : \exists (\lambda, \mu) \in E\Lambda(x^0, y^0) \text{ such that } (D1) \text{ and } (D2) \text{ are satisfied}\}$$

and the edge set

$$\begin{aligned}
\mathcal{E} = \quad & \{(I, J) \in \mathcal{V} \times \mathcal{V} : \|I\| - |J|\| = 1 \text{ and either } I \subseteq J \text{ or } J \subseteq I\} \cup \\
& \{(I, J) \in \mathcal{V} \times \mathcal{V} : |I| = |J| \text{ and there exist indices} \\
& \qquad\qquad i, j \text{ such that } J = I \setminus \{i\} \cup \{j\}\}.
\end{aligned}$$

Here, $|A|$ denotes the number of elements in the set A. For each of the sets $I \in \mathcal{V}$ there exists one continuously differentiable selection function $x^I(\cdot)$ of $x(\cdot)$, i.e. one essential element of the generalized derivative of $x(\cdot)$. Each of the cones T^I used in the proof of Lemma 4.1 coincide with one set of directions on which the directional derivative $x'(y^0; \cdot)$ reduces to a linear function. By parametric quadratic optimization, two of these cones have common elements iff they are related as given in the edge set $\mathcal{E}$ of the polytope graph. Hence, the edges of this graph give exactly the family of all pairs of sets $\mathcal{N}$ which are needed to construct the above representation of $x'(y^0; r)$. It is easy to verify that a quasidifferential of the i-th component function $x_i(\cdot)$ coincides with the pair (V_i, W_i), where V_i and $-W_i$ are the subdifferentials of the convex functions

$$u_i(r) = \sum_{(I,J) \in \mathcal{E}} |(\nabla x_i^I(y^0) - \nabla x_i^J(y^0))r|$$

and

$$w_i(r) = x_i'(y^0; r) + \sum_{(I,J) \in \mathcal{E}} |(\nabla x_i^I(y^0) - \nabla x_i^J(y^0))r|$$

at zero. The subdifferential of the convex function $u_i(r)$ at zero is given by

$$\partial u_i(0) = \text{conv} \sum_{(I,J) \in \mathcal{E}} \{\nabla x_i^I(y^0) - \nabla x_i^J(y^0), \nabla x_i^J(y^0) - \nabla x_i^I(y^0)\}$$

(cf. [28]). It is difficult to give a closed formula for the subdifferential of the second function $w_i(r)$ at zero, but we can use Theorem 23.2 in [28] to formulate an algorithm computing it. This Theorem implies that

$$\partial w_i(0) = \text{conv} \left\{ v : \exists r \in \mathbb{R}^m \text{ such that} \right.$$
$$\left. \langle v, r \rangle = x_i'(y^0; r) + \sum_{(I,J) \in \mathcal{E}} |(\nabla x_i^I(y^0) - \nabla x_i^J(y^0))r|\right\}.$$

Hence, we can proceed as follows: Take any vertex I in $\mathcal{V}$. Let

$$\mathcal{N}(I) := \{J \in \mathcal{V} : (I, J) \in \mathcal{E} \text{ or } (J, I) \in \mathcal{E}\}$$

denote its family of neighboring sets in the polytope graph. Next, we have to determine if

$$|(\nabla x_i^I(y^0) - \nabla x_i^J(y^0))r| = (\nabla x_i^I(y^0) - \nabla x_i^J(y^0))r$$

for some $r \in T^I$ or not. If this equation is possible, then $v_J = 1$:

$$v_J = \begin{cases} 1, & \exists \ r \in T^I : \langle \nabla x^I(y^0), r \rangle < \langle \nabla x^J(y^0), r \rangle, \\ 0, & \text{else.} \end{cases}$$

In the other case if the equation

$$|(\nabla x_i^I(y^0) - \nabla x_i^J(y^0))r| = -(\nabla x_i^I(y^0) - \nabla x_i^J(y^0))r$$

is (also) possible, we set $w_J = 1$:

$$w_J = \begin{cases} 1, & \exists \ r \in T^I : \langle \nabla x^I(y^0), r \rangle > \langle \nabla x^J(y^0), r \rangle, \\ 0, & \text{else.} \end{cases}$$

For the computation of the values of v_J, w_J we simply can use linear programming:

$$(\nabla x_i^I(y^0) - \nabla x_i^J(y^0))r \to \min / \max \tag{12}$$

subject to the Karush-Kuhn-Tucker conditions of the problem $(QP(r))$ for a fixed set I, a fixed vertex $(\lambda^0, \mu^0) \in E\Lambda(x^0, y^0)$ with $\{j : \lambda_j^0 > 0\} \subseteq I$ and variable vectors d, r, ν, η:

$$\begin{aligned} &\nabla_{xx}^2 L(x^0, y^0, \lambda^0, \mu^0)d + \nabla_{xy}^2 L(x^0, y^0, \lambda^0, \mu^0)r + \\ &\qquad + \nabla_x^T g_I(x^0, y^0)\nu + \nabla_x^T h(x^0, y^0)\eta = 0, \\ &\nabla_x g_i(x^0, y^0)d + \nabla_y g_i(x^0, y^0)r \begin{cases} = 0 & \text{if } i \in I, \\ \leq 0 & \text{if } i \in I^0 \setminus I \end{cases} \\ &\nabla_x h_j(x^0, y^0)d + \nabla_y h_j(x^0, y^0)r = 0, \text{ for } j = 1, \ldots, q \\ &\nu = 0, \ i \in I^0 \setminus I, \ \nu_i \geq 0, \ i \in I \setminus \{j : \lambda_j^0 > 0\} \\ &-1 \leq r_i \leq 1, \ i = 1, \ldots, m. \end{aligned} \tag{13}$$

Then,

$$\partial w_i(0) = \text{conv} \ \{a(I) : I \in \mathcal{V}\},$$

where

$$a(I) = \{\nabla x_i^I(y^0)\} + \sum_{J \in \mathcal{N}(I)} b_J \{\nabla x_i^I(y^0) - \nabla x_i^J(y^0)\}, \tag{14}$$

$$b_J = \begin{cases} [-1, 1], & \text{if } v_J = w_J = 1, \\ 1, & \text{if } v_J = 1, \ w_J = 0, \\ -1, & \text{if } v_J = 0, \ w_J = 1, \end{cases}$$

and (14) is defined as a Minkowski sum.

By applying this procedure for each i simultaneously we compute the whole subdifferential of the function $w(\cdot)$ at zero. Thus we have shown

Theorem 4.2 *If the assumptions (C), (MFCQ), (SSOC), (CRCQ), (E) are satisfied for the problem (2), then the solution function $x(\cdot)$ is quasidifferentiable and $(\partial w(0), -\partial u(0))$ is one representation of a quasidifferential of $x(\cdot)$ at $y = y^0$.*

For the proof, the above results can be used in combination with [9].

5 Conclusion

In this paper we have considered the concepts of Clarke as well as of Dem'yanov and Rubinov for defining generalizations of the Jacobian of (local) solution functions of a smooth parametric optimization problem. We have shown that this function is differentiable in both generalized senses provided that besides the strong sufficient optimality condition of second order and the Mangasarian-Fromowitz constraint qualification also the constant rank constraint qualification are satisfied. For the computation of the generalized Jacobian (and some quasidifferential) of $x(\cdot)$ we need one additional assumption. An example shows that this assumption cannot be dropped in general. Formulae for the computation of both the generalized Jacobian and the quasidifferential have been given.

The method for the computation of some representation of the quasidifferential seems to be very expensive. But, we hope that the generalized Jacobian of the function $x(\cdot)$ contains only a very few elements. Then, the polytope graph used is only very small and the computational burden needed for computing the quasidifferential is not too large. It was not our intention to compute a representation of the quasidifferential which is as small as possible. For an idea for reducing it, the reader is referred to [7].

Acknowledgement: The author would like to thank Prof. D. Pallaschke for many fruitful discussions not only on the topic of quasidifferentiability. Part of the work presented was based on research supported by a promotion award of the Deutsche Akademie der Naturforscher Leopoldina and financial aid from the Bundesministerium für Bildung, Wissenschaft, Forschung und Technologie, Germany.

References

[1] B. Bank, J. Guddat, D. Klatte, B. Kummer, and K. Tammer. *Non-Linear Parametric Optimization.* Akademie-Verlag, Berlin, 1982.

[2] J. Bonnans. A semistrong sufficiency condition for optimality in nonconvex programming and its connection to the perturbation problem. *Journal of Optimization Theory and Applications*, 60:7–18, 1989.

[3] J. Bonnans and A. Sulem. Pseudopower expansion of solutions of generalized equations and constrained optimization problems. *Mathematical Programming*, 70:123–148, 1995.

[4] F. Clarke. *Optimization and Nonsmooth Analysis.* J. Wiley & Sons, New York et al., 1983.

[5] S. Dempe. Directional differentiability of optimal solutions under Slater's condition. *Mathematical Programming*, 59:49–69, 1993.

[6] S. Dempe. On generalized differentiability of optimal solutions and its application to an algorithm for solving bilevel optimization problems. In D.-Z. Du, L. Qi, and R.S. Womersley, editor, *Recent advances in nonsmooth optimization*, pages 36–56. World Scientific Publishers, 1995.

[7] S. Dempe and D. Pallaschke. Quasidifferentiability of optimal solutions in parametric nonlinear optimization. *Optimization.* to appear.

[8] S. Dempe and H. Schmidt. On an algorithm solving two-level programming problems with nonunique lower level solutions. *Computational Optimization and Applications*, 6:227–249, 1996.

[9] V. Dem'yanov and A. Rubinov. *Quasidifferential Calculus.* Optimization Software Inc., Publ. Division, New York, 1986.

[10] M. Dyer and L. Proll. An algorithm for determining all extreme points of a convex polytope. *Mathematical Programming*, 12:81–96, 1985.

[11] A. Fiacco. *Introduction to Sensitivity and Stability Analysis in Nonlinear Programming.* Academic Press, New York, 1983.

[12] A. Fiacco and G. McCormick. *Nonlinear Programming: Sequential Unconstrained Minimization Techniques.* J. Wiley & Sons, New York et al., 1968.

[13] V. Gorokhovik and O. Zorko. Piecewise affine functions and polyhedral sets. *Optimization*, 31:209–221, 1994.

[14] J. Guddat, F. G. Vasquez, and H. Jongen. *Parametric Optimization: Singularities, Pathfollowing and Jumps.* J. Wiley, Chichester et al. and B.G. Teubner, Stuttgart, 1990.

[15] W. Hager. Lipschitz continuity for constrained processes. *SIAM Journal on Control and Optimization*, 17:321–328, 1979.

[16] K. Jittorntrum. Solution point differentiability without strict complementarity in nonlinear programming. *Mathematical Programming Study*, 21:127–138, 1984.

[17] D. Klatte. Nonlinear optimization problems under data perturbations. In W. Krabs and J. Zowe, editors, *Modern Methods of Optimization*, pages 204–235. Springer Verlag, 1992.

[18] D. Klatte. On quantitative stability for non-isolated minima. *Control and Cybernatics*, 23:183–200, 1994.

[19] D. Klatte and K. Tammer. On second-order sufficient optimality conditions for $C^{1,1}$-optimization problems. *Optimization*, 19:169–179, 1988.

[20] M. Kojima. Strongly stable stationary solutions in nonlinear programs. In S. Robinson, editor, *Analysis and Computation of Fixed Points*, pages 93–138. Academic Press, New York, 1980.

[21] M. Kojima and S. Shindo. Extension of Newton and quasi-Newton methods to systems of PC^1 equations. *Journal of the Operational Research Society of Japan*, 29:352–375, 1986.

[22] B. Kummer. Newton's method for non-differentiable functions. In *Advances in Mathematical Optimization*, volume 45 of *Mathematical Research*, pages 114–125. Akademie-Verlag, Berlin, 1988.

[23] L. Kuntz. Topological aspects of nonsmooth optimization. Technical report, Institut für Statistik und Mathematische Wirtschaftstheorie, Universität Karlsruhe, Germany, 1994.

[24] J. Liu. Sensitivity analysis in nonlinear programs and variational inequalities via continuous selections. *SIAM Journal on Control and Optimization*, 33:1040–1060, 1995.

[25] K. Malanowski. Differentiability with respect to parameters of solutions to convex programming problems. *Mathematical Programming*, 33:352–361, 1985.

[26] J. Outrata. On the numerical solution of a class of Stackelberg problems. *ZOR - Methods and Models of Operations Research*, 34:255–277, 1990.

[27] D. Ralph and S. Dempe. Directional derivatives of the solution of a parametric nonlinear program. *Mathematical Programming*, 70:159–172, 1995.

[28] R. Rockafellar. *Convex analysis*. Princeton University Press, Princeton, 1970.

[29] S. Scholtes. Introduction to piecewise differentiable equations. Technical report, Universität Karlsruhe, Institut für Statistik und Mathematische Wirtschaftstheorie, 1994. No. 53/1994.

[30] A. Shapiro. private communication.

[31] A. Shapiro. Perturbation theory of nonlinear programs when the set of optimal solutions is not a singleton. *Applied Mathematics and Optimization*, 18:215–229, 1988.

[32] A. Shapiro. Sensitivity analysis of nonlinear programs and differentiability properties of metric projections. *SIAM Journal on Control and Optimization*, 26:628–645, 1988.

[33] L. Veselý and L. Zajíček. Delta-convex mappings between Banach spaces and applications. *Dissertationes Mathematicae*, 289, 1989.

Characterizations of Lipschitzian Stability in Nonlinear Programming

A. L. DONTCHEV
Mathematical Reviews, Ann Arbor, MI 48107
and
R. T. ROCKAFELLAR
Dept. of Math., Univ. of Washington, Seattle, WA 98195

Abstract

Nonlinear programming problems are analyzed for Lipschitz and upper-Lipschitz behavior of their solutions and stationary points under general perturbations. Facts from a diversity of sources are put together to obtain new characterizations of several local stability properties.

1. INTRODUCTION

In this paper we consider the following nonlinear programming problem with canonical perturbations:

$$\text{minimize } g_0(w, x) + \langle v, x \rangle \text{ over all } x \in C(u, w), \tag{1}$$

where $C(u, w)$ is given by the constraints

$$g_i(w, x) - u_i \begin{cases} = 0 & \text{for } i \in [1, r], \\ \leq 0 & \text{for } i \in [r+1, m], \end{cases} \tag{2}$$

for $\mathcal{C}^2$ functions $g_i : \mathbb{R}^d \times \mathbb{R}^n \to \mathbb{R}$, $i = 0, 1, \ldots, m$. The vectors $w \in \mathbb{R}^d$, $v \in \mathbb{R}^n$ and $u = (u_1, \ldots, u_m) \in \mathbb{R}^m$ are parameter elements. Consolidating them as $p = (v, u, w)$, we denote by $X(p)$ the set of local minimizers of (1) and refer to the map $p \mapsto X(p)$ as the *solution map*. An element $x \in X(p)$ is *isolated* if $X(p) \cap U = \{x\}$ for some neighborhood U of x. To fit with this notational picture, we write $C(p)$ for the set of feasible solutions, even though this only depends on the (u, w) part of p; the map $p \mapsto C(p)$ is the *constraint map*.

This work was supported by National Science Foundation grants DMS 9404431 for the first author and DMS 9500957 for the second.

In terms of the basic Lagrangian function

$$L(w, x, y) = g_0(w, x) + y_1 g_1(w, x) + \cdots + y_m g_m(w, x),$$

the Karush-Kuhn-Tucker (KKT) system associated with problem (1) has the form:

$$\begin{cases} v + \nabla_x L(w, x, y) = 0, \\ -u + \nabla_y L(w, x, y) \in N_Y(y) \end{cases} \quad \text{for } Y = \mathbb{R}^r \times \mathbb{R}_+^{m-r}, \tag{3}$$

where $N_Y(y)$ is the normal cone to the set Y at the point y. For a given $p = (v, u, w)$ the set of solutions (x, y) of the KKT system (the set of the KKT pairs) is denoted by $S_{KKT}(p)$; the map $p \mapsto S_{KKT}(p)$ is called the *KKT map*. We write $X_{KKT}(p)$ for the set of stationary points; that is, $X_{KKT}(p) = \{x \mid$ there exists y such that $(x, y) \in S_{KKT}(p)\}$; the map $p \mapsto X_{KKT}(p)$ is the *stationary point map*. The set of Lagrange multiplier vectors associated with x and p is $Y_{KKT}(x, p) = \{y \mid (x, y) \in S_{KKT}(p)\}$.

Recall that the *Mangasarian-Fromovitz condition* holds at (p, x) if $y = 0$ is the only vector satisfying

$$y = (y_1, y_2, \ldots, y_m) \in N_{\mathcal{K}}\big(g_1(w, x) - u_1, \ldots, g_m(w, x) - u_m\big)$$
$$\text{and } y_1 \nabla_x g_1(w, x) + \cdots + y_m \nabla_x g_m(w, x) = 0,$$

where $\mathcal{K}$ is the convex and closed cone in $\mathbb{R}^m$ with elements whose first r components are zeros and the remaining $m - r$ components are nonpositive numbers. Under the Mangasarian-Fromovitz condition, the KKT system (3) represents a necessary condition for a feasible point x for (1) to be locally optimal, see e.g. [37].

In this paper we study Lipschitz-type properties of the maps S_{KKT} and X_{KKT}. We complement and unify a number of results scattered in the literature by putting together facts from a diversity of sources and exploiting the canonical form of the perturbations in our model (1). In Section 2 we discuss the robustness of the local upper-Lipschitz property with respect to higher-order perturbations and give a characterization of a stronger version of this property for the KKT map S_{KKT}. Section 3 is devoted to the stationary point map X_{KKT}. In Theorem 3.1 we present a characterization of the local upper-Lipschitz property of this map by utilizing a condition for its proto-derivative. Theorem 3.3 complements a known result of Kojima; we prove that if the map X_{KKT} is locally single-valued and Lipschitz continuous with its values locally optimal solutions, then both the Mangasarian-Fromovitz condition and the strong second-order condition hold. The converse is true under the constant rank condition for the constraints. Further, in the line of our previous paper [8], we show in Theorem 3.6 that the combination of the Mangasarian-Fromovitz condition and the Aubin continuity of the map X_{KKT} is equivalent to the local single-valuedness and Lipschitz continuity of this map, provided that the values of X_{KKT} are locally optimal solutions. In Section 4 we present a sharper version of the characterization of the local Lipschitz continuity of the solution-multiplier pair obtained in [8].

The literature on stability of nonlinear programming problems is enormous, and even a short survey would be beyond the scope of the present paper. We refer here to papers that are explicitly related to the results presented. For recent surveys also on other aspects of the subject, see [4] and [15].

Throughout we denote by $\mathbb{B}_a(x)$ the closed ball with center x and radius a. The ball $\mathbb{B}_1(0)$ is denoted simply by $\mathbb{B}$. For a (potentially set-valued) map Γ from $\mathbb{R}^m$

to $I\!R^n$ we denote by $\operatorname{gph}\Gamma$ the set $\{(u,x)\,|\,u\in I\!R^m,\ x\in\Gamma(u)\}$. We associate with any point $(u_0,v_0,w_0,x_0,y_0)\in\operatorname{gph}S_{\mathrm{KKT}}$ the index sets I_1, I_2, I_3 in $\{1,2,\ldots,m\}$ defined by

$$
\begin{aligned}
I_1 &= \{i\in[r+1,m]\,|\,g_i(w_0,x_0)-u_{0i}=0,\ y_{0i}>0\}\cup\{1,\ldots,r\},\\
I_2 &= \{i\in[r+1,m]\,|\,g_i(w_0,x_0)-u_{0i}=0,\ y_{0i}=0\},\\
I_3 &= \{i\in[r+1,m]\,|\,g_i(w_0,x_0)-u_{0i}<0,\ y_{0i}=0\}.
\end{aligned}
$$

Recall that the *strict Mangasarian-Fromovitz condition* holds at a point (p_0,x_0) if there is a Lagrange multiplier vector $y_0\in Y_{\mathrm{KKT}}(x_0,p_0)$ such that:
(a) the vectors $\nabla_x g_i(w_0,x_0)$ for $i\in I_1$ are linearly independent;
(b) there is a vector $z\in I\!R^n$ such that

$$
\begin{aligned}
\nabla_x g_i(w_0,x_0)^\top z &= 0 \quad \text{for all } i\in I_1\\
\nabla_x g_i(w_0,x_0)^\top z &< 0 \quad \text{for all } i\in I_2.
\end{aligned}
$$

It is known that the strict Mangasarian-Fromovitz condition holds at (p_0,x_0) if and only if there is a unique multiplier vector y_0 associated with (p_0,x_0); that is, $Y_{\mathrm{KKT}}(x_0,p_0)=\{y_0\}$ (cf. Kyparisis [21]).

Let (x_0,y_0) satisfy the KKT conditions (3) for a given $p_0=(v_0,u_0,w_0)$. In the notation $A=\nabla^2_{xx}L(w_0,x_0,y_0)$, $B=\nabla^2_{yx}L(w_0,x_0,y_0)$, the *linearization* of (3) at (u_0,v_0,w_0,x_0,y_0) is the linear variational inequality:

$$
\begin{cases}
v+\nabla_x L(w_0,x_0,y_0)+A(x-x_0)+B^\top(y-y_0)=0,\\
-u+g(w_0,x_0)+B(x-x_0)\in N_Y(y).
\end{cases}
\tag{4}
$$

We denote by L_{KKT} the map assigning to each (u,v) the set of all pairs (x,y) that solve (4).

2. THE LOCAL UPPER-LIPSCHITZ PROPERTY

Robinson [31] introduced the following definition. The set-valued map $\Gamma:I\!R^n\to I\!R^m$ is *locally upper-Lipschitz* at y_0 with modulus M if there is a neighborhood V of y_0 such that

$$
\Gamma(y)\subset\Gamma(y_0)+M\|y-y_0\|I\!\!B \quad \text{for all } y\in V.
\tag{5}
$$

In [31] he proved that if $F:I\!R^n\mapsto I\!R^m$ is a set-valued map whose graph is a (possibly nonconvex) polyhedron, then F is locally upper-Lipschitz at every point y in $I\!R^n$, moreover with a modulus M that is independent of the choice of y.

The upper-Lipschitz property is not a completely local property of the graph of a map, so for the sake of investigating stability under local perturbations we work with the following variant.

Definition 2.1. *The map $\Gamma:I\!R^n\to I\!R^m$ is locally upper-Lipschitz with modulus M at a point (y_0,x_0) in its graph if there exist neighborhoods U of x_0 and V of y_0 such that*

$$
\Gamma(y)\cap U\subset\{x_0\}+M\|y-y_0\|I\!\!B \quad \text{for all } y\in V.
\tag{6}
$$

The local upper-Lipschitz property at a point $(y_0,x_0)\in\operatorname{gph}\Gamma$ implies that $\Gamma(y_0)\cap U=\{x_0\}$ for some neighborhood U of x_0. Conversely, if $\Gamma(y_0)\cap U=$

$\{x_0\}$ for some neighborhood U of x_0, then the local upper-Lipschitz property at y_0 in the Robinson's sense implies the local upper-Lipschitz property at the point $(y_0, x_0) \in \text{gph}\,\Gamma$; and the latter is in turn equivalent to the local upper-Lipschitz property of $\Gamma \cap U$ holding at y_0 with respect to some neighborhood U of x_0. Note that $\Gamma(y) \cap U$ might be empty for some y near y_0. Of course, if Γ is single-valued and locally upper-Lipschitz at $(y_0, \Gamma(y_0))$, it need not be Lipschitz continuous in a neighborhood of y_0.

Bonnans [2] studied a version of the local upper-Lipschitz property in Definition 2.1 for solution mappings of variational inequalities under the name "semistability". Levy [22] called it the "local upper-Lipschitz property at y_0 for x_0." Pang [27], in the context of the linear complementarity problem, introduced a stronger property of a map Γ; in addition to the condition that Γ is locally upper-Lipschitz at the point (y_0, x_0) in its graph (Definition 2.1) he also requires that there exist neighborhoods U of x_0 and V of y_0 such that

$$\Gamma(y) \cap U \neq \emptyset \text{ for all } y \in V.$$

The latter is equivalent to the openness of the inverse Γ^{-1} at (x_0, y_0). Throughout we call such maps locally nonempty-valued and upper-Lipschitz at the point in the graph.

We show first that, for maps defined by solutions to generalized equations, the local upper-Lipschitz property at a point in the graph is "robust under higher-order perturbations." Note that the local openness at a point, and hence the property introduced by Pang, are not robust in this sense.

Let $P = \mathbb{R}^d \times \mathbb{R}^m$ and consider the map Σ from P to the subsets of $\mathbb{R}^n$ defined by

$$\Sigma(p) = \{x \in \mathbb{R}^n \mid y \in f(w, x) + F(w, x)\} \quad \text{for } p = (w, y), \tag{7}$$

where $f : \mathbb{R}^d \times \mathbb{R}^n \to \mathbb{R}^m$ is a function and $F : \mathbb{R}^d \times \mathbb{R}^n \to \mathbb{R}^m$ is a possibly set-valued map. Assume that $x_0 \in \Sigma(p_0)$ for some $p_0 = (w_0, y_0) \in P$ and that the function $f(w_0, \cdot)$ is differentiable at x_0 with Jacobian matrix $\nabla_x f(w_0, x_0)$. As in (7), consider the map obtained by the linearization of f:

$$\Lambda(p) = \{x \in \mathbb{R}^n \mid y \in f(w_0, x_0) + \nabla_x f(w_0, x_0)(x - x_0) + F(w, x)\}. \tag{8}$$

The following result was established by Dontchev [6] in a more abstract setting.

Theorem 2.2. *Suppose there exist neighborhoods U of x_0 and W of w_0 along with a constant l such that, for every $x \in U$ and $w \in W$,*

$$\|f(w, x) - f(w_0, x)\| \leq l\|w - w_0\|. \tag{9}$$

Then the following are equivalent:
   (i) *Λ is locally upper-Lipschitz at the point (p_0, x_0) in its graph;*
   (ii) *Σ is locally upper-Lipschitz at the point (p_0, x_0) in its graph.*

We note that a result closely related to the implication (i) $\Rightarrow$ (ii), but in a different setting, is proved in Robinson [32], Theorem 4.1. If the map F is polyhedral and independent of w, we obtain the following fact by combining Theorem 2.2 with Robinson's result in [31] mentioned at the beginning of this section.

Corollary 2.3. *Let the assumptions of Theorem 2.2 be fulfilled, and let $F : \mathbb{R}^n \to \mathbb{R}^m$ be a polyhedral map. Then the following are equivalent:*

(i) *there exists a neighborhood U of x_0 such that*

$$[f(w_0, x_0) + \nabla f(w_0, x_0)(\cdot - x_0) + F(\cdot)]^{-1}(y_0) \cap U = \{x_0\};$$

(ii) *The map Σ is locally upper-Lipschitz at the point (p_0, x_0) in its graph.*

Proof. The map $\Lambda = [f(w_0, x_0) + \nabla f(w_0, x_0)(\cdot - x_0) + F(\cdot)]^{-1}$ is polyhedral, hence from [31] it is locally upper-Lipschitz in $\mathbb{R}^m$. Then (i) implies that Λ is locally upper-Lipschitz at the point (y_0, x_0) in the graph. Applying Theorem 2.2, Σ is locally upper-Lipschitz at the point (p_0, x_0) in its graph. The converse implication follows again from Theorem 2.2. $\qquad\square$

Bonnans in [2], Theorem 3.1(a), showed that the local upper-Lipschitz property at a point in the graph of the solution map of a variational inequality over a polyhedral set is equivalent to the requirement that the reference point be an isolated solution of the linearized variational inequality. This conclusion follows immediately from Corollary 2.3. The precise result, specialized for the KKT system (3) (where N_Y is a polyhedral set) is as follows.

Corollary 2.4. *The following are equivalent:*

(i) (x_0, y_0) *is an isolated point of the set $L_{\mathrm{KKT}}(p_0)$;*

(ii) *The map S_{KKT} is locally upper-Lipschitz at $(p_0, x_0, y_0) \in \mathrm{gph}\, S_{\mathrm{KKT}}$.*

We note that Corollary 2.4 can be also deduced by the characterization of the local upper-Lipschitz property at a point in the graph of a map in terms of its graphical derivative, see Section 3 for an application of this result to the stationary point map.

Recall that a set-valued map Γ from $\mathbb{R}^m$ to the subsets of $\mathbb{R}^n$ has the *Aubin property*[2] at $(y_0, x_0) \in \mathrm{gph}\,\Gamma$ with constant M if there exist neighborhoods U of x_0 and V of y_0 such that

$$\Gamma(y_1) \cap U \subset \Gamma(y_2) + M\|y_1 - y_2\|\mathbb{B} \text{ for all } y_1, y_2 \in V.$$

The following lemma is a particular case of Theorem 1 of Klatte [14], see also Theorem 4.3 in Robinson [33]; for completeness we present a short proof.

Lemma 2.5. *Suppose x_0 is an isolated local minimizer of (1) for $p = p_0$, and let the Mangasarian-Fromovitz condition hold at (p_0, x_0). Then the map X is lower semicontinuous at (p_0, x_0); that is, for every neighborhood U of x_0 there exists a neighborhood V of p_0 such that for every $p \in V$ the set $X(p) \cap U$ is nonempty.*

Proof. The constraint map C defined by (2) has the Aubin property at (w_0, u_0, x_0) if and only if the Mangasarian-Fromovitz condition holds at (w_0, u_0, x_0), see e.g., [26], Corollary 4.5. Let a, b and γ be the constants in the definition of the Aubin property of the map C; that is, for $p_1, p_2 \in \mathbb{B}_b(p_0)$,

$$C(p_1) \cap \mathbb{B}_a(x_0) \subset C(p_2) + \gamma(\|p_1 - p_2\|)\mathbb{B}.$$

---

[2] In [1], J.-P. Aubin used the name "pseudo-Lipschitz continuity". Following [8], we prefer to call this concept the Aubin property.

Let U be an arbitrary neighborhood of x_0. Choose $\alpha \in (0, a)$ in such a way that x_0 is the unique minimizer in $I\!B_\alpha(x_0)$ of (1) with $p = p_0$ and $I\!B_\alpha(x_0) \subset U$.

For this fixed α and for $p \in I\!B_b(p_0)$ consider the map

$$p \mapsto C_\alpha(p) = \left\{ x \in C(p) \mid \quad \|x - x_0\| \le \alpha + \gamma\|p - p_0\| \right\}.$$

It is clear that the map C_α is upper semicontinuous at $p = p_0$. Let us show that it is lower semicontinuous at $p = p_0$ as well. Take $x \in C_\alpha(p_0) = C(p_0) \cap I\!B_\alpha(x_0)$. From the Aubin property of the map C, for any p near p_0 there exists $x_p \in C(p)$ such that $\|x_p - x\| \le \gamma\|p - p_0\|$. Then $\|x_p - x_0\| \le \|x_p - x\| + \|x - x_0\| \le \alpha + \gamma\|p - p_0\|$. Thus $x_p \in C_\alpha(p)$ and $x_p \to x$ as $p \to p_0$. Hence C_α is lower semicontinuous at $p = p_0$.

The problem

$$\text{minimize } g_0(w, x) + \langle x, v \rangle \text{ in } x \text{ subject to } x \in C_\alpha(p) \tag{10}$$

has a solution for every p near p_0, because $C_\alpha(p)$ is nonempty and compact. Moreover, because of the choice of α, x_0 is the unique minimizer of this problem for $p = p_0$. From the Berge theorem (see e.g. Chapter 9, Theorem 3, in [9]), the solution map X_α giving the argmin in (10) is upper semicontinuous at $p = p_0$; in other words, for any $\delta > 0$ there exists $\eta \in (0, b)$ such that for any $p \in I\!B_\eta(p_0)$ the set of (global) minimizers of (10) is nonempty and included within $I\!B_\delta(x_0)$. Since $X_\alpha(p_0) = \{x_0\}$, the map X_α is actually continuous at p_0. Let δ' be such that $0 < \delta' < \alpha$. Then there exists $\eta' > 0$ such that for every $p \in I\!B_{\eta'}(p_0)$ any solution $x \in X_\alpha(p)$ satisfies $\|x - x_0\| \le \delta' < \alpha + \gamma\|p - p_0\|$. Hence for $p \in I\!B_{\eta'}(p_0)$ the constraint $\|x - x_0\| \le \alpha + \gamma\|p - p_0\|$ is inactive in the problem (10). Then for every $p \in I\!B_{\eta'}(p_0)$ we have $X_\alpha(p) \subset X(p) \cap I\!B_{\delta'}(x_0)$. The proof is now complete. □

Recall that the *second-order sufficient condition* holds at $(p_0, x_0, y_0) \in \text{gph } S_{\text{KKT}}$ if

$$\langle x', \nabla^2_{xx} L(w_0, x_0, y_0)x' \rangle > 0 \text{ for all } x' \ne 0 \text{ in the cone}$$

$$D = \left\{ x' \mid \nabla_x g_i(w_0, x_0)x' = 0 \text{ for } i \in I_1, \ \nabla_x g_i(w_0, x_0)x' \le 0 \text{ for } i \in I_2 \right\}.$$

Theorem 2.6. *The following are equivalent:*

(i) *The map S_{KKT} is upper-Lipschitz at the point (p_0, x_0, y_0) in its graph and is locally nonempty-valued there, and x_0 is a locally optimal solution to problem (1) for p_0;*

(ii) *The strict Mangasarian-Fromovitz condition and the second-order sufficient condition for local optimality hold for (p_0, x_0, y_0).*

Proof. Suppose that (i) holds. Then y_0 is an isolated point in $Y_{\text{KKT}}(x_0, p_0)$. Noting that $Y_{\text{KKT}}(x_0, p_0)$ is convex, we get $Y_{\text{KKT}}(x_0, p_0) = \{y_0\}$. Hence the strict Mangasarian-Fromovitz condition holds, see [21] for instance. Further, from Corollary 2.4, there is no (x, y) close to (x_0, y_0) such that $(x, y) \in L_{\text{KKT}}(p_0)$. Without loss of generality, suppose that $I_1 = \{1, 2, \dots, m_1\}$ and $I_2 = \{m_1 + 1, \dots, m_2\}$ and denote by B_1, B_2 the submatrices of B corresponding to the indices I_1, I_2, respectively. Then the vector $(x, y) = (0, 0)$ is an isolated solution of the variational system

$$Ax + B^\top y = 0,$$
$$B_1 x = 0,$$
$$B_2 x \le 0, \quad y_i \ge 0, \quad y_i(Bx)_i = 0 \quad \text{for } i \in [m_1 + 1, m_2]. \tag{11}$$

Note that there is no restriction here on the sign of y_i for $i \in I_1$, since $y_{0i} > 0$ for $i \in I_1$. As a matter of fact, $(0,0)$ is the unique solution to (11), because the set of solutions to (11) is a cone. Applying the second-order necessary condition for local optimality of x_0 we get

$$\langle x', Ax' \rangle \geq 0 \quad \text{for all} \quad x' \neq 0 \quad \text{in} \quad D.$$

All we need is to show that this inequality is $>$. Suppose to the contrary that there exists a nonzero vector $x' \in D$ with $Ax' = 0$. Then the nonzero vector $(x', 0)$ is a solution to (11), a contradiction.

If (ii) holds, then it is known that x_0 is an isolated local solution of (1) for $p = p_0$ and y_0 is the corresponding unique multiplier vector. Suppose that the index set I_1 associated with (p_0, x_0) is nonempty and $\mathcal{U}$ and $\mathcal{W}$ are neighborhoods of x_0 and w_0 respectively such that the vectors $\nabla_x g_i(w, x)$ for $i \in I_1$ are linearly independent for all $x \in \mathcal{U}$ and $w \in \mathcal{W}$. From Lemma 2.5, $X(p) \cap \mathcal{U} \neq \emptyset$ for p near p_0. Then for all p near p_0 and $x(p) \in X(p)$ near x_0 there exist $y_i(p)$ for $i \in I_1$ which are close to the values y_{0i} for $i \in I_1$ and such that

$$v + \nabla_x g_0(w, x(p)) + \sum_{i \in I_1} y_i(p) \nabla_x g_i(w, x(p)) = 0.$$

Note that $y_i(p) > 0$ for all $i \in I_1$. Taking $y_i(p) = 0$ for $i \in I_2 \cup I_3$, we obtain that the vector $y(p) = (y_1(p), \ldots, y_m(p))$ is a Lagrange multiplier for the perturbed problem which is close to y_0. Hence, if U is a neighborhood of (x_0, y_0) and p is sufficiently close to p_0, then $S_{\text{KKT}}(p) \cap U \neq \emptyset$.

If $I_1 = \emptyset$, then $y_0 = 0 = Y_{\text{KKT}}(x_0, p_0)$. From Lemma 2.5, $X(p) \cap \mathcal{U} \neq \emptyset$ for any neighborhood $\mathcal{U}$ of x_0 provided that p is sufficiently close to p_0. Further, the Mangasarian-Fromovitz condition yields that for p near p_0 and x near x_0 the set of Lagrange multipliers $Y_{\text{KKT}}(x, p)$ is nonempty and contained in a bounded set, see e.g. Theorem 2.3 in [32]. Suppose that there exist $\alpha > 0$, a sequence $p_k \to p_0$ and a sequence $x_k \to x_0$ such that $\|y\| \geq \alpha$ for all $y \in Y_{\text{KKT}}(x_k, p_k)$, $k = 1, 2, \ldots$. Take a sequence $y_k \in Y_{\text{KKT}}(x_k, p_k)$; this sequence is bounded, hence it has an accumulation point, say $\bar{y}$, and then $\bar{y} \neq 0$. Passing to the limit with k in the KKT system we obtain that $\bar{y} \in Y_{\text{KKT}}(x_0, p_0)$ which means that $Y_{\text{KKT}}(x_0, p_0)$ is not a singleton. This contradicts the strict Mangasarian-Fromovitz condition. Hence, for any neighborhood $\mathcal{Y}$ of $y_0 = 0$, $Y_{\text{KKT}}(x, p) \cap \mathcal{Y} \neq \emptyset$ when p is sufficiently close to p_0 and $x \in X(p)$ is sufficiently close to x_0. Thus, for a neighborhood U of (x_0, y_0) and for p close to p_0, $S_{\text{KKT}}(p) \cap U \neq \emptyset$ also in this case.

Assume that the map S_{KKT} is not locally upper-Lipschitz at the point (p_0, x_0, y_0) in its graph. Then, from Corollary 2.4, the system (11) has a nonzero solution (x', y') and this solution can be taken as close to $(0,0)$ as desired. With a slight abuse of the notation, suppose that $y' \in \mathbb{R}^m$ with $y'_i = 0$ for $i \in I_3$. If $x' = 0$, then $y' \neq 0$. Note that if $y'_i \neq 0$ for some $i \in I_2$, then $y'_i > 0$. Since $y_{0i} > 0$ for $i \in I_1$, and y' is close to zero, the vector $y_0 + y'$ is a Lagrange multiplier for x_0 and p_0. This contradicts the strict Mangasarian-Fromovitz condition. Hence $x' \neq 0$. But $x' \in D$. Multiplying the first equality in (11) by x', we obtain $\langle x', Ax' \rangle = 0$, a contradiction. This proves the theorem. $\square$

Theorem 2.6 can be also derived by combining the equivalence between the strict Mangasarian-Fromovitz condition and the uniqueness of the Lagrange multiplier with Proposition 6.2 of Bonnans [2], where it is assumed that (x_0, y_0) is an isolated solution of (3) and x_0 is a local solution to (1) for $p = p_0$, and then it is shown that the local upper-Lipschitz property of S_{KKT} at (p_0, x_0, y_0) is equivalent to the second-order sufficient condition at (p_0, y_0, x_0). In a different setting, Pang [28] considered the KKT system for a variational inequality and proved (roughly speaking, see Theorem 5 in [28]) that under the strict Mangasarian-Fromovitz condition and a second-order necessary optimality condition, the map S_{KKT} is locally nonempty-valued and upper-Lipschitz at the point in its graph if and only if the second-order sufficient optimality condition holds. For results relating upper-Lipschitz properties of local minimizers to growth conditions for the objective function, see Klatte [15].

3. THE STATIONARY POINT MAP

One can get a characterization of the local upper-Lipschitz property at a point of the stationary point map X_{KKT} in terms of the proto-derivative of this map by combining results from Levy [22] and Levy and Rockafellar [24]. Namely, it was shown in King and Rockafellar [13], Proposition 2.1, and Levy [22], Proposition 4.1, that a map has the local upper-Lipschitz property at a point in its graph if and only if its graphical (contingent) derivative at that point has image $\{0\}$ at 0. In our case the graphical derivative is actually the proto-derivative of X_{KKT}, a formula for which is given in Theorems 3.1 and 3.2 of Levy and Rockafellar [24] (cf. also Theorem 5.1 in Levy [22]).

Theorem 3.1. *Let $x_0 \in X_{\mathrm{KKT}}(p_0)$, $p_0 = (v_0, u_0, w_0)$, be such that the Mangasarian-Fromovitz condition is fulfilled. Then the following condition is necessary and sufficient for the map X_{KKT} to be locally upper-Lipschitz at (p_0, x_0): there exists no vector $x' \neq 0$ which for some choice of*

$$y_0 \in \operatorname{argmax}\left\{\langle x', \nabla_{xx}^2 L(w_0, x_0, y) x'\rangle \;\middle|\; y \text{ with } (x_0, y) \in S_{\mathrm{KKT}}(p_0)\right\}$$

satisfies the KKT conditions for the subproblem having objective function $h_0(x') = \langle x', \nabla_{xx}^2 L(w_0, x_0, y_0) x'\rangle$ and constraint system

$$\begin{cases} \langle \nabla_x g_0(w_0, x_0) - v_0, x'\rangle = 0, \\ \langle \nabla_x g_i(w_0, x_0), x'\rangle = 0 \text{ for } i \in [1, r], \\ \langle \nabla_x g_i(w_0, x_0), x'\rangle \leq 0 \text{ for } i \in [r+1, m] \text{ with } g_i(w_0, x_0) - u_{0i} = 0. \end{cases}$$

Proof. According to the cited Theorems 3.1 and 3.2 of Levy and Rockafellar [24] as specialized to this situation, the vectors x' that satisfy the KKT conditions for one of the subproblems in question form the image of 0 under the proto-derivative mapping associated with X_{KKT} at (p_0, x_0). (The first of the cited theorems establishes the proto-differentiability.) Applying Proposition 2.1 of King and Rockafellar [13], we see that the nonexistence of a vector $x' \neq 0$ in this set is equivalent to the property we wish to characterize. $\qquad\square$

On the basis of our Lemma 2.5 and Theorem 3.1 we now are able to get the following.

Corollary 3.2. *Let x_0 be an isolated local minimizer of (1) for $p_0 = (v_0, u_0, w_0)$. Suppose that the Mangasarian-Fromovitz condition holds for (p_0, x_0), and let the* *condition in Theorem 3.1 hold as well. Then the solution map X of (1) is locally nonempty-valued and upper-Lipschitz at the point (p_0, x_0).*

Note that both the local optimality of x_0 and the condition in Theorem 3.1 are satisfied if the second-order sufficient condition holds at (p_0, x_0, y_0) for *every* choice of a Lagrange multiplier vector y_0. Under this stronger form of the second-order sufficient condition, the property of the solution map X obtained in Corollary 3.2 can be derived by combining Theorem 3.2 and Corollary 4.3 of Robinson [32]; for more recent results in this direction see [3], [16], [36] and [38].

Recall that a map Σ from $\mathbb{R}^m$ to the subsets of $\mathbb{R}^n$ with $(y_0, x_0) \in \text{gph}\,\Sigma$ is *locally single-valued and (Lipschitz) continuous around* (y_0, x_0) if there exist neighborhoods U of x_0 and V of y_0 such that the map $y \mapsto \Sigma(y) \cap U$ is single-valued and (Lipschitz) continuous on V. Our next result is related to Theorem 7.2 in Kojima [18]. Kojima showed that, under the Mangasarian-Fromovitz condition and for C^2 perturbations of the functions in the problem, as long as the reference point is a local minimizer, the stationary point map is locally single-valued and continuous if and only if the strong second-order sufficient condition holds. Note that for the case when the perturbations are represented by parameters, the continuity of the stationary point map does not imply the strong second-order condition (consider the example of $\min_x \{x^4 + vx \mid x \in \mathbb{R}\}$, $v \in \mathbb{R}$).

The theorem below complements the "only if" part of Kojima's theorem in the following way: we use a narrower class of perturbations represented by parameters in canonical form and show that a stronger condition, namely of the stationary point map being locally single-valued and *Lipschitz* continuous with its values locally optimal solutions, implies both the Mangasarian-Fromovitz condition and the strong second-order condition for local optimality.

Theorem 3.3. *Suppose that $x_0 \in X_{\text{KKT}}(p_0)$, $p_0 = (v_0, u_0, w_0)$ and assume that the stationary point map X_{KKT} is locally single-valued and Lipschitz continuous around (p_0, x_0), moreover with the property that for all $(p, x) \in \text{gph}\,X_{\text{KKT}}$, $p = (v, u, w)$ in some neighborhood of (p_0, x_0), x is a locally optimal solution to the nonlinear programming problem (1) for p. Then the following conditions hold:*

(i) The Mangasarian-Fromovitz condition holds at (p_0, x_0);

(ii) The strong second-order sufficient condition for local optimality holds at (p_0, x_0); that is, for every $y_0 \in Y_{\text{KKT}}(x_0, p_0)$, if I_1 is the set of indices with positive y_{0i}, then

$$\langle x', \nabla^2_{xx} L(w_0, x_0, y_0) x' \rangle > 0 \quad \text{for all } x' \neq 0 \tag{12}$$

in the subspace $M = \{ x' \in \mathbb{R}^n \mid x' \perp \nabla_x g_i(w_0, x_0) \text{ for all } i \in I_1 \}$.

In the proof of the theorem we use the following general result from [6]. Let Γ be a set-valued map from $\mathbb{R}^n$ to the subsets of $\mathbb{R}^m$ and let $x_0 \in \Gamma(y_0)$. The function $s : Y \to X$ is said to be a local selection of Γ around (y_0, x_0) if $x_0 = s(y_0)$ and there exists a neighborhood V of y_0 such that $s(y) \in \Gamma(y)$ for all $y \in V$.

Let us consider the maps Σ and Λ defined in (7) and (8) under the following two conditions:

(A) *f is differentiable with respect to x with Jacobian matrix $\nabla_x f(w, x)$ depending continuously on (w, x) in a neighborhood of (w_0, x_0).*

(B) *f is Lipschitz continuous in w uniformly in x around (w_0, x_0); that is, there exist neighborhoods U of x_0 and V of w_0 and a number $l > 0$ such that $\|f(w_1, x) - f(w_2, x)\| \leq l\|w_1 - w_2\|$ for all $x \in U$ and $w_1, w_2 \in V$.*

The following result, proved in Dontchev [6], Theorem 4.1 (see also [7], Theorem 2.4) shows that, similarly to the local upper-Lipschitz property (and also to the Aubin property, see [6], Theorem 2.4) the existence of a Lipschitz continuous local selection is robust under (non)linearization:

Theorem 3.4. *Consider the maps Σ and Λ defined in (7) and (8) respectively, let $x_0 \in \Sigma(p_0)$ for some $p_0 = (w_0, y_0) \in P$ and let the conditions (A) and (B) hold. Then the following are equivalent:*

(i) *Λ has a Lipschitz continuous local selection around (p_0, x_0);*
(ii) *Σ has a Lipschitz continuous local selection around (p_0, x_0).*

Proof of Theorem 3.3. The assumption that the stationary point map X_{KKT} is locally single-valued and Lipschitz continuous in x around (p_0, x_0) implies that the feasible map $p \mapsto C(p)$ has a Lipschitz continuous local selection around (p_0, x_0). Then, from Theorem 3.4 it follows that for every u near u_0 there exists a solution of the linearized system of constraints,

$$- u_i + g_i(w_0, x_0) + \nabla_x g_i(w_0, x_0)(x - x_0) \in \mathcal{K} \tag{13}$$

where $\mathcal{K}$, as in the introduction, denotes the convex and closed cone in $\mathbb{R}^m$ of vectors whose first r components are zeros and the remaining $m - r$ components are nonpositive numbers. The map $T : \mathbb{R}^n \mapsto \mathbb{R}^m$, defined as

$$x \mapsto T(x) := u_0 - g(w_0, x_0) - \nabla_x g(w_0, x_0)(x - x_0) + \mathcal{K},$$

has convex and closed graph and the condition (13) means that $0 \in \mathbb{R}^m$ is in the interior of the range of T. Then we can apply the Robinson-Ursescu theorem, see e.g. [6], Theorem 2.2, obtaining that the constraint map C is Aubin continuous at (w_0, u_0, x_0); the latter is in turn equivalent to the Mangasarian-Fromovitz condition. Thus (i) is established. Note that in obtaining (i) the only condition we use is that X_{KKT} has a Lipschitz continuous local selection. Actually, it is sufficient to assume that X_{KKT} has a set-valued local selection which is Aubin continuous.

Let us prove (ii). It is known that, under the Mangasarian-Fromovitz condition, the set $Y_{\mathrm{KKT}}(x_0, p_0)$ is a nonempty polyhedron, and moreover, if y_0 is any extreme point of $Y_{\mathrm{KKT}}(x_0, p_0)$ and if I_i for $i = 1, 2, 3$, are the sets of indices associated with (p_0, x_0, y_0), then the gradients $\nabla_x g_i(x_0, w_0)$, $i \in I_1$, must be linearly independent. Consider the variational system (3) with the following value of the parameter vector $p = (v, u, w)$ denoted by p_ε: $v = v_0$, $u_i = u_{0i}$ for $i \in I_1 \cup I_3$, $u_i = u_{0i} + \varepsilon$ for $i \in I_2$, and $w = w_0$, where ε is a positive number. For a sufficiently small $\varepsilon > 0$ there exist neighborhoods V of p_ε and U of x_0 such that for every $p \in V$ the set $X_{\mathrm{KKT}}(p) \cap U$ is a singleton and the map $p \mapsto X_{\mathrm{KKT}}(p) \cap U$ is Lipschitz continuous in V. Choose U and V smaller, if necessary, so if $p \in V$ and $x \in U$, one has $-u_i + g_i(w, x) < 0$ for $i \in I_2 \cup I_3$; moreover the vectors $\nabla_x g_i(w, x)$, $i \in I_1$, are linearly independent, and if y_i, $i \in I_1$, satisfy

$$v + \nabla_x g_0(w, x) + \sum_{i \in I_1} y_i \nabla_x g_i(w, x) = 0, \tag{14}$$

then $y_i > 0$ for $i \in I_1$ (the latter being true by the linear independence of the vectors $\nabla_x g_i(w, x)$ for $i \in I_1$ and the fact that $y_{0i} > 0$ for $i \in I_1$).

If $p \in V$ and $x \in X_{\mathrm{KKT}}(p) \cap U$, then every associated multiplier vector y for (3) must satisfy $y_i = 0$ for $i \in I_2 \cup I_3$; furthermore, y_i for $i \in I_1$ must satisfy (14), so that $y_i > 0$ for $i \in I_1$. Denoting by W a neighborhood of y_0 such that $y_i > 0$ for $i \in I_1$ whenever $y \in W$, we obtain that $(x, y) \in U \times W$ is a solution of the variational system:

$$v + \nabla_x g_0(w, x) + \sum\nolimits_{i \in I_1} y_i \nabla_x g_i(w, x) = 0,$$
$$-u_i + g_i(w, x) = 0, \text{ for } i \in I_1,$$
$$-u_i + g_i(w, x) \le 0, \ y_i = 0 \text{ for } i \in I_2 \cup I_3. \tag{15}$$

Further, if $(x, y) \in U \times W$ is a solution of (15) for some $p \in V$, then $x \in X_{\mathrm{KKT}}(p) \cap U$ and, because of the linear independence of $\nabla_x g_i(w, x)$ for $i \in I_1$, y is uniquely defined and the function $p \mapsto y(p)$ is Lipschitz continuous in V. Thus, the solution map of (15) is locally single-valued and Lipschitz continuous around $(p_\varepsilon, x_0, y_0)$. Observe that (x_0, y_0) is a locally unique solution of (15) and x_0 is a locally optimal solution of (1), both for $p = p_\varepsilon$. In particular, the map $p \mapsto (x(p), y(p))$ is locally nonempty-valued and upper-Lipschitz at the point $(p_\varepsilon, x_0, y_0)$. From Theorem 2.6 it follows that the second-order sufficient condition holds at $(p_\varepsilon, x_0, y_0)$. Noting that the set I_2 associated with $(p_\varepsilon, x_0, y_0)$ is empty, and $\nabla^2_{xx} L$ does not depend on ε and is affine in y, we see that (12) holds for every $y_0 \in Y_{\mathrm{KKT}}(x_0, p_0)$. $\qquad\square$

Observe that, from the above proof, Theorem 3.3 remains valid for a smaller class of perturbations $p = (v, u, w)$ where $w = w_0$ is kept constant.

As indicated by a counterexample of Robinson [32], the statement converse to Theorem 3.3 is false, in general. It has been noted recently, see Liu [25] and Ralph and Dempe [29], that, under the constant rank condition for the set of constraints at the reference point, the converse statement holds; that is the combination of the Mangasarian-Fromovitz condition and the strong second-order sufficient condition implies that both the solution map and the stationary point map are locally single-valued and Lipschitz continuous (and B-differentiable). Recall that the set of constraints (2) satisfies the *constant rank condition* if there exists a neighborhood $\mathcal{W}$ of (w_0, x_0) such that for every $I \subset I_1 \cup I_2$, rank$\{\nabla_x g_i(w, x) \mid i \in I\}$ is constant for every $(w, x) \in \mathcal{W}$. By combining the above mentioned result with Theorem 3.3 we obtain the following characterization of the Lipschitzian stability of the stationary point map:

Corollary 3.5. *Suppose that $x_0 \in X_{\mathrm{KKT}}(p_0)$ and the constant rank condition holds. Then the following are equivalent:*

(i) The stationary point map X_{KKT} is locally single-valued and Lipschitz continuous around (p_0, x_0), moreover with the property that for all $(p, x) \in \mathrm{gph}\, X_{\mathrm{KKT}}$ in some neighborhood of (p_0, x_0), x is a locally optimal solution to the nonlinear programming problem (1) for p;

(ii) The Mangasarian-Fromovitz condition and the strong second-order sufficient condition for local optimality (12) hold for (p_0, x_0).

In our previous paper [8] we proved that if the solution map, say Σ, of a variational inequality over a convex polyhedral set has the Aubin property at $(y_0, x_0) \in \mathrm{gph}\, \Sigma$, it must be locally single-valued and Lipschitz continuous around (y_0, x_0) (for more recent related results see [11] and [23]). In particular, if the KKT map S_{KKT} has the Aubin property, it is locally single-valued and Lipschitz continuous. Here we present an analogue of this result for the stationary point map X_{KKT}, but under the additional assumptions that the values of this map are locally optimal solutions.

Theorem 3.6. *Let* $(p_0, x_0) \in \mathrm{gph}\, X_{\mathrm{KKT}}$, $p_0 = (v_0, u_0, w_0)$, *and suppose that if* $(p, x) \in \mathrm{gph}\, X_{\mathrm{KKT}}$ *in some neighborhood of* (p_0, x_0) *then* x *is a locally optimal solution to (1) for* p, *that is* $(p, x) \in \mathrm{gph}\, X$. *Then the following are equivalent:*

(i) *The Mangasarian-Fromovitz condition holds for* (p_0, x_0) *and the map* X_{KKT} *is Aubin continuous at* (p_0, x_0);

(ii) *The map* X_{KKT} *is locally single-valued and Lipschitz continuous around* (p_0, x_0).

Proof. The implication (ii) $\Rightarrow$ (i) follows from Theorem 3.3 (note that the local optimality is not needed in this direction). If we prove that (i) implies the strong second-order condition for local optimality (12), then from Theorem 7.2 in Kojima [18] it follows that the map X_{KKT} is locally single-valued and continuous, hence Lipschitz continuous, and we obtain (ii).

Invoking the argument given in the proof of Theorem 3.3, we choose an extreme point y_0 of $Y_{\mathrm{KKT}}(x_0, p_0)$ and consider the KKT system (3) for $p = (v, u, w)$ in a neighborhood of the value p_ε whose components are: $v = v_0$, $u_i = u_{0i}$ for $i \in I_1 \cup I_3$, $u_i = u_{0i} + \varepsilon$ for $i \in I_2$, and $w = w_0$. For a sufficiently small $\varepsilon > 0$ the map X_{KKT} is Aubin continuous at (p_ε, x_0), by the very definition of the Aubin continuity at (p_0, x_0). Choose neighborhoods V of p_ε and U of x_0 such that for every $p \in V$ and $x \in U$, one has $-u_i + g_i(w, x) < 0$ for $i \in I_2 \cup I_3$; moreover the vectors $\nabla_x g_i(w, x)$, $i \in I_1$, are linearly independent, and if y_i, $i \in I_1$, satisfy (14), then $y_i > 0$ for $i \in I_1$. From the definition of the Aubin continuity one can find neighborhoods $U' \subset U$ of x_0 and $V' \subset V$ of p_ε such that if $p', p'' \in V'$ and $x' \in X_{\mathrm{KKT}}(p') \cap U'$, then there exists $x'' \in X_{\mathrm{KKT}}(p'')$ with $\|x'' - x'\| \leq M\|p' - p''\|$; moreover, U' and V' can be chosen so small that $x'' \in U$ for every choice of $p', p'' \in V'$ and $x' \in U'$. Denoting by W a neighborhood of y_0 such that $y_{0i} > 0$ for $i \in I_1$, let $(x', y') \in U' \times W$ be a KKT point for (1) for p', that is, (x', y') solves (3) for p'. Then (x', y') must satisfy the system (15) for p' and the Lagrange multiplier y' is unique, from the linear independence of $\nabla_x g_i(w', x')$, $i \in I_1$. Let $x'' \in X_{\mathrm{KKT}}(p'')$ be such that $\|x'' - x'\| \leq M\|p' - p''\|$, and let y'' be an associate Lagrange multiplier. Then, from the choice of the neighborhoods U' and V', y'' must satisfy (15) and then $y_i'' > 0$ for $i \in I_1$ and $y_i'' = 0$ for $i \in I_2 \cup I_3$. Further, since y', y'' satisfy (14) for (p', x') and $(p,'' x'')$, respectively, and because of the linear independence of $\nabla_x g_i(w, x)$, $i \in I_1$ for $p \in V'$, $x \in U'$, there exists a constant $c > 0$ such that

$$\|y'' - y'\| \leq c(\|x'' - x'\| + \|p'' - p'\|) \leq c(M+1)\|p'' - p'\|.$$

This means that the KKT map S_{KKT} of the problem (1) is Aubin continuous at $(p_\varepsilon, x_0, y_0)$.

Applying the characterization of the Aubin property from Theorem 5 in [8] with $I_1' = I_1$, $I_3' = I_2 \cup I_3$, see Theorem 4.1 in the following section, we obtain that

$$\langle x', \nabla_{xx}^2 L(w_0, x_0, y_0) x' \rangle \neq 0 \text{ for all } x' \neq 0, x \in M. \tag{16}$$

Since every $y_0 \in Y_{\mathrm{KKT}}(x_0, p_0)$ can be represented as a convex combination of the extreme points of $Y_{\mathrm{KKT}}(x_0, p_0)$ and $\nabla_{xx}^2 L$ is affine in y, we get (16) for every choice of the Lagrange multiplier y_0. Taking into account the second-order necessary conditions for local optimality of x_0 for (1) with p_ε, we complete the proof. □

4. STRONG REGULARITY

In Robinson's terminology [30], the KKT system (3) is *strongly regular* if the map L_{KKT} defined by the linearization (4) of the KKT system (3) is locally single-valued and Lipschitz continuous around (q_0, x_0), $q_0 = (v_0, u_0)$. In [30] it is proved that if the KKT system (3) is strongly regular, then the KKT map S_{KKT} is locally single-valued and Lipschitz continuous; moreover, for the problem (1), the strong second-order sufficient condition for local optimality together with the linear independence of the gradients of the active constraints implies the strong regularity.

From the result in [8] mentioned in the preceding section it follows that the map S_{KKT} has the Aubin property if and only if this map is locally single-valued and Lipschitz continuous. In [8] we obtained a characterization of the strong regularity in a form of a "critical face condition." The precise result is as follows.

Theorem 4.1 ([8], Theorem 5). *The KKT system (3) is strongly regular for (p_0, x_0, y_0), $p_0 = (v_0, u_0, w_0)$ if and only if the following two requirements are fulfilled:*

(a) The vectors $\nabla_x g_i(w_0, x_0)$ for $i \in I_1 \cup I_2$ are linearly independent;

(b) For each partition of $\{1, 2, \ldots, m\}$ into index sets I'_1, I'_2, I'_3 with $I_1 \subset I'_1 \subset I_1 \cup I_2$ and $I_3 \subset I'_3 \subset I_3 \cup I_2$, the cone $K(I'_1, I'_2) \subset \mathbb{R}^n$ consisting of all the vectors x' satisfying

$$\langle \nabla_x g_i(w_0, x_0), x' \rangle \begin{cases} = 0 & \text{for } i \in I'_1, \\ \leq 0 & \text{for } i \in I'_2, \end{cases}$$

should be such that

$$x' \in K(I'_1, I'_2), \ \nabla^2_{xx} L(w_0, x_0, y_0) x' \in K(I'_1, I'_2)^* \quad \Longrightarrow \quad x' = 0.$$

Here K^* denotes the polar to K. For other characterizations of strong regularity and a detailed discussion of related results, see Klatte and Tammer [17] and Kummer [19]. In particular, Kummer's condition is based on a general implicit-function theorem for nonsmooth functions, while we use the equivalence of the Aubin property and the strong regularity established in [8] and then apply Mordukhovich's characterization of the Aubin property. It is not clear to us how one could derive the equivalence between these various conditions directly.

Relying on the above result, in [8], Theorem 6 we showed that, under canonical perturbations, the combination of the strong second-order sufficient condition for local optimality with the linear independence of the gradients of the active constraints is actually equivalent to the requirement the map S_{KKT} be locally single-valued and Lipschitz continuous with the x-component being a locally optimal solution. Below, we present a refinement of this theorem with a short proof.

Theorem 4.2. *The following are equivalent:*

(i) The map $p \mapsto S(p) = \{(x, y) \in S_{\text{KKT}}(p) \mid x \in X(p)\}$ is locally single-valued and Lipschitz continuous around (p_0, x_0, y_0), $p_0 = (u_0, v_0, w_0)$;

(ii) The map S_{KKT} is locally single-valued and Lipschitz continuous around (p_0, x_0, y_0), moreover with the property that for all $(p, x, y) \in \text{gph} S_{KKT}$ in some

neighborhood of $(u_0, v_0, w_0, x_0, y_0)$, x *is a locally optimal solution to the nonlinear programming problem (1) for* $p = (u, v, w)$;

(iii) x_0 *is an isolated local minimizer of (1) and the associated KKT system (3) is strongly regular around* (p_0, x_0, y_0);

(iv) *The constraint gradients* $\nabla_x g_i(w_0, x_0)$ *for* $i \in I_1 \cup I_2$ *are linearly independent and the strong second-order sufficient condition for local optimality holds for* (p_0, x_0, y_0): *one has*

$$\langle x', \nabla^2_{xx} L(w_0, x_0, y_0) x' \rangle > 0 \text{ for all } x' \neq 0 \text{ in the subspace}$$

$$M = \left\{ x' \mid x' \perp \nabla_x g_i(w_0, x_0) \text{ for all } i \in I_1 \right\}.$$

Proof. Let (i) hold. Since $S(p) \subset S_{\text{KKT}}(p)$, the KKT map S_{KKT} has a Lipschitz continuous local selection around (p_0, x_0, y_0). Consider the Kojima map associated with the KKT system (3):

$$F(x, y) = \begin{pmatrix} \nabla_x g_0(w_0, x) + \sum_{i=1}^{r} y_i \nabla_x g_i(w_0, x) + \sum_{i=r+1}^{m} y_i^+ \nabla_x g_i(w_0, x) \\ -g_1(w_0, x) \\ \cdot \\ \cdot \\ \cdot \\ -g_r(w_0, x) \\ -g_{r+1}(w_0, x) + y_{r+1}^- \\ \cdot \\ \cdot \\ \cdot \\ -g_m(w_0, x) + y_m^- \end{pmatrix},$$

where $y^+ = \max\{0, y\}$ and $y^- = \min\{0, y\}$. Let $G(x, y) = F(x, y + g(w_0, x))$, where $g = (g_1, \cdots, g_m)$. Then every $(x, y) \in S_{\text{KKT}}(p)$ for $p = (u, v, w_0)$ is a solution of the equation

$$G(x, y) = \begin{pmatrix} -v \\ -u \end{pmatrix}.$$

The continuous map $G : \mathbb{R}^{n+m} \mapsto \mathbb{R}^{n+m}$ has the property that its inverse G^{-1} has a Lipschitz continuous local selection around (p_0, x_0, y_0). Then the map G^{-1} must be locally single-valued, see Lemma 1 in Kummer [20]. Hence S_{KKT} is locally single-valued and Lipschitz continuous and thus (ii) is established.

The equivalence (ii) $\Leftrightarrow$ (iii) follows from Proposition 2 in [8]. The implication (iii) $\Rightarrow$ (iv) is a consequence of Theorems 3.3 and 4.1. Finally, (iv) $\Rightarrow$ (i) is proved in Robinson [30]; for other proofs see [9], p. 370 and [17]. $\square$

Under the linear independence of the gradients of the active constraints, the equivalence of (iii) and (iv) follows from a combination of earlier results of Kojima [18] and Jongen et al. [12]; alternative proofs of this equivalence have been furnished recently by Bonnans and Sulem [5] and Pang [28].

At the end, as a consequence of Theorem 4.2 we present a characterization of the continuous differentiability of the pair solution-Lagrange multiplier which complements a basic result due to Fiacco [10]. Using the implicit function theorem, Fiacco showed that the combination of the linear independence of the gradient of the active constraints, the second-order sufficient condition and the strict complementary slackness implies that the map S is locally single-valued and continuously differentiable around the reference point (p_0, x_0, y_0). The following result shows that, for canonical perturbations, these conditions are also necessary for the latter property. Recall that strict complementary slackness condition holds if there are no zero Lagrange multipliers associated with active constraints at the reference point; that is, $I_2 = \emptyset$.

Corollary 4.3. *The following are equivalent:*

(i) *The map $p \mapsto S(p) = \{(x, y) \in S_{KKT}(p) \mid x \in X(p)\}$ is locally single-valued and continuously differentiable around (p_0, x_0, y_0), $p_0 = (u_0, v_0, w_0)$;*

(ii) *The strict complementary slackness (i.e., $I_2 = \emptyset$), the linear independence of the gradients of the active constraint, and the second-order sufficient condition for local optimality hold for (p_0, x_0, y_0).*

Proof. All we need to prove is that (i) implies the strict complementarity; the rest follows from Theorem 4.2 and Fiacco's theorem. On the contrary, assume that $I_2 \neq \emptyset$ and (i) holds. Let $i \in I_2$ and consider the problem (1) with the following values of the parameter p denoted p_ε: $w = w_0$, $v = v_0$, $u_i = u_{0i} + \varepsilon$ and $u_j = u_{0j}$ for $j \neq i$, where ε is a real parameter from a neighborhood of zero. As already noted in the previous proof, since $y_{0i} = 0$, for every nonnegative and sufficiently small ε and for some neighborhood W of (x_0, y_0) the only element in $S(p_\varepsilon) \cap W$ is (x_0, y_0). Then the derivative of x must satisfy

$$\frac{d}{d\varepsilon} x(p_\varepsilon)|_{\varepsilon=0} = 0. \tag{17}$$

On the other hand, for $\varepsilon < 0$ the corresponding solution $x(p_\varepsilon)$ is feasible, that is,

$$-u_{0i} - \varepsilon + g_i(w_0, x(p_\varepsilon)) \leq 0.$$

Combining this inequality with the assumed equality $-u_{0i} + g_i(w_0, x_0) = 0$, we obtain

$$-1 + \frac{1}{\varepsilon}[g(w_0, x(p_\varepsilon)) - g(w_0, x_0)] \geq 0.$$

Passing to zero with ε and using (17) results in $-1 \geq 0$, a contradiction. This completes the proof. □

References

[1] J.-P. Aubin, Lipschitz behavior of solutions to convex minimization problems, *Math. Oper. Res.* **9**, 1984, 87–111.

[2] J. F. Bonnans, Local analysis of Newton-type methods for variational inequalities and nonlinear programming, *Appl. Math. and Optim.* **29** (1994), 161–186.

[3] J. F. Bonnans, A. D. Ioffe, Quadratic growth and stability in convex programming problems with multiple solutions, *J. Convex Analysis* **2** (1995) 41-57.

[4] J. F. Bonnans, A. Shapiro, Optimization problems with perturbations, a guided tour, INRIA preprint 2872, Théme 4, INRIA Rocquencourt, April 1996.

[5] J. F. Bonnans, A. Sulem, Pseudopower expansion of solutions of generalized equations and constrained optimization problems, *Math. Programming* **70** (1995), 123–148.

[6] A. L. Dontchev, Characterizations of Lipschitz stability in optimization; in *Recent Developments in Well-Posed Variational Problems*, R. Lucchetti and J. Revalski (eds.), Kluwer, 1995, 95–116.

[7] A. L. Dontchev, Implicit function theorems for generalized equations, *Math. Programming* **70** (1996) 91-106.

[8] A. L. Dontchev, R. T. Rockafellar, Characterizations of strong regularity for variational inequalities over polyhedral convex sets, *SIAM J. Optim.*, accepted 1995.

[9] A. L. Dontchev, T. Zolezzi, Well-Posed Optimization Problems, Lecture Notes in Mathematics, vol. **1543**, Springer-Verlag, 1993.

[10] A. V. Fiacco, Sensitivity analysis for nonlinear programming using penalty methods, *Math. Programming* **10** (1976), 287–311

[11] M. Seetharama Gowda, R. Sznajder, On the pseudo-Lipschitzian behavior of the inverse of a piecewise affine function, In *Proc. International Conference on Complementarity Problems*, Baltimore, Nov. 1995, to appear.

[12] H. Th. Jongen, T. Möbert, J. Rückmann, K. Tammer, Implicit functions and sensitivity of stationary points, *Linear Algebra and Appl.* **95** (1987), 97–109.

[13] A. J. King, R. T. Rockafellar, Sensitivity analysis for nonsmooth generalized equations, *Math. Programming* **55** (1992), 341–364.

[14] D. Klatte, On the stability of local and global optimal solutions in parametric problems of nonlinear programming, Part I: Basic results, *Seminarbericht 75 der Sektion Mathematik der Humboldt-Universität zu Berlin*, Berlin, 1985, 1–21.

[15] D. Klatte, On quantitative stability for non-isolated minima, *Control Cybernet.* **23** (1994), 183–200.

[16] D. Klatte, On quantitative stability for $C^{1,1}$ programs; in *Proceeding of the French-German Conference on Optimization*, R. Durier and C. Michelot (eds.), Lecture Notes in Economics and Mathematical Systems, vol. **429**, Springer-Verlag, 1995, 214–230.

[17] D. Klatte, K. Tammer, Strong stability of stationary solutions and Karush-Kuhn-Tucker points in nonlinear optimization, *Annals of Oper. Research* **27** (1990), 285–308.

[18] M. Kojima, Strongly stable stationary solutions in nonlinear programming; in *Analysis and computation of fixed points*, S. M. Robinson (ed.), Academic Press, New York, 1980, 93–138.

[19] B. Kummer, Lipschitzian inverse functions, directional derivatives and applications in $C^{1,1}$ optimization, *J. Optim. Theory Appl.* **70** (1991), 561–582.

[20] B. Kummer, Lipschitzian and pseudo-Lipschitzian inverse functions and applications to nonlinear optimization, preprint, Dept. of Mathematics, Humboldt University, Berlin 1996.

[21] J. Kyparisis, On uniqueness of Kuhn-Tucker multipliers in nonlinear programming, *Math. Programming* **32** (1985), 242–246.

[22] A. B. Levy, Implicit multifunction theorems for the sensitivity analysis of variational conditions, *Math. Programming*, **74** (1996) 333–350.

[23] A. B. Levy, R. A. Poliquin, Characterizations of the local single-valuedness of multifunctions, preprint, Dept. of Mathematics, Bowdoin College, 1996.

[24] A. B. Levy, R. T. Rockafellar, Sensitivity of solutions in nonlinear programming problems with nonunique multipliers, in *Recent Advances in Nonsmooth Optimization*, D. Z. Du, L. Qi and R. S. Womersley (eds.), World Scientific Press, 1995, 215–223.

[25] Jiming Liu, Sensitivity analysis in nonlinear programs and variational inequalities via continuous selections, *SIAM J. Control Optim.* **33** (1995) 1040-1060.

[26] B. Mordukhovich, Lipschitzian stability of constraint systems and generalized equations, *Nonlinear analysis*, **22** (1994) 173–206.

[27] J.-S. Pang, Convergence of splitting and Newton methods for complementarity problems: an application of some sensitivity results, *Math. Programming* **58** (1993), 149–160.

[28] J.-S. Pang, Necessary and sufficient conditions for solution stability of parametric nonsmooth equations, in *Recent Advances in Nonsmooth Optimization*, D. Z. Du, L. Qi and R. S. Womersley (eds.), World Scientific Press, 1995, 261–288.

[29] D. Ralph, S. Dempe, Directional derivatives of the solution of a parametric nonlinear program, *Math. Programming* **70** (1995) 159–172.

[30] S. M. Robinson, Strongly regular generalized equations, *Math. Oper. Res.* **5** (1980), 43–62.

[31] S. M. Robinson, Some continuity properties of polyhedral multifunctions, *Math. Programming Study* **14** (1981), 206–214.

[32] S. M. Robinson, Generalized equations and their solutions, Part II: Applications to nonlinear programming, *Math. Programming Study* **19** (1982), 200–221.

[33] S. M. Robinson, Local epi-continuity and local optimization, *Mathematical Programming* **37** (1987), 208–223.

[34] R. T. Rockafellar, Proto-differentiability of set-valued mappings and its applications in optimization, *Ann. Inst. H. Poincaré: Analyse Non Linéaire* **6** (1989), suppl., 449–482.

[35] S. Scholtes, Introduction to piecewise differentiable equations, preprint 53/1994, Institut für Statistik und Mathematische Wirtschaftstheorie, Universität Karlsruhe, May 1994.

[36] A. Shapiro, Sensitivity analysis of nonlinear programs and differentiability properties of metric projections, *SIAM J. Control Optim.* **26** (1988), 628–645.

[37] G. Still, M. Streng, Optimality conditions in smooth nonlinear programming, *J. Optim. Theory Appl.* **90** (1996) 483–515.

[38] D. E. Ward, Characterizations of strict local minima and necessary conditions for weak sharp minima, *J. Optim. Theory Appl.* **80** (1994) 551-571.

On Second Order Sufficient Conditions for Structured Nonlinear Programs in Infinite-Dimensional Function Spaces

J. C. DUNN
Mathematics Department, Box 8205,
North Carolina State University, Raleigh, NC 27695-8205

Abstract. In this expository article, we examine several recently proved second order sufficiency theorems for nonlinear programs with smooth structured nonconvex cost functions and affine constraints in infinite-dimensional function spaces. Sufficient conditions for two nonequivalent species of local optimality are described in this setting, and compared with earlier sufficient conditions for general nonlinear programs in infinite- dimensional normed vector spaces. The structure and smoothness assumptions invoked in the new theorems are satisfied by constrained minimization problems in the calculus of variations and optimal control, and the new sufficient conditions are not far removed from necessary conditions for local optimality.

1. Introduction. Second order sufficient conditions are key hypotheses in sensitivity theorems and local convergence theorems for iterative algorithms (cf.,Fiacco and McCormick [20] [21] [42], Bertsekas and Gafni [7]–[9], [22], Luenberger [32] [33], Robinson [45], Hager, Dontchev, et. al. [12] [13] [27] [28], Maurer, Zowe, et. al. [39] [38], Dunn and Gawande [14] [23] [24], AlJazzaf [1], Tian and Dunn [15] [16] [47] [18] [19] [48], Alt and Malanowski [2]–[4], [34]–[37], Ito and Kunisch [31] and others). For the classic smooth nonlinear program in $\mathbb{R}_n$, the gap between the standard necessary conditions and sufficient conditions is not large, and the corresponding sensitivity and convergence theorems are therefore widely applicable. On the other hand, this gap can open substantially for normed vector space extensions of the finite-dimensional optimality conditions (Maurer and Zowe [39], Ioffe [30]) and when this happens, the scope of the sensitivity and convergence theories is limited accordingly. Since norm equivalence is lost in infinite-dimensional spaces, the hypotheses and conclusions in the principal theorems are now also norm-dependent, and there may be no single norm with respect to which all the required hypotheses hold simultaneously (Ioffe [29], Maurer [38], Malanowski [36]). These points are developed here for general and special nonlinear programs with affine constraints. For such problems, the basic optimality conditions assume a particularly simple aspect, free of explicit representations for the feasible set and attendant regularity conditions, multiplier vectors and Lagrangians.

Investigation supported by NSF Research Grant #DMS-9500908

In the following sections, a map F from a normed vector space $\mathbb{Z}$ to a normed vector space $\mathbb{U}$ is said to be *twice directionally differentiable* at $\underline{z}$ *iff* the following limits exist for all ξ, η in $\mathbb{Z}$:

$$F'(\underline{z})(\eta) \stackrel{def}{=} \lim_{s \to 0} \frac{F(\underline{z} + s\eta) - F(\underline{z})}{s}$$

$$F''(\underline{z})(\xi, \eta) \stackrel{def}{=} \lim_{s \to 0} \frac{F'(\underline{z} + s\xi)(\eta) - F'(\underline{z})(\eta)}{s}.$$

A twice directionally differentiable map F is *twice Gâteaux differentiable* at $\underline{z}$ *iff* the associated functions $F'(\underline{z})(\cdot) : \mathbb{Z} \to \mathbb{U}$ and $F''(\underline{z})(\cdot, \cdot) : \mathbb{Z} \oplus \mathbb{Z} \to \mathbb{U}$ are linear and bilinear respectively, and bounded in the sense that

$$\|F'(\underline{z})\| \stackrel{def}{=} \sup_{\|\eta\|_{\mathbb{Z}}=1} \|F'(\underline{z})(\eta)\|_{\mathbb{U}} < \infty$$

and

$$\|F''(\underline{z})\| \stackrel{def}{=} \sup_{\|\xi\|_{\mathbb{Z}}=1} \sup_{\|\eta\|_{\mathbb{Z}}=1} \|F''(\underline{z})(\xi, \eta)\|_{\mathbb{U}} < \infty.$$

For a linear map $F'(\underline{z})(\cdot)$, boundedness is equivalent to continuity relative to the norms $\| \cdot \|_{\mathbb{U}}$ and $\| \cdot \|_{\mathbb{Z}}$. Similarly, for a bilinear map $F''(\underline{z})(\cdot, \cdot)$, boundedness is equivalent to continuity relative to $\| \cdot \|_{\mathbb{U}}$ and any of the standard induced norms on the direct sum $\mathbb{Z} \oplus \mathbb{Z}$ (e.g., $\|(\xi, \eta)\|_{\mathbb{Z} \oplus \mathbb{Z}} = (\|\xi\|_{\mathbb{Z}}^2 + \|\eta\|_{\mathbb{Z}}^2)^{\frac{1}{2}}$). A twice Gâteaux differentiable map F is *twice Fréchet differentiable* at $\underline{z}$ *iff*

$$\|F(z) - F(\underline{z}) - F'(\underline{z})(z - \underline{z})\|_{\mathbb{U}} = o(\|z - \underline{z}\|_{\mathbb{Z}})$$

and

$$\sup_{\|\eta\|_{\mathbb{Z}}=1} \|F'(z)(\eta) - F'(\underline{z})(\eta) - F''(\underline{z})(z - \underline{z}, \eta)\|_{\mathbb{U}} = o(\|z - \underline{z}\|_{\mathbb{Z}}).$$

Finally, F is *twice continuously Fréchet differentiable* at $\underline{z}$ *iff* F is twice Fréchet differentiable near $\underline{z}$, and

$$\lim_{\|z - \underline{z}\|_{\mathbb{Z}} \to 0} \sup_{\|\eta\|_{\mathbb{Z}}=1} \|F'(z)(\eta) - F'(\underline{z})(\eta)\|_{\mathbb{U}} = 0$$

and

$$\lim_{\|z - \underline{z}\|_{\mathbb{Z}} \to 0} \sup_{\|\xi\|_{\mathbb{Z}}=1} \sup_{\|\eta\|_{\mathbb{Z}}=1} \|F''(z)(\xi, \eta) - F''(\underline{z})(\xi, \eta)\|_{\mathbb{U}} = 0.$$

When $\mathbb{Z}$ and $\mathbb{U}$ are finite-dimensional spaces, the foregoing differentiability properties are *norm-invariant*, i.e., they hold for all combinations of norms $\| \cdot \|_{\mathbb{Z}}$ and $\| \cdot \|_{\mathbb{U}}$ *iff* they hold for some such combination of norms. However, directional differentiability generally depends on the norm assigned to $\mathbb{U}$, while Gâteaux and Fréchet differentiability depend on the norms $\| \cdot \|_{\mathbb{Z}}$ and $\| \cdot \|_{\mathbb{U}}$. When $\mathbb{U} = \mathbb{R}_1$, it is understood that $\|u\|_{\mathbb{U}} = |u|$. When $\mathbb{U} = \mathbb{R}_1$ and F is twice continuously Fréchet differentiable at $\underline{z}$, it follows immediately from Taylor's formula in $\mathbb{R}_1$ that

$$F(z) = F(\underline{z}) + F'(\underline{z})(z - \underline{z}) + \frac{1}{2}F''(\underline{z})(z - \underline{z}, z - \underline{z}) + r_2(\underline{z}, z),$$

with

$$\lim_{\|z - \underline{z}\|_{\mathbb{Z}} \to 0} \frac{r_2(\underline{z}, z)}{\|z - \underline{z}\|_{\mathbb{Z}}^2}.$$

If $\mathbb{Z}$ is equipped with an inner product $\langle \cdot, \cdot \rangle_{\mathbb{Z}}$, and if $J : \mathbb{Z} \to \mathbb{R}_1$ is twice Gâteaux differentiable at $\underline{z}$, then the first and second order differentials of J often have useful representations of the form,

$$J'(\underline{z})(w) = \langle \nabla J(\underline{z}), w \rangle_{\mathbb{Z}}$$

and

$$J''(\underline{z})(v, w) = \langle v, \nabla^2 J(\underline{z})w \rangle_{\mathbb{Z}},$$

where $\nabla J(\underline{z})$ is a vector in $\mathbb{Z}$ called the *gradient* of J at $\underline{z}$, and $\nabla^2 J(\underline{z})$ is a bounded linear operator on $\mathbb{Z}$ called the *Hessian* of J at $\underline{z}$. Gradients and Hessians are always unique. Moreover, if $\mathbb{Z}$ is metrically complete with respect to the inner product induced norm, $\|z\| = \sqrt{\langle z, z \rangle_{\mathbb{Z}}}$, then the existence of $\nabla J(\underline{z})$ and $\nabla^2 J(\underline{z})$ follows automatically from the Riesz-Fréchet representation theorem in Hilbert spaces. On the other hand, the existence of $\nabla J(\underline{z})$ and $\nabla^2 J(\underline{z})$ can often be verified directly for specific problem classes in incomplete inner product spaces. Finally, if $\mathbb{Z} = \mathbb{R}_m$ and $\langle \cdot, \cdot \rangle_{\mathbb{Z}}$ is the standard Euclidean inner product, $\langle \xi, \eta \rangle = \sum_{i=1}^{m} \xi_i \eta_i$, then $\nabla J(\underline{z})$ and $\nabla^2 J(\underline{z})$ automatically exist and have the familiar matrix representors with entries $\frac{\partial J}{\partial z_i}(\underline{z})$ and $\frac{\partial^2 J}{\partial z_i \partial z_j}(\underline{z})$ respectively.

2. Optimality Conditions for Nonlinear Programs With Affine Constraints. The standard nonlinear program in $\mathbb{R}_n$ is subsumed by the general constrained minimization problem,

(2.1a)
$$\min_{z \in C} J(z)$$

(2.1b)
$$C = G^{-1}(\mathcal{K}) = \{z \in \mathbb{Z} : G(z) \in \mathcal{K}\},$$

where J and G are twice continuously Fréchet differentiable functions from a normed vector space $\mathbb{Z}$ to $\mathbb{R}_1$ and a normed vector space $\mathbb{Y}$ respectively, and $\mathcal{K}$ is a closed convex cone in $\mathbb{Y}$ with vertex at 0. More specifically, the finite-dimensional nonlinear program is obtained from (2.1) by setting $\mathbb{Z} = \mathbb{R}_n$, $\mathbb{Y} = \mathbb{R}_q \oplus \mathbb{R}_r$, $\mathcal{K} = \mathcal{K}_g \times \mathcal{K}_h$, $\mathcal{K}_g = \{\eta \in \mathbb{R}_q : \eta \leq 0\}$, $\mathcal{K}_h = \{0\}$, and $G(z) = (g(z), h(z))$, where g and h map $\mathbb{R}_n$ to $\mathbb{R}_q$ and $\mathbb{R}_r$ respectively. Formal extensions of the finite-dimensional first and second order necessary conditions for local optimality are proved for the general problem (2.1) by Luenberger [32], Guignard [26], Robinson [44][45], Maurer and Zowe [39][38], Ioffe [29][30], Ben-Tal and Zowe [5][6], Cominetti [11] and others. Analogous formal extensions of the finite-dimensional second order sufficient conditions do *not* imply local optimality for (2.1) [39]; however, valid single norm and multinorm sufficient conditions are obtained in [39] and [38] by replacing the standard positive-definiteness condition on a Lagrangian second differential $L''(z) = J''(z) - \lambda(G''(z))$ by a coercivity condition, and then enlarging the cone on which the coercivity condition is enforced. When G is a continuous affine map, the set $C = G^{-1}(\mathcal{K})$ is closed and convex, and the fundamental necessary conditions and sufficient conditions in [39] and [38] are subsumed by the following elementary representation-free theorems which apply to any local minimizer in any convex set C, without regularity conditions.

At each point z in a convex set C, let $\mathcal{F}_C$ denote the *conical hull* of $C - z$; equivalently,

$$(2.2) \qquad \mathcal{F}_C(z) = \{w \in \mathbb{Z} : \exists \alpha > 0 \quad z + \alpha w \in C\}.$$

Since C is convex, it follows that $z + tw \in C$ when $w \in \mathcal{F}_C(z)$ and t is positive and sufficiently small. Thus $\mathcal{F}_C(z)$ is the *cone of feasible directions* at z, and an application of one-dimensional calculus yields well-known rudimentary first and second order necessary conditions for optimality in convex sets.

THEOREM 2.1. *Suppose that* $J : \mathbb{Z} \to \mathbb{R}_1$ *is twice directionally differentiable at a point* $\underline{z}$ *in a convex set* C, *and that* $\underline{z}$ *is a local minimizer for* J *in* C. *Then for all* w,

$$(2.3a) \qquad w \in \mathcal{F}_C(\underline{z}) \Rightarrow J'(\underline{z})(w) \geq 0$$

and

$$(2.3b) \qquad w \in \mathcal{F}_C(\underline{z}) \text{ and } J'(\underline{z})(w) = 0 \Rightarrow J''(\underline{z})(w, w) \geq 0$$

The second order sufficient conditions in [39] and [38] also have representation-free counterparts that are easily proved in any convex feasible set C. We first state and prove a simple single-norm second order sufficient condition related to the analysis in [39].

THEOREM 2.2. *Let* J *be twice continuously Fréchet differentiable at a point* $\underline{z}$ *in a convex set* $C \subset \mathbb{Z}$, *and suppose that for some* $\mu > 0$ *and* $\beta > 0$, *and all* w,

$$(2.4a) \qquad w \in \mathcal{F}_C(\underline{z}) \Rightarrow J'(\underline{z})(w) \geq 0$$

and

$$(2.4b) \qquad w \in \mathcal{F}_C(\underline{z}) \text{ and } J'(\underline{z})(w) \leq \beta\|w\|_{\mathbb{Z}} \Rightarrow J''(\underline{z})(w, w) \geq \mu\|w\|_{\mathbb{Z}}^2.$$

Then $\underline{z}$ *is a strict local minimizer for* J *in* C, *and for each* $\mu' \in (0, \mu)$, *there is a* $\delta > 0$ *such that for all* z,

$$(2.5) \qquad z \in C \text{ and } \|z - \underline{z}\|_{\mathbb{Z}} < \delta \Rightarrow J(z) - J(\underline{z}) \geq \frac{1}{2}\mu'\|z - \underline{z}\|_{\mathbb{Z}}^2.$$

Proof. Since J is twice continuously Fréchet differentiable at $\underline{z}$, we have

(2.6a) $J(z) - J(\underline{z}) = J'(\underline{z})(z - \underline{z}) + \frac{1}{2}J''(\underline{z})(z - \underline{z}, z - \underline{z}) + r_2(\underline{z}, z),$

with

(2.6b) $$\lim_{\|z - \underline{z}\|_{\mathbb{Z}} \to 0} \frac{r_2(\underline{z}, z)}{\|z - \underline{z}\|_{\mathbb{Z}}^2} = 0$$

and

(2.6c) $$M \overset{def}{=} \sup_{\|w\|_{\mathbb{Z}}=1} |J''(\underline{z})(w, w)| < \infty.$$

Fix μ' in $(0, \mu)$ and choose $\delta > 0$ so that

(2.7) $$\beta \geq \frac{1}{2}(M + \mu)\delta,$$

and for all z

(2.8) $$\|z - \underline{z}\|_{\mathbb{Z}} < \delta \Rightarrow |r_2(\underline{z}, z)| \leq \frac{1}{2}(\mu - \mu')\|z - \underline{z}\|_{\mathbb{Z}}^2.$$

Suppose that $z \in C$ and $\|z - \underline{z}\|_{\mathbb{Z}} < \delta$. Since C is convex, it follows that $z - \underline{z} \in \mathcal{F}_C(\underline{z})$ and therefore $J'(\underline{z})(z - \underline{z}) \geq 0$, by (2.4a). If $0 \leq J'(\underline{z})(z - \underline{z}) \leq \beta \|z - \underline{z}\|_{\mathbb{Z}}$, then (2.4b), (2.6) and (2.8) imply that

$$\begin{aligned} J(z) - J(\underline{z}) &\geq 0 + \frac{1}{2}\mu\|z - \underline{z}\|_{\mathbb{Z}}^2 - \frac{1}{2}(\mu - \mu')\|z - \underline{z}\|_{\mathbb{Z}}^2 \\ &= \frac{1}{2}\mu'\|z - \underline{z}\|_{\mathbb{Z}}^2. \end{aligned}$$

On the other hand, if $\beta\|z - \underline{z}\|_{\mathbb{Z}} < J'(\underline{z})(z - \underline{z})$, then (2.6)–(2.8) imply that

$$\begin{aligned} J(z) - J(\underline{z}) &\geq [\frac{\beta}{\delta} - \frac{1}{2}M - \frac{1}{2}(\mu - \mu')]\|z - \underline{z}\|_{\mathbb{Z}}^2 \\ &\geq \frac{1}{2}\mu'\|z - \underline{z}\|_{\mathbb{Z}}^2. \end{aligned}$$

$\square$

Next, we formulate a variant of Theorem 2 that imposes differentiability and coercivity hypotheses in different, typically non-equivalent norms on $\mathbb{Z}$. This result is related to the two-norm sufficient conditions in [38], which are directly applicable to an important class of Bolza ODE optimal control problems in $\mathbb{L}^p$ function spaces. For such problems, the coercivity condition (2.4b) typically holds in an $\mathbb{L}^p$ norm only if $1 \leq p \leq 2$, whereas J and G are often Fréchet differentiable only relative to the $\mathbb{L}^\infty$ norm (which dominates all the nonequivalent $\mathbb{L}^p$ norms for $1 \leq p < \infty$).

THEOREM 2.3. *Let $\mathbb{Z}$ be equipped with an auxiliary norm $\|\cdot\|_a$ which is dominated by the principal norm $\|\cdot\|_{\mathbb{Z}}$ on $\mathbb{Z}$, i.e.,*

(2.9) $$\sup_{z \neq 0} \frac{\|z\|_a}{\|z\|_{\mathbb{Z}}} < \infty.$$

Suppose that J is twice Gâteaux differentiable relative to the norm $\|\cdot\|_a$ at a point $\underline{z}$ in a convex set $C \subset \mathbb{Z}$, with

$$(2.10a) \qquad J(z) - J(\underline{z}) = J'(\underline{z})(z - \underline{z}) + \frac{1}{2}J''(\underline{z})(z - \underline{z}, z - \underline{z}) + r_2(\underline{z}, z),$$

and

$$(2.10b) \qquad \lim_{\|z - \underline{z}\|_{\mathbb{Z}} \to 0} \frac{r_2(\underline{z}, z)}{\|z - \underline{z}\|_a^2} = 0,$$

In addition, suppose that for some $\mu > 0$ and $\beta > 0$, and all w,

$$(2.11a) \qquad w \in \mathcal{F}_C(\underline{z}) \Rightarrow J'(\underline{z})(w) \geq 0$$

and

$$(2.11b) \qquad w \in \mathcal{F}_C(\underline{z}) \text{ and } J'(\underline{z})(w) \leq \beta\|w\|_a \Rightarrow J''(\underline{z})(w, w) \geq \mu\|w\|_a^2.$$

Then $\underline{z}$ is a strict local minimizer for J in C, and for each $\mu' \in (0, \mu)$, there is a $\delta > 0$ such that for all z,

$$(2.12) \qquad z \in C \text{ and } \|z - \underline{z}\|_{\mathbb{Z}} < \delta \Rightarrow J(z) - J(\underline{z}) \geq \frac{1}{2}\mu'\|z - \underline{z}\|_a^2.$$

The proof for Theorem 2.3 is similar to the proof for Theorem 2.2. Note that the two-norm remainder hypothesis (2.10b) is typically stronger than the conventional single-norm property (2.6b), but weaker than (2.6b) with $\|\cdot\|_{\mathbb{Z}}$ replaced by $\|\cdot\|_a$. Note also that readily tested conditions which imply (2.10b) in practice, generally also imply that J is twice continuously Fréchet differentiable relative to $\|\cdot\|_{\mathbb{Z}}$ (cf., §4).

If G is a continuous affine map and C is the closed convex set $G^{-1}(\mathcal{K})$, then the differential $G''(\underline{z})$ vanishes, the feasible direction cone $\mathcal{F}_C(\underline{z})$ coincides with the so-called *linearizing cone*,

$$\mathcal{L}_C(\underline{z}) = \{w \in \mathbb{Z} : \exists \alpha > 0 \quad G(\underline{z}) + G'(\underline{z})(\alpha w) \in \mathcal{K}\},$$

and the principal Lagrangian necessary conditions and sufficient conditions in [39] and [38] reduce to (2.3), (2.4) and (2.11). The latter conditions are said to be *representation-free* since G does not appear in them explicitly. When $dim\ \mathbb{Y} = \infty$ and G is allowed to be any continuous affine map, the corresponding class of convex sets $C = G^{-1}(\mathcal{K})$ is very large and the optimality conditions (2.3), (2.4) and (2.11) can be quite coarse. More specifically, it can easily happen that the intersection of the cone $\mathcal{F}_C(\underline{z})$ and the closed subspace $nul\ J'(\underline{z}) \stackrel{def}{=} \{w \in \mathbb{Z} : J'(\underline{z})(w) = 0\}$ is the singleton set $\{0\}$ at generic local minimizers in the boundary of C. The second order necessary condition (2.3b) is met trivially at all such $\underline{z}$, while the second order sufficient conditions may hold only for a narrow atypical subclass of boundary point minimizers. On the other hand, if $dim\ \mathbb{Z} < \infty$, $dim\ \mathbb{Y} < \infty$, G is affine, and $\mathcal{K}$ is a polyhedral convex cone, then $G^{-1}(\mathcal{K})$ is a polyhedral convex set, the coercivity conditions (2.4b) and (2.11b) are equivalent [1], the gap between (2.4b) and (2.3b)

---

[1] since any two norms on $\mathbb{Z}$ are equivalent

is small, and local minimizers that satisfy (2.4b) are in some sense generic. In the balance of this section, we examine (2.3), (2.4) and (2.11) more closely in polyhedral convex subsets $Z \subset \mathbb{R}_m$ and in sets C of vector-valued $\mathbb{L}^p$ functions on the interval $[0, 1]$ with range in a polyhedral convex set $Z \subset \mathbb{R}_m$, i.e.,

$$(2.13) \qquad C = \{z \in \mathbb{L}_m^p[0,1] : z(t) \overset{a.e.}{\in} Z\}.$$

Roughly speaking, these *quasi-polyhedral* convex sets are infinite-dimensional limits of finite Cartesian products of polyhedral convex sets; as such, they inherit some but not all of the properties of finite-dimensional polyhedral convex sets. Recall that Z is a polyhedral convex set in $\mathbb{R}_m$ *iff* Z is the intersection of k half-spaces $\{\zeta \in \mathbb{R}_m : \langle a_i, \zeta \rangle + b_i \leq 0\}$ with $a_i \in \mathbb{R}_m$ and $b_i \in \mathbb{R}_1$ for $i = 1, \cdots, k$. Thus, every quasi-polyhedral convex set (2.13) has a representation $C = G^{-1}(\mathcal{K})$ with $\mathbb{Z} = \mathbb{L}_m^p[0,1]$, $\mathbb{Y} = \mathbb{L}_k^p[0,1]$, $\mathcal{K} = \{y \in \mathbb{Y} : \forall i = 1, \cdots, k \quad y_i(t) \overset{a.e.}{\leq} 0\}$, and a continuous affine map $G : \mathbb{Z} \to \mathbb{Y}$ of the special form,

$$G(z)(t) \overset{a.e.}{=} (\langle a_1, z(t) \rangle + b_1, \cdots, \langle a_k, z(t) \rangle + b_k).$$

If Z is a polyhedral convex set in $\mathbb{R}_m$, and the first order necessary condition (2.3a) is satisfied at $\underline{z} \in Z$, then the coercivity condition (2.4b) automatically holds provided $J''(\underline{z})$ is merely positive-definite on the cone $\mathcal{F}_Z(\underline{z}) \cap nul\ J'(\underline{z})$. The proof for this claim depends on a simple variant of Theorem 5.2 in [39].

THEOREM 2.4. *Suppose that C is a convex set in a finite-dimensional normed vector space $\mathbb{Z}$, and let J be twice Gâteaux differentiable at a point $\underline{z} \in C$ where (2.3a) holds. Then the coercivity condition (2.4b) is implied by the positive-definiteness condition,*

$$(2.14) \qquad \forall w \in \overline{\mathcal{F}_C(\underline{z})} \cap nul\ J'(\underline{z}) \quad (w \neq 0 \Rightarrow J''(\underline{z})(w,w) > 0).$$

Proof. We employ the indirect proof technique of [39]. Suppose that (2.4b) is false. Then there is a sequence of vectors $w_n \in \mathcal{F}_C(\underline{z}) \cap \{w \in \mathbb{Z} : \|w\|_{\mathbb{Z}} = 1\}$ such that $0 \leq J'(\underline{z})(w_n) \leq 1/n$ and $J''(\underline{z})(w_n, w_n) \leq 1/n$ for all positive integers n. Since $dim\ \mathbb{Z} < \infty$, the unit sphere in $\mathbb{Z}$ is compact. Hence there is a subsequence $\{w_{n_k}\}$ and a vector w such that $\|w_{n_k} - w\|_{\mathbb{Z}} \to 0$ as $k \to \infty$, and therefore $\|w\|_{\mathbb{Z}} = 1$ and $w \in \overline{\mathcal{F}_C(\underline{z})}$. Moreover, the maps $J'(\underline{z})(\cdot)$ and $J''(\underline{z})(\cdot, \cdot)$ are continuous and therefore $J'(\underline{z})(w) = 0$ and $J''(\underline{z})(w, w) \leq 0$. This proves that (2.14) is false, and hence that (2.14) $\Rightarrow$ (2.4b). $\square$

COROLLARY 2.5. *Suppose that Z is a polyhedral convex set in $\mathbb{R}_m$, and let J be twice Gâteaux differentiable at a point $\underline{z} \in Z$ where (2.3a) holds. Then the coercivity condition (2.4b) is equivalent to the positive-definiteness condition,*

$$(2.15) \qquad \forall w \in \mathcal{F}_C \cap nul\ J'(\underline{z}) \quad (w \neq 0 \Rightarrow J''(\underline{z})(w,w) > 0).$$

Proof. The polyhedral convex set Z has a representation of the form,

$$Z = \{z \in \mathbb{R}_m : \forall i = 1, \cdots, k \quad \langle a_i, z \rangle + b_i \leq 0\}.$$

It follows easily that at each $\underline{z} \in Z$,

$$\mathcal{F}_Z(\underline{z}) = \{w \in \mathbb{R}_m : \forall i \in \mathcal{I}(\underline{z}) \quad \langle a_i, w \rangle \leq 0\},$$

where $\mathcal{I}(\underline{z})$ is the set of *active constraint indices* at $\underline{z}$, i.e.,

$$\mathcal{I}(\underline{z}) = \{i \in \{1, \cdots k\} : \langle a_i, \underline{z} \rangle + b_i = 0\}.$$

The cone $\mathcal{F}_Z(\underline{z})$ is therefore closed.
□

Since the positive-definiteness condition (2.15) is not far-removed from the second order necessary condition (2.3b), the sufficient conditions (2.4) in Theorem 2.2 are now seen to be quite sharp in polyhedral convex sets $Z \subset \mathbb{R}_m$. However, for the quasi-polyhedral convex sets in (2.13), the cone $\mathcal{F}_C(\underline{z})$ is typically not closed in $\mathbb{L}_m^p[0, 1]$, and the corresponding cone $\overline{\mathcal{F}_C(\underline{z})} \cap nul\ J'(\underline{z})$ can be larger than its counterpart $\mathcal{F}_C(\underline{z}) \cap nul\ J'(\underline{z})$ in the necessary condition (2.3b). Moreover, since the unit sphere in $\mathbb{L}_m^p[0, 1]$ is not compact, it seems likely that (2.4b) is not implied by (2.14), much less (2.15). In fact, the following example shows that for quasi-polyhedral convex sets, (2.4b) is not implied by the *coercivity* condition,

$$(2.16) \qquad \exists \mu_0 > 0\ \forall w \in \overline{\mathcal{F}_C(\underline{z})} \cap nul\ J'(\underline{z}) \quad J''(\underline{z})(w, w) \geq \mu_0 \|w\|_{\mathbb{Z}}^2.$$

Note that if J is twice Gâteaux differentiable at $\underline{z}$, condition (2.16) is equivalent to (2.14) when $dim\ \mathbb{Z} < \infty$, but stronger than (2.14) when $dim\ \mathbb{Z} = \infty$.

EXAMPLE 2.1. Let $\mathbb{Z} = \mathbb{L}^2[0, 1]$ and $C = \{z \in \mathbb{Z} : z(t) \overset{a.e}{\geq} 0\}$. Then C is a quasi-polyhedral convex set, and $C = G^{-1}(\mathcal{K})$ for $\mathbb{Y} = \mathbb{Z}$, $G = I$, and $\mathcal{K} = \{y \in \mathbb{Y} : y(t) \overset{a.e.}{\geq} 0\}$. Define $J : \mathbb{Z} \to \mathbb{R}_1$ and $\underline{z} \in C$ by

$$J(z) = \int_0^1 [2az - (sgn\ a)z^2]dt$$

and

$$\underline{z} = \max\{0, -a\},$$

with

$$a(t) = 1 - 2t.$$

We will show that $\underline{z}$ satisfies the first order necessary condition (2.3a) and the coercivity condition (2.16), but is not a local minimizer for J in C and therefore cannot satisfy the sufficient conditions (2.4).

Relative to the norm $\| \cdot \|_2$ on $\mathbb{Z}$ and $\mathbb{Y}$, the convex cone $\mathcal{K}$ is closed, the map G is continuous, and the quadratic functional J is twice continuously Fréchet differentiable, with

$$J'(\underline{z})(w) = \int_0^1 2[a - (sgn\ a)\underline{z}]wdt = \int_0^{\frac{1}{2}} 2awdt$$

and

$$J''(\underline{z})(w,w) = \int_0^1 -2(sgn\ a)w^2 dt.$$

Moreover, the feasible direction cone at $\underline{z}$ and its $\mathbb{L}^2$ closure are prescribed by,

$$
\begin{aligned}
\mathcal{F}_C(\underline{z}) &= \{w \in \mathbb{Z} : \exists \alpha > 0\ \ \underline{z}(t) + \alpha w(t) \overset{a.e}{\geq} 0\} \\
&= \bigcup_{\alpha > 0} \{w \in \mathbb{Z} : w(t) \overset{a.e.}{\geq} -\alpha^{-1}\underline{z}(t)\}
\end{aligned}
$$

and

$$\overline{\mathcal{F}_C(\underline{z})} = \{w \in \mathbb{Z} : w(t) \overset{a.e.}{\geq} 0\ \ t \in [0, \tfrac{1}{2}]\}.$$

Hence,

$$\overline{\mathcal{F}_C(\underline{z})} \cap nul\ J'(\underline{z}) = \{w \in \mathbb{Z} : w(t) \overset{a.e.}{=} 0\ \ t \in [0, \tfrac{1}{2}]\}.$$

Conditions (2.3a) and (2.16) are now easily verified at $\underline{z}$; however, $\underline{z}$ is not $\mathbb{L}^2$ locally optimal, and therefore (2.4b) can not hold at $\underline{z}$. In fact, $\underline{z}$ is not even locally optimal with respect to perturbations that are small in the *essential sup* norm $\| \cdot \|_\infty$. To see this, let χ_ϵ denote the characteristic function of the interval $[\tfrac{1}{2} - \epsilon, \tfrac{1}{2}]$ for $\epsilon \in (0, \tfrac{1}{2})$, let $z_\epsilon = \underline{z} + 3\epsilon\chi_\epsilon$, and consider that $z_\epsilon \in C$, $\|\underline{z} - z_\epsilon\|_\infty = 3\epsilon$, and

$$
\begin{aligned}
J(z_\epsilon) - J(\underline{z}) &= \int_{\frac{1}{2}-\epsilon}^{\frac{1}{2}} [6\epsilon(1 - 2t) - 9\epsilon^2]dt \\
&= -3\epsilon^3 < 0
\end{aligned}
$$

◇

Example 2.1 demonstrates that the small gap between necessary conditions and sufficient conditions for local optimality in finite-dimensional polyhedral convex sets is not preserved for the quasi-polyhedral convex sets (2.13). In this example, the feasible direction cones are not closed, the coercivity condition (2.16) is substantially stronger than the second order necessary condition (2.3b), and yet (2.16) is still not strong enough to imply (2.4b) or local optimality. Our next example shows that the gap between the sufficient conditions (2.4) or (2.11) and the local optimality growth properties (2.5) or (2.12) can also be quite large in quasi-polyhedral convex sets.

EXAMPLE 2.2. Let $\mathbb{Z} = \mathbb{L}^p[0, 1]$ and $C = \{z \in \mathbb{Z} : z(t) \overset{a.e.}{\in} [-1, 1]\}$, with $p = 2$ or ∞. Note that C is a quasi-polyhedral convex set, and $C = G^{-1}(\mathcal{K})$ for $\mathbb{Y} = \mathbb{L}_2^p[0, 1]$, and $\mathcal{K}$ and G prescribed by,

$$\mathcal{K} = \{y \in \mathbb{Y} : y_1(t) \overset{a.e}{\leq} 0\ \ y_2(t) \overset{a.e.}{\leq} 0\}$$

and

$$G(z)(t) \overset{a.e.}{=} (-z(t) - 1, z(t) - 1).$$

Define $J : \mathbb{Z} \to \mathbb{R}_1$ by

$$J(z) = -2 \int_0^{\frac{1}{2}} (z + z^2) dt + \int_{\frac{1}{2}}^1 z^2 dt.$$

Let $\| \cdot \|_p$ denote the standard $\mathbb{L}^p$ norms on $\mathbb{Z}$ and $\mathbb{Y}$, and let $\| \cdot \|_2$ serve as an auxiliary norm on $\mathbb{Z}$. When $p = \infty$, the norms $\| \cdot \|_p$ and $\| \cdot \|_2$ are not equivalent on $\mathbb{Z}$, and $\| \cdot \|_2$ is dominated by $\| \cdot \|_p$. Relative to the norms $\| \cdot \|_p$ on $\mathbb{Z}$ and $\mathbb{Y}$, the convex cone $\mathcal{K}$ is closed, the affine map G is continuous, and the quadratic functional J is twice continuously Fréchet differentiable. Define $\underline{z} \in C$ by

$$\underline{z}(t) = \begin{cases} 1 & t \in [0, \frac{1}{2}] \\ 0 & t \in (\frac{1}{2}, 1] \end{cases}$$

It is easily seen that $\underline{z}$ is the unique global minimizer of J in C, and satisfies the growth condition,

$$\forall z \in C \quad J(z) - J(\underline{z}) \geq \|z - \underline{z}\|_2^2.$$

Moreover,

$$J'(\underline{z})(w) = -6 \int_0^{\frac{1}{2}} w \, dt$$

and

$$J''(\underline{z})(w, w) = -4 \int_0^{\frac{1}{2}} w^2 dt + 2 \int_{\frac{1}{2}}^1 w^2 dt.$$

We will now show that the coercivity condition (2.16) is satisfied at $\underline{z}$ relative to $\| \cdot \|_2$, but (2.11b) does *not* hold at $\underline{z}$ for $\| \cdot \|_a = \| \cdot \|_2$ (or *a fortiori*, for $\| \cdot \|_a = \| \cdot \|_\infty$). Note first that

$$\begin{aligned} \mathcal{F}_C(\underline{z}) &= \{w \in \mathbb{Z} : \exists \alpha > 0 \quad \underline{z}(t) + \alpha w(t) \overset{a.e.}{\in} [-1,1]\} \\ &= \bigcup_{\alpha > 0} \{w \in \mathbb{Z} : -\frac{1}{\alpha}(1 + \underline{z}(t)) \overset{a.e.}{\leq} w(t) \overset{a.e.}{\leq} \frac{1}{\alpha}(1 - \underline{z}(t))\} \\ &= \bigcup_{\alpha > 0} \{w \in \mathbb{Z} : -\frac{2}{\alpha} \overset{a.e.}{\leq} w(t) \overset{a.e.}{\leq} 0 \quad t \in [0, \frac{1}{2}] \; ; \; |w(t)| \overset{a.e.}{\leq} \frac{1}{\alpha} \quad t \in (\frac{1}{2}, 1]\} \end{aligned}$$

Relative to the norm $\| \cdot \|_2$, the closure of $\mathcal{F}_C(\underline{z})$ is then

$$\overline{\mathcal{F}_C(\underline{z})} = \{w \in \mathbb{Z} : w(t) \overset{a.e.}{\leq} 0 \quad t \in [0, \frac{1}{2}]\}$$

and thus

$$\overline{\mathcal{F}_C(\underline{z})} \cap nul \, J'(\underline{z}) = \{w \in \mathbb{Z} : w(t) \overset{a.e.}{=} 0 \quad t \in [0, \frac{1}{2}]\}.$$

Condition (2.16) is now readily checked at $\underline{z}$ for the norm $\| \cdot \|_2$. On the other hand, let χ_ϵ be the characteristic function of $[0, \epsilon]$ for $\epsilon \in (0, \frac{1}{2}]$, let $w_\epsilon = -\epsilon^{-\frac{1}{2}}$, and consider that $w_\epsilon \in \mathcal{F}_C(\underline{z})$ and $\|w_\epsilon\|_2 = 1$, with

$$J'(\underline{z})(w_\epsilon) = 6 \int_0^\epsilon \epsilon^{-\frac{1}{2}} dt = 6\epsilon^{\frac{1}{2}} \|w_\epsilon\|_2$$

and

$$J''(\underline{z})(w_\epsilon, w_\epsilon) = -4 \int_0^\epsilon \epsilon^{-1} dt = -4\|w_\epsilon\|_2^2.$$

In the limit as $\epsilon \to 0^+$, this shows that (2.11b) can not hold at $\underline{z}$ relative to $\|\cdot\|_a = \|\cdot\|_2$.

◇

3. Alternative Optimality Conditions In Quasi-Polyhedral Convex Sets. We have seen that a potentially large gap exists between the necessary conditions (2.3) and the sufficient conditions (2.4) or (2.11) in quasi-polyhedral convex sets (2.13). In this section, we describe recently proved alternative optimality conditions that exploit the simple product structure inherent in (2.13). The new sufficient conditions incorporate a strengthened version of the first order necessary condition (2.3a) that is closely related to strict complementarity in finite-dimensional nonlinear programming. Further refinements of the results in this section are discussed in §4 for an important special class of cost functionals J.

We begin with some basic facts about normal cones and optimality conditions, first for general convex sets C in a real inner product space $\mathbb{Z}$, and then for k-fold Cartesian products $C_k = Z \times \cdots \times Z$ in the k-fold direct sum $\mathbb{R}_{km} = \mathbb{R}_1 \times \cdots \times \mathbb{R}_m$, with Z a polyhedral convex set in $\mathbb{R}_m$. The products C_k are in some sense finite-dimensional approximations to the quasi-polyhedral convex sets in (2.13).

Let C be a convex set in an inner product space $\{\mathbb{Z}, \langle\cdot,\cdot\rangle_{\mathbb{Z}}, \|\cdot\|_{\mathbb{Z}}\}$. At $\underline{z} \in C$, the exterior normals to C make a closed convex cone,

$$(3.17) \qquad \mathcal{N}_C(\underline{z}) = \{v \in \mathbb{Z} : \forall z \in C \quad \langle v, z - \underline{z}\rangle_{\mathbb{Z}} \le 0\}.$$

This *normal cone* is related to the feasible direction cone by the formula,

$$(3.18) \qquad \mathcal{N}_C(\underline{z}) = \mathcal{F}_C(\underline{z})^* \stackrel{def}{=} \{v \in \mathbb{Z} : \forall w \in \mathcal{F}_C(\underline{z}) \quad \langle v, w\rangle_{\mathbb{Z}} \le 0\}.$$

If $\mathbb{Z}$ is complete, it follows that

$$(3.19) \qquad \mathcal{N}_C(\underline{z})^* = \mathcal{F}_C(\underline{z})^{**} = \overline{\mathcal{F}_C(\underline{z})}.$$

If $\mathcal{F}_C(\underline{z})$ is also closed in $\mathbb{Z}$ then $\overline{\mathcal{F}_C(\underline{z})} = \mathcal{F}_C(\underline{z})$, and $\mathcal{N}_C(\underline{z})$ and $\mathcal{F}_C(\underline{z})$ are *dual cones* in $\mathbb{Z}$.

Suppose that J is twice Gâteaux differentiable at $\underline{z} \in C$, and that $\nabla J(\underline{z})$ and $\nabla^2 J(\underline{z})$ exist (cf., §1). In view of (3.18), the first order necessary condition (2.3a) is equivalent to

$$(3.20a) \qquad -\nabla J(\underline{z}) \in \mathcal{N}_C(\underline{z}),$$

and the second order necessary condition (2.3b) can be restated as

$$(3.20b) \qquad \forall w \in \mathcal{F}_C(\underline{z}) \cap \nabla J(\underline{z})^\perp \quad \langle w, \nabla^2 J(\underline{z})w\rangle_{\mathbb{Z}} \ge 0,$$

where

$$\nabla J(\underline{z})^\perp = \{w \in \mathbb{Z} : \langle \nabla J(\underline{z}), w \rangle_{\mathbb{Z}} = 0\} = nul\ J(\underline{z}).$$

Let $N_C(\underline{z})$ and $T_C(\underline{z})$ denote the closed linear hull of $\mathcal{N}_C(\underline{z})$ and its orthogonal complement in $\mathbb{Z}$, i.e.,

(3.21a)
$$N_C(\underline{z}) = \overline{span\ \mathcal{N}_C(\underline{z})}$$

and

(3.21b)
$$T_C(\underline{z}) = N_C(\underline{z})^\perp = \mathcal{N}_C(\underline{z})^\perp$$

Let $ri\ \mathcal{N}_C(\underline{z})$ denote the *relative interior* of the normal cone $\mathcal{N}_C(\underline{z})$, i.e., the interior of $\mathcal{N}_C(\underline{z})$ relative $N_C(\underline{z})$. If $\mathbb{Z}$ is complete and $\underline{z}$ satisfies the first order necessary condition (3.20a), it can be seen that (3.18) and (3.19) imply

(3.22)
$$\overline{\mathcal{F}_C(\underline{z})} \cap \nabla J(\underline{z})^\perp \supset T_C(\underline{z}).$$

If $\mathbb{Z}$ is complete and $\underline{z}$ satisfies the *geometric strict complementarity condition*,

(3.23)
$$-\nabla J(\underline{z}) \in ri\ \mathcal{N}_C(\underline{z}),$$

(cf., [14], [15] and [10]) then (3.18) and (3.19) imply

(3.24)
$$\overline{\mathcal{F}_C(\underline{z})} \cap \nabla J(\underline{z})^\perp = T_C(\underline{z}).$$

Hence if $\mathbb{Z}$ is complete and $\mathcal{F}_C(\underline{z})$ is closed, then for all $\underline{z}$ in C,

$$-\nabla J(\underline{z}) \in \mathcal{N}_C(\underline{z}) \Rightarrow \overline{\mathcal{F}_C(\underline{z})} \cap \nabla J(\underline{z})^\perp \supset T_C(\underline{z})$$

and

$$-\nabla J(\underline{z}) \in ri\ \mathcal{N}_C(\underline{z}) \Rightarrow \overline{\mathcal{F}_C(\underline{z})} \cap \nabla J(\underline{z})^\perp = T_C(\underline{z}).$$

In a general convex set C, the cones $\mathcal{F}_C(\underline{z})$ are typically not closed and $\overline{\mathcal{F}_C(\underline{z})} \cap \nabla J(\underline{z})^\perp$ can be much larger than $\mathcal{F}_C(\underline{z}) \cap \nabla J(\underline{z})^\perp$. On the other hand, in the polyhedral convex product set C_k, the cones $\mathcal{F}_{C_k}(\underline{z})$ are always closed. Moreover, the finite-dimensional space $\mathbb{R}_{km}$ is automatically complete and the normal cones $\mathcal{N}_{C_k}$ inherit the product structure of C_k, i.e., at each $\underline{z} = (\underline{z}_1, \cdots, \underline{z}_k)$ in C_k,

(3.25)
$$\mathcal{N}_{C_k}(\underline{z}) = \mathcal{N}_Z(\underline{z}_1) \times \cdots \times \mathcal{N}_Z(\underline{z}_k),$$

and therefore

(3.26)
$$ri\ \mathcal{N}_{C_k}(\underline{z}) = ri\ \mathcal{N}_Z(\underline{z}_1) \times \cdots \times ri\ \mathcal{N}_Z(\underline{z}_k),$$

(3.27)
$$N_{C_k}(\underline{z}) = N_Z(\underline{z}_1) \times \cdots \times N_Z(\underline{z}_k),$$

and

(3.28)
$$T_{C_k}(\underline{z}) = T_Z(\underline{z}_1) \times \cdots \times T_Z(\underline{z}_k).$$

Accordingly, if $\nabla J(\underline{z})_i$ denotes the ith component of the vector-valued $k-tuple$ $\nabla J(\underline{z}) = (\nabla J(\underline{z})_1, \cdots, \nabla J(\underline{z})_k)$, then the first order necessary condition (3.20a) and the geometric strict complementarity condition (3.23) can be expressed by analogous *component-wise* conditions on $\nabla J(\underline{z})_i$ and $\mathcal{N}_Z(\underline{z}_i)$.

The foregoing observations, Theorems 2.1 and 2.2, and Corollary 2.5 now yield basic optimality conditions in the product sets C_k.

THEOREM 3.1. *Let Z be a polyhedral convex set in $\mathbb{R}_m$, let C_k be the corresponding k-fold Cartesian product $Z \times \cdots \times Z$ in $\mathbb{R}_{km}$, and suppose that J is twice Gâteaux differentiable at $\underline{z} = (\underline{z}_1, \cdots, \underline{z}_k) \in C_k$. If $\underline{z}$ is a local minimizer of J in C_k, then $\underline{z}$ satisfies the (component-wise) first order necessary condition,*

$$(3.29a) \qquad -\nabla J(\underline{z})_i \in \mathcal{N}_Z(\underline{z})_i \quad i = 1, \cdots, k$$

and the second order necessary condition,

$$(3.29b) \qquad \forall w \in T_{C_k}(\underline{z}) \quad \langle w, \nabla^2 J(\underline{z})w \rangle \geq 0.$$

Conversely, let J be twice continuously Fréchet differentiable at $\underline{z} \in C_k$, and suppose that $\underline{z}$ satisfies the (component-wise) geometric strict complementarity condition,

$$(3.30a) \qquad -\nabla J(\underline{z})_i \in ri\, \mathcal{N}_Z(\underline{z})_i \quad i = 1, \cdots, k$$

and the coercivity condition,

$$(3.30b) \qquad \exists \mu > 0 \,\forall w \in T_{C_k}(\underline{z}) \quad \langle w, \nabla^2 J(\underline{z})w \rangle \geq \mu \|w\|^2.$$

Then $\underline{z}$ is a strict local minimizer for J in C_k, and for each $\mu' \in (0, \mu)$ there is a $\delta > 0$ such that,

$$(3.31) \qquad \forall z \in C_k \quad \|z - \underline{z}\| < \delta \Rightarrow J(z) - J(\underline{z}) \geq \frac{1}{2}\mu' \|z - \underline{z}\|^2.$$

Moreover, the coercivity condition (3.30b) is equivalent to the positive-definiteness condition,

$$(3.32) \qquad \forall w \in T_{C_k} \quad w \neq 0 \Rightarrow \langle w, \nabla^2 J(\underline{z})w \rangle > 0.$$

Our goal now is to formulate extensions of the necessary conditions (3.29) and sufficient conditions (3.30) for quasi- polyhedral convex sets,

$$(3.33) \qquad C = \{z \in \mathbb{L}_m^\infty[0,1] : z(t) \overset{a.e.}{\in} Z\}$$

in the vector space $\mathbb{Z} = \mathbb{L}_m^\infty[0,1]$ [2]. In this development, $\mathbb{L}_m^\infty[0,1]$ is equipped with the usual *essential sup* norm $\|\cdot\|_\infty$ and the auxiliary norm $\|z\|_2 = \sqrt{\langle z, z \rangle_2}$ induced by the standard $\mathbb{L}^2$ inner product $\langle z_1, z_2 \rangle_2 = \int_0^1 \langle z_1(t), z_2(t) \rangle dt$. The inner product

---

[2] The equivalence of coercivity and positive-definiteness does not survive the passage from $\mathbb{R}_{km}$ to $\mathbb{L}_m^\infty[0,1]$, since the unit sphere for any norm on the infinite-dimensional space $\mathbb{L}_m^\infty[0,1]$ is not compact.

space $\{\mathbb{Z}, \langle \cdot, \cdot \rangle_2, \| \cdot \|_2\}$ is incomplete; however, this will not matter in the following analysis, which relies heavily on the product structure in (3.33) and the related normal cones $\mathcal{N}_C(\underline{z})$ and subspaces $N_C(\underline{z})$ and $T_C(\underline{z})$.

The quasi-polyhedral convex set C in (3.33) has decomposition formulas analogous to (3.25), (3.27) and (3.28) for the product set C_k, namely,

$$(3.34) \qquad \mathcal{N}_C(\underline{z}) = \{v \in \mathbb{L}_m^\infty[0,1] : v(t) \overset{a.e.}{\in} \mathcal{N}_Z(\underline{z}(t))\},$$

$$(3.35) \qquad N_C(\underline{z}) = \{v \in \mathbb{L}_m^\infty[0,1] : v(t) \overset{a.e.}{\in} N_Z(\underline{z}(t))\},$$

and

$$(3.36) \qquad T_C(\underline{z}) = \{w \in \mathbb{L}_m^\infty[0,1] : w(t) \overset{a.e.}{\in} T_Z(\underline{z}(t))\}.$$

On the other hand, the point-wise counterpart of the decomposition formula (3.26) is *false* for $ri\,\mathcal{N}_C(\underline{z})$, relative to either of the norms $\| \cdot \|_2$ or $\| \cdot \|_\infty$ on $\mathbb{L}_m^\infty[0,1]$. In fact, $ri\,\mathcal{N}_C(\underline{z})$ is typically *empty* relative to $\| \cdot \|_2$, while $v \in ri\,\mathcal{N}_C(\underline{z})$ relative to $\| \cdot \|_\infty$ *iff* the distance from $v(t)$ to the relative boundary $rb\,\mathcal{N}_Z(\underline{z}(t))$ is essentially bounded away from zero on $[0,1]$.

The gradients and Hessians in $\mathbb{R}_{km}$ also have natural analogues in $\mathbb{L}_m^\infty[0,1]$ (cf., §1). If $J : \mathbb{L}_m^\infty[0,1] \to \mathbb{R}_1$ is twice Gâteaux differentiable at $\underline{z}$ relative to the norm $\| \cdot \|_2$, then $\nabla J(\underline{z})$ is a vector in $\mathbb{L}_m^\infty[0,1]$ such that

$$\forall w \in \mathbb{L}_m^\infty[0,1] \quad J'(\underline{z})(w) = \langle \nabla J(\underline{z}), w \rangle_2,$$

and $\nabla^2 J(\underline{z})$ is a linear operator on $\mathbb{L}_m^\infty[0,1]$ such that

$$\forall v, w \in \mathbb{L}_m^\infty[0,1] \quad J''(\underline{z})(v,w) = \langle v, \nabla^2 J(\underline{z})w \rangle_2,$$

and

$$\sup_{\|w\|_2=1} \|\nabla^2 J(\underline{z})w\|_2 < \infty.$$

Since the inner product space $\{\mathbb{L}_m^\infty[0,1], \langle \cdot, \cdot \rangle_2, \| \cdot \|_2\}$ is incomplete, a twice Gâteaux differentiable functional does not automatically have gradients or Hessians. In our development, we *assume* that $\nabla J(\underline{z})$ and $\nabla^2 J(\underline{z})$ exist, and this assumption is often directly verifiable without recourse to the Riesz-representation theorem.

The first and second order necessary conditions in Theorem 3.1 have the following obvious formal counterparts in quasi-polyhedral convex sets (3.33):

$$(3.37a) \qquad -\nabla J(\underline{z})(t) \overset{a.e.}{\in} \mathcal{N}_Z(\underline{z}(t))$$

and

$$(3.37b) \qquad \forall w \in T_C(\underline{z}) \quad \langle w, \nabla J(\underline{z})w \rangle_2 \geq 0.$$

Equations (3.18) and (3.34) and Theorem 2.1 immediately show that the pointwise first order condition (3.37a) is actually valid in quasi-polyhedral convex sets (3.33). On the other hand, the formal second order condition (3.37b) requires more scrutiny since the crucial inclusion, $T_C(\underline{z}) \subset \mathcal{F}_C(\underline{z}) \cap \nabla J(\underline{z})^\perp$, that yields (3.29b) in the polyhedral convex set C_k, is generally false in nonpolyhedral convex sets, even in finite-dimensional spaces. The key to the proof of (3.37b) is provided by Lemma 5.1 in [17], which establishes that $T_C(\underline{z})$ is the $\mathbb{L}^2$ closure of the union of a nondecreasing sequence of subspaces $T^{(n)}$ in the feasible direction cone $\mathcal{F}_C(\underline{z})$. If the first order necessary condition (3.37a) holds at $\underline{z}$, it is easily seen that the subspaces $T^{(n)}$ must lie in $\mathcal{F}_C(\underline{z}) \cap \nabla J(\underline{z})^\perp$. Since the linear map $\nabla^2 J(\underline{z})$ is $\mathbb{L}^2$ continuous by hypothesis, condition (2.3b) in Theorem 2.1 extends to $T_C(\underline{z})$ and it is now possible to state the following variant of Theorem 5.2 in [17].

THEOREM 3.2. *Let C be a quasi-polyhedral convex set (3.33), let $J : \mathbb{L}_m^\infty[0,1] \to \mathbb{R}_1$ be twice Gâteaux differentiable at $\underline{z} \in C$ relative to the norm $\| \cdot \|_2$ on $\mathbb{L}_m^\infty[0,1]$, and suppose that J has a gradient and Hessian at $\underline{z}$. If $\underline{z}$ is a local minimizer of J in C with respect to any norm on $\mathbb{L}_m^\infty[0,1]$, then $\underline{z}$ satisfies the first and second order necessary conditions (3.37).*

The sufficient conditions (3.30) also have obvious formal extensions in quasi-polyhedral sets (3.33), namely,

$$(3.38a) \qquad -\nabla J(\underline{z})(t) \overset{a.e.}{\in} ri \, \mathcal{N}_Z(\underline{z}(t)))$$

and

$$(3.38b) \qquad \exists \mu > 0 \; \forall w \in T_C(\underline{z}) \quad \langle w, \nabla^2 J(\underline{z}) w \rangle_2 \geq \mu \|w\|_2^2.$$

In the polyhedral convex product set C_k of Theorem 3.1, the component-wise strict complementarity condition (3.30a) is *equivalent* to the geometric strict complementarity condition, $-\nabla J(\underline{z}) \in \mathcal{N}_{C_k}(\underline{z})$, and the latter condition implies that $\langle \nabla J(\underline{z}), (z - \underline{z}) \rangle$ grows like $\|z - \underline{z}\|$ in the exterior of some sufficiently small conical neighborhood of the subspace $T_{C_k}(\underline{z})$ where $\langle z - \underline{z}, \nabla^2 J(\underline{z})(z - \underline{z}) \rangle$ increases like $\|z - \underline{z}\|^2$ in accordance with (3.30b). These facts support a sufficiency proof for (3.30) that does not rely on compactness of the unit sphere in $\mathbb{R}_{km}$ and might therefore transfer to (3.38) in the infinite-dimensional inner product space $\{\mathbb{L}_m^\infty[0,1], \langle \cdot, \cdot \rangle_2, \| \cdot \|_2\}$. Unfortunately, this proof strategy fails precisely because (3.38a) does not imply that $-\nabla J(\underline{z}) \in ri \, \mathcal{N}_C(\underline{z})$ in the $\mathbb{L}^2$ sense. Moreover, Example 2.1 demonstrates that the formal sufficient conditions (3.38) are actually not sufficient for local optimality in quasi-polyhedral convex sets. In this example, Z is the closed interval $[0, \infty) \subset \mathbb{R}_1$, and it is easily seen that

$$[-\nabla J(\underline{z})(t), ri \, \mathcal{N}_Z(\underline{z}(t))] \overset{a.e.}{=} \begin{cases} [-2(1 - 2t), (-\infty, 0]] & t \in [0, \frac{1}{2}] \\ [0, \{0\}] & t \in (\frac{1}{2}, 1] \end{cases}$$

and

$$T_C(\underline{z}) = \{w \in \mathbb{L}_m^\infty[0,1] : w(t) \overset{a.e.}{=} 0 \quad t \in [0, \frac{1}{2}]\}.$$

Hence, conditions (3.38) clearly hold at this $\underline{z}$, and yet $\underline{z}$ is not a local minimizer *even in the weak* $\mathbb{L}^\infty$ *sense.*

A variant of the proof technique sketched above is used in [17] to establish a key sufficiency lemma for $\mathbb{L}^\infty$ local optimality in quasi-polyhedral convex sets. In this lemma, the coercivity condition (3.38b) is extended to a larger subspace $\hat{T} \supset T_C(\underline{z})$, and (3.38a) is modified accordingly to insure suitable complementarity in the growth properties of the first and second order terms $\langle \nabla J(\underline{z}), (z - \underline{z}) \rangle_2$ and $\langle z - \underline{z}, \nabla^2 J(\underline{z})(z - \underline{z}) \rangle_2$ near $\underline{z}$ in C. More specifically, suppose that $\underline{z}$ satisfies the first order necessary condition (3.37a) and let $\Delta(\underline{z})(t)$ denote the distance from $-\nabla J(\underline{z})(t)$ to the relative boundary $rb \, \mathcal{N}_Z(\underline{z}(t))$, or equivalently,

(3.39)

$$\Delta(\underline{z})(t) \overset{a.e.}{=} \sup\{\rho \geq 0 : \forall \eta \in N_Z(\underline{z}(t)) \quad \|\eta\| \leq \rho \Rightarrow -\nabla J(\underline{z})(t) + \eta \in \mathcal{N}_Z(\underline{z}(t))\}.$$

Since polyhedral convex sets Z have finitely many distinct normal cones, the set-valued function $\mathcal{N}(\underline{z}(\cdot))$ has a finite range and $\Delta(\underline{z})(\cdot)$ is seen to be Lebesgue measurable with values in the extended half-line $[0, \infty]$ [3]. Moreover, $\Delta(\underline{z})(t) \overset{a.e.}{>} 0$ when the pointwise strict complementarity condition (3.38a) holds. On the other hand, (3.38a) does not imply that $\Delta(\underline{z})(t)$ is bounded away from zero on $[0, 1]$. In fact, for commonly encountered cases where $\underline{z}$ satisfies (3.38a) and $\underline{z}$ and $\nabla J(\underline{z})$ are *continuous* functions in $\mathbb{L}_m^\infty[0, 1]$, the quantity $\Delta(\underline{z})(t)$ *must* approach zero as t approaches a point τ where $\underline{z}(t)$ passes from the relative interior of one polyhedral face of Z to the relative interior of a contiguous face. When conditions (3.38) hold and $\Delta(\underline{z})(t)$ is not essentially bounded away from zero, the vector $-\nabla J(\underline{z})$ does not fall in the $\mathbb{L}^\infty$ relative interior of $\mathcal{N}_C(\underline{z})$, and complementarity in the growth properties of $\langle \nabla J(\underline{z}), (z - \underline{z}) \rangle_2$ and $\langle z - \underline{z}, \nabla^2 J(\underline{z})(z - \underline{z}) \rangle_2$ is correspondingly deficient, even for $\mathbb{L}^\infty$-local optimality. Accordingly, reference [17] considers modifications of (3.38) that entail complementary orthogonal subspace pairs $(\hat{N}, \hat{T})$ of the form,

(3.40a) $$\hat{N} = \{w \in \mathbb{L}_m^\infty[0, 1] : w(t) \overset{a.e.}{\in} \hat{N}(t)\}$$

and

(3.40b) $$\hat{T} = \{w \in \mathbb{L}_m^\infty[0, 1] : w(t) \overset{a.e.}{\in} \hat{T}(t)\},$$

where $\hat{N}(t)$ and $\hat{T}(t)$ are complementary orthogonal subspaces in $\mathbb{R}_m$ such that

(3.40c) $$\hat{N}(t) \subset N_Z(\underline{z}(t)) \qquad \hat{T}(t) \supset T_Z(\underline{z}(t)).$$

By construction,

(3.41) $$\hat{N} \subset \mathcal{N}_C(\underline{z}) \qquad \hat{T} \supset T_C(\underline{z}),$$

and $\hat{N}$ and $\hat{T}$ are *pointwise orthogonal* in the sense that

(3.42) $$\forall v \in \hat{N} \, \forall w \in \hat{T} \quad \langle v(t), w(t) \rangle \overset{a.e.}{=} 0.$$

---

[3] Note that $\Delta(\underline{z})(t) = \infty \Leftrightarrow rb \, \mathcal{N}_Z(\underline{z}(t)) = \emptyset \Leftrightarrow \mathcal{N}_Z(\underline{z}(t)) = \{0\} \Leftrightarrow \underline{z}(t) \in int \, Z.$

Moreover, the counterpart of the quantity $\Delta(\underline{z})(t)$ for $\hat{N}$ is

(3.43)

$$\hat{\Delta}(t) \overset{a.e.}{=} \sup\{\rho \geq 0 : \forall \eta \in \hat{N}(t) \quad \|\eta\| \leq \rho \Rightarrow -\nabla J(\underline{z})(t) + \eta \in \mathcal{N}_Z(\underline{z}(t))\},$$

with

(3.44) $$\hat{\Delta}(t) \overset{a.e.}{\geq} \Delta(\underline{z}(t)) \overset{a.e.}{\geq} 0$$

when the first order necessary condition (3.37a) holds at $\underline{z}$. The following version of Lemma 6.3 in [17] assumes that for suitably constructed $\hat{N}$ and $\hat{T}$, the majorant $\hat{\Delta}(t)$ is essentially bounded away from zero and the coercivity condition (3.38b) extends from $T_C(\underline{z})$ to the larger subspace $\hat{T}$. This lemma also assumes that a two-norm Taylor remainder estimate similar to (2.10b) in Theorem 2.3 holds at $\underline{z}$. In §4, the remainder hypothesis is connected to structure/continuity conditions for an important class of cost functionals J which are twice continuously Fréchet differentiable relative to the norm $\|\cdot\|_2$, but possibly not relative $\|\cdot\|_2$. For this class of functionals, we shall also see that the requisite subspace pairs $(\hat{N}, \hat{T})$ exist when certain mild regularity conditions are added to (3.38).

LEMMA 3.3. *Let C be a quasi-polyhedral convex set (3.33). Suppose that J is twice Gâteaux differentiable at $\underline{z}$ relative to the norm $\|\cdot\|_2$, that J has a gradient and Hessian at $\underline{z}$, and that*

(3.45a)

$$J(z) - J(\underline{z}) = \langle \nabla J(\underline{z}), z - \underline{z}\rangle_2 + \frac{1}{2}\langle z - \underline{z}, \nabla^2 J(\underline{z})(z - \underline{z})\rangle_2 + r_2(\underline{z}, z),$$

and

(3.45b) $$\lim_{\|z - \underline{z}\|_\infty \to 0} \frac{r_2(\underline{z}, z)}{\|z - \underline{z}\|_2^2} = 0.$$

In addition, assume that the following conditions hold for some pair of complementary pointwise orthogonal subspaces $(\hat{N}, \hat{T})$ in (3.40):

(3.46a) $$\exists \epsilon > 0 \quad \hat{\Delta}(t) \overset{a.e.}{\geq} \epsilon$$

(3.46b) $$\exists \hat{\mu} > 0 \; \forall w \in \hat{T} \quad \langle w, \nabla^2 J(\underline{z})w\rangle_2 \geq \hat{\mu}\|w\|_2^2.$$

Then $\underline{z}$ is a strict $\mathbb{L}^\infty$- local minimizer for J in C, and for each $\mu' \in (0, \hat{\mu})$, there is a $\delta > 0$ such that for all z,

(3.47) $$z \in C \text{ and } \|z - \underline{z}\|_\infty < \delta \Rightarrow J(z) - J(\underline{z}) \geq \frac{1}{2}\mu'\|z - \underline{z}\|_2^2.$$

If conditions (3.38) hold at $\underline{z}$ and if $\Delta(\underline{z})(t)$ is essentially bounded away from zero on $[0,1]$, then conditions (3.46) are automatically satisfied for $\hat{N} = N_C(\underline{z})$ and $\hat{T} = T_C(\underline{z})$. This happens in Example 2.2[4], where $Z = [-1,1]$ and

$$(\underline{z}(t), \nabla J(\underline{z})(t), (\nabla^2 J(\underline{z})w)(t)) = \begin{cases} (1, -6, -4w(t)) & t \in [0, \tfrac{1}{2}] \\ (0, 0, 2w(t)) & t \in (\tfrac{1}{2}, 1] \end{cases}$$

$$(\mathcal{N}_Z(\underline{z}(t)), \Delta(\underline{z})(t)) = \begin{cases} ([0,\infty], 6) & t \in [0, \tfrac{1}{2}] \\ (\{0\}, \infty) & t \in (\tfrac{1}{2}, 1] \end{cases}$$

The quantity $\Delta(\underline{z})(t)$ is bounded away from zero in this example precisely because discontinuities in $\underline{z}$ and $\nabla J(\underline{z})$ occur just where $\underline{z}(t)$ passes from $\{0\}$ to $(-1,1)$ in Z. However, we have already observed that $\Delta(\underline{z})(t)$ is usually not essentially bounded away from zero on $[0,1]$, and in such cases it is important to know when subspaces $\hat{N}$ and $\hat{T}$ with the requisite properties in Lemma 3.3 are likely to exist. This question is addressed in §4 for specially structured cost functionals J.

4. Refined Optimality Conditions in Quasi-Polyhedral Convex Sets.
We now consider functionals $J : \mathbb{L}_m^\infty[0,1] \to \mathbb{R}_1$ that meet the following requirements.

STRUCTURE/CONTINUITY CONDITIONS.
  (I) At each $z \in \mathbb{L}_m^\infty[0,1]$, J is twice Gâteaux differentiable relative to the norm $\|\cdot\|_2$ and has a gradient and Hessian. Moreover, there are essentially bounded $m \times m$ matrix-valued functions $S(z) \in \mathbb{L}_{m\times m}^\infty[0,1]$ and $K(z) \in \mathbb{L}_{m\times m}^\infty([0,1] \times [0,1])$ such that for all $w \in \mathbb{L}_m^\infty[0,1]$,

(4.48) $(\nabla^2 J(z)w)(t) \overset{a.e.}{=} S(z)(t)w(t) + \int_0^1 K(z)(t,s)w(s)ds.$

  (II) The maps $z \to S(z)$ and $z \to K(z)$ are continuous in the sense that at each $\underline{z} \in \mathbb{L}_m^\infty[0,1]$,

(4.49a) $\lim_{\|z-\underline{z}\|_\nu \to 0} \|S(z) - S(\underline{z})\|_\infty = 0$

and

(4.49b) $\lim_{\|z-\underline{z}\|_\nu \to 0} \|K(z) - K(\underline{z})\|_2 = 0,$

where ν is either ∞ or 2, and

$$\|S(z)\|_\infty = ess \sup_{t\in[0,1]} \|S(z)(t)\|_{\mathbb{M}_{m\times m}}$$

and

$$\|K(z)\|_2^2 = \int_0^1 \int_0^1 \|K(z)(t,s)\|_{\mathbb{M}_{m\times m}}^2 dt ds.$$

---

[4] Recall that the sufficient conditions (2.11) in Theorem 2.2 do *not* hold relative to the norms $\|\cdot\|_{\mathbb{Z}} = \|\cdot\|_\infty$ and $\|\cdot\|_a = \|\cdot\|_2$ in Example 2.2.

If conditions (I) and (II) are met for $\nu = \infty$, then J is twice continuously Fréchet differentiable relative to $\|\cdot\|_\infty$ and the two-norm Taylor remainder estimate (3.45b) holds. If (I) and (II) hold for $\nu = 2$, then J is twice continuously Fréchet differentiable relative to $\|\cdot\|_2{}^5$.

The balance of this section describes a lemma and three theorems specifically tied to quasi-polyhedral convex sets C and cost functionals J that satisfy the structure/continuity conditions. Lemma 4.1 provides technical but frequently satisfied conditions that guarantee the existence of complementary subspace pairs $(\hat{N}, \hat{T})$ meeting the requirements of Lemma 3.3. The $\mathbb{L}^\infty$- local optimality sufficient conditions in Theorem 6.4 of [17] is immediately recovered from Lemmas 3.3 and 4.1, and is restated here in Theorem 4.2. Theorem 4.3 is essentially Theorem 5.4 in [17], which establishes a necessary condition for $\mathbb{L}^2$-local optimality closely related to the Pontryagin Minimum Principle. Finally, Theorem 4.4 is a restatement of Theorem 4.1 in [18], which asserts that conditions (3.38) and a strengthened version of the $\mathbb{L}^2$-local optimality necessary condition imply $\mathbb{L}^2$-local optimality. Detailed proofs for all the results in this section are supplied in [17] and [18]. We note that precursors of Lemma 3.3 and Theorems 4.2–4.4 are proved by Dunn and Tian in [19], and a Lagrangian variant of Theorem 4.2 is established by Dontchev, Hager et. al. in [13] for ODE optimal control problems with terminal state constraints and pointwise nonlinear constraints on control inputs. The relationship between the $\mathbb{L}^\infty$-local optimality sufficient conditions in [17] and [13] is discussed at length in [17]. We also note related contributions to the theory of $\mathbb{L}^\infty$-local optimality second order sufficient conditions for optimal control problems in the work of Zeidan and Orrel [43][49][50], Goldberg and Tröltzsch [25], Maurer and Pickenhain [41], and in the previously cited papers of Alt, Malanowski, and Maurer et.al.

A few additional observations and definitions are needed at the outset. In [46] it is shown that each polyhedral convex set Z in $\mathbb{R}_m$ has finitely many distinct *polyhedral faces* $F_1, \cdots, F_d$ such that

$$Z = \bigcup_{j=1}^{d} ri\, F_j$$

and

$$ri\, F_j \cap ri\, F_k = \emptyset$$

for $j \neq k$. Moreover, the normal cone $\mathcal{N}_Z(\zeta)$ (and hence the associated subspaces $N_Z(\zeta)$ and $T_Z(\zeta)$) are *invariant* for ζ in each set $ri\, F_j$, and we may therefore write

$$\mathcal{N}_Z(\zeta) = \mathcal{N}_j \qquad N_Z(\zeta) = N_j \qquad T_Z(\zeta) = T_j$$

for all $\zeta \in ri\, F_j$. For $z \in \mathbb{L}_m^\infty[0,1]$ and $j \in \{1, \cdots, d\}$, define the corresponding pairwise disjoint measurable sets,

$$\alpha_j(z) = z^{-1}(ri\, F_j) = \{t \in [0,1] : z(t) \in ri\, F_j\},$$

---

^{5} The first case occurs for ODE optimal control problems with Bolza cost functionals; The second case occurs for a narrower but still significant subclass of Bolza problems with control-quadratic Hamiltonians [48] [18] and note that the set $[0,1] \setminus \cup_{j=1}^{d} \alpha_j(z)$ has Lebesgue measure zero. Let $\partial \alpha_j(z)$ and *int* $\alpha_j(z)$ denote the boundary and interior of $\alpha_j(z)$ *relative to the closed interval* $[0,1]$, and note that every point τ in the set,

$$(4.50) \qquad \gamma(z) = [0,1] \setminus \cup_{j=1}^{d} int\ \alpha_j(z)$$

lies in the boundary of at least one (and possibly many) of the sets $\alpha_j(z)$. While the structure of $\gamma(z)$ can be highly complex, it is often quite simple at local minimizers or stationary points z where the first order necessary condition (3.37a) holds. For many such z, $\gamma(z)$ is finite or denumerable, each $\alpha_j(z)$ is the union of a finite or denumerable set of pairwise disjoint subintervals of $[0,1]$, z and $\nabla J(z)$ are continuous at $\tau \in \gamma(z)$, and each $\tau \in \gamma(z)$ lies in the boundary of *exactly two* sets $\alpha_{j_1}(z)$ and $\alpha_{j_2}(z)$, with $rb\ F_{j_1} \subset ri\ F_{j_2}$ and $N_{j_2} \subset N_{j_1}$. For a still larger class of stationary points z, the set $\gamma(z)$ has Lebesgue measure zero and all of its points meet the requirements in the following definition.

DEFINITION 4.1. *Let z satisfy the first order necessary condition (3.37a) and let $\gamma(z)$ be the corresponding set in (4.50). A point $\tau \in \gamma(z)$ is regular if and only if:*

(i) The collection of subspaces $\{N_j\}_{\tau \in \partial \alpha_j(z)}$ has a (unique) minimal element, i.e.,

$$(4.51) \qquad \exists \nu_\tau\ \forall j \quad \tau \in \partial \alpha_{\nu_\tau}(z)\ and\ (\tau \in \partial \alpha_j(z) \Rightarrow N_{\nu_\tau} \subset N_j)$$

(ii) The set $\alpha_{\nu_\tau}(z)$ has a nonempty interior in $[0,1]$ and is contained in the closure of its interior, i.e.,

$$(4.52) \qquad cl\ int\ \alpha_{\nu_\tau}(z) \supset \alpha_{\nu_\tau}(z) \supset int\ \alpha_{\nu_\tau}(z) \neq \emptyset$$

Suppose that the pointwise strict complementarity condition (3.38a) holds at $z \in C$, let τ be a regular point of $\gamma(z)$, and put

$$(4.53)$$

$$\Delta_\tau(z)(t) \stackrel{a.e.}{=} \sup\{\rho \geq 0 : \forall \eta \in N_{\nu_\tau}\quad \|\eta\| \leq \rho \Rightarrow -\nabla J(z)(t) + \eta \in \mathcal{N}_Z(z(t))\}.$$

By construction,

$$\Delta_\tau(z)(t) \stackrel{a.e.}{\geq} \Delta(z)(t) \stackrel{a.e}{>} 0,$$

for t near τ in $[0,1]$. We have seen that the quantity $\Delta(z)(t)$ is often not essentially bounded away from zero near points $\tau \in \gamma(z)$; however, in many cases, $\Delta(z)(t)$ is essentially bounded away from zero *on each compact subset of* $[0,1] \setminus \gamma(z) = \cup_{j=1}^{d} int\ \alpha_j(z)$, while all points of $\gamma(z)$ are regular and $\Delta_\tau(z)(t)$ is essentially bounded away from zero near its corresponding point $\tau \in \gamma(z)$. When this happens, it is possible to construct subspaces $\hat{N} \subset N_C(z)$ meeting requirement (3.46a) in Lemma 3.3; moreover, if the multiplication operator kernel $S(z)(\cdot)$ in the structure/continuity conditions is continuous at each $\tau \in \gamma(z)$, then the coercivity condition (3.38b) automatically extends to (3.46b) on the orthogonal complement $\hat{T} \supset T_C(z)$ of some such $\hat{N}$ in $\mathbb{L}_m^\infty[0,1]$. These conclusions and their implications for $\mathbb{L}^\infty$-local optimality are summarized in the following lemma and theorem. The lengthy technical proofs for these results and Lemma 3.3 are provided in Section 6 of [17], along with additional analysis and examples pertaining to the set $\gamma(\underline{z})$ and Definition 4.1.

LEMMA 4.1. *Let C be a quasi-polyhedral convex set (3.33), let J satisfy the structure/continuity condition (I) and suppose that the pointwise strict complementarity condition (3.38a) and the $\mathbb{L}^2$ coercivity condition (3.38b) hold at $\underline{z} \in C$. In addition, assume that each point of the set $\gamma(\underline{z})$ in (4.50) is a regular point and a point of continuity for $S(\underline{z})(\cdot)$. Finally, suppose that the distance $\Delta(\underline{z})(t)$ in (3.39) is essentially bounded away from zero on every compact subset of $[0,1] \setminus \gamma(\underline{z})$, and for each $\tau \in \gamma(\underline{z})$ the quantity $\Delta_\tau(\underline{z})(t)$ in (4.53) is essentially bounded away from zero for t near τ. Then conditions (3.46) in Lemma 3.3 hold for some pair of complementary orthogonal subspaces $N \subset N_C(\underline{z})$ and $\hat{T} \supset T_C(\underline{z})$ in $\mathbb{L}_m^\infty[0,1]$.*

THEOREM 4.2. *Let the hypotheses of Lemma 4.1 hold and in addition, suppose that J satisfies the structure/continuity condition (II) for $\nu = \infty$. Then $\underline{z}$ is a strict $\mathbb{L}^\infty$-local minimizer of J in C, and the two-norm growth condition (3.47) holds near $\underline{z}$.*

The norms $\|\cdot\|_\infty$ and $\|\cdot\|_2$ are not equivalent on $\mathbb{L}_m^\infty[0,1]$, and local optimality relative to $\|\cdot\|_\infty$ is a weaker property than its counterpart for $\|\cdot\|_2$. In general, the conditions in Lemma 3.3 and Theorem 4.2 do not imply local optimality in the $\mathbb{L}^2$ sense; however, sufficient conditions for $\mathbb{L}^2$-local optimality in quasi-polyhedral sets (3.33) are provided in Theorems 6.6 of [17] and Theorem 4.1 of [18] for cost functions J that satisfy the structure/continuity conditions for $\nu = 2$, and a strengthened form of the necessary condition for $\mathbb{L}^2$-local optimality in Theorem 5.4 of [17]. The main conclusion in the latter theorem is restated below.

THEOREM 4.3. *Let C be a quasi-polyhedral convex set (3.33), let J satisfy the structure/continuity conditions for $\nu = 2$, and suppose that $\underline{z} \in C$ is a local minimizer of J in C relative to the norm $\|\cdot\|_2$. Then*

$$(4.54) \qquad \inf_{\zeta \in Z} \left[\langle \nabla J(\underline{z})(t), \zeta - \underline{z}(t)\rangle + \frac{1}{2}\langle \zeta - \underline{z}(t), S(\underline{z})(t)(\zeta - \underline{z}(t))\rangle \right] \stackrel{a.e.}{=} 0.$$

If the hypotheses on C and J in Theorem 4.3 hold, and if there is a positive number c such that the condition,

$$(4.55)$$

$$\forall \zeta \in Z \quad \langle \nabla J(\underline{z})(t), \zeta - \underline{z}(t)\rangle + \frac{1}{2}\langle \zeta - \underline{z}(t), S(\underline{z})(t)(\zeta - \underline{z}(t))\rangle \geq c\|\zeta - \underline{z}(t)\|^2,$$

holds almost everywhere in $[0,1]$, then every $\mathbb{L}^\infty$- local minimizer satisfying the two-norm growth condition (3.47) in Lemma 3.3 and Theorem 4.2 is automatically an $\mathbb{L}^2$-local minimizer (Theorem 6.6, [17]). Hence, sufficient conditions for $\mathbb{L}^2$-local optimality are obtained from the $\mathbb{L}^\infty$-local optimality conditions in Lemma 3.3 or Theorem 4.2, in conjunction with the structure/continuity conditions for $\nu = 2$, and condition (4.55). However, the analysis in [18] shows that the structural features of the set $\gamma(\underline{z})$ and the behavior of $\nabla J(\underline{z})(t)$ near points $\tau \in \gamma(\underline{z})$ are essentially irrelevant when (4.55) is satisfied, and that $\mathbb{L}^2$-local optimality of $\underline{z}$ then follows directly from the pointwise strict complementarity condition (3.38a), the $\mathbb{L}^2$ coercivity condition (3.38b), and the structure/continuity conditions for $\nu = 2$. We restate the relevant Theorem 4.1 of [18] here.

THEOREM 4.4. *Let C be a quasi-polyhedral convex set (3.33) and let J satisfy the structure/continuity conditions for $\nu = 2$. In addition, suppose that conditions (3.38) and (4.55) hold at $\underline{z} \in C$. Then $\underline{z}$ is a strict $\mathbb{L}^2$- local minimizer for J in C, and for each $\mu' \in (0, \hat{\mu})$, there is a $\delta > 0$ such that for all z,*

$$(4.56) \qquad z \in C \text{ and } \|z - \underline{z}\|_2 < \delta \Rightarrow J(z) - J(\underline{z}) \geq \frac{1}{2}\mu'\|z - \underline{z}\|_2^2.$$

Note that in Example 2.2, the structure/continuity conditions hold trivially for $\nu = 2$, with $K(z) = 0$ and

$$S(z)(t) = \begin{cases} -4 & t \in [0, \frac{1}{2}] \\ 2 & t \in (\frac{1}{2}, 1] \end{cases}$$

for all $z \in \mathbb{L}_m^\infty[0, 1]$. Condition (4.55) is also readily verified here, and in this special case immediately implies *global* optimality of $\underline{z}$ and quadratic growth of J in the norm $\| \cdot \|_2$ (since J is quadratic and $K(z)$ is identically zero.) The remaining conditions invoked in Theorem 4.4 have already been verified for Example 2.2.

5. Optimal Control Problems. Continuous-time input-constrained Bolza optimal control problems are specially structured nonlinear programs (2.1) with feasible sets,

$$C = \{z \in \mathbb{L}_m^\infty[0, 1] : z(t) \overset{a.e.}{\in} Z\},$$

and objective functions,

$$J(z) = P(x(z)(1)) + \int_0^1 f^0(t, z(t), x(z)(t))dt,$$

where Z is a given nonempty set in $\mathbb{R}_m$, and $x(z)(\cdot)$ is the solution of an ODE initial value problem,

$$(5.57a) \qquad \frac{dx}{dt}(t) \overset{a.e.}{=} f(t, z(t), x(t))$$

$$(5.57b) \qquad x(t_0) = x^0,$$

corresponding to $z \in \mathbb{L}_m^\infty[0, 1]$. In this setting, $x(t)$ represents the state of some natural or artificial system at time t, and $z(t)$ is the instantaneous control input[6].

We assume that $z(t)$ and $x(t)$ are vectors in $\mathbb{R}_m$ and $\mathbb{R}_n$ respectively, that the second partial derivatives of $P(\xi)$ are continuous, and that $f^0(t, \zeta, \xi)$, $f(t, \zeta, \xi)$ and the associated second partial derivatives with respect to ζ and ξ are continuous. Under certain additional nonlocal Lipschitz continuity conditions on f, f^0

---

[6] The examples in §2 are readily seen as rudimentary Bolza optimal control problems.

and their partial derivatives with respect to ξ and ζ, it can be shown that the state equations (5.57) have a unique solution $x(z)$ for each z in $\mathbb{L}_m^\infty[0,1]$, that the corresponding function-valued map $z \to x(z)$ and objective functional J are twice continuously Fréchet differentiable relative to the norm $\|\cdot\|_\infty$ on $\mathbb{L}_m^\infty[0,1]$, that the structure/continuity conditions of §4 are satisfied for $\nu = \infty$, and that $\nabla J(z)$, $S(z)$ and $K(z)$ are readily constructed from partial gradients and Hessians of the system Hamiltonian,

$$H(t,\zeta,\xi,\psi) = f^0(t,\zeta,\xi) + \langle \psi, f(t,\zeta,\xi) \rangle,$$

and solutions of the affine *equations of variation*,

$$\frac{dy}{dt}(t) \overset{a.e.}{=} \frac{\partial f}{\partial \xi}(t, z(t), x(z)(t))\, y(t) + \frac{\partial f}{\partial \zeta}(t, z(t), x(z)(t))\, v(t)$$

$$y(t_0) = 0.$$

More specifically, it can be shown that

$$\nabla J(z)(t) \overset{a.e.}{=} \nabla_\zeta H(t, z(t), x(z)(t), p(z)(t)),$$

and

$$S(z)(t) \overset{a.e.}{=} \nabla^2_{\zeta\zeta} H(t, z(t), x(z)(t), p(z)(t)),$$

where $p(z)(\cdot)$ is the unique solution of the affine adjoint *costate equations*,

$$\frac{dp}{dt}(t) \overset{a.e.}{=} -\nabla_\xi H(t, z(t), x(z)(t), p(t))$$

$$p(t_1) = \nabla P(x(z)(1)).$$

Under more stringent conditions on f and f^0, it is possible to establish analogous Fréchet differentiability properties for J relative to the inner product induced norm $\|\cdot\|_2$ on $\mathbb{L}_m^\infty[0,1]$, and to show that the structure continuity conditions of §4 hold for $\nu = 2$. The stronger hypotheses require in part that the system Hamiltonian $H(t,\zeta,\xi,\psi)$ is quadratic in its control input argument ζ, and hence that $\nabla_{\zeta\zeta} H(t,\zeta,\xi,\psi)$ depends only on (t,ξ,ψ). These conditions are met by an important class of regulator problems with affine or nonaffine state equations and nonconvex nonquadratic objective functions. For additional discussion and supporting references, see [48].

REFERENCES

[1] M. Aljazzaf, *Multiplier Methods with Partial Elimination of Constraints for Nonlinear Programming*, North Carolina State University Ph.D. Thesis (1989).

[2] W. Alt, Stability of solutions for a class of nonlinear cone constrained optimization problems, Part I: Basic theory, *Numer. Funct. Anal. Optim.*, 10 (1989), 1053–1064.

[3] W. Alt, The Lagrange-Newton method for nonlinear optimal control problems, *Comp. Optimiz. Applic.*, 2 (1993), 77–100.

[4] W. Alt and K. Malanowski, The Lagrange-Newton method for nonlinear optimal control problems, *Comp. Optimiz. and Applic. 2* (1993), 77–100.

[5] A. Ben-Tal, Second-order theory of extremum problems, in *Extremal Methods and Systems Analysis*, Springer-Verlag, New York, (1980).

[6] A. Ben-Tal and J. Zowe, A unified theory of first and second order conditions for extremum problems in topological spaces, *Mathematical Programming Study 19* (1982), 39–76.

[7] D.P. Bertsekas, "On the Goldstein-Levitin-Polyak gradient projection method", Proc. 1974 IEEE CDC, Phoenix, AZ, 47–52, and *IEEE Trans. Auto. Control*, AC-10 (1976), 174–184.

[8] D.P. Bertsekas, *Constrained Optimization and Lagrange Multiplier Methods*, Academic Press, NY, (1982).

[9] D. P. Bertsekas, Projected Newton methods for optimization problems with simple constraints, *SIAM J. Control Optimiz., 20* (1982), 221–246.

[10] J.V. Burke and J.J. Moré, On the identification of active constraints, *SIAM J. Numer. Anal. 25, 5* (1988) 1197–1211.

[11] R. Cominetti, Metric regularity, tangent sets, and second-order optimality conditions, , *Applied Math. Optimiz., 21* (1990), 265–287.

[12] A.L. Dontchev and W.W Hager, Lipschitzian stability in nonlinear control and optimization, *SIAM J. Control Optimiz., 31* (1993), 569–603.

[13] A.L. Dontchev, W.W. Hager, A.B. Poore, and B. Yang, Optimality, stability and convergence in nonlinear control, *Appl. Math. Optimiz. 31* (1995), 297–326.

[14] J.C. Dunn, On the convergence of projected gradient processes to singular critical points, *J. Optimiz. Th. Applic., 55, 2* (1987), 203–215.

[15] J. C. Dunn, A subspace decomposition principle for scaled gradient projection methods: local theory, SIAM J. Control, Optimiz., 31, 1 (1993), 219–246.

[16] J. C. Dunn, *Gradient-related constrained minimization algorithms in function spaces: convergence properties and computational implications*, in Large Scale Optimization: State of the Art, Kluwer Academic Publishers, Dordrecht,(1994).

[17] J.C. Dunn, Second order optimality conditions in sets of $\mathbb{L}^\infty$ functions with range in a polyhedron, *SIAM J. Control Optimiz., 33, 5* (1995), 1603–1635.

[18] J.C. Dunn, On $\mathbb{L}^2$ sufficient conditions and the gradient projection method for optimal control problems, *SIAM J. Control Optimiz., 34, 4* (1996), 1270–1290.

[19] J.C. Dunn and T.Tian, Variants of the Kuhn-Tucker sufficient conditions in cones of non-negative functions, *SIAM J. Control Optimiz., 30, 6* (1992), 1361–1384.

[20] A.V. Fiacco, and G.P. McCormick, *Nonlinear Programming: Sequential Unconstrained Minimization Techniques*, John Wiley, New York, (1968).

[21] A.V. Fiacco, *Introduction to Sensitivity and Stability Analysis in Nonlinear Programming*, Academic Press, New York, (1983).

[22] E.M. Gafni and D.P. Bertsekas, Two-metric projection methods for constrained minimization, *SIAM J. Control Optimiz., 22 ,6* (1984), 936–964.

[23] M. Gawande, *Projection Algorithms for Specially Structured Constrained Minimization Problems*, North Carolina State University Ph.D. Thesis, (1986).

[24] M. Gawande and J. C. Dunn, Variable metric gradient projection processes in convex feasible sets defined by nonlinear inequalities, *Appl. Math. Optimiz., 17* (1988), 103–119.

[25] H. Goldberg and F. Tröltzsch, Second-order sufficient optimality conditions for a class of nonlinear parabolic boundary control problems, *SIAM J. Control Optimiz., 31, 4* (1993), 1007–1025.

[26] M. Guignard, Generalized Kuhn-Tucker conditions for mathematical programming problems in a Banach space, *SIAM J. Control, 7* (1969) 232–241.

[27] W.W. Hager, Lipschitz continuity for constrained processes, *SIAM J. Control Optimiz. 17* (1979), 321–338.

[28] W.W. Hager, Multiplier methods for nonlinear optimal control," *SIAM J. Numer. Anal., 27* (1990), 1061–1080.

[29] A.D. Ioffe, Necessary and sufficient conditions for a local minimum 3: second order conditions and augmented duality, *SIAM J. Control Optimiz., 17* (1979), 266–288.

[30] A.D. Ioffe, On some recent developments in the theory of second order optimality

conditions, in *Optimization*, S. Doleki (ed.), Lecture Notes in Mathematics, Springer-Verlag, New York, (1989).

[31] K. Ito and K. Kunisch, Sensitivity analysis of solutions to optimization probems in Hilbert spaces with applications to optimal control and estimation, *J. Diff. Eqs., 99* (1992), 1–40.

[32] D. G. Luenberger, *Optimization by Vector Space Methods*, Wiley, NY, (1969).

[33] D. G. Luenberger, *Linear and Nonlinear Programming, 2nd ed.*, Addison Wesley, Reading, MA (1984).

[34] K. Malanowski, Sensitivity analysis of optimization problems in Hilbert space, with application to optimal control, , *Applied Math. Optimiz., 21* (1990), 1-20

[35] K. Malanowski, Second order conditions and constraint qualifications in stability and sensitivity analysis of solutions of optimization problems in Hilbert spaces, *Appl. Math. Optimiz. 25* (1992), 51–79.

[36] K. Malanowski, Two-norm approach in stability and sensitivity analysis of optimization and optimal control problems, *Adv. Math. Sci. Applic., 2* (1993), 397–443.

[37] K. Malanowski, Stability and sensitivity of solutions to nonlinear optimal control problems, *Appl. Math. Optimiz. 32* (1995), 111-141.

[38] H. Maurer, First and second order sufficient optimality conditions in mathematical programming and optimal control, *Mathematical Programming Study, 14* (1981), 163–177.

[39] H. Maurer and J. Zowe, First and second-order necessary and sufficient optimality conditions for infinite-dimensional programming problems, *Math. Programming, 16, 1*(1979), 98–110.

[40] H. Maurer and H-J. Pesch, Solution differentiability for parametric nonlinear control problems with control-state constraints, *Control, Cyber., 23* (1994), 201–227.

[41] H. Maurer and S. Pickenhain, Second order sufficient conditions for control problems with mixed control-state constraints, *J. Optimiz. Th. Applic. 86, 3* (1995), 649-667.

[42] G.P. McCormick, Second order conditions for constrained minima, *SIAM J. Appl. Math., 15* (1967), 641–652.

[43] D. Orrell and V. Zeidan, Another Jacobi sufficiency criterion for optimal control with smooth constraints, *J. Optimiz. Th., Applic., 58* (1988), 283–300.

[44] S.M. Robinson, First order conditions for general nonlinear optimization, *SIAM J. Appl. Math., 30* (1976), 597–607.

[45] S.M. Robinson, Strongly regular geberalized equations, *Math Oper. Res. 5, 1* (1980), 43–62.

[46] R. T. Rockafellar, *Convex Analysis*, Princeton University Press, Princeton, N. J. , (1970).

[47] T. Tian, *Convergence Analysis of a Projected Gradient Method for a Class of Optimal Control Problems*, Ph. D. Dissertation, North Carolina State University, (1992).

[48] T. Tian and J. C. Dunn , On the gradient projection method for optimal control problems with non-negative $\mathbb{L}^2$ inputs, *SIAM J. Control Optimiz., 32, 2* (1994), 517–537.

[49] V. Zeidan, Sufficient conditions for the generalized problem of Bolza, *Trans. Amer. Math Soc. 275*, (1983), 561-586.

[50] V. Zeidan, Sufficiency criteria via focal points and via coupled points, *SIAM J. Control Optimiz. 30* (1992), 82–98.

Algorithmic Stability Analysis for Certain Trust Region Methods

URSULA FELGENHAUER   Technical University of Cottbus (BTU), Institute of Mathematics,   Karl-Marx-Str.17, D - 03044 Cottbus, Germany

1   INTRODUCTION

The stability of the solution set with respect to small data perturbation is a general necessary condition for the successful numerical treatment of a nonlinear programming problem. The design of algorithms, however, has to take into account the specifics of error propagation during the approximation so that the user will be enabled to estimate the accuracy of the final numerical results.

In the paper there will be given an anlysis of the asymptotic behavior of some trust region type minimization algorithms under data perturbations. As a particular error source there are considered gradient approximation errors. They are of particular interest when the algorithm uses only derivative information up to order one (for example, when the quasi-NEWTON approach is applied). Starting from the direct LYAPUNOV method we will formulate basic definitions and then point out the central role of descent properties for the stability of an iteration.

For two particular trust region methods there will be estimated least descent bounds, and the question will be discussed how to control the iteration parameters in the case of noisy data. The main result consists in the estimation of the critical point set approximation, namely, the tolerances turn out to be of the same order as the gradient approximation accuracy (provided the latter is sufficiently small).

The relation between decrease and global convergence of minimization procedures has been widely investigated in the literature in the context of line search criteria, for example see POWELL [24]; further WERNER et al.

[31], also [1], [13]. In [22] and [26] there has been considered the role of descent bounds for certain trust region methods.

The stability of iterative methods is a closely connected question. Investigations in this field mainly had been influenced and suggested by a number of publications of EVTUSHENKO et al., see for ex. [8], [7]. Further, impulses for the analysis of quasi-NEWTON formulas with perturbed gradients were given particularly by DENNIS and WALKER [5], [6], see also [12], and the paper [15]. In [11] there was outlined the stabilizing role of descent line search criteria in quasi-NEWTON algorithms, and for unconstrained optimization some first sensitivity estimates had been derived. Similiar ideas were used by ZAVRIEV, [32], to analyze wider classes of algorithms for their numerical (or algorithmic) stability.

2   PROBLEM. BASIC DEFINITIONS

Let be given a function $f : I\!R^n \to I\!R$ and suppose that for some given x_o the level set

$$D = \{ x \in I\!R^n : \quad f(x) \leq f(x_o)\}$$

is a compact set. Assume f to be differentiable on D with locally LIPSCHITZ continuous gradient $g = \nabla f$. We will consider the minimization problem

$$f(x) \to \min \quad s.t. \quad x \in D \cap B, \tag{1}$$

where the set B may be **(i)** – the space $I\!R^n$ (unconstrained case), or **(ii)** – the set
$B = \{x \in I\!R^n : l \leq x \leq u\}$ (bound constrained case). By f_* there will be denoted the minimal function value in (1), by X_* – the solution set; when f is strictly convex, we have $X_* = \{x_*\}$.

In general iterative procedures for solving optimization problems can be written in the form

$$x_{k+1} \quad = \quad \Phi_k(x_k) \quad = \quad x_k + \phi_k(x_k) \tag{2}$$

where the iteration functions Φ_k essentially determine the approximation properties of $\{x_k\}$, i.e. asymptotic behavior, convergence rates and numerical stability. In some cases it is possible to formulate a related "continuous version" of (2), i.e.

$$\dot{x} = \phi(x, t) \quad , \quad t > 0 \quad . \tag{3}$$

The iteration formulas then may be interpreted as explicit one-step discretization methods for (3), and their behavior (at least for small step size parameters) is mainly determined by the asymptotic properties of solution trajectories of (3). This idea has been used by a number of authors for stability investigations; let's remark namely EVTUSHENKO and ZHADAN (for ex. [8]) and KOVÁCS / VASIL'EV [18], [19], [20]; further comp. also [30], [23]. The main advantage of the "continuous version approach" has to be seen in the convenient manner the LYAPUNOV's Theorems may be applied. On the other hand, not for all iteration functions there exist "natural" continuations, for instance the characteristic behavior of quasi-NEWTON (or secant) methods hardly can be reproduced this way. Thus we will prefer the discrete approach where starting from the theory of ODE and related discretization methods (see [29] e.g.; also cf. [32]) stability results for iterative (discrete) procedures are proved directly.

Main facts we will refer to are the following:

Definition 1 (Stability in the whole) *A sequence $\{x_k\}$ given by (2) is called stable in the whole (with respect to a solution x_*) if for all $\epsilon > 0$ there exist $\delta > 0$ such that for $\{\tilde{x}_k\}$ with*

$$\tilde{x}_{k+1} = \Phi_k(x_k) + \xi_{k+1} \quad , \quad \tilde{x}_0 = x_0 + \xi_0$$

it holds that

$$\|\xi_k\| \leq \delta \quad \forall k \quad \Longrightarrow \quad \limsup_{k \to \infty} \|\tilde{x}_k - x_*\| < \epsilon \quad .$$

This property may be interpreted as a least stability condition we need for numerical purposes, but its direct proof is mostly impossible. A stronger but easier to verify condition is given by:

Definition 2 (Exponential stability) *A sequence $\{x_k\}$ is called exponentially stable (with respect to a solution x_*) if there exist positive numbers c , δ and μ such that*
for $\| x_0 - x_* \| < \delta$ it holds that

$$\|x_k - x_*\| < c \, \exp(-\mu \, k) \quad .$$

The connection between both Definitions has been enlightened by HAHN (see [29], chap.2.3; also [16]) in the following

THEOREM 1 (HAHN) *If $\{x_k\}$ is exponentially stable with resp. to x_*, then it is stable in the whole as well.*

The main instrument in LYAPUNOV's theory for proving stability (in the sense of the attraction to the solution set for our problem (1)) will be the descent analysis of an appropriate indicator function, i.e. a scalar merit function representing in some sense the distance to the set X_*. Its role is described in Theorem 2 (cf. [29], chap.2.3), a discrete version of well-known classical results:

THEOREM 2 (LJAPUNOV) *A sequence $\{x_k\}$ (or the solution x_* resp.) is exponentially stable **iff** there exists a function $v : \mathrm{R}^n \times \mathrm{R} \to \mathrm{R}$ with*

$$v(\cdot, t) \in \mathrm{C}^1(\mathrm{R}^n) \quad ; \quad v(\xi, \cdot) \in Lip(\mathrm{R}_+)$$

and a constant $r > 0$ such that for $\|x\| < r$ the following holds: There exist constants $a_1, a_2, a_3 > 0$ so that

(i)
$$a_1 \|x - x_*\|^2 \leq v(x, k) \leq a_2 \|x - x_*\|^2$$

(ii)
$$\begin{aligned} \Delta v_k : &= v(\Phi_k(x), k+1) - v(x, k) \\ &\leq -a_3 \|x - x_*\|^2 \end{aligned} \qquad \forall k \geq 0$$

To illustrate how this Theorem may be applied to minimization algorithms, assume for the moment that f is a strictly convex function on D and $B = I\!\!R^n$. If the indicator function v is chosen as $v = f - f_*$ or $v = \|g\|^2$, then condition (i) can be satisfyed under standard assumptions. Condition (ii) then has to be treated as a descent property of Φ_k; for example for line search methods we just get a so-called "efficient" step length rule (cf. [31],[24]; [28]) (for example in the form:)

$$f(x_{k+1}) - f(x_k) \leq -c \|g(x_k)\|^2 \quad , \tag{4}$$

i.e. conditions which are usually associated with global convergence results for descent algorithms. The connection between step length choice and stability more detailed has been considered in [11].

In the following we therefore will spend our attention to similiar properties of trust region type algorithms designed for the solution of (1) when f is not (or not strictly) convex.

3 DESCENT PROPERTIES OF CERTAIN TRUST REGION METHODS

In 1982 J.J.MORÉ [22], and in parallel also M.J.D. POWELL [26], considered global convergent trust region type algorithms. The approach in [22] consists in a steepest descent prediction step for minimizing a quadratic local model function of the objective f and a descent-preserving correction combined with ("usual") trust region parameter control.

For the case of problems with simple bounds this method had been modified later by CONN, GOULD and TOINT [2]; an implementation now is part of [4] (see also [3]).

The main ideas may be shortly described as folllows:

Let the k-th iterate be x_k, Δ_k – the actual trust region radius, then the trust region T (in terms of the maximum-norm) is given by

$$T = T_k = \{x \, : \, \|x - x_k\|_\infty \leq \Delta_k\} \quad .$$

Consider the quadratic model function

$$m_k(x) = f_k + g_k^T (x - x_k) + 0.5 \, (x - x_k)^T B_k (x - x_k) \, ,$$

where $f_k = f(x_k)$, $g_k = g(x_k)$ and the matrix B_k stands for an appropriate symmetric approximation of the HESSE matrix of f in x_k. In general, B will be determined by some quasi-NEWTON formula with an update rule $\mathcal{Q}$. Throughout the paper we will make the assumption that (i) B_k cannot be the zero matrix and (ii) $\|B_k\|$ are uniformly bounded w.r.t. k.

To evaluate the quality of the quadratic model (and to control the choice of the trust region parameter) define the ratio

$$\rho(x) = (f(x) - f(x_k))/(m_k(x) - f(x_k)) \quad . \tag{5}$$

This term will be used for the trust region adaptation throughout the iteration.

The algorithm will start with the minimization of m_k over T along some search path

$$p(t) = \Pi_B(x_k - t \, g_k) \, , \quad t > 0 \, ,$$

with Π_B – the orthogonal projector onto B if $B \neq \mathbb{R}^n$ and $\Pi_B = I$ else. Denote

$$t_C = \arg\min_{t > 0}\{m_k(p(t)) \, | \, p(t) \in T\} \, , \quad x_C = p(t_C) \, . \tag{6}$$

If $B = \mathbb{R}^n$ then the point x_C is called the CAUCHY point (cf. [22]). In the case $B \neq \mathbb{R}^n$ the point $x_C = p(t_C)$ will be called a generalized CAUCHY point, [2].

The path p is a piecewise linear function in t. The vector $\dot{p}(0+)$ is the projected (anti-)gradient in x_k with respect to the feasible set B and will be denoted by $(-z_k)$. In the solution point obviously z is the zero vector. If we denote

$$s(t) = x_k - t \, z_k \quad \text{and} \quad t_B = \arg\min_{t > 0}\{m_k(s(t)) \, | \, s(t) \in T \cap B\} \, ,$$

then $x_B = s(t_B)$ is the first break point of $p(t)$ (or coincides with x_C). For the constrained problem in general $x_B \neq x_C$ and $m_k(x_C) \leq m_k(x_B)$.

To unify in the following the notation of active bounds we will assume that the restrictions defining B are enumerated, e.g. $b_i = [x]_i - u_i \leq 0$; $b_{n+i} = l_i - [x]_i \leq 0$; $i = 1, \ldots, n$. For an arbitrary point $x \in B$ then the active index set is defined by $I(x) = \{j \, : \, b_j(x) = 0\}$.

Now let's describe two model minimization algorithms:

TR: Trust Region Algorithm [22]; $(B = I\!\!R^n)$

initialization: Choose positive constants with: $\alpha \leq 1 \leq \beta; \eta_1 < \eta_2 < 1; \gamma_1 < 1 < \gamma_2;$
further let be given $x_o \in I\!\!R^n; \Delta_o > 0; B_o \in I\!\!R^{n \times n}$ – a symmetric, positive definite matrix. Set $k = 0$.

while $z(x_k) \neq 0$ **do**:

prediction: (line search)
Determine the CAUCHY point x_C according to (6).

correction: (local improvement)
Find x_k^+ such that
(i) $m_k(x_k^+) - f_k \leq \alpha(m_k(x_C) - f_k)$, (ii) $\|x_k^+ - x_k\| \leq \beta \Delta_k$,

trust region test: Calculate $\rho_k = \rho(x_k^+)$;
if $\rho_k < \eta_1$ then $x_{k+1} := x_k$; $\Delta_{k+1} := \gamma_1 \Delta_k$;
else: $x_{k+1} := x_k^+$ and (if $\rho_k < \eta_2$ then $\Delta_{k+1} := \Delta_k$, else $\Delta_{k+1} := \gamma_2 \Delta_k$).

matrix update: $B_{k+1} = Q_k(x_{k+1}, x_k, g_{k+1}, g_k, B_k, ...)$; $k := k + 1$

end.

For the bound constrained case consider the

MTR: Modified Trust Region Algorithm [2]; $(B \neq I\!\!R^n)$
which differs from (**TR**) in the following parts:

prediction: (path search)
Determine the generalized CAUCHY point x_C according to (6).

correction: (local improvement)
Find x_k^+ such that
(i) and (ii) as in (**TR**) hold and in addition
(iii) $I(x_k^+) \supseteq I(x_k)$.

For shortness in the following we will denote by $\{x_k\}$ the subsequence of iterates corresponding to "successful" steps (with $\rho_k \geq \eta_1$ and $x_{k+1} \neq x_k$) only. With this new enumeration the above algorithms guarantee the descent of $\{f_k\}$ with the following least descent bounds:

$$
\begin{aligned}
f(x_{k+1}) - f(x_k) &\leq -\eta_1 \, \alpha \, (m_k(x_C) - f_k) \\
&\leq -\eta_1 \, \alpha \, (m_k(x_B) - f_k) \\
&\leq -\eta_1 \, \alpha \, (0.5 \, t_B \, \|z_k\|^2) \,,
\end{aligned} \tag{7}
$$

where for the last estimate we made usage of the fact that $\phi(t) = m_k(p(t))$ is a quadratic function on $[0, t_B]$ and $\dot\phi(t_B) \leq 0$. This inequality in essence is of the same structure as the abstract descent bound (4).

In fact for the convergence proofs of the above algorithms there had been derived the stronger descent estimates

- for **(TR)**,[22]:

$$
f(x_{k+1}) - f(x_k) \leq -c \, \|g_k\| \, \min\{ \, \|B_k\|^{-1} \, \|g_k\| \,, \Delta_k\} \tag{8}
$$

- for **(MTR)** (compare [2]):

$$
f(x_{k+1}) - f(x_k) \leq -c \frac{h_k}{\|z_k\|} \, h_k \, \min\{ \, (1 + \|B_k\|)^{-1} \, h_k \,, \Delta_k\} \tag{9}
$$

with $h = \|\Pi_B(x - g) - x\|$.

Before we come to detailed conclusions on the algorithmic stability let's make some remarks on the error sources to be considered and how they will affect the descent properties of our iteration.

Suppose that the gradient values used within the iteration are inexact, for example due to rounding errors or finite difference approximation, but may be obtained with arbitrary given accuracy at step k. We assume that additional rounding errors (in the function values of f particularly) can be neglected in comparison to the gradient errors.

Denote the gradient approximation by $\tilde{g}$, the error level in a current iteration point x_k let be given by the absolute error bound θ_k or alternatively by the relative error τ_k, i.e.

$$
\|\tilde{g}_k - g(x_k)\| \leq \theta_k \leq \theta \quad \forall \, k \,; \tag{10}
$$

$$
\|\tilde{g}_k - g(x_k)\| \leq \tau_k \, \|g(x_k)\| \,, \tag{11}
$$

In the algorithms **(TR)**, **(MTR)** then instead of g the inexact values $\tilde{g}$ will be used what particularly means that we have to do with a perturbed local quadratic model function,

$$
\tilde{m}_k(x) = f_k + (x - x_k)^T \tilde{g}_k + 0.5(x - x_k)^T \tilde{B}_k(x - x_k) \,,\cdot
$$

(Remark that besides of the direct effects on the gradients in quasi-NEWTON type methods we have an indirect error accumulation due to the matrix secant update.)

When in analogy to the previous notation by $\tilde{z}$ we denote the projected (inexact) gradient obtained by truncation of $\tilde{g}$, then the algorithm will guarantee at least a generalized function value descent

$$f(x_{k+1}) - f(x_k) \leq -c\,\tilde{t}_B\,\|\tilde{z}_k\|^2\,,$$

or – when we take into account that $\|\tilde{z}\| \geq |\,\|z\|\, - \theta\,|$

$$f(x_{k+1}) - f(x_k) \leq -c'\,\tilde{t}_B\,\|z_k\|^2 + c''\theta^2\,.$$

Descent properties of this kind then will allow to estimate the asymptotical behavior of the distance to the solution set under appropriate regularity conditions on g resp. f (cf. [11] and [23]). Main results concerning the trust region approach will be proved in the next two sections.

For illustration , however, let's shortly consider the situation of an unconstrained optimization problem where f – twice continuously differentiable and strictly convex in D. We have $z = g$, and there exist constants μ, M so that

$$\mu\,\|x - x'\| \leq \|\,g(x') - g(x)\,\| \leq L\,\|x - x'\| \quad \forall\, x, x' \in D\,. \tag{12}$$

If now there exists $t_o : 0 < t_o \leq \tilde{t}_B(k)$ for all k, then we obtain

$$
\begin{aligned}
f(x_{k+1}) - f_* \;&\leq\; f(x_k) - f_* - c'\,\tilde{t}_B\,\|z_k\|^2 + c''\theta^2 \\
&\leq\; \left(1 - \frac{\mu}{L}c't_o\right)(f_k - f_*) + c''\theta^2 \\
&=\; (1 - \nu)\,(f_k - f_*) + c''\theta^2\,; \\
f(x_{k+1}) - f_* - \sigma\theta^2 \;&\leq\; (1 - \nu)\,(f_k - f_* - \sigma\theta^2)\,,
\end{aligned}
$$

when $\sigma\nu = c''$. Decreasing c' if necessary we get $\nu < 1$, so that

$$\lim sup(f_k - f_* - \sigma\theta^2) \leq 0\,.$$

In the result we have $\lim sup\|x_k - x_*\| = O(\theta)$, i.e. the solution will be approximated with essentially the same accuracy as that of the input data, the gradient values $\tilde{g}_k$.

4   STABILITY OF TRUST REGION METHODS FOR UNCONSTRAINED MINIMIZATION

To begin with we will confirm that for the method (**TR**) with perturbed gradients a descent bound of the form (8) holds. The algorithm for successful steps ensures (cf. (7))

$$f_{k+1} - f_k \leq \eta_1\,(\tilde{m}_k(x_{k+1}) - f_k) \leq \alpha\,\eta_1\,(\tilde{m}_k(\tilde{x}_C) - f_k)\,, \tag{13}$$

where $\tilde{x}_C$ is the first minimum point of $\tilde{m}_k$ along $(x_k - t\tilde{g}_k)$. There are two cases to be distinguished:

(a) $\tilde{x}_C \in T^o$ (inner point);

$$\Rightarrow \quad \dot{\phi}(t_C) = \frac{d}{dt}\, m_k(p(t))|_{t=t_C} = 0\,,$$

$$t_C = \|\tilde{g}_k\|^2 / \tilde{g}_k^T \tilde{B}_k \tilde{g}_k \geq \|\tilde{B}_k\|^{-1}\,;$$

so that

$$\tilde{m}_k(\tilde{x}_C) - f_k \leq -0.5\, t_C \|\tilde{g}_k\|^2 \leq -0.5\,\|\tilde{B}_k\|^{-1}\|\tilde{g}_k\|^2\,.$$

(b) $\tilde{x}_C \in \partial T$ (trust region boundary);

$$\Rightarrow \quad \dot{\phi}(t_C) \leq 0\,, \quad t_C = \Delta_k/\|\tilde{g}_k\| \leq \|\tilde{g}_k\|^2 / \tilde{g}_k^T \tilde{B}_k \tilde{g}_k\,,$$

and thus

$$\tilde{m}_k(\tilde{x}_C) - f_k \leq -0.5\, t_C \|\tilde{g}_k\|^2 = -0.5\,\Delta_k\,\|\tilde{g}_k\|\,.$$

Both bounds together may be written in the unique form

$$\tilde{m}_k(\tilde{x}_C) - f_k \leq -0.5\,\|\tilde{g}_k\|\,\min\{\Delta_k, \|\tilde{B}_k\|^{-1}\|\tilde{g}_k\|\} \leq 0\,. \qquad (14)$$

Based on this generalized descent bound we may prove the following Lemma:

LEMMA 1 *Let be L a positive constant such that*

$$\|g(x') - g(x)\| \leq L\,\|x - x'\| \quad \forall\, x, x' \in D\,.$$

Suppose the matrices $\tilde{B}_k$ to be uniformly bounded: $\|\tilde{B}_k\| \leq b$. Further assume that the accuracy of the gradient approximation satisfies the relations

$$\theta_k \leq \kappa\,\Delta_k \quad\quad and \quad\quad \theta_k \leq \theta \qquad (15)$$

for all k and some positive constants κ, θ. Then,

$$\lim\,inf_{k\to\infty} \|\tilde{g}_k\| = 0\,, \quad\quad \lim\,sup_{k\to\infty} \|\tilde{g}_k\| \leq 2\,\theta\,.$$

Remark that the proof is based on ideas from [22] (and partly [2]) and there are only added modifications connected with gradient error accumulation where necessary.

Proof: From the smoothness assumptions on f (see sect.1) we know that f is bounded below on the set D; the minimal function value denote by f_*. As a consequence

$$\sum_{k=0}^{n} (f_{k+1} - f_k) = f_{n+1} - f_o \geq f_* - f_o > -\infty$$

for all n, so that from (13) and (14) it follows that $\lim_{k\to\infty} (f_{k+1} - f_k) = 0$. Let us assume first that for all $k:\ \Delta_k \geq \|\tilde{B}_k\|^{-1}\|\tilde{g}_k\|$, i.e.

$$f_{k+1} - f_k \leq -c\,\|\tilde{g}_k\|^2 \leq 0 \,.$$

In this case $\lim_{k\to\infty} \|\tilde{g}_k\| = 0$ and the assertion of the Lemma is true.

On the contrary assume now that there exists a subsequence for which $\Delta_k < \|\tilde{B}_k\|^{-1}\|\tilde{g}_k\|$. Under the descent condition for f_k it follows that

$$\lim\inf{}_{k\to\infty}\Delta_k\,\|\tilde{g}_k\| = 0 \,. \tag{16}$$

Suppose there exists $\epsilon > 0$ such that $\|\tilde{g}_k\| \geq \epsilon\ \forall\, k$. Consider the term $\rho(x)$:

$$\rho(x) = (f(x) - f_k)/(\tilde{m}_k(x) - f_k) \,;$$
$$|\rho(x) - 1| = |\tilde{m}_k(x) - f(x)|\,/\,|\tilde{m}_k(x) - f_k|$$

with

$$
\begin{aligned}
|\tilde{m}_k(x) - f(x)| &\leq |f(x) - f_k - g_k^T(x - x_k)| + 0.5\,|(x - x_k)^T B_k(x - x_k)| \\
&\quad + |(\tilde{g}_k - g_k)^T(x - x_k)| \\
&\leq (L + 0.5\,b)\|x - x_k\|^2 + \kappa\,\Delta_k\,\|x - x_k\|
\end{aligned}
$$

under the assumptions of the Lemma. Consider the point $x = x_{k+1}$: from the algorithm **(TR)** we know that $\|x_{k+1} - x_k\| \leq \beta\,\Delta_k$; on the other hand the inequality (14) leads to

$$|\tilde{m}_k(x) - f_k| \geq 0.5\,\alpha\,\epsilon\,\Delta_k$$

if only $\Delta_k \leq \epsilon\,b^{-1} \leq \|\tilde{B}_k\|^{-1}\|\tilde{g}_k\|$.
But then it follows that

$$|\rho(x) - 1| \leq 2\,\beta\,((L + 0.5\,b)\,\beta + \kappa)\,\Delta_k\,/\,(\alpha\,\epsilon) =: c_1\,\Delta_k \,,$$

what yields $|\rho_k - 1| < 1 - \eta_2$ whenever $\Delta_k \leq \Delta' = \min\{\epsilon b^{-1}, (1 - \eta_2)c^{-1}\}$. Thus the trust region parameter control in **(TR)** would guarantee that for all steps $\Delta_k \geq \gamma_1\Delta'$ what contradicts the assumption (16). Consequently, $\lim\inf{}_{k\to\infty} \|\tilde{g}_k\| = 0$.

Now it remains to estimate the $\lim sup\|\tilde{g}_k\|$:

Assume there exists $\epsilon > 0$ and a subsequence $\{k_j\}$ such that

$$\|\tilde{g}_{k_j}\| \geq \epsilon_1 = 2\theta(1+\epsilon) .$$

From the first part of the proof we know that for arbitrary $\epsilon_2 < \epsilon_1$ there exists another sequence, say $\{l_j\}$, with

$$\|\tilde{g}_k\| \geq \epsilon_2 \quad \text{for } k_j \leq k < l_j, \quad \|\tilde{g}_{l_j}\| < \epsilon_2 .$$

if l_j – sufficiently large.

Choose $\epsilon_2 = \epsilon\theta$. With $K = \{k : k_j \leq k < l_j, \, j = 1, \ldots\}$ we obtain

$$f_k - f_{k+1} \geq c\epsilon_2 \min\{b^{-1}\epsilon_2, \Delta_k\} > 0 \quad \text{for } k \in K ;$$

i.e. $\lim_{k \in K} \Delta_k = 0$. For sufficiently large j, $l_j > k \geq k_j$, then: $f_k - f_{k+1} \geq c\epsilon_2\Delta_k$.

Consider the term $d_j := \|x_{k_j} - x_{l_j}\|$:

$$
\begin{aligned}
d_j &\leq \sum_{i=k_j}^{l_j-1} \|x_{i+1} - x_i\| \\
&\leq \beta \sum_{i=k_j}^{l_j-1} \Delta_i \leq \beta/(c\epsilon_2) \sum_{i=k_j}^{l_j-1} (f_i - f_{i+1}) \\
&\leq \beta/(c\epsilon_2) \sum_{i=k_j}^{\infty} (f_i - f_{i+1}) \longrightarrow 0 \quad \text{for } j \to \infty .
\end{aligned}
$$

From the last relation may be concluded

$$
\begin{aligned}
\|\tilde{g}_{k_j} - \tilde{g}_{l_j}\| &\leq \|g_{k_j} - g_{l_j}\| + \theta_{k_j} + \theta_{l_j} \\
&\leq 2\theta + L\,d_j < \theta(2+\epsilon)
\end{aligned}
$$

for large j. But this contradicts the choice of $\{k_j\}, \{l_j\}$ and $\epsilon_{1,2}$ yielding

$$\|\tilde{g}_{k_j} - \tilde{g}_{l_j}\| \geq \left| \|\tilde{g}_{k_j}\| - \|\tilde{g}_{l_j}\| \right| \geq \theta(2(1+\epsilon) - \epsilon) ,$$

and thus the proof of the Lemma is complete. **q.e.d.**

The condition (15) played an important role in the above proof; it shows that the parameters of an algorithm has to be chosen with special care when it is run with perturbed data. Nevertheless, the first inequality in (15) seems to be a quite natural restriction in the context of **TR** methods because the idea of a "trust" region itself consists in the guarantee of a sufficient good approximation of the objective function by the local quadratic model on the set T, and this should be particularly true then for the linear part of it.

We will come to the formulation of a stability result for the unconstrained minimization algorithm (**TR**). Remember that for the solution set of problem (1) there was used the notation X_* and define the critical point set $\bar{X}$ by

$$\bar{X} = \{x \in D : g(x) = 0\}; \text{ then } X_* = \{x \in D : f(x) = \min_D f = f_*\} \subseteq \bar{X} .$$

Further denote by U the open unit ball in $\mathbb{R}^n$; $(x + \rho U)$ consequently stands for the ρ-neighborhood of a given point x.

THEOREM 3 *Suppose the gradient function g to be* LIPSCHITZ*ian invertible on the set $X_\rho = D \cap (\bar{X} + \rho U)$ for some $\rho > 0$, i.e $\exists \mu > 0$:*

$$x, y \in X_\rho \Rightarrow \|g(x) - g(y)\| \geq \mu \|x - y\| .$$

Then there exists $\theta_o > 0$:

$$\theta \leq \theta_o \Rightarrow \lim sup_{k \to \infty} (dist \{x_k, \bar{X}\}) \leq 4\theta / \mu .$$

*If the sequence $\{x_k\}$ from (**TR**) does not terminate, then for every accumulation point $\bar{x}$ there exists a local minimizer of f on D, $\bar{x}_* \in X_*$, such that $\|\bar{x} - \bar{x}_*\| = O(\theta)$.*

In the case that f in addition is a twice continuously differentiable and convex function on D, and if $\Delta_k \geq \underline{\Delta} > 0$, the attractor set has the form $(x_* + O(\theta) U)$ and is at least exponentially stable, i.e. there exist $\nu_1, \nu_2 > 0, 1 > q > 0$ and k_o:

$$\|x_k - x_*\|^2 \leq \nu_1 q^k + \nu_2 \theta^2 \qquad \forall k > k_o . \tag{17}$$

Proof: At the beginning we remark that the invertibility condition on g particularly leads to the conclusion that $\bar{X}$ consists of a finite number of isolated points while D – compact.

For the first estimate consider the auxiliary set $V = D \backslash X_{\rho/2}$, from our assumption we conclude that

$$\delta_o = (1/3) \min_V \|g\| > 0 .$$

It follows that for all $\delta < \delta_o$

$$\bar{X} \subseteq \{x \in D : \|g\| < 2\delta\} \subseteq X_{\rho/2} .$$

On the other hand, from Lemma 1 and the tolerance restriction on the gradient we have

$$\lim sup_{k \to \infty} \|g_k\| \leq \lim sup_{k \to \infty} \|\tilde{g}_k\| + \theta \leq 3\theta$$

so that for $\theta \leq \theta_o = \delta_o / 2$ we get $x_k \in X_{\rho/2}$ if only k sufficiently large (say greater than k_1). Taking into account the structure of $\bar{X}$ we can find then for every $k > k_1$ a unique point $\bar{x}(k) \in \bar{X}$ such that

$$dist\{x_k, \bar{X}\} = dist\{x_k, \bar{x}(k)\} \leq \|g_k\|/\mu \leq 4\theta/\mu \quad \text{for } k \geq k_2 > k_1 .$$

The result in [14], Theor.1, on the connection between injectivity of the gradient and strict convexity of a function for our situation yields that any element in $\bar{X}$ is either a strict local minimizer (convex situation) or a local maximizer of f. Taking into account the monotonicity of $\{f_k\}$ (according to (14) and (13)) it follows that the iteration cannot accumulate to a maximum point; if it is not terminating after a finite number of steps the algorithm thus can omly approximate local minimum points of f.

Last we prove the attraction property (17):
Due to the boundedness of f on D and the monotonicity of $\{f_k\}$ it follows that $(f_{k+1} - f_k) \to 0$ so that under the assumption that $\{\Delta_k\}$ – bounded below we obtain: $\lim_{k\to\infty} \|\tilde{g}_k\| = 0$ and

$$f_{k+1} - f_k \leq -c\, b^{-1} \|\tilde{g}_k\|^2 .$$

Then we have

$$
\begin{aligned}
f_{k+1} - f_k &\leq -c\, b^{-1} \|\tilde{g}_k\|^2 \\
&\leq -c/b\, (\mu \|x_k - x_*\| - \theta)^2 \\
&\leq -c'\, (f_k - f_*) + c''\theta^2
\end{aligned}
$$

(c', c'' depend on the LIPSCHITZ constant L of g, of μ and of b. Without loss of generality we require $c' < 1$.)
Then it follows (comp. the end of sect 3) that

$$
\begin{aligned}
f(x_{k+1}) - f_* - \nu\theta^2 &\leq q\, (f_k - f_* - \nu\theta^2) \\
&\leq q^k (f(x_o) - f_* - \nu\theta^2) ;
\end{aligned}
$$

with $q = 1 - c'$; $\nu(1 - q) = c''$; the conclusion about the iteration points then is an immediate consequence of the convexity of the function f on D. **q.e.d.**

5 STABILITY IN THE BOUND CONSTRAINED CASE

In the case of box constraints for proving a descent bound of the form (9) it is necessary to analyze the piecewise linear search path $p(t) = \Pi_B(x_k - t\,\tilde{g}_k)$ first. Let $T = \{t_i : i = 1, \ldots, r\}$ denote the set of break points of p, $(t > 0)$, and assume that they are enumerated so that $0 = t_o < t_1 < \ldots < t_r$. The derivative $\dot{p}(t)$ exists for all positive $t \notin T$ and may be given in terms of reduced gradient vectors:

$$t < t_1 : \dot{p}(t) = -\tilde{z}_k; \quad t_m < t < t_{m+1} : \dot{p}(t) = -\tilde{z}_k(m) \qquad (18)$$

where we define $\tilde{z}_k(m)$ for $m > 0$ componentwise

$$[\tilde{z}_k(m)]_j = \begin{cases} [\tilde{z}_k]_j & \text{if } j \notin I(\Pi_B(x_k - t_m \tilde{g}_k)) \\ 0 & \text{else} \end{cases}$$

Using these relations we may get a local descent bound for $\tilde{m}$ along p.

LEMMA 2 *Consider the function* $\phi(t) = \tilde{m}_k(p(t)) - f_k$, $t > 0$.
There exists $t_\phi > 0$ *such that*

$$\phi(t) \leq -t\,\tilde{h}_k^2 / 8 \qquad \text{for} \quad 0 < t < t_\phi .$$

Proof: For simplicity define $\dot{p}(t_m) = \dot{p}(t_m + 0)$. The above formulas for the derivative vectors show **(i)** $\delta(t) = \|p(t) - x_k\|$ is a continuous monotone increasing function and **(ii)** δ is piecewise constant and decreasing. Further $\delta(0) = 0$; $\delta(1) = \tilde{h}_k$; thus $\exists\, t' \in (0, 1)$ with $\delta(t') < 0.5\,\tilde{h}_k$, particularly one can take $t' = \tilde{h}_k/(2\|\tilde{z}_k\|)$ (cf. (18)). Then we get

$$\begin{aligned}
\|\dot{p}(t')\| &\geq (1 - t')\,\|\dot{p}(t')\| \\
&\geq \|p(1) - p(t')\| \\
&\geq \|p(1) - x_k\| - \|p(t') - x_k\| \geq 0.5\,\tilde{h}_k .
\end{aligned}$$

and consequently $\|\dot{p}(t)\| \geq 0.5\,\tilde{h}_k$ for all $t < t'$.
Now consider the derivative of ϕ :

$$\begin{aligned}
\dot{\phi}(t) &= \tilde{g}_k^T \dot{p}(t) + (p(t) - x_k)^T \tilde{B}_k \dot{p}(t) \\
&\leq -\|\dot{p}(t)\|^2 + t\,\|\tilde{z}_k\|\,\|\tilde{B}_k\|\,\|\dot{p}(t)\| \\
&\leq -\|\dot{p}(t)\| \left[\|\dot{p}(t)\| - t\,\|\tilde{B}_k\|\,\|\tilde{z}_k\| \right] \quad .
\end{aligned}$$

If now $t < t'$ and $t < t'' = \tilde{h}_k/(4\,\|\tilde{B}_k\|\,\|\tilde{z}_k\|)$ then $\dot{\phi}(t) \leq -(1/8)\,\tilde{h}_k^2 < 0$; so that the assertion of the Lemma follows after integration immediately with $t_\phi = \min\{t', t''\}$.   **q.e.d.**

After this rather technical proof it is easy to conclude that a descent bound of the kind (9) holds: Indeed, there are two possible cases for the position of $p(t_\phi)$:
(a) $p(t_\phi) \in T$; then $\tilde{m}_k(x_C) \leq \tilde{m}_k(p(t_\phi))$;

$$\tilde{m}_k(x_C) - f_k \leq -(\tilde{h}_k^2/8)\,t_\phi$$

due to Lemma 2.

(b) $p(t_\phi) \notin T$; then $\delta(t_\phi) = \|p(t_\phi) - x_k\| > \Delta_k$; on the other hand: take $\bar{t} = \Delta_k / \|\tilde{z}_k\|$ we have from $\|\dot{p}\| \leq \|\tilde{z}_k\|$ that $\delta(\bar{t}) \leq \Delta_k$; it follows then from the monotonicity of δ that $\bar{t} \leq t_\phi$; $\dot{\phi}(\bar{t}) < 0$. But in this case

$$\tilde{m}_k(x_C) - f_k \leq -(\tilde{h}_k^2/8)\,\bar{t}\,.$$

From (a), (b) and Lemma 2 then follows

$$\tilde{m}_k(x_C) - f_k \leq -(\tilde{h}_k^2/8)\,\min\{t', t'', \bar{t}\}\,,$$

(where t' and t'' are the same as in the proof of the above Lemma) or in other terms

$$f(x_{k+1}) - f(x_k) \leq -c\frac{\tilde{h}_k}{\|\tilde{z}_k\|}\,\tilde{h}_k\,\min\{\,(1 + \|\tilde{B}_k\|)^{-1}\,\tilde{h}_k\,,\,\Delta_k\}\,. \qquad (19)$$

As a first result on the asymptotic behavior of the algorithm **(MTR)** under gradient perturbation notice the following:

LEMMA 3 *Let the assumptions of Lemma 1 hold, particularly suppose that the matrices $\tilde{B}_k$ are uniformly bounded and assume that the accuracy of the gradients satisfies (15), i.e.*

$$\theta_k \leq \kappa\,\Delta_k \qquad and \qquad \theta_k \leq \theta$$

Then the iteration will approximate nearly-critical points in the sense

$$\lim\inf{}_{k\to\infty}\,\tilde{h}_k = 0\,, \qquad \lim\sup{}_{k\to\infty}\,\tilde{h}_k \leq 2\theta\,.$$

The proof in essence repeats the conclusions used for proving Lemma 1 but based on the estimate (19) now; it will be omitted here (for comparison see also [2], Theor.8).

In the presence of constraints, however, it is not sufficient to analyze the terms h which can be seen as some measure for the "criticality" of a point:

$$x \quad \text{is critical} \quad \Longleftrightarrow \quad h = 0\,.$$

A not less important question is whether the algorithm determines the active face where a solution point is located. One main advantage of the algorithm **(MTR)** of CONN, GOULD and TOINT consists in the fact that under the strict complementarity slackness condition (s.c.s.) the set of active bounds in the limit point is identified after a finite number of steps. To prove a similiar result for the algorithm under gradient perturbation, the complementarity slackness condition has to be strengthened to get a stable condition. Before we give an appropriate formulation define the "nearly"-active index set $I_\epsilon(x)$ for positive ϵ and $x \in B$;

$$I_\epsilon(x) = \{ i \in \{1, \ldots, 2n\} : [x]_i < l_i + \epsilon \text{ or } [x]_{i-n} > u_{i-n} - \epsilon \} . \quad (20)$$

In the following we will always assume that the problem data satisfy the condition

stable s.c.s.: There exist $\epsilon_o, \gamma_o : \gamma_o > \epsilon_o > 0$ such that for all $\epsilon < \epsilon_o$:

$$h(x) < \epsilon \quad \text{and} \quad i \in I_\epsilon(x) \quad \Longrightarrow \quad |[g(x)]_i| > \gamma_o . \quad (21)$$

Remark that then particularly for x with $h(x) < \epsilon$ the signs of the gradient components related to ϵ-active bounds are determined by

$$[x]_i > u_i - \epsilon \Rightarrow [g(x)]_i < -\gamma_0 < 0; \qquad [x]_i < l_i + \epsilon \Rightarrow [g(x)]_i > \gamma_0 > 0 . \quad (22)$$

Concerning the gradient approximation error it will be additionally required that

$$\theta_k \le \theta < 1/4\epsilon \quad ; \quad (23)$$

further, the algorithm has to guarantee, that the iterate x_{k+1} is "not worse" than the generalized CAUCHY point x_C in the sense that ϵ-active bounds of x_C remain to be ϵ-active in the new test point. An adequate modification has to be inserted in the correction step, condition (iii), of **(MTR)**.
Under these assumptions there will hold the following identification theorem:

THEOREM 4 *Let be given a minimzation problem (1) with $f \in C^{1,1}$ and $B = \{x : l \le x \le u\}$. We assume that the strict complementarity slackness condition is fulfilled in its stable form (21), and that in the algorithm (MTR) the sets I are replaced by I_ϵ, (20), with (fixed) $\epsilon \in (0, \epsilon_o)$. Further, let the following accuracy restrictions hold:*

$$\| \tilde{g}_k - g(x_k) \| \le \theta_k \le \theta < 1/4\epsilon \quad ; \quad \theta_k \le \kappa \Delta_k$$

for all $k \ge 0$. Then, $\exists\, k_2$ so that:

$$I_\epsilon(x_k) = I_\epsilon(x') \qquad \forall\, k > k_2 ,$$

where x' is an arbitrary accumulation point of the iteration sequence $\{x_k\}$ from (MTR) when perturbed gradients are used within the iteration.

In analogy to [2] there has to be proved an auxiliary result on the sequence $\{I_\epsilon(x_k)\}$ first:

LEMMA 4 *Consider (MTR) using inexact gradient data for which the assumptions of Theorem 4 hold. Then*

$$\exists\, k_1 > 0 : \quad I_\epsilon(x_k) \subseteq I_\epsilon(x_C^{(k)}) \subseteq I_\epsilon(x_{k+1}) \qquad \forall\, k \ge k_1 .$$

Proof: To begin with let be given an arbitrary accumulation point x' of the iteration sequence.

From Lemma 3 we obtain: $\lim \sup h_k \leq \lim \sup \tilde{h}_k + \theta \leq 3\theta < (3/4)\,\epsilon$, consequently x' is "nearly"-critical and the (stable) Strict Complementarity Slackness Condition (21) holds.

Consider x with $\|x - x'\| < r_o = \epsilon/(4L+8)$, where L denotes the LIPSCHITZ constant for g on D:

$$|h(x)-h(x')| \leq \|\Pi_B(x-g(x)) - \Pi_B(x'-g(x'))\| + \|x-x'\| \leq (L+2)\,\|x-x'\| ,$$

so that $h(x) \leq h(x') + (L+2)\,r_o < \epsilon$.

We prove next that then $I_\epsilon(x) \subseteq I_\epsilon(x')$:

Assume the contrary, i.e. there exists $j \in I_\epsilon(x) \setminus I_\epsilon(x')$, then due to (21)

$$|[g(x)]_j| > \gamma_o > \epsilon_o ;$$

at the same time we have from (23) and Lemma 3

$$|[g(x')]_j| \leq h(x') < (3/4)\,\epsilon_o$$

so that

$$
\begin{aligned}
\|x - x'\| &\geq L^{-1}\,\|g(x) - g(x')\| \\
&\geq L^{-1}\,|[g(x)]_j - [g(x')]_j| \\
&> L^{-1}\,(\gamma_o - (3/4)\,\epsilon_o) > \epsilon_o/(4\,L) > r_o .
\end{aligned}
$$

This last inequality stands in contradiction to the choice of r_o.

Let's now return to the iteration $\{x_k\}$: The descent of $\{f_k\}$ guarantees that all iterates belong to the (compact) set D. Denote by X' the set of all accumulation points of $\{x_k\}$. Then for arbitrary $r > 0$ there exists $k = k(r)$ such that

$$dist\,(x_k, X') \leq r \qquad \forall\,k \geq k(r) .$$

Take $k_o = k(r_o)$, x_k such that $k \geq k_o$, then there exists a point $x'_{(k)} \in X'$ with

$\|x_k - x'_{(k)}\| \leq r_o$. But as we have shown in the first part for these points there are valid the relations

$$h(x_k) < \epsilon,\ h(x'_{(k)}) < \epsilon;\quad I_\epsilon(x_k) \subseteq I_\epsilon(x'_{(k)}) .$$

Remember now the construction of x_C: starting from the quadratic model function we ask for the minimum along the path $p(t)$, $t > 0$. Condition (22) together with the accuracy restriction leads to the conclusion that on this path the distance to ϵ-active bounds can become only smaller so that $I_\epsilon(x_k) \subseteq I_\epsilon(x_C)$ follows.

In the "correction step" of the algorithm (**MTR**) for the perturbed case there will be further accepted only such new iterates for which $I_\epsilon(x_C) \subseteq I_\epsilon(x_{k+1})$ (see the remark on the modification above) so that the monotonicity of the ϵ-active index sets will hold if only k is sufficiently large.   **q.e.d.**

Proof of Theor.4: From the last Lemma it follows that $\{I_\epsilon(x_k)\}$ and $\{I_\epsilon(x_k^C)\}$ tend to a (common) limit set I'.

We consider an arbitrary accumulation point x' of $\{x_k\}$ and denote by $\{k_l\}$ an index sequence with $\lim_{l\to\infty} x_{k_l} = x'$. Then as was shown above there exists l_o such that $I_\epsilon(x') \supseteq I_\epsilon(x_{k_l})$ for all $l > l_o$.

Suppose $I_\epsilon(x') \setminus I' = T \neq \emptyset$. We have

$$h(x') \leq 3/4\,\epsilon \quad ; \quad |\,[g(x')]_i| > \gamma_o \quad \text{for } i \in T\,.$$

Without loss of generality, let $i \in T$ be an index corresponding to an upper bound u_i, then $[g(x')]_i < -\gamma_o < -\epsilon$.

Due to the continuouity of g and h there exists an index l_1 such that for all $l > \max\{l_o, l_1\}$

$$\| x_{k_l} - x'\| \leq \epsilon/16\,; \quad [g(x_{k_l})]_i < -15/16\,\epsilon\,; \quad h(x_{k_l}) < 7/8\,\epsilon\,.$$

On the other hand, $i \notin I_\epsilon(x_{k_l})$ when l (resp. l_1) is sufficiently large; thus

$$u_i - \epsilon \geq [x_{k_l}]_i \geq [x']_i - \| x_{k_l} - x'\| > u_i - 17/16\,\epsilon\,;$$
$$[x_{k_l}]_i - [g_{k_l}]_i > u_i - 1/8\,\epsilon\,;$$
$$[\Pi_B(x_{k_l} - g_{k_l}) - x_{k_l}]_i > (u_i - 1/8\,\epsilon) - (u_i - \epsilon) = 7/8\,\epsilon\,;$$

and we obtain

$$h_{k_l} = \| \Pi_B(x_{k_l} - g_{k_l}) - x_{k_l}\| \geq [\Pi_B(x_{k_l} - g_{k_l}) - x_{k_l}]_i > 7/8\,\epsilon\,,$$

what contradicts the above inequality for h when only $l > l_1$.

Thus $I' = I_\epsilon(x')$ for arbitrary $x' \in X'$; taking into account that $\{I_\epsilon(x_k)\}$ is monotone increasing and all sets $I_\epsilon(x)$ are finite the desired result follows. **q.e.d.**

At the end we come to the stability result for the trust region algorithm in the bound constrained case. For shortness define the projection operator Π_ϵ,

$$[\Pi_\epsilon(x)]_j = \begin{cases} b_j & \text{if } j \in I_\epsilon(x) \\ [x]_j & \text{else} \end{cases}$$

where b_j is the ϵ-active bound corresponding to $j \in I_\epsilon(x)$. Further we will use the notations $z(x)$, $z_\epsilon(x)$ for projected gradients corresponding to I, I_ϵ as follows:

$$[z(x)]_j = \begin{cases} 0 & \text{if } j \in I(x) \\ [g_k]_j & \text{else} \end{cases} \quad ; \quad [z_\epsilon(x)]_j = \begin{cases} 0 & \text{if } j \in I_\epsilon(x) \\ [g_k]_j & \text{else} \end{cases} \quad ;$$

remark that this notation coincides with the previous definition of z if the point x is "nearly"-critical in the sense of (21).

Define $\bar{X}_B = \{ x \in D \cap B : h(x) = 0 \}$ – the set of the critical points related to problem (1). Obviously $X_* \subseteq \bar{X}_B$; $h(x_*) = \|z(x_*)\| = \|z_\epsilon(x_*)\| = 0$ for all local minimizers x_* of f on the set $D \cap B$.

THEOREM 5 *Suppose the function z to be* LIPSCHITZ*ian invertible on the set $\bar{X}_B$ in the following sense: there exist ρ, $\mu > 0$ such that*

$$\bar{x} \in \bar{X}_B \ \Rightarrow \ \|z(x) - z(\bar{x})\| \geq \mu \|x - \bar{x}\| \ .$$

for all x with $\|x - \bar{x}\| < \rho$ and $I(x) = I(\bar{x})$.
Then there exist $\theta_1 > 0$, $\epsilon_1 > 0$ with the property:
If $\epsilon < \epsilon_1$ and $\theta < \min\{\theta_1, \epsilon/4\}$ then every accumulation point x' of $\{x_k\}$ together with its projection $p' = \Pi_\epsilon(x')$ satisfy
(i) $\qquad f(p') \leq f(x')$;
(ii) $\qquad \exists \, \bar{x} \in \bar{X}_B : \qquad I(p') = I(\bar{x}) \quad and \quad \|p' - \bar{x}\| = O(\theta)$.
If the sequence $\{x_k\}$ from **(MTR)** *does not terminate, then for every accumulation point x' there exists a local minimizer of f on D, $\bar{x}_* \in X_*$, such that $\|p' - \bar{x}_*\| = O(\theta)$.*

Proof:
   part (i). For $p, x \in D$ the TAYLOR formula yields:

$$f(p) - f(x) = g(x)^T (p - x) + \int_0^1 (g(x + t(p - x)) - g(x))^T (p - x) \, dt \ .$$

Consider $p = p'$, $x = x'$ and denote $\eta = \max_{i \in I_\epsilon(x')} \{| b_i - [x']_i |\}$; then there are valid the relations (L – LIPSCHITZ constant of g):

$$\|x' - p'\|^2 = \sum_{i \in I_\epsilon} ([x']_i - b_i)^2 \leq n \, \eta^2 ;$$

$$\|g(x' + t(p' - x')) - g(x')\|^2 \leq L^2 \, n \, \eta^2$$

$$g(x')^T (p' - x') = \sum_{i \in I_\epsilon} [g(x')]_i \, (b_i - [x']_i) \leq -\gamma_0 \, \eta \, ,$$

Therefore, the function values satisfy

$$f(p') - f(x') \leq \eta \, (-\gamma_0 + n \, L \, \epsilon) < 0$$

if only $\epsilon < \epsilon' = \gamma_0/(nL)$.

   part (ii). Without loss of generality assume: $\rho < \gamma_0/L$. First, taking θ_1 sufficiently small, we can guarantee $x' \in \bar{X}_B + 0.5 \, \rho \, U$: indeed, if

$W = D \cap B \backslash (\bar{X}_B + 0.5\,\rho\, U)$ then $\min_W h \geq h_o > 0$, so that due to Lemma 3 $x' \notin W$ if $\theta < h_o/4$.

Thus there $\exists\, \bar{x} \in \bar{X}_B$ with $\| x' - \bar{x} \| < 0.5\,\rho$.

Secondly, the set $\bar{X}_B$ under the smoothness assumptions on f and for D – compact is a compact set; due to the strict complementarity slackness condition (21) then there exists a constant $R > 0$ such that for every element of $\bar{X}_B$ the distance to inactive bounds is greater R. Let $\rho < R$, then the distance of x' to bounds inactive in $\bar{x}$ is greater than $(R - 0.5\rho) > 0.5\rho$. Thus, if we suppose $\epsilon < \epsilon'' = 0.5\,\rho$, it follows that $I_\epsilon(x') \subseteq I(\bar{x})$.

Following the arguments of the proof of Theor. 4 we can show that they coincide (indeed, for $j \in I(\bar{x}) \backslash I_\epsilon(x')$ we have: $|[\bar{g}]_j| > \gamma_o$; $|[g']_j| < h' < \epsilon$, what due to the LIPSCHITZ continuity of g and the choice of ρ leads to a contradiction).

Consequently, $I(p') = I_\epsilon(x') = I(\bar{x})$.

Using the just found index set relation we have $\| p' - \bar{x} \| \leq \| \bar{x} - x' \| < 0.5\,\rho$; it follows that

$$\mu \| p' - \bar{x} \| \leq \| z(p') - z(\bar{x}) \| = \| z(p') \| .$$

Further,

$$
\begin{aligned}
\| z(p') \| = \| z_\epsilon(p') \| &\leq \| z_\epsilon(x') \| + \| z_\epsilon(p') - z_\epsilon(x') \| \\
&\leq \| z_\epsilon(x') \| + \| g(p') - g(x') \| \qquad (24) \\
&\leq \| z_\epsilon(x') \| + L \| p' - x' \| .
\end{aligned}
$$

Consider the vector $y(x') = (\Pi_B(x' - g(x')) - x')$:

For $i \notin I_\epsilon(x')$:

$$| [x']_i - b_i | > 0.5\,\rho ; \qquad | [y(x')]_i | \leq \| y(x') \| = h(x') < 3\,\epsilon/4 ,$$

so that

$$[\Pi_B(x' - g')]_i = [x']_i - [g']_i ; \quad | [y(x')]_i | = | [g(x')]_i | = | [z_\epsilon(x')]_i | .$$

For $i \in I_\epsilon(x')$:

$$| [x']_i - b_i | < \epsilon , \qquad | [g']_i | > \gamma_o > \epsilon ,$$

taking into account that the sign of the gradient component further must coincide (for ρ - small enough) with its sign in $\bar{x}$, we get

$$[\Pi_B(x' - g')]_i = b_i ; \qquad [y(x')]_i = [p' - x']_i .$$

As a result we come to the formula

$$h^2(x') = \| y(x') \|^2 = \sum_{i \notin I_\epsilon} \left([g(x')]_i \right)^2 + \sum_{i \in I_\epsilon} \left([x']_i - b_i \right)^2$$

$$= \| z_\epsilon(x') \|^2 + \| p' - x' \|^2$$

from which we obtain $\| z_\epsilon(x') \| \le h(x') \le 3\theta$ and $\| p' - x' \| \le h(x') \le 3\theta$. Now from (24) we may conclude that

$$\| p' - \bar{x} \| \le (3/\mu)(1 + L)\theta,$$

i.e. the assertion **(ii)** of the Theorem is valid.

The fact that the projected point $\Pi_\epsilon(x')$ represents an approximation of an critical point not worse than x' leads to the idea to consider in the final phase the sequence $\{\Pi_\epsilon(x_k)\}$: For sufficiently large k these points all are located on the same face of B as their accumulation point p'; further due to the descent of the function values $\{f_k\}$ and the property **(i)** the projected sequence may not approximate a local maximum point of f w.r.t. the ϵ-active face related to $\{x_k\}$, x' resp. On the other hand the LIPSCHITZian invertibility of the projected gradient vectors z yields that x' may be either a local (subspace-) minimizer or a local maximizer; cf. [14]. In analogy to the proof of Theor. 3 the second case may be excluded for the algorithm under consideration.   **q.e.d.**

If we assume that the algorithm works with asymptotically exact data $\theta_k \to 0$ and $\epsilon \to 0$ the (real) active index set of the limit point is obtained. The final phase of the algorithm then may be considered as an unconstrained optimization method with respect to a subspace of lower dimension. For this situation the attraction speed may be estimated in accordance to Theorem 3: if f is a smooth function and locally convex near the limit point the algorithm is exponentially stable (cf. sect.4).

Acknowlegdement: First of all I would like to thank Professor A. V. Fiacco for his attention to algorithmic stability topics. I am also grateful to A.Griewank for his comments concerning several forms of convexity assumptions.

REFERENCES

[1] Byrd R.H., Nocedal J., Yuan Ya-Xiang Global convergence of quasi-Newton methods on convex problems, *SIAM J.Numer.Anal.*, *24(5)*:1171-1190 (1987).

[2] Conn A.R., Gould N.I.M., Toint Ph.L. Global convergence of a class of trust region algorithms for optimization problems with simple bounds, *SIAM J. Numer.Anal.*, *25*:433-460 (1988).

[3] Conn A.R., Gould N.I.M., Toint Ph.L. Testing a class of methods for solving minimization problems with simple bounds on the variables, *Math. Comp.*, *50(182)*:399-430 (1988).

[4] Conn A.R., Gould N.I.M., Toint Ph.L. LANCELOT: A Fortran Package for large-scale nonlinear optimization (Release A), *Series in Computat. Mathematics 17* Springer-V., New York (1992).

[5] Dennis jun. J.E., Walker H.F. Inaccuracy in quasi-Newton methods: local improvement theorems, *Math. Program. Study 22*:70-85 (1984).

[6] Dennis jun. J.E., Walker H.F. Least change sparse secant update methods with inaccurate secant conditions , *SIAM J.Numer.Anal. 22(4)*:760-778 (1985).

[7] Evtushenko Yu.G. Metody reshenia ekstremal'nykh zadach i ikh primenenie v sistemakh optimizacii, Moskva, Nauka (1982).

[8] Evtushenko Yu.G., Zhadan V.G. Primenenie metoda funkcij Ljapunova dl'a issledovania skhodimosti chislennykh metodov, *Zhurnal vych. matem.i matem fiziki 15(1)*:101-112 (1975).

[9] Felgenhauer U. Influence of gradient approximation errors on the convergence rate of quasi-Newton algorithms, *Preprint Math 2/88, TU Magdeburg* (1988).

[10] Felgenhauer U. On the stable global convergence of particular quasi-Newton methods, *optimization, 26*:97-113 (1992).

[11] Felgenhauer U. Stabilizing properties of descent step lengths, *Proc. "Parametric Optimization and Related Topics III", Güstrow 1991, eds. F.Deutsch, B.Brosowski, J.Guddat; Ser. Approximation and Optimization*, Peter Lang Publ. House Frankfurt; 3:135-146 (1993).

[12] Fontecilla R. Inexact secant methods for nonlinear constrained optimization, *SIAM J. Numer. Anal., 27(1)*:154-165 (1990).

[13] Griewank A. The "global"' convergence of Broyden-like methods with a suitable line search, *J.Austr.Math.Soc., Ser. B, 28*:75-92 (1986).

[14] Griewank A., Jongen H.Th., Kwong M.K. The equivalence of strict convexity and injectivity of the gradient in bounded level sets, *Math. Progr. 51*:273-278 (1991).

[15] Griewank A., Toint Ph. L. Local convergence analysis for partitioned quasi-NEWTON updates , *Numer. Math. 39*:429-448 (1982).

[16] Hahn W. Stability of motion, Springer-V., Berlin-Heidelberg-New York (1967).

[17] Hiriart-Urruty J.-B., Lemarechal C. Convex analysis and minimization algorithms I, *Grundlehren der mathemat. Wissensch. 305*, Springer-V., Berlin-Heidelberg-New York (1993).

[18] Kovács M., Vasil'ev F.P. Some convergence theorems on nonstationary minimization processes, *optimization, 15(2)*:203-210 (1984).

[19] Kovács M., Vasil'ev F.P. The convergence rate of the continuous versions of the regularized gradient method, *optimization, 18(5)*:689-696 (1987).

[20] Kovács M., Vasil'ev F.P. Convergence rate of the regularized barrier function method, *optimization, 22(3):*427-438 (1991).

[21] Lukšan L. Inexact trust region method for large sparse systems of nonlinear equations, *J. Optim. Theory Applic., 18(3):*569-590 (1994).

[22] Moré J.J. Recent developments in algorithms and software for trust region methods, *Mathematical Programming. The State of the Art. Bonn 1982* eds.: Bachem A., Grötschel M., Korte B., Springer-V. Berlin 258-287 (1983).

[23] Pol'jak B.T. Skhodimost i skorost skhodimosti iterativnykh stokhasticheskikh algoritmov, *Avtomatika i telemekhanika, part I: 12:*83-94 (1976); *part II: 4:*71-80 (1977).

[24] Powell M.J.D. Some global convergence properties of variable metric algorithms without exact line search, *Nonlinear Programming* ed. Cottle R.W., Lemke C.E., SIAM-AMS Proc., IX, AMS Providence, (1976).

[25] Powell M.J.D. Convergence properties of a class of optimization algorithms, *Nonlinear Programming 2,* eds. Mangasarian O.L., Meyer R.R., Robinson S.M., Academic Press (1975).

[26] Powell M.J.D. On the global convergence of trust region algorithms for unconstrained optimization, *Math. Programming 29:*297-303 (1984).

[27] Rockafellar R.T. Convex analysis, Princeton University Press (1970).

[28] Schwetlick H. Numerische Lösung nichtlinearer Gleichungen , Dtsch. Verlag d. Wissensch., Berlin (1979).

[29] Stetter H.J. Analysis of discretization methods for ordinary differential equations, Springer-Verlag Berlin -Heidelberg - New York, (1973); (in Russ.: Moskva, Mir, (1978)).

[30] Venec V.I., Rybashov M.V. Metod funkcij Ljapunova v issledovanii nepreryvnykh algoritmov matematicheskogo programmirovania, *Zhurnal vych. matem.i mat. fiziki, 17(3):*622-633 (1977).

[31] Warth W., Werner J. Efficiente Schrittweitenfunktionen bei unrestringierten Optimierungsaufgaben, *Computing 19:*59-72 (1977).

[32] Zavriev S.K. On the global optimization properties of finite-difference local descent algorithms, *Journal of Global Optimization, 3:*67-78 (1993).

A Note on Using Linear Knowledge to Solve Efficiently Linear Programs Specified with Approximate Data

SHARON FILIPOWSKI The Boeing Company, PO Box 3707, MS 7L-20, Seattle, WA 98124-2207

Abstract. An algorithm that solves symmetric linear programs specified with approximate data is given. The algorithm uses knowledge that the coefficients of the actual (unknown) instance satisfy a system of linear equations to reduce the data precision necessary to solve the instance in question. In some cases, problem instances that would otherwise require perfect precision to solve, called *ill-posed* problem instances, can now be solved without perfect precision because of the use of the knowledge. The algorithm is computationally efficient and, for certain types of linear knowledge, requires not much more data accuracy than the minimum amount necessary to solve the actual instance. Furthermore, for these certain types of linear knowledge, the complexity bounds can be expressed in terms of a condition measure and an angle between the subspace of problem instances that satisfy the linear knowledge and a certain subspace of problem instances. In some cases, this certain subspace corresponds to the set of ill-posed problem instances.

1. Introduction

Renegar [4, 5] developed a complexity theory that allows for real and approximate data while still maintaining finite precision calculations. The efficiency of an algorithm in solving a problem instance is measured in terms of a *condition measure*, a quantity that reflects the difficulty of the problem instance to be solved. Vera [6, 7] also has contributed to the development of this new complexity theory.

Our work in [1, 2, 3] took a step forward by allowing for the use of *knowledge* in this new complexity theory. An example of knowledge, and one that is considered in [3], is that it is known that the linear program to be solved is feasible before computations begin. Furthermore, another type of knowledge that was considered in [2] is that some of the constraint matrix coefficients are known to be equal to zero before computations begin. (We briefly discuss the work in [2] as well as that of Renegar [5] and Vera [6] at the end of this section.) In this paper we assume it is known that the coefficients of the actual instance satisfy linear knowledge (i.e., a system of *linear* equations) before computations begin.

We now discuss what it means to *solve* a linear program specified with approximate data and known to satisfy linear knowledge before computations begin. We consider linear programs in symmetric form so that the actual (unknown) instance can be written:

$$\max \ c^T x$$
$$Ax \leq b$$
$$x \geq 0,$$

where $A \in I\!\!R^{m \times n}$, $b \in I\!\!R^m$, $c \in I\!\!R^n$, and $x \in I\!\!R^n$. The dual linear program can be written:

$$\min \ b^T y$$
$$A^T y \geq c$$
$$y \geq 0,$$

where $y \in I\!\!R^m$.

Let $d = (A, b, c) \in I\!\!R^{mn+m+n}$ denote the data vector of the actual (unknown) instance. Because the solution set of a linear program is invariant under positive scaling, we assume that $\|d\| = 1$, where

$$\|d\| \equiv \max_i (|d_i|),$$

as throughout this paper.

We also assume that a rational approximation $\bar{d} = (\bar{A}, \bar{b}, \bar{c})$ to the actual instance that satisfies the linear knowledge is given along with a rational error bound $\bar{\delta}$. Thus we have *approximate data* $(\bar{d}, \bar{\delta})$ satisfying both $\|\bar{d} - d\| < \bar{\delta}$ and the linear knowledge (i.e., both d and $\bar{d}$ satisfy $Bd = B\bar{d} = 0$ for some matrix $B \in I\!\!R^{k \times (mn+m+n)}$, for some $k \geq 0$).

An algorithm *solves a linear program specified with approximate data and known to satisfy linear knowledge before computations begin* if, given the approximate data $(\bar{d}, \bar{\delta})$ and the linear knowledge, it correctly replies with one of the following statements:

- The actual instance is infeasible.

- The actual instance is unbounded.

- The actual instance has an optimal solution, and $\bar{x} \in I\!\!R^n$ and $\bar{\epsilon} \in [0, \infty)$ are guaranteed to satisfy $\bar{x} \in \{ \tilde{x} : \|\tilde{x} - x^*\| \leq \bar{\epsilon}$ where x^* solves $\max c^T x$ such that $Ax \leq b$, $x \geq 0 \}$. (We define $\bar{x}$ to be an $\bar{\epsilon}$−approximate solution.)

- Better data accuracy is needed.

  (The algorithm cannot determine that the actual instance is infeasible, determine that the actual instance is unbounded, or produce an $\bar{\epsilon}$-approximate solution $\bar{x}$ for any $\bar{\epsilon} \in [0, \infty)$, given the approximate data $(\bar{d}, \bar{\delta})$ and the linear knowledge.)

In this approximate data setting, for an algorithm to be able to reply that the actual instance is infeasible, it must be able to guarantee that all instances $\tilde{d}$ satisfying both $\|\bar{d} - \tilde{d}\| < \bar{\delta}$ and the linear knowledge are infeasible. A similar statement holds for an algorithm to be able to reply that the actual instance is unbounded. For an algorithm to be able to give $\bar{x}$ as an $\bar{\epsilon}$-approximate solution, it must be able to guarantee that all instances $\tilde{d}$ satisfying both $\|\bar{d} - \tilde{d}\| < \bar{\delta}$ and the linear knowledge have an optimal solution and that $\bar{x}$ serves as an $\bar{\epsilon}$-approximate solution for all such instances.

Because an algorithm is considering fewer instances when using the linear knowledge, it is hoped that the use of the knowledge will lessen the amount of data accuracy necessary for an algorithm to be able to solve the instance in question. Also, in some cases, problem instances that would otherwise require prefect precision to solve, called *ill-posed* problem instances, can now be solved with approximate data because of the use of the linear knowledge. (The use of the linear knowledge to solve ill-posed problem instances is discussed in more detail in Section 4.2)

We now discuss what it means for an algorithm to be *efficient* in solving a linear program specified with approximate data and known to satisfy linear knowledge before computations begin. We refer the reader to the paper by Renegar [5], where the foundation for this new complexity theory is given, for a more thorough discussion.

As first discussed by Renegar [5], an algorithm is said to be *fully efficient* if it is both *computationally efficient* and *data efficient*. An algorithm is said to be computationally efficient if it runs in polynomial-time in the bit length of the rational approximate data $(\bar{d}, \bar{\delta})$. An algorithm is said to be data efficient if it uses *nearly minimal data precision*. We now discuss data efficiency.

For each instance d, desired solution accuracy $\epsilon \in [0, \infty)$, and linear knowledge LK, there is a minimum perturbation size necessary such that no algorithm is able correctly to reply that the actual instance is infeasible, reply that the actual instance is unbounded, or provide an ϵ-approximate solution when given any approximate data that has error bound that is strictly larger than this minimum perturbation size. Denote this minimum perturbation size by $\delta(d, \epsilon, LK)$.

In defining what it means for an algorithm to be data efficient, Renegar [5] used this minimum perturbation size and introduced the notion of a *condition measure*. The condition measure for the instance d, desired solution accuracy ϵ, and linear knowledge LK is

$$C(d, \epsilon, LK) \equiv \frac{\|d\|}{\delta(d, \epsilon, LK)} = \frac{1}{\delta(d, \epsilon, LK)},$$

assuming that $\|d\| = 1$. Roughly, any algorithm will require at least $\log[C(d, \epsilon, LK)]$ relative bits of accuracy to solve the instance d. Therefore, this condition measure reflects the difficulty of correctly replying that the actual instance is infeasible, replying that the actual instance is unbounded, or providing an ϵ-approximate solution.

Finally, an algorithm is said to be data efficient if there exist polynomials $p(m, n)$, $q(m, n)$, $r(m, n)$, and $t(m, n)$ in the variables m and n that are independent of the problem instance, desired solution accuracy, and linear knowledge such that the algorithm is guaranteed correctly to reply that the actual instance is infeasible, reply that the actual instance is unbounded, or give a $q(m, n)\epsilon$-approximate solution, when provided with approximate data that has error bound satisfying

$$\bar{\delta} \leq (\frac{1}{C(d,\epsilon,LK)})^{r(m,n)}(\frac{1}{p(m,n)^{t(m,n)}}) = \frac{(\delta(d,\epsilon,LK))^{r(m,n)}}{p(m,n)^{t(m,n)}}.$$

Therefore, the algorithm is guaranteed to stop with a correct answer when provided with only linearly more, in terms of $\log[C(d,\epsilon,LK)]$, bits of accuracy than the minimum amount necessary. (It is assumed that all polynomials $k(m,n)$, in the variables m and n, in this paper satisfy $k(m,n) \geq 1$ for all $m,n \geq 1$.)

We now discuss the importance of linear knowledge. In this new complexity theory, transformations of linear programs do not exist such that one algorithm for one form of a linear program provides a fully efficient algorithm for a linear program in any form. As a result, an algorithm needs to be constructed for each type of linear program.

Consider the transformation in traditional complexity theory that converts a linear program in standard form

$$\max \ c^T x$$
$$Ax = b$$
$$x \geq 0,$$

into a linear program in symmetric form. After the transformation, the new linear program to be solved is

$$(\tilde{P}) \ \max \ c^T x$$
$$\tilde{A}x \leq \tilde{b}$$
$$x \geq 0$$

where $\tilde{b} \in I\!\!R^{2m}$ and $\tilde{A} \in I\!\!R^{2m \times n}$ satisfy:

$$\tilde{b} \equiv \begin{bmatrix} b \\ -b \end{bmatrix} \quad \tilde{A} \equiv \begin{bmatrix} A \\ -A \end{bmatrix}.$$

If x^* is a solution to $(\tilde{P})$, it is also a solution to the original linear program.

At first it might seem that this transformation can be used in this new complexity theory as well. However, when given approximate data, to consider only those linear programs considered when using the standard form, we have to consider only perturbations of $(\tilde{P})$ where the new constraint matrix $\hat{A}$ has the same structure as $\tilde{A}$ and where the new right-hand side vector $\hat{b}$ has the same structure as $\tilde{b}$. In particular, we need to consider only those perturbations such that $(\hat{b}_i + \Delta \hat{b}_i) + (\hat{b}_{m+i} + \Delta \hat{b}_{m+i}) = 0$ for $1 \leq i \leq m$ and $(\hat{a}_{i,j} + \Delta \hat{a}_{i,j}) + (\hat{a}_{i+m,j} + \Delta \hat{a}_{i+m,j}) = 0$ for $1 \leq i \leq m$ and $1 \leq j \leq n$.

Therefore, we will have to consider only those perturbations that satisfy *linear knowledge* of the form $Bd = 0$ where B is a matrix describing the linear knowledge that the actual instance satisfies. In particular, to describe the linear knowledge used for $(\tilde{P})$, the first row of B would consist of two 1's in the appropriate places to describe that $(\tilde{b}_1 + \Delta \tilde{b}_1) + (\tilde{b}_{m+1} + \Delta \tilde{b}_{m+1}) = 0$.

This example provides some reasons why the study of linear knowledge is important: it might provide a more unified approach toward solving problems in this new complexity theory, and it might give some insight into the reason why transformations of linear programs do not exist such that one algorithm provides a fully efficient for all linear programs.

In this paper we consider only special cases of linear knowledge. (These special cases are discussed in more detail in the next section.) In particular, we do not consider the linear knowledge that characterizes the transformation of a linear program in standard form to a linear program in symmetric form, as discussed above. As a result, the work in this paper gives only a start to the consideration of all types of linear knowledge.

Finally, we briefly discuss some previous work. We refer the reader to the papers by Renegar [4, 5] as well as to the introductions in [6, 7] and [1] for a more thorough introduction to and discussion of this new complexity theory.

In [5] Renegar considers solving symmetric linear programs specified with approximate data in the case that there is no knowledge. Solving linear programs specified with approximate data when there is no knowledge is the same as when there is linear knowledge; however, an algorithm must work differently with the approximate data. For example, because there is no knowledge, for an algorithm to be able to reply correctly that the actual instance is infeasible when given approximate data $(\bar{d}, \bar{\delta})$, it must be able to guarantee that *all* instances $\tilde{d}$ satisfying $\|\bar{d} - \tilde{d}\| < \bar{\delta}$ are infeasible. A similar statement holds for an algorithm to be able to reply that the actual instance is unbounded and for an algorithm to be able to provide an ϵ-approximate solution.

Similar to the case when there is linear knowledge, for each instance d and desired solution accuracy $\epsilon \in [0, \infty)$, there is a minimum perturbation size necessary such that no algorithm is able correctly to reply that the actual instance is infeasible, reply that the actual instance is unbounded, or provide an ϵ-approximate solution when provided with approximate data that has error bound that is strictly larger than this minimum perturbation size. Denote this minimum perturbation size by $\delta(d, \epsilon)$.

Finally, Renegar [5] gave a computationally efficient algorithm that is guaranteed correctly to reply that the actual instance is infeasible, reply that the actual instance is unbounded, or give an ϵ-approximate solution when provided with approximate data that has error bound satisfying

$$\bar{\delta} \leq \frac{\delta(d, \epsilon)^6}{\tilde{p}(m, n)},$$

where $\tilde{p}(m, n)$ is a polynomial in the variables m and n that is independent of the actual instance and desired solution accuracy.

Vera [6, 7] extended Renegar's results to provide a fully efficient algorithm that solves linear programs in inequality form:

$$\max c^T x$$
$$Ax \leq b$$

and to provide a fully efficient algorithm that solves linear programs in standard form:

$$\max c^T x$$

$$Ax = b$$
$$x \geq 0.$$

We now discuss the work in [2] where some of the constraint matrix coefficients of the actual instance are known to be equal to zero before computations begin. We call this *sparsity knowledge*. Again, solving the actual instance is the same as when there is linear knowledge. Furthermore, for each instance d, desired solution accuracy $\epsilon \in [0, \infty)$, and sparsity knowledge s, there is a minimum perturbation size necessary such that no algorithm is able correctly to reply that the actual instance is infeasible, reply that the actual instance is unbounded, or give an ϵ-approximate solution when provided with any approximate data that has error bound that is strictly larger than this minimum perturbation size. Denote this minimum perturbation size by $\delta(d, \epsilon, s)$.

The main theorem is that there exist a computationally efficient algorithm and polynomials $p_s(m, n)$ and $t_s(m, n)$ in the variables m and n that are independent of the actual instance, desired solution accuracy, and sparsity knowledge such that the algorithm is guaranteed correctly to reply that the actual instance is infeasible, reply that the actual instance is unbounded, or give a 2ϵ-approximate solution when provide with approximate data that has error bound satisfying

$$\bar{\delta} \leq \frac{(\delta(d, \epsilon, s))^{6mn}}{p_s(m, n)^{t_s(m, n)}}.$$

2.   The Algorithm

The algorithm will be given after some definitions are made and some lemmas are stated. For a particular instance $d = (A, b, c)$, let Feas(d) denote the feasible region for the instance d, let DualFeas(d) denote the feasible region for the dual of the instance d, let Opt(d) denote the set of optimal solutions for the instance d, and let DualOpt(d) denote the set of optimal solutions for the dual of the instance d. Also, if d has an optimal solution, let $k(d)$ denote the optimal objective function value for the instance d. That is,

$$\text{Feas}(d) \equiv \{ x : Ax \leq b, \ x \geq 0 \},$$

$$\text{DualFeas}(d) \equiv \{ y : A^T y \geq c, \ y \geq 0 \},$$

$$\text{Opt}(d) \equiv \{ x^* : x^* \in \text{Feas}(d) \text{ and } c^T x^* \geq c^T x \text{ for all } x \in \text{Feas}(d) \},$$

$$\text{DualOpt}(d) \equiv \{ y^* : y^* \in \text{DualFeas}(d) \text{ and } b^T y^* \leq b^T y \text{ for all } y \in \text{DualFeas}(d) \},$$

and

$$k(d) \equiv \max\{ c^T x : Ax \leq b, \ x \geq 0 \}.$$

As mentioned in the introduction, let $B \in \mathbb{R}^{k \times (mn+m+n)}$ denote a matrix that defines the linear knowledge that the actual instance d satisfies, for some $k \geq 0$. Also, let $e_i \in \mathbb{R}^{mn+m+n}$ be the ith unit vector for i satisfying $1 \leq i \leq mn + m + n$, and consider the following $mn + m + n$ linear programs, in the variables $\Delta d \in \mathbb{R}^{mn+m+n}$, for i satisfying $1 \leq i \leq mn + m + n$:

$$(P_i) \quad \max \quad e_i^T \Delta d$$
$$B \Delta d = 0$$
$$-1 \leq \Delta d_j \leq 1 \text{ for } 1 \leq j \leq mn + m + n.$$

Let $\bar{\Delta}d \equiv (\bar{\Delta}d_A, \bar{\Delta}d_b, \bar{\Delta}d_c) \in (\mathbb{R}^{mn}, \mathbb{R}^m, \mathbb{R}^n)$, where $\bar{\Delta}d_A \in \mathbb{R}^{mn}$ consists of the optimal objective function values for the mn linear programs corresponding to the coefficients for the constraint matrix A, $\bar{\Delta}d_b \in \mathbb{R}^m$ consists of the optimal objective function values for the m linear programs corresponding to the coefficients for the right-hand side vector b, and $\bar{\Delta}d_c \in \mathbb{R}^n$ consists of the optimal objective function values for the n linear programs corresponding to the coefficients for the objective function vector c.

Then, given approximate data $(\bar{d}, \bar{\delta})$ and the linear knowledge, define

$$\bar{d}^+ \equiv (\bar{A} - \bar{\delta}\bar{\Delta}A, \ \bar{b} + \bar{\delta}\bar{\Delta}d_b, \ \bar{c} + \bar{\delta}\bar{\Delta}d_c)$$

$$\bar{d}^- \equiv (\bar{A} + \bar{\delta}\bar{\Delta}A, \ \bar{b} - \bar{\delta}\bar{\Delta}d_b, \ \bar{c} - \bar{\delta}\bar{\Delta}d_c),$$

where $\bar{\Delta}A \in \mathbb{R}^{m \times n}$ is the constraint matrix corresponding to $\bar{\Delta}d_A$.

The algorithm uses the following lemmas that are similar to results of Renegar [4].

Lemma 2.1. *Given approximate data $(\bar{d}, \bar{\delta})$, $Feas(\bar{d}^-) \subseteq Feas(\tilde{d}) \subseteq Feas(\bar{d}^+)$ for all $\tilde{d}$ satisfying both $\|\tilde{d} - \bar{d}\| \leq \bar{\delta}$ and the linear knowledge.*

Because of the symmetry of the linear programs, a similar result holds for the feasible regions for the dual linear programs.

Lemma 2.2. *Given approximate data $(\bar{d}, \bar{\delta})$, $DualFeas(\bar{d}^+) \subseteq DualFeas(\tilde{d}) \subseteq DualFeas(\bar{d}^-)$ for all $\tilde{d}$ satisfying both $\|\tilde{d} - \bar{d}\| \leq \bar{\delta}$ and the linear knowledge.*

The following lemma claims that given approximate data $(\bar{d}, \bar{\delta})$, if $\bar{d}^-$ has an optimal solution, then a portion of the feasible region for the instance $\bar{d}^+$ contains all optimal solutions for all instances $\tilde{d}$ satisfying both $\|\tilde{d} - \bar{d}\| \leq \bar{\delta}$ and the linear knowledge.

Lemma 2.3. *Assume that $\bar{d}^-$ has an optimal solution. If $\tilde{d}$ satisfies both $\|\tilde{d}-\bar{d}\| \leq \bar{\delta}$ and the linear knowledge, then $\tilde{x} \in \mathrm{Opt}(\tilde{d})$ implies that $\tilde{x} \in \{ x : (\bar{c}+\bar{\delta}\bar{\Delta}d_c)^T x \geq k(\bar{d}^-)\} \cap Feas(\bar{d}^+)$.*

Furthermore, let $e_i \in \mathbb{R}^n$ denote the ith unit vector in $\mathbb{R}^n$ for i satisfying $1 \leq i \leq n$. Assuming that the feasible region for an instance $\tilde{d}$ is bounded, let $cen_\infty(Feas(\tilde{d}))$ denote the *infinity center*, defined in terms of the infinity norm, of the feasible region for the instance $\tilde{d}$. That is,

$$(\mathrm{cen}_\infty(\mathrm{Feas}(\tilde{d})))_i \equiv \frac{1}{2}(v_i^+(\tilde{d}) + v_i^-(\tilde{d})),$$

with

$$v_i^+(\tilde{d}) \equiv \max\{ e_i^T x \ : \ \tilde{A}x \leq \tilde{b}, \ x \geq 0 \}$$

and

$$v_i^-(\tilde{d}) \equiv \min\{\ e_i^T x \ : \ \tilde{A}x \le \tilde{b}, \ x \ge 0 \ \}.$$

Also, if the feasible region for an instance $\tilde{d}$ is bounded, let $rad_\infty(Feas(\tilde{d}))$ denote the *infinity radius*, measured again in terms of the infinity norm, of the feasible region for the instance $\tilde{d}$. That is,

$$\mathrm{rad}_\infty(\mathrm{Feas}(\tilde{d})) \equiv \max\{\ \frac{1}{2}(v_i^+(\tilde{d}) - v_i^-(\tilde{d})) \ : \ 1 \le i \le n\}.$$

Finally, if the feasible region of an instance $\tilde{d}$ is unbounded, the infinity center of that region is undefined, and we define the infinity radius of that region to be ∞.

We now give the algorithm. The algorithm is similar to Algorithm 2.4 in [2], where a sketch can be found.

Algorithm 2.4.

(0) *The algorithm assumes that $(\bar{d}, \bar{\delta})$ and the linear knowledge are given and that $\bar{d}$ satisfies the linear knowledge.*

(1) *Check if $\bar{d}^+$ is infeasible. If so, **STOP**; the actual instance is infeasible.*

(2) *Check if $\bar{d}^-$ is feasible. If not, **GOTO** (6).*

(3) *Check if $\bar{d}^-$ is dual infeasible. If so, **STOP**; the actual instance is unbounded.*

(4) *Check if $\bar{d}^+$ is dual feasible. If not, **GOTO** (6).*

(5) *Check if $\bar{\epsilon} = \mathrm{rad}_\infty(Feas(\bar{d}^+) \cap \{\ x : (\bar{c} + \bar{\delta}\bar{\Delta}d_c)^T x \ge k(\bar{d}^-)\}) < \infty$. If so, **STOP**; $cen_\infty(Feas(\bar{d}^+) \cap \{\ x : (\bar{c} + \bar{\delta}\bar{\Delta}d_c)^T x \ge k(\bar{d}^-)\})$ is an $\bar{\epsilon}-$approximate solution.*

(6) *Better data accuracy is needed.*

3. Efficiency of the Algorithm

The algorithm is computationally efficient because it relies just on linear programming (i.e., we assume that the algorithm uses a polynomial-time linear programming algorithm to solve all linear programs). The remainder of this section is devoted to showing that the algorithm is data efficient and, hence, fully efficient.

Before proving that the algorithm is data efficient, we need some definitions. Assuming that the actual instance is primal feasible. let $\delta_p'(d, LK)$ denote the *linear knowledge distance* (i.e., the distance using only perturbations that satisfy the given linear knowledge) of the actual instance to the set of primal infeasible instances. This distance can be written:

$$\delta_p'(d, LK) \equiv \sup\{\ \delta : \|d - \tilde{d}\| < \delta \text{ and } \tilde{d} \text{ satisfies the linear knowledge imply that } \mathrm{Feas}(\tilde{d}) \ne \emptyset\}.$$

If d is primal infeasible, let $\delta_p'(d, LK)$ denote the linear knowledge distance of the actual instance to the set of primal feasible instances. Also, if d is dual feasible, let $\delta_d'(d, LK)$ denote the linear knowledge distance of the actual instance to the set of dual infeasible instances, and if d is dual infeasible, let $\delta'_d(d, LK)$ denote the linear knowledge distance of the actual instance to the set of dual feasible instances. These distances can be written in a similar way. Finally, let

$$\delta'(d, LK) \equiv \min\{ \delta'_p(d, LK),\ \delta'_d(d, LK)\}.$$

Assuming that the actual instance d has an optimal solution, for a solution accuracy $\epsilon \in [0, \infty)$, there is a minimum perturbation size necessary such that there does not exist an ϵ-approximate solution for all instances $\tilde{d}$ satisfying both $\|\tilde{d} - d\| < \delta$ and the linear knowledge for any δ strictly larger than this minimum perturbation size. This minimum perturbation size can be written:

$$\delta(d, \epsilon, LK) \equiv \sup\{ \delta : \exists\, \bar{x} \in I\!\!R^n \text{ such that } \|d - \tilde{d}\| < \delta \text{ and } \tilde{d} \text{ satisfies the linear knowledge}$$
$$\text{imply that } \exists\, \tilde{x} \in \text{Opt}(\tilde{d}) \text{ satisfying } \|\bar{x} - \tilde{x}\| \le \epsilon\}.$$

Assuming that the actual instance is infeasible, there also is a minimum perturbation size necessary such that not all instances $\tilde{d}$ satisfying both $\|\tilde{d} - d\| < \delta$ and the linear knowledge are infeasible for any δ strictly larger than this minimum perturbation size. This minimum perturbation size is independent of ϵ and is also denoted by $\delta(d, \epsilon, LK)$. Clearly, in this case, $\delta(d, \epsilon, LK) = \delta'_p(d, LK)$.

Furthermore, assuming that the actual instance is unbounded, there also is a minimum perturbation size necessary such that not all instances $\tilde{d}$ satisfying both $\|\tilde{d} - d\| < \delta$ and the linear knowledge are unbounded for any δ strictly larger than this minimum perturbation size. This minimum perturbation size is independent of ϵ, is also denoted by $\delta(d, \epsilon, LK)$, and can be written:

$$\delta(d, \epsilon, LK) \equiv \sup\{ \delta : \|d - \tilde{d}\| < \delta \text{ and } \tilde{d} \text{ satisfies the linear knowledge imply that } \tilde{d} \text{ is}$$
$$\text{unbounded }\}.$$

For the remainder of this paper, we consider only certain types of linear knowledge. First, we assume that the linear knowledge is *separate* for the constraint matrix A, the right-hand side vector b, and the objective function vector c. By separate we mean that the matrix B describing the linear knowledge has the following form:

$$B = \begin{bmatrix} B_A & 0 & 0 \\ 0 & B_b & 0 \\ 0 & 0 & B_c \end{bmatrix},$$

where $B_A \in I\!\!R^{k_A \times mn}$, $B_b \in I\!\!R^{k_b \times m}$, and $B_c \in I\!\!R^{k_c \times n}$, for some $k_A \ge 0$, $k_b \ge 0$, and $k_c \ge 0$. For linear knowledge of this form, we consider two different cases:

- There is no *sparsity* knowledge.

- There is *sparsity* knowledge in the constraint matrix only.

By *sparsity knowledge* it is meant that some of the coefficients of the instance in question are known to be equal to zero before computations begin. We consider these cases separately because the analysis for the case where there is sparsity knowledge in the constraint matrix is more complicated and does not simplify to the case where there is no sparsity knowledge. (Sparsity knowledge is a special case of linear knowledge and was considered in [2].)

Let $\Delta d = (\Delta d_A, \Delta d_b, \Delta d_c) \in (I\!R^{mn}, I\!R^m, I\!R^n)$ and let $e \in I\!R^{mn+m+n}$, $e_A \in I\!R^{mn}$, $e_b \in I\!R^m$, and $e_c \in I\!R^n$ denote vectors of all ones with the specified dimensions. Finally, consider the following three linear programs:

$$(P_A) \min \|e_A - \Delta d_A\|$$
$$B_A \Delta d_A = 0$$
$$\Delta d_A \leq e_A,$$

$$(P_b) \min \|e_b - \Delta d_b\|$$
$$B_b \Delta d_b = 0$$
$$\Delta d_b \leq e_b,$$

$$(P_c) \min \|e_c - \Delta d_c\|$$
$$B_c \Delta d_c = 0$$
$$\Delta d_c \leq e_c.$$

Let Δd_A^* be an optimal solution to (P_A), let Δd_b^* be an optimal solution to (P_b), and let Δd_c^* be an optimal solution to (P_c). Δd_A^* can be thought of as the *normalized projection*, using the infinity norm, of e_A onto the null space of B_A (i.e., $\{ d_A : B_A d_A = 0 \}$). Δd_b^* and Δd_c^* have similar interpretations.

Let δ_A^* be the optimal objective function value for (P_A), let δ_b^* be the optimal objective function value for (P_b), and let δ_c^* be the optimal objective function value for (P_c). Also, let

$$\delta_A \equiv 1 - \delta_A^*, \quad \delta_b \equiv 1 - \delta_b^*, \text{ and } \delta_c \equiv 1 - \delta_c^*.$$

Moreover, let

$$\delta_{A,b} \equiv \min\{ \delta_A, \delta_b\}, \quad \delta_{A,c} \equiv \min\{ \delta_A, \delta_c\}, \text{ and } \delta_{A,b,c} \equiv \min\{ \delta_A, \delta_b, \delta_c\}.$$

Furthermore, if there is sparsity knowledge in the constraint matrix (i.e., $(\bar{\Delta} d_A)_i = 0$ for some i satisfying $1 \leq i \leq mn$), let

$$\delta_A^s \equiv \min\{ (\Delta d_A^*)_i : 1 \leq i \leq mn, (\bar{\Delta} d_A)_i > 0\}.$$

Furthermore, let

$$\delta_{A,b}^s \equiv \min\{ \delta_A^s, \delta_b\}, \quad \delta_{A,c}^s \equiv \min\{ \delta_A^s, \delta_c\}, \text{ and } \delta_{A,b,c}^s \equiv \min\{ \delta_A^s, \delta_b, \delta_c\}.$$

Note that $\delta_A \geq 0$, $\delta_b \geq 0$, $\delta_c \geq 0$, and $\delta_A^s \geq 0$.

The algorithm can be shown to be data efficient when linear knowledge appears in the constraint matrix only and when $\delta_A > 0$ if there is no sparsity knowledge or $\delta_A^s > 0$ if there is sparsity knowledge.

Theorem 3.1. *Assume that $\|d\| = 1$, that there is linear knowledge in the constraint matrix only, that there is no sparsity knowledge, and that $\delta_A > 0$. There exists a polynomial $p_1(m, n)$ in the variables m and n that is independent of the actual instance, desired solution accuracy, and linear knowledge, such that Algorithm 2.4 is guaranteed correctly to reply that the actual instance is infeasible, reply that the actual instance is unbounded, or give a 2ϵ-approximate solution when provided with approximate data that has error bound satisfying*

$$\bar{\delta} \leq \frac{(\delta_A)^6 (\delta(d, \epsilon, LK))^6}{p_1(m, n)}.$$

Theorem 3.2. *Assume that $\|d\| = 1$, that there linear knowledge in the constraint matrix only, and that $\delta_A^s > 0$. There exist polynomials $p_2(m, n)$ and $t(m, n)$ in the variables m and n such that Algorithm 2.4 is guaranteed correctly to reply that the actual instance is infeasible, reply that the actual instance is unbounded, or give a 2ϵ-approximate solution when provided with approximate data that has error bound satisfying*

$$\bar{\delta} \leq \frac{(\delta_A^s)^{6mn} (\delta(d, \epsilon, LK))^{6mn}}{p_2(m, n)^{t(m,n)}}.$$

Before we prove the theorems, we give an example that shows that Algorithm 2.4 is not data efficient when linear knowledge is not restricted to the constraint matrix only. Consider the following linear program:

$$\max\ x_1 + x_2$$
$$x_1 + x_2 \leq 1$$
$$x \geq 0.$$

Assume that the linear knowledge is of the following form:

$$a_{11} = a_{12}$$
$$c_1 = c_2.$$

Then, $\delta_{A,b,c} = 1$ and $\text{Opt}(\tilde{d}) = \{\ (x_1, x_2)\ :\ x_1 + x_2 = \tilde{b}/\tilde{a}_{11}\}$ for all $\tilde{d}$ satisfying both $\|d - \tilde{d}\| < 1$ and the linear knowledge. Therefore, even though $\delta(d, \epsilon, LK) > 0$ and $\delta_{A,b,c} > 0$ for all $\epsilon \in [0, \infty)$, Algorithm 2.4 cannot provide an approximate solution for any $\epsilon \in [0, 1]$ without perfect precision. As a result, we consider linear knowledge in the constraint matrix only when showing that Algorithm 2.4 is data efficient.

The proofs of the theorems use bounds on optimal primal solutions, optimal dual solutions, and the change in the optimal objective function value for perturbations in the data that satisfy the linear knowledge. The proofs of the bounds and the theorems follow from previous results in [5] and [2] and relationships between the linear knowledge distances and the distances when there is no linear knowledge to the set of ill-posed problem instances.

To express these relationships, we need the following definitions. Assuming that d is feasible, let $\delta_p'(d)$ denote the distance of the instance d to the set of infeasible instances. This distance can be written:

$$\delta'_p(d) \equiv \sup\{ \ \delta : \|d - \tilde{d}\| < \delta \text{ implies that } \mathrm{Feas}(\tilde{d}) \neq \emptyset\}.$$

If d is infeasible, let $\delta'_p(d)$ denote the distance of the instance d to the set of feasible instances. Furthermore, if d is dual feasible, let $\delta'_d(d)$ denote the distance of the instance d to the set of dual infeasible instances, and if d is dual infeasible, let $\delta'_d(d)$ denote the distance of the instance d to the set of dual feasible instances. These distances can be written in a similar way. Finally, let

$$\delta'(d) \equiv \min\{ \ \delta'_p(d), \delta'_d(d)\}.$$

We now give similar definitions in the case that there is sparsity knowledge. Assuming that d is feasible, let $\delta'_p(d, s)$ denote the sparse distance (i.e., the distance using only perturbations that satisfy the sparsity knowledge) of the instance d to the set of infeasible instances. This distance can be written:

$$\delta'_p(d, s) \equiv \sup\{ \ \delta : \|d - \tilde{d}\| < \delta \text{ and } \tilde{d} \text{ satisfies the sparsity knowledge imply that } \mathrm{Feas}(\tilde{d}) \neq \emptyset\}.$$

If d is infeasible, let $\delta'_p(d, s)$ denote the sparse distance of the instance d to the set of feasible instances. Finally, $\delta'_d(d, s)$ has a similar definition, and we let

$$\delta'(d, s) \equiv \min\{ \ \delta'_p(d, s), \delta'_d(d, s)\}.$$

We have the following lemma.

Lemma 3.3. *In the case that there is no sparsity knowledge,*

$$\delta'_p(d) \geq \delta_{A,b}\delta'_p(d, LK), \tag{3.1}$$

and

$$\delta'_d(d) \geq \delta_{A,c}\delta'_d(d, LK). \tag{3.2}$$

Furthermore, in the case that there is sparsity knowledge,

$$\delta'_p(d, s) \geq \delta^s_{A,b}\delta'_p(d, LK). \tag{3.3}$$

and

$$\delta'_d(d, s) \geq \delta^s_{A,c}\delta'_d(d, LK). \tag{3.4}$$

Proof: First, assume that d is primal feasible. Using the definitions of $\delta'_p(d, LK)$ and $(\Delta d^*_A, \Delta d^*_b, \Delta d^*_c)$, we have that the instances

$$\tilde{d} \equiv d + \gamma(\Delta d^*_A, \Delta d^*_b, 0)$$

for $|\gamma| < \delta'_p(d, LK)$ are feasible. As a result, using the definition of $\delta_{A,b}$, the instances

$$\tilde{d} \equiv d + \gamma(e_A, e_b, 0)$$

for $|\gamma| < \delta_{A,b}\delta'_p(d, LK)$ are feasible. Therefore, (3.1) follows. It can be shown in a similar way that (3.2), (3.3), and (3.4) hold. □

We also have the following result that gives a relationship between $\delta(d, \epsilon, LK)$ and $\delta(d, \epsilon)$ and a relationship between $\delta(d, \epsilon, LK)$ and $\delta(d, \epsilon, s)$, where

$$\delta(d, \epsilon) \equiv \sup\{ \delta : \exists \bar{x} \in I\!R^n \text{ such that } \|d - \tilde{d}\| < \delta \text{ implies that } \exists \tilde{x} \in \text{Opt}(\tilde{d}) \text{ satisfying} \\ \|\bar{x} - \tilde{x}\| \le \epsilon\},$$

and where $\delta(d, \epsilon, s)$ has a similar definition to $\delta(d, \epsilon, LK)$ except that *sparsity knowledge* replaces *linear knowledge*. These relationships will be used to prove the theorems.

Lemma 3.4. *In the case that there is no sparsity knowledge,*

$$\delta(d, \epsilon) \ge \delta_{A,b,c}\delta(d, \epsilon, LK). \tag{3.5}$$

Furthermore, in the case that there is sparsity knowledge,

$$\delta(d, \epsilon, s) \ge \delta^s_{A,b,c}\delta(d, \epsilon, LK). \tag{3.6}$$

Proof: Using the definitions of $\delta(d, \epsilon, LK)$ and $(\Delta d^*_A, \Delta d^*_b, \Delta d^*_c)$, we have that all instances

$$\tilde{d} = d + \gamma(\Delta d^*_A, \Delta d^*_b, \Delta d^*_c)$$

for $|\gamma| < \delta(d, \epsilon, LK)$ are infeasible, are unbounded, or have a common ϵ-approximate solution. Using the definition of $\delta_{A,b,c}$, we have that all instances

$$\tilde{d} = d + \gamma(e_A, e_b, e_c)$$

for $|\gamma| < \delta_{A,b,c}\delta(d, \epsilon, LK)$ are infeasible, are unbounded, or have a common ϵ-approximate solution. Therefore, (3.5) holds. It can be shown in a similar way that (3.6) holds. □

We now use Lemma 3.3 along with results from [2] and [5] to give bounds on a feasible point, optimal primal and dual solutions, and the optimal objective function value in terms of the linear knowledge distances to primal and dual infeasibilities and the linear knowledge parameters δ_A, δ_b, δ_c, and δ^s_A. For these bounds, we consider linear knowledge in all coefficients with sparsity knowledge in the constraint matrix only. We also assume that $\delta_{A,b,c} > 0$ or $\delta^s_{A,b,c} > 0$. In Section 4.1, it is shown that if $\delta_{A,b,c} = 0$, $\delta^s_{A,b,c} = 0$, or sparsity knowledge is not restricted to the contraint matrix only then there exist instances that have optimal solutions whose norms have no finite bound even though $\delta'(d, LK) > 0$.

We first give bounds when there is no sparsity knowledge. The case when there is sparsity knowledge is considered separately from the case when there is no sparsity knowledge because the two bounds do not generalize.

Proposition 3.5. *Assume that $\|d\| = 1$, that linear knowledge is separate for A, b, and c, that there is no sparsity knowledge, that $\delta'_d(d, LK) > 0$, and that $\delta_{A,c} > 0$. Furthermore, assume that d has an optimal solution. If $x^* \in Opt(d)$, then*

$$\|x^*\|_1 \le \frac{\max\{ 1, -k(d)\}}{(\delta_{A,c})(\delta'_d(d, LK))}.$$

where $\|x^*\|_1 \equiv \sum_{j=1}^{n} |x_j^*|$.

Proof: When knowledge is not considered, Proposition 2.1 in [5] gives the following bound:

$$\|x^*\|_1 \leq \frac{\max\{\, 1, -k(d)\}}{\delta_d'(d)}.$$

Using (3.2), the result follows. $\square$

Proposition 3.6. *Assume that $\|d\| = 1$, that the linear knowledge is separate for A, b, and c, that there is no sparsity knowledge, that $\delta_p'(d, LK) > 0$, and that $\delta_{A,b} > 0$. Furthermore, assume that d has an optimal solution. If $y^* \in \mathrm{DualOpt}(d)$, then*

$$\|y^*\|_1 \leq \frac{\max\{\, 1, k(d)\}}{(\delta_{A,b})(\delta_p'(d, LK))},$$

where $\|y^*\|_1 \equiv \sum_{i=1}^{m} |y_i^*|$.

Proof: Because of the symmetry of the linear programs, the proof is similar to the proof of Proposition 3.5. $\square$

Proposition 3.7. *Assume that $\|d\| = 1$, that the linear knowledge is separate for A, b, and c, that there is no sparsity knowledge, that $\delta'(d, LK) > 0$, and that $\delta_{A,b,c} > 0$. Furthermore, assume that d has an optimal solution. Then,*

$$-\frac{1}{(\delta_{A,b})(\delta_p'(d, LK))} \leq k(d) \leq \frac{1}{(\delta_{A,c})(\delta_d'(d, LK))}.$$

Proof: If $k(d) \geq 0$, then the proof follows from Proposition 3.5 and that $\|c\| \leq 1$. Otherwise, the proof follows from Proposition 3.6 and that $\|b\| \leq 1$. $\square$

Proposition 3.8. *Assume that $d = (A, b)$ is feasible and satisfies $\|d\| = 1$. Furthermore, assume that the linear knowledge is separate for A and b, that there is no sparsity knowledge, that $\delta_p'(d, LK) > 0$, and that $\delta_{A,b} > 0$. Then, there exists an $x \in \mathrm{Feas}(d)$ satisfying*

$$\|x\| \leq \frac{1}{(\delta_{A,b})(\delta_p'(d, LK))}.$$

Proof: The proof is similar to the proof of Proposition 3.8 in [2]. $\square$

We now give bounds on a feasible point, primal and dual optimal solutions, and the optimal objective function value assuming that, along with the linear knowledge in possibly all coefficients, there is sparsity in the constraint matrix only.

Proposition 3.9. *Assume that* $\|d\| = 1$, *that the linear knowledge is separate for A, b, and c, that there is sparsity knowledge in the constraint matrix only, that* $\delta'_d(d, LK) > 0$, *and that* $\delta^s_{A,c} > 0$. *Furthermore, assume that d has an optimal solution and that* $k(d) \geq 0$. *If* $x^* \in Opt(d)$, *then*

$$\|x^*\| \leq \frac{(2n)^n}{(\delta^s_{A,c})^n (\delta'_d(d, LK))^n}.$$

Proof: Proposition 3.2 in [2] give the following bound on a optimal solution when there is sparsity knowledge in the constraint matrix only:

$$\|x^*\| \leq \frac{(2n)^n}{(\delta'_d(d, s))^n}.$$

Finally, using (3.4), the result follows. □

Proposition 3.10. *Assume that* $\|d\| = 1$, *that the linear knowledge is separate for A, b, and c, that there is sparsity knowledge in the constraint matrix only, that* $\delta'_d(d, LK) > 0$, *and that* $\delta^s_{A,c} > 0$. *Furthermore, assume that d has an optimal solution and that* $k(d) \geq 0$. *Then*

$$k(d) \leq \frac{n(2n)^n}{(\delta^s_{A,c})^n (\delta'_d(d, LK))^n}.$$

Proof: The proof follows from Proposition 3.9 and that $\|c\| \leq 1$. □

Proposition 3.11. *Assume that* $\|d\| = 1$, *that the linear knowledge is separate for A, b, and c, that there is sparsity knowledge in the constraint matrix only, that* $\delta'_p(d, LK) > 0$, *and that* $\delta^s_{A,b} > 0$. *Furthermore, assume that d has an optimal solution and that* $k(d) \leq 0$. *If* $y^* \in DualOpt(d)$, *then*

$$\|y^*\| \leq \frac{(2m)^m}{(\delta^s_{A,b})^m (\delta'_p(d, LK))^m}.$$

Proof: Because of the symmetry of the linear programs, the proof is similar to the proof of Proposition 3.9. □

Proposition 3.12. *Assume that* $\|d\| = 1$, *that the linear knowledge is separate for A, b, and c, that there is sparsity knowledge in the constraint matrix only, that* $\delta'_p(d, LK) > 0$, *and that* $\delta^s_{A,b} > 0$. *Furthermore, assume that d has an optimal solution. Then,*

$$k(d) \geq -\frac{m(2m)^m}{(\delta^s_{A,b})^m (\delta'_p(d, LK))^m}.$$

Proof: The proof follows from Proposition 3.11 and that $\|b\| \leq 1$. □

Proposition 3.13. *Assume that* $\|d\| = 1$, *that the linear knowledge is separate for* A, b, *and* c, *that there is sparsity knowledge in the constraint matrix only, that* $\delta'(d, LK) > 0$, *and that* $\delta^s_{A,b,c} > 0$. *Furthermore, assume that* d *has an optimal solution and that* $k(d) \leq 0$. *If* $x^* \in Opt(d)$, *then*

$$\|x^*\| \leq \max\{ \frac{(2m)^{2m+2}}{(\delta^s_{A,b})^{2m}(\delta'_p(d, LK))^{2m}}, \frac{(2n)^n}{(\delta^s_{A,c})^n(\delta'_d(d, LK))^n} \}.$$

Proof: The proof follows from Proposition 3.7 in [2] and (3.3) and (3.4). $\square$

Proposition 3.14. *Assume* $\|d\| = 1$, *that the linear knowledge is separate for* A, b *and* c, *that there is sparsity knowledge in the constraint matrix only, that* $\delta'(d, LK) > 0$, *and that* $\delta^s_{A,b,c} > 0$. *Furthermore, assume that* d *has an optimal solution and that* $k(d) \geq 0$. *If* $y^* \in DualOpt(d)$, *then*

$$\|y^*\| \leq \max\{ \frac{(2n)^{2n+2}}{(\delta^s_{A,c})^{2n}(\delta'_d(d, LK))^{2n}}, \frac{(2m)^m}{(\delta^s_{A,b})^m(\delta'_p(d, LK))^m} \}.$$

Proof: Because of the symmetry of the linear programs, the proof is similar to the proof to Proposition 3.13. $\square$

Proposition 3.15. *Assume that* $d = (A, b)$ *is feasible and that* $\|d\| = 1$. *Furthermore, assume that the linear knowledge is separate for* A *and* b, *that there is sparsity knowledge in the constraint matrix only, that* $\delta'_p(d, LK) > 0$, *and that* $\delta^s_{A,b} > 0$. *Then, there exists an* $x \in Feas(d)$ *satisfying*

$$\|x\| \leq \frac{m(2m)^m}{(\delta^s_{A,b})^m(\delta'_p(d, LK))^m}.$$

Proof: The proof is similar to the proof of Proposition 3.8. $\square$

Using the previous propositions, we give a bound on the change in the optimal objective function value for changes in the data that satisfy the linear knowledge. We first give a bound on the change in the optimal objective function value when there is no sparsity knowledge.

Proposition 3.16. *Assume that* $\|d\| = 1$, *that the linear knowledge is separate for* A, b, *and* c, *that there is no sparsity knowledge, that* $\delta'(d, LK) > 0$, *and that* $\delta_{A,b,c} > 0$. *Also, assume that* d *has an optimal solution and that* Δd *satisfies both the linear knowledge and* $\|\Delta d\| \leq \delta'(d, LK)/2$. *Then,*

$$|k(d + \Delta d) - k(d)| \leq \|\Delta d\| \frac{9}{((\delta_{A,b}\delta_{A,c})^2\delta_b))((\delta'_p(d,LK))^2\delta'_d(d,LK))^2)}$$

$$\leq \|\Delta d\| \frac{9}{(\delta_{A,b,c})^5(\delta'(d,LK))^4}.$$

Proof: The proof is similar to the proof of Corollary 3.11 in [2] and uses Lemma 3.3. □

We now provide similar bounds in the case that, along with the linear knowledge, there is sparsity knowledge in the constraint matrix only.

Proposition 3.17. *Assume that* $\|d\| = 1$, *that the linear knowledge is separate for A, b, and c, that there is sparsity knowledge in the constraint matrix only, that* $\delta'(d, LK) > 0$, *and that* $\delta^s_{A,b,c} > 0$. *Also, assume that* Δd *satisfies both the linear knowledge and* $\|\Delta d\| \leq \delta'(d, LK)/2$. *Then*

$$|k(d + \Delta d) - k(d)| \leq \|\Delta d\| \frac{(6mn)^{7mn}}{(\delta^s_{A,b,c})^{5mn}(\delta'(d, LK))^{4mn}}.$$

Proof: The proof is similar to the proof of Corollary 3.11 in [2] and uses Lemma 3.3. □

We now prove Theorems 3.1 and 3.2. The proof of Theorem 3.1 follows from Lemma 3.3 and Renegar's proof of Proposition 4.2 in [5], where a data efficient algorithm that solves symmetric linear programs specified with approximate data when there is no knowledge. Furthermore, the proof of Theorem 3.2 follows Lemma 3.3 and Theorem 3.1 in [2] where there is a data efficient algorithm that solves symmetric linear programs specified with approximate data when there is sparsity knowledge (not general linear knowledge) in the constraint matrix.

Proof of Theorem 3.1: In the proof of Proposition 4.2 in [5], Renegar shows that his algorithm is guaranteed correctly to reply that the actual instance is infeasible, reply that the actual instance is unbounded, or give an ϵ-approximate solution when provided with approximate data that has error bound satisfying

$$\bar{\delta} \leq \frac{(\delta(d, \epsilon))^6}{\bar{p}(m, n)}, \tag{3.7}$$

where $\bar{p}(m, n)$ is a polynomial in the variables m and n that is independent of the actual instance and desired solution accuracy. Because every region considered in Algorithm 2.4 is a subset of the corresponding region considered in Renegar's algorithm (i.e., each coefficient in Renegar's algorithm is perturbed in absolute value by $\bar{\delta}$ while each coefficient in Algorithm 2.4 is perturbed in absolute value by at most $\bar{\delta}$), Algorithm 2.4 is guaranteed to stop once the approximate data error bound satisfies (3.7). Using (3.5), the results of the lemma follow. □

Proof of Theorem 3.2: The proof is similar to the proof of Theorem 3.1 except that it uses Theorem 3.1 in [2] and (3.6). □

4. Comments

4.1. Examples

We first give examples indicating that the dependence on the parameters $\delta_{A,b,c}$ and $\delta^s_{A,b,c}$ in the bounds on optimal primal and dual solutions are necessary without relying on parameters in addition to the linear knowledge distances to infeasibilities. We give examples just for primal linear programs and note that because of the symmetry there are similar examples for dual linear programs.

Consider the following linear program:

$$\max \quad x_1$$
$$tx_1 \;+\; x_2 \;\leq\; 1$$
$$x \;\geq\; 0,$$

where $0 < t \leq 1$. Let x^* denote and optimal solution, and assume that the linear knowledge is $a_{11} = ta_{12}$. We have that $x^* = (1/t, 0)$, $\delta_p'(d, LK) = 1$, $\delta_d'(d, LK) = 1$, and $\|x^*\| = 1/t$. Furthermore, $\delta_A = t$. Therefore the factor of $1/\delta_A$ is needed to bound an optimal solution in the case when there is no sparsity knowledge.

Now, consider the following linear program:

$$\max \quad -tx_1 \;-\; x_2$$
$$-tx_1 \;-\; x_2 \;\leq\; -1$$
$$x \;\geq\; 0,$$

where $0 < t \leq 1$. Let x^* denote an optimal solution, and assume that the linear knowledge is $c_1 = tc_2$. We have that $x^* = (1/t, 0)$ is an optimal solution, $\delta_p'(d, LK) = 1$, $\delta_d'(d, LK) = 1$, and $\|x^*\| = 1/t$. Furthermore, $\delta_c = t$. Therefore the factor of $1/\delta_c$ is needed to bound an optimal solution.

We now consider sparsity knowledge in the constraint matrix only along with linear knowledge in all coefficients. First, assume that sparsity knowledge in the constraint matrix is the only type of linear knowledge considered. Then, it was shown in [2] that the factors $1/(\delta_d'(d, LK))^n$ and $1/(\delta_p'(d, LK))^m$ are needed to bound a primal optimal solution without relying on additional parameters. We now give examples to show that the factors $1/(\delta_A)^n$ and $1/(\delta_c)^n$ are needed to bound a primal optimal solution without relying on additional parameters when general linear knowledge is considered.

Consider the following linear program:

$$
\begin{array}{rcrcrcrcr}
\max & & & & & & x_m \\
& x_1 & & & & & & \leq 1 \\
& -x_1 & + & tx_2 & & & & \leq 0 \\
& & & -x_2 & + & tx_3 & & \leq 0 \\
& & & & & \vdots & & \\
& & & & -x_{m-1} & + & tx_m & \leq 0 \\
& & & & & & x & \geq 0,
\end{array}
$$

where $0 < t \leq 1$ and $2 \leq m \leq n$. Assume that all coefficients in the constraint matrix equal to zero are known to be zero, assume that the linear knowledge is

$$a_{22} = ta_{11},$$
$$a_{33} = ta_{11},$$
$$\vdots$$
$$a_{mm} = ta_{11},$$

and let x^* denote an optimal solution. We have that $\delta'_d(d, LK) = 1$, $\delta'_p(d, LK) = 1$, and $\|x^*\| = 1/t^{m-1}$. Furthermore, $\delta^s_A = t$. Therefore, the factor $1/(\delta^s_A)^m$, with possibly $m = n$, is needed to bound a primal optimal solution.

Next, consider the following linear program:

$$
\begin{array}{rcccccccccl}
\max & -x_1 & - & tx_2 & \cdots & & - & tx_k & - & tx_{k+1} & \cdots & - & tx_n & \\
& -x_1 & & & & & & & & & & & & \leq -1 \\
& x_1 & - & tx_2 & & & & & & & & & & \leq 0 \\
& & & x_2 & - & tx_3 & & & & & & & & \leq 0 \\
& & & & & \vdots & & & & & & & & \\
& & & & & & x_{m-1} & - & tx_m & & & & & \leq 0 \\
& & & & & & & & & & & x & \geq 0,
\end{array}
$$

where $0 < t \leq 1$ and $2 \leq m \leq n$. Again, assume that all constraint matrix coefficients equal to zero are known to be zero. Furthermore, let x^* denote an optimal solution and let

$$
c_2 = tc_1,
$$
$$
c_3 = tc_1,
$$
$$
\vdots
$$
$$
c_n = tc_1,
$$

and

$$
a_{22} = ta_{11},
$$
$$
a_{33} = ta_{11},
$$
$$
\vdots
$$
$$
a_{mm} = ta_{11},
$$

be the linear knowledge. Then, $\delta'_d(d, LK) = 1$, $\delta'_p(d, LK) = 1$, and $\|x^*\| = 1/t^{m-1}$. Furthermore, $\delta_c = \delta^s_A = t$ so that the extra factor of $1/\delta^m_c = 1/(\delta^s_A)^m$, with possibly $m = n$, is needed to bound an optimal solution.

Finally, consider the following linear program:

$$
\begin{array}{rccccccccl}
\max & & & & & & & x_m & & \\
& x_1 & & & & & & & & \leq 1 \\
& -x_1 & + & x_2 & & & & & & \leq 1 \\
& & & & \vdots & & & & & \\
& -x_1 & - & x_2 & - & x_3 & \cdots & -x_{m-1} & + & x_m & \leq 1 \\
& & & & & & & & x & \geq 0,
\end{array}
$$

where $2 \leq m \leq n$. Again, assume that all coefficients in the constraint matrix equal to zero are known to be zero and let x^* denote an optimal solution. We have $\delta'_d(d, LK) = \delta'_p(d, LK) = 1$ and $\|x^*\| = 2^{m-1}$. Therefore, the factor of 2^m or possibly 2^n is needed to bound a primal optimal solution. Furthermore, it is a factor like this that leads to $t(m, n)$

not being equal to 1 in Theorem 3.2.

We now give examples to show that the bounds on optimal solutions, in terms of the linear knowledge distances to primal and dual infeasibilities and the linear knowledge parameter $\delta_{A,b,c}$ or $\delta^s_{A,b,c}$ do not exist if sparsity is not restricted to be in the constraint matrix only. Consider the following linear program:

$$\begin{array}{rcccc} \max & x_1 & & & \\ & x_1 & + & 0x_2 & \leq & 1/2 \\ & x_1 & & & \leq & 1 \\ & & & x & \geq & 0. \end{array}$$

Assume that c_2 and a_{21} are known to be equal to zero. Now, $x^* = (1/2, \gamma)$ for any $\gamma \geq 0$ is an optimal solution. However, $\delta'_p(d, LK) = 1/2$, $\delta'_d(d, LK) = 1$, and $\delta^s_{A,b,c} = 1$. Therefore, an additional parameter is needed to bound all optimal primal solutions.

We now show that if $\delta_{A,b,c} = 0$ or $\delta^s_{A,b,c} = 0$ then there exist instances that have optimal solutions that are infinite in norm even though $\delta'(d, LK) > 0$. Consider the following example:

$$\begin{array}{rcccc} \max & x_1 & - & x_2 & \\ & x_1 & - & x_2 & \leq & 1/2 \\ & x_1 & - & x_2 & \leq & 1 \\ & & & x & \geq & 0. \end{array}$$

Assume that the linear knowledge is of the form:

$$c_1 + c_2 = 0,$$
$$a_{21} + a_{22} = 0.$$

The set of optimal solutions is given by $X^* \equiv \{ x^* : x_1^* - x_2^* = 1/2, \; x^* \geq 0 \}$. Furthermore, $\delta'_p(d, LK) = \delta'_d(d, LK) = 1$, and $\delta_{A,b,c} = 0$.

4.2. Comparison to the Case When there is no Knowledge

We first discuss how the algorithm and the complexity of the algorithm compare with Renegar's algorithm and the complexity of his algorithm for symmetric linear programs where there is no knowledge [5]. First, Renegar's algorithm is guaranteed correctly to reply that the actual instance is infeasible, reply that the actual instance is unbounded, or provide an ϵ-approximate solution when provided with approximate data that has error bound satisfying

$$\bar{\delta} \leq \frac{(\delta(d, \epsilon))^6}{\tilde{p}(m, n)},$$

where $\tilde{p}(m, n)$ is polynomial in the variables m and n that is independent of the actual instance and desired solution accuracy and where

$\delta(d, \epsilon) \equiv \sup\{ \delta : \exists \, \bar{x} \in I\!\!R^n$ such that $\|d - \tilde{d}\| < \delta$ implies that $\exists \, \tilde{x} \in \mathrm{Opt}(\tilde{d})$ satisfying $\|\bar{x} - \tilde{x}\| \leq \epsilon\}$.

At first it might seem that Renegar's complexity result is stronger because $6 \leq 6mn$ whenever $m, n \geq 1$. However, this is not necessarily the case. First, when there is no linear knowledge, Algorithm 2.4 is the same as Renegar's algorithm. In this case the instances used by Algorithm 2.4, $\bar{d}^+$ and $\bar{d}^-$, are the same as those used by Renegar (i.e., each coefficient is perturbed by either $-\bar{\delta}$ or $\bar{\delta}$).

Second, it might be the case that $\delta(d, \epsilon) < \delta(d, \epsilon, LK)$ and even the case that $\delta(d, \epsilon) <<$ $\delta(d, \epsilon, LK)$. In particular, consider the following infeasible linear program:

$$
\begin{array}{rlll}
\max & c_1 x_1 & + & c_2 x_2 \\
& x_1 & & \leq 1/2 \\
& -x_1 & & \leq -1 \\
& & x & \geq 0,
\end{array}
$$

for any objective function coefficients, where the constraint matrix coefficients equal to be zero are known to be zero. Here, $\delta(d, \epsilon, LK) = 1/7$ while $\delta(d, \epsilon) = 0$ (i.e., let $a_{12} = -\rho$ for any $\rho > 0$). Therefore,

$$
\frac{(\delta(d, \epsilon))^6}{\tilde{p}(m, n)} << \frac{(\delta_A^s)^{5mn} (\delta(d, \epsilon, LK))^{6mn}}{p(m, n)^{16mn}} .
$$

Finally, both the algorithm and the complexity results reduce to those given in [2] where sparsity in the constraint matrix only is considered.

4.3. Angles

In this section we give a geometric interpretation of the complexity results given in Theorem 3.1. The presence of the condition measure in the complexity bound given in Theorem 3.1 already has a geometric interpretation: the closer the problem instance is to the set of ill-posed problem instances, the more precision the algorithm is given to solve the instance in question. In this section we show that there is a relationship between the linear knowledge parameters and an angle between the subspace of problem instances that satisfy the linear knowledge and a certain subspace of problem instances. We also show that in some cases, this certain set of problem instances is the set of ill-posed problem instances. In particular, in some cases, the relationship that will be given expresses that the smaller the angle is between the subspace of instances that satisfy the linear knowledge and the set of ill-posed problem instances, the more precision the algorithm is given to solve the problem instance in question.

In this section, we consider the problem of determining the feasibility or infeasibility of a system of linear inequalities. Assume that $d = (A, b)$ is feasible. The results extend to the case that d is infeasible and to solving linear programs.

Recall that $\Delta d_A^* \in I\!\!R^{mn}$ solves the following linear program:

$$
\begin{array}{rl}
(P_A) \min & \|e_A - \Delta d_A\| \\
& B_A \Delta d_A = 0 \\
& \Delta d_A \leq e_A
\end{array}
$$

and that $\Delta d_b^* \in I\!\!R^m$ solves the following linear program:

$$(P_b) \ \min \ \|e_b - \Delta d_b\|$$
$$B_b \Delta d_b = 0$$
$$\Delta d_b \leq e_b.$$

We consider sparsity in all coefficients so that $\delta_A^s \equiv \min\{\ (\Delta d_A^*)_i : (\bar{\Delta} d_A)_i > 0\}$, $\delta_b^s \equiv \min\{\ (\Delta d_b^*)_i : (\bar{\Delta} d_b)_i > 0\}$, and $\delta_{A,b}^s \equiv \min\{\ \delta_A^s, \delta_b^s\}$.

Let

$S \equiv \{\ (\tilde{d}_A, \tilde{d}_b) : (\tilde{d}_A)_i = 0 \text{ if } (\bar{\Delta} d_A)_i = 0, \ (\tilde{d}_A)_i > d_A + \delta_p'(d,s) \text{ if } (\bar{\Delta} d_A)_i > 0, \ (\tilde{d}_b)_i = 0 \text{ if } (\bar{\Delta} d_b)_i = 0, \ (\tilde{d}_b)_i < d_b - \delta_p'(d,s) \text{ if } (\bar{\Delta} d_b)_i > 0\}.$

Therefore, every instance in S is guaranteed to be infeasible. Furthermore, let $\bar{S}$ be the closure of S. Finally, assuming that $\delta_{A,b}^s > 0$, let

$\tilde{S} \equiv \bar{S} \cap \{\ (\tilde{d}_A, \tilde{d}_b) : (\tilde{d}_A)_i = (d_A)_i + \delta_p'(d,s) \text{ if } (\Delta d_A^*)_i = \delta_{A,b}^s, \ (\tilde{d}_b)_i = (d_b)_i - \delta_p'(d,s) \text{ if } (\Delta d_b^*)_i = \delta_{A,b}^s\}.$

For certain problem instances, $\tilde{S}$ describes the set of ill-posed problem instances. In particular, for certain problem instances, if $(\Delta d_A)_i < \delta_p'(d,s)$ or $(\Delta d_b)_i > -\delta_p'(d,s)$ for a particular i, then $d + \Delta d = (d_A + \Delta d_A, d_b + \Delta d_b)$ is feasible. For example, consider the following instance:

$$x_1 - t x_2 \leq -1$$
$$-x_2 \leq 1$$
$$x \geq 0$$

where $0 < t \leq 1$. Then, $\delta_p'(d,s) = t$ (i.e., adding t to the second constraint matrix coefficient a_{12} gives an infeasible instance). Furthermore, as long as the second constraint matrix coefficient is perturbed by less than t, the system remains feasible. Assuming that the linear knowledge is of the following form:

$$a_{12} = t a_{22},$$

$\tilde{S}$ describes the set of ill-posed problem instances.

This example can be extended to allow sparsity knowledge. For example, a_{21} can be known to be equal to zero. This example also can be extended to instances with any number of variables (i.e., let the constraint matrix coefficients for any new variable be known to be equal to zero, for example) and any number of constraints (i.e., duplicate the constraints, for example).

Assuming that $\delta_{A,b}^s > 0$, we now give a relationship between $\delta_{A,b}^s$ and an *angle* between $\tilde{S}$ and the subspace of instances that satisfy the linear knowledge:

$$S_{LK} \equiv \{\ (\tilde{d}_A, \tilde{d}_b) : B_A \tilde{d}_A = 0, \ B_b \tilde{d}_b = 0\}.$$

We consider first a simpler case for the problem instance and linear knowledge considered and then the general case.

Proposition 4.1. *Assume that $d = (A, b)$ is feasible, that $\|d\| = 1$, that the linear knowledge is separate for A and b, and that $\delta_{A,b}^s > 0$. Furthermore, assume that the rank of B_A is equal to $mn - 1$ and that the rank of B_b is equal to $m - 1$. Then,*

$$\frac{\sin(\theta)}{\sqrt{mn + m}} \leq \delta_{A,b}^s \leq \sqrt{mn + m}\sin(\theta),$$

where θ is the minimum angle between S_{LK} and $\tilde{S}$.

Proof: From the assumptions about the ranks of B_A and B_b,

$$S_{LK} = \{ (\tilde{d}_A, \tilde{d}_b) : (\tilde{d}_A, \tilde{d}_b) = (\gamma_A \Delta d_A^*, \gamma_b \Delta d_b^*), \gamma_A, \gamma_b \in I\!\!R \}.$$

Let

$$d(\gamma) \equiv (d_A + \gamma \Delta d_A^*, d_b - \gamma \Delta d_b^*).$$

Then, $d(\gamma)$ does not intersect $\tilde{S}$ until $\gamma = \delta_p'(d, s)/\delta_{A,b}^s$. Let $d_{I_0} \equiv (d_A + \frac{\delta_p'(d,s)}{\delta_{A,b}^s}\Delta d_A^*, d_b - \frac{\delta_p'(d,s)}{\delta_{A,b}^s}\Delta d_b^*)$ be where $d(\gamma)$ first intersects $\tilde{S}$. We can express $\tilde{S}$ by

$$\tilde{S} \equiv \{ \Delta d_{I_0} : \Delta d_{I_0} = d + \delta_p'(d, s)\bar{e} + \sum_{j=1, j \notin I_0 \cup \hat{S}}^{mn+m} \gamma_j e_j, \gamma_j \in I\!\!R \text{ for } 1 \leq j \leq mn + m, j \notin I_0 \cup \hat{S} \},$$

where $e_j \in I\!\!R^{mn+m}$ is the jth unit vector, where $j \in I_0$ if $(\Delta d_A^*, \Delta d_b^*)_j = \delta_{A,b}^s$, where $\bar{e} \in I\!\!R^{mn+m}$ satisfies $\bar{e}_i = 1$ if $i \in I_0$ and i corresponds to a constraint matrix coefficient, where $\bar{e}_i = -1$ if $i \in I_0$ and i corresponds to a right-hand side coefficient, and where $\bar{e}_i = 0$ otherwise. Finally, $j \in \hat{S}$ if d_j is known to be equal to zero.

Now, the angle between S_{LK} and any vector $y \in \tilde{S}$ with $\|y\|_2 = 1$ is given by

$$\angle (y, S_{LK}) \equiv \cos^{-1}\left(\frac{(\Delta d^*)^T y}{\|\Delta d^*\|_2}\right), \tag{4.1}$$

where $\Delta d^* \equiv (\Delta d_A^*, \Delta d_b^*)$. Clearly, the minimum angle between S_{LK} and $\tilde{S}$ is achieved by the normalized projection of Δd^* onto $\tilde{S}$. That is,

$$\bar{y} \equiv \frac{\text{Proj}_{\tilde{S}}(\Delta d^*)}{\|\text{Proj}_{\tilde{S}}(\Delta d^*)\|_2}.$$

We now give an expression for $\bar{y}$. Because $\{ e_j \}_{j=1, j \notin I_0 \cup \hat{S}}^{mn+m}$ is a basis for $\tilde{S}$,

$$\text{Proj}_{\tilde{S}}(\Delta d^*) = C(C^T C)^{-1}C^T(\Delta d^*),$$

where $C \in I\!\!R^{(mn+m) \times (mn+m-|I_0 \cup \hat{S}|)}$ has the unit vectors $e_j \in I\!\!R^{mn+m}$, for $j \notin I_0 \cup \hat{S}$, as its columns. (Here $|I_0 \cup \hat{S}|$ denotes the number of elements in $I_0 \cup \hat{S}$.) Therefore,

$$\begin{aligned}
\text{Proj}_{\tilde{S}}(\Delta d^*) &= CC^T(\Delta d^*), \\
&= \sum_{j=1, j \notin I_0 \cup \hat{S}}^{mn+m} e_j e_j^T(\Delta d^*) \\
&= \Delta d^* - \sum_{j \in I_0} e_j (\Delta d^*)_j.
\end{aligned}$$

Therefore,

$$\bar{y} = \frac{\Delta d^* - \sum_{j \in I_0} e_j \delta_{A,b}}{\|\Delta d^* - \sum_{j \in I_0} e_j \delta_{A,b}\|_2}.$$

Finally,

$$\cos(\theta) \equiv \cos(\angle \ (\bar{y}, S_{LK})) = \frac{(\Delta d^*)^T \bar{y}}{\|\Delta d^*\|_2}$$

$$= \frac{\sqrt{\|\Delta d^*\|_2^2 - |I_0|\delta_{A,b}^2}}{\|\Delta d^*\|_2}.$$

Therefore,

$$(\cos(\theta))^2 \|\Delta d^*\|_2^2 = \|\Delta d^*\|_2^2 - |I_0|\delta_{A,b}^2$$

so that

$$|I_0|\delta_{A,b}^2 = \|\Delta d^*\|_2^2 (1 - (\cos(\theta))^2)$$

$$= \|\Delta d^*\|_2^2 (\sin(\theta))^2.$$

Finally, because $\|\Delta d^*\| = 1$ and $0 \le \theta \le 45$,

$$\sin(\theta) \le \sqrt{|I_0|}\delta_{A,b} \le \sqrt{mn + m}\sin(\theta)$$

so that

$$\frac{\sin(\theta)}{\sqrt{mn + m}} \le \delta_{A,b} \le \sqrt{mn + m}\sin(\theta). \square$$

We now extend the results to the case that the rank of B_A is not necessarily equal to $mn - 1$ and that rank of B_b is not necessarily equal to $m - 1$.

Proposition 4.2. *Assume that $d = (A, b)$ is feasible, that $\|d\| = 1$, that the linear knowledge is separate for A and b, and that $\delta_{A,b}^s > 0$. Then,*

$$\frac{\sin(\theta)}{\sqrt{mn + m}} \le \delta_{A,b}^s \le \sqrt{mn + m}\sin(\theta),$$

where θ is the minimum angle between $\{ \ (\tilde{d}_A, \tilde{d}_b) = \gamma(\Delta \tilde{d}_A, -\Delta \tilde{d}_b), \ \gamma \in I\!R\}$ and $\tilde{S}$, where $\Delta \tilde{d} = (\Delta \tilde{d}_A, \Delta \tilde{d}_b) \in S_{LK}$ approaches $\tilde{S}$ from d first.

Proof: From the definitions,

$$d + \frac{\delta_p'(d, s)}{\delta_{A,b}^s}\Delta d^* = d + \frac{\delta_p'(d, s)}{\delta_{A,b}^s}(\Delta d_A^*, \Delta d_b^*)$$

is where S_{LK} first intersects $\tilde{S}$ from d. For, let $\Delta \tilde{d} = (\Delta \tilde{d}_A, \Delta \tilde{d}_b)$ satisfy $B_A \Delta \tilde{d}_A = 0$ and $B_b \Delta \tilde{d}_b = 0$ and $(\Delta \tilde{d}_A, \Delta \tilde{d}_b) \ge 0$. Because Δd_A^* and Δd_b^* solve (P_A) and (P_b) respectively, we have that

$$0 \le \min_i (\Delta \tilde{d})_i \le \min_i (\Delta d^*)_i,$$

where $\Delta d^* = (\Delta d_A^*, \Delta d_b^*)$. Therefore, $d + \gamma \Delta d^*$ will intersect $\tilde{S}$ at least as fast as $d + \gamma \Delta \tilde{d}$ does.

Therefore, using the analysis in the proof of the previous proposition, the result of the proposition holds. □

Similar results hold if d is infeasible.

Finally, consider the following algorithm, which essentially consists of the first two steps of Algorithm 2.4:

Algorithm 4.3.

(0) *The algorithm assumes that $(\bar{d}, \bar{\delta})$ and the linear knowledge are given and that $\bar{d}$ satisfies the linear knowledge.*

(1) *Check if $\bar{d}^+ \equiv (\bar{A} - \bar{\delta}\bar{\Delta}A, \ \bar{b} + \bar{\delta}\bar{\Delta}b)$ is infeasible. If so, STOP; the actual instance is infeasible.*

(2) *Check if $\bar{d}^- \equiv (\bar{A} + \bar{\delta}\bar{\Delta}A, \ \bar{b} - \bar{\delta}\bar{\Delta}b)$ is feasible. If so, STOP; the actual instance is feasible.*

(3) *Better data accuracy is needed.*

Using Theorem 3.1 and the relationship between $\delta_{A,b}^s$ and an angle between the subspace of instances satisfying the linear knowledge and $\tilde{S}$ given in Propositions 4.1 and 4.2, we have the following theorem.

Theorem 4.4. *Assume that $d = (A, b)$ satisfies $\|d\| = 1$ and that $\delta_{A,b} > 0$. Then, Algorithm 4.3 is guaranteed to reply correctly that the actual instance is feasible or infeasible when provided with approximate data that has error bound satisfying*

$$\bar{\delta} \leq \frac{\sin \theta}{\sqrt{mn + m}} \frac{\delta_p'(d, LK)}{2},$$

where θ is the minimum angle between $\{ \ (\tilde{d}_A, \tilde{d}_b) = \gamma(\Delta \tilde{d}_A, -\Delta \tilde{d}_b), \ \gamma \in I\!R\}$ and $\tilde{S}$, where $\Delta \tilde{d} = (\Delta \tilde{d}_A, \Delta \tilde{d}_b) \in S_{LK}$ approaches $\tilde{S}$ from d first.

Proof: From the proof of Theorem 3.1, Algorithm 4.3 is guaranteed to reply correctly that the actual instance is feasible or infeasible when provided with approximate data that has error bound satisfying

$$\bar{\delta} \leq \frac{\delta_{A,b}^s \delta_p'(d, LK)}{2}.$$

Therefore, using the results of Propositions 4.1 and 4.2, the results of the theorem hold. □

We finally note that the ideas extend to solving linear programs instead of just determining feasibility or infeasibility. In this case, S_{LK} and $\tilde{S}$ have similar definitions except that the objective coefficients are included and the the definitions vary slightly, depending on if d is infeasible, unbounded, or has an optimal solution.

5. Acknowledgments

I thank Jim Renegar and Adrian Lewis for discussions that led me to consider linear knowledge. I also thank a referee for many helpful comments. Finally, most of the work on this paper was done while I was a faculty member in the Department of Industrial and Manufacturing Systems Engineering at Iowa State University.

References

[1]   S. Filipowski, "On the Complexity of Solving Feasible Systems of Linear Inequalities Specified with Approximate Data," *Mathematical Programming* 71 (1995) 259-288.

[2]   S. Filipowski, "On the Complexity of Solving Sparse Symmetric Linear Programs Specified with Approximate Data," to appear in *Mathematics of Operations Research*.

[3]   S. Filipowski, "On the Complexity of Solving Feasible Linear Programs Specified with Approximate Data," preprint.

[4]   J. Renegar, "Some Perturbation Theory for Linear Programming," *Mathematical Programming* 65 (1994) 73-92.

[5]   J. Renegar, "Incorporating Condition Measures into the Complexity Theory of Linear Programming," *SIAM Journal on Optimization* 5 (1995) 506-524.

[6]   J. Vera, "Ill-posedness and the Complexity of Deciding Existence of Solutions to Linear Programs," *SIAM Journal on Optimization* 6 (1996) 549-569.

[7]   J. Vera, "Ill-posedness and the Computation of Solutions to Linear Programs with Approximate Data," preprint, Department of Industrial Engineering, University of Chile (1992), *jvera@dii.uchile.cl*.

On the Role of the Mangasarian–Fromovitz Constraint Qualification for Penalty-, Exact Penalty-, and Lagrange Multiplier Methods

JÜRGEN GUDDAT Humboldt University, Berlin, Germany

FRANCISCO GUERRA Univ.de las Américas, Puebla, Mexico

DIETER NOWACK Humboldt University, Berlin, Germany

dedicated to the 65. birthday of Prof. Dr. A. Fiacco

Abstract

In this paper we consider three embeddings (one-parametric optimization problems) motivated by penalty, exact penalty and Lagrange multiplier methods. We give an answer to the question under which conditions these methods are successful with an arbitrarily chosen starting point. Using the theory of one-parametric optimization (the local structure of the set of stationary points and of the set of generalized critical points, singularities, topological stability, pathfollowing and jumps) the so-called Mangasarian-Fromovitz condition and its extension play an important role. The analysis shows us that the class of optimization problems for which we can surely find a stationary point using a pathfollowing procedure for the penalty and exact penalty embedding is much larger than the class where the Lagrange multiplier embedding is successful. For the first class, the objective may be a "really non-convex" function, but for the second one we are restricted to convex optimization problems. This fact was a surprise at least for the authors.

1 INTRODUCTION

We consider a nonlinear optimization problem with the following standard structure:

(P) $\min\{f(x) \mid x \in M\}$,

where

$$M := \{x \in \mathbb{R}^n \mid h_i(x) = 0,\ i \in I,\ g_j(x) \leq 0,\ j \in J\}, \tag{1.0}$$

[1]This research was supported by the Deutsche Forschungsgemeinschaft under Grant Gu304/1-4

$I := \{1, \ldots, m\}$, $m < n$, $J := \{1, \ldots, s\}$, and $f, h_i, g_j \in C^2(\mathbb{R}^n, \mathbb{R})$, $i \in I$, $j \in J$.

For some of the results presented we need a higher degree of differentiability. We recall the well-known concept of embedding (cf. e.g. [5,8,13]). Consider a one-parametric optimization problem

$$P(t) \quad \min\{f(y, t) \mid y \in M(t)\}, \ t \in [0, 1], \tag{1.1}$$

where

$$M(t) := \{y \in \mathbb{R}^{\bar{n}} \mid h_i(y, t) = 0, \ i \in I, \ g_j(y, t) \leq 0, \ j \in \bar{J}\}$$

$n \leq \bar{n}$, $\bar{J}$ is a finite index set with $J \subseteq \bar{J}$, with at least the following properties:

(A1) A stationary point of $P(0)$ is known (and the corresponding Lagrange multipliers are known or easy to compute);

(A2) $P(t)$ has a global minimizer for every $t \in [0, 1]$;

(A3) $P(1)$ is equivalent to (P) in a certain sense (to be specified below).

Let us mention 4 papers that are related to our work: The use of continuation methods for smooth penalty functions is given in [20], with extensions and numerics in [19]. A bifurcation analysis for quadratic penalty, augmented Lagrangian, and log barrier methods can be found in [17], [18].

In this paper we consider three embeddings (cf. [6,2,3]) motivated by penalty, exact penalty and Lagrange multiplier methods and we ask the following question: Are there conditions for finding a discretization of the interval $[0, 1]$:

$$0 = t_0 < \cdots < t_i < \cdots < t_N = 1$$

and corresponding stationary points $y(t_i)$ of $\tilde{P}(t_i)$, $i = 1, \ldots, N$ starting with an arbitrarily chosen stationary point $y(t_0)$. We will see that the so-called Mangasarian-Fromovitz Constraint Qualification and its extension play an important role. First we introduce the following modified penalty embedding (we refer to [6])

$$\tilde{P}_1(t) : \min\{F_1(x, v, w, t) \mid (x, v, w) \in \tilde{M}_1(t)\}, \ t \in [0, 1]$$

where

$$F_1(x, v, w, t) := F_{(A,B,C,x^0,v^0,w^0)}(x, v, w, t) =$$
$$= tf(x) + (1-t)(x-x^0)^T A(x-x^0) + (v-v^0)^T B(v-v^0) + (w-w^0)^T C(w-w^0)$$

(here A is an (n, n) matrix, B is an (m, m) matrix and C is an (s, s) matrix)

$$\tilde{M}_1(t) := M_{(x^0,v^0,w^0,w^1,b,p,q)} = \{(x, v, w) \in \mathbb{R}^n \times \mathbb{R}^m \times \mathbb{R}^s \mid th_i(x) + (1-t)(v_i - v_i^0) = 0, i \in I,$$
$$tg_j(x) + (1-t)(w_j - w_j^1) \leq 0, j \in J$$
$$\|x - x^0\|^2 + b^T x - p \leq 0$$
$$\|v - v^0\|^2 + \|w - w^0\|^2 - q \leq 0\}.$$

In distinction to $\tilde{P}_4(t)$ in [6] we consider two points w^0 and w^1 in a suitable manner. Then the starting point (x^0, v^0, w^0) has better properties (cf. Theorem 1.1 below).

In [6] it is explained how we obtain such a kind of model for the penalty method. Let us consider this here only briefly. The model for the original quadratic penalty method reads

$$\min\{f(x) + \Big(\frac{t}{1-t}\Big)^2 \Big(\sum_{i \in I}(h_i(x))\Big)^2 + \sum_{j \in J}(\max\{g_j(x), 0\})^2 \mid x \in I\!R^n\}, t \in [0, 1).$$

We observe that the penalty parameter $c(t) := \frac{t}{1-t}$ tends to $+\infty$ if t tends to 1. This one-parametric optimization problem has the following disadvantages: The problem is not defined for $t = 1$, the objective function is exactly once continuously differentiable (i.e., the results of parametric optimization presented in [7,8,9,10,11,12,13,15,16] – a short summary is given in Chapter 2 – are not applicable); we do not know any starting point for $t = 0$. It is easy to see that these disadvantages will not appear for $P_1(t)$. Moreover, there are further important properties of $P_1(t)$ (cf. Theorem 1.1). The term $(1 - t)(x - x^0)^T A(x - x^0) + (v - v^0)^T B(v - v^0) + (w - w^0)^T C(w - w^0)$ is a regularization term (cf. Theorem 3.5(i)) which does not have any influence on the solution of (P) ((P) is equivalent to $\tilde{P}_1(1)$ (cf. Theorem 1.1(iii))).

Now we assume that

(B1) the feasible set M of the original problem (P) is non-empty and compact.

We denote $E(x^0, b, p) := \{x \in I\!R^n \mid \|x - x^0\|^2 + b^T x - p \leq 0\}$. Then, for a fixed x^0 and a fixed b there exists a $\tilde{p} > 0$ such that $M \subseteq E(x^0, b, p)$ for all $p \geq \tilde{p}$. Furthermore, there exists a $\tilde{q} > 0$ such that $\tilde{M}_1(t) \neq \emptyset$ for all $t \in [0, 1]$ and all $p \geq \tilde{p}$ and $q > \tilde{q}$.

We assume that

(B2) $p > \tilde{p}$ and $q > \tilde{q}$.

Now it is easy to see:

THEOREM 1.1 (cf. [6], too). *$\tilde{P}_1(t)$ has the following properties:*

(i) *Choosing $w^0, w^1 \in I\!R^s$ with $w_j^0 < w_j^1, j \in J$, $x^0, b \in I\!R^n$ with $b^T x^0 - p < 0$, $v^0 \in I\!R^m$, A, B and C positive definite, $p > 0$ and $q > 0$ sufficiently large, then (x^0, v^0, w^0) is the global minimizer and the only stationary point for $\tilde{P}_1(0)$. Furthermore, $(x^0, v^0, w^0, 0)$ is a non-degenerate stationary point (non-degenerate cf. the conditions (2.4)).*

(ii) *Assume that (B1) and (B2) are satisfied. Then $\psi_{\mathrm{glob}}(t) \neq \emptyset$ for all $t \in [0, 1]$, where $\psi_{\mathrm{glob}}(t)$ is the set of all global minimizers for $\tilde{P}_1(t)$.*

(iii) *$\tilde{P}_1(1)$ is equivalent to (P) in the following sense:*

    (a) *If (x, v, w) is a stationary point for $\tilde{P}_1(1)$, then x is a stationary point for (P).*

    (b) *If x is a stationary point for (P), then there exist vectors $v \in I\!R^m, w \in I\!R^s$, such that (x, v, w) is a stationary point for $\tilde{P}_1(1)$.*   □

Secondly, we consider the so-called exact penalty methods proposed e.g. in [14] for problems without equality constraints (i.e., $I = \emptyset$). We propose the following embedding (cf. [2])

$$\tilde{P}_2(t): \ \min\{F_2(x, w, t) \mid (x, w) \in \tilde{M}_2(t)\}, \ t \in [0, 1]$$

where

$$F_2(x, w, t) := F_{(A, B, x^0, w^0)}(x, w, t) = tf(x) + (1 - t)\sum_{j \in J} \mu_j(x)(w_j - w_j^0) +$$
$$+ (1 - t)(x - x^0)^T A(x - x^0) + (w - w^0)^T [\mathrm{diag}(\alpha_j - g_j(x))]^{-1} B(w - w^0),$$

A is an (n, n) matrix, B is an (s, s) matrix,

$\alpha^T = (\alpha_1, \ldots, \alpha_s)$ with $\alpha_j > g_j(x^0)$, $j = 1, \ldots, s$,
$\mu(x) = -[Dg(x)Dg(x)^T + \gamma^2 G(x) \cdot G(x)]^{-1} Dg(x)Df(x)^T$
$g(x) = (g_1(x), \ldots, g_s(x))$, $Dg(x)$ is the Jacobian, $\gamma \neq 0$,
$G(x) = \mathrm{diag}\,(g_j(x))$,

$$\tilde{M}_2(t) := M_{(x^0, w^0, w^1, b, p, q)}(t) := \{(x, w) \in I\!\!R^n \times I\!\!R^s \mid tg_j(x) - (1 - t)(\omega_j - w_j^1) \leq 0, j \in J,$$
$$\|x - x^0\|^2 + b^T x - p \leq 0,$$
$$\|w - w^0\|^2 - q \leq 0\}$$

Then we have

THEOREM 1.2 (cf. [2], too). *$\tilde{P}_2(t)$ has the following properties*

(i) *Choosing $w^0, w^1 \in I\!\!R^s$ with $w_j^1 < w_j^0$, $j \in J$, $x^0, b \in I\!\!R^n$ with $b^T x^0 - p < 0$, $\alpha \in I\!\!R^s$ with $\alpha_j > g_j(x^0)$, $j \in J$, $p > 0$ and $q > 0$ sufficiently large, the matrices A and B positive definite in such a way that the Hessian $D^2 F_2(x^0, w^0, 0)$ is positive definite, then (x^0, w^0) is the global minimizer and the only stationary point for $\tilde{P}_2(0)$. Furthermore, (x^0, w^0) is a non-degenerate stationary point.*

(ii) *Analogously to Theorem 1.1(ii).*

(iii) *Analogously to Theorem 1.1(iii).*   □

Thirdly, we introduce the following Lagrange multiplier embedding (cf. [3]), for the original method cf. e.g. [1].

$$\tilde{P}_3(t): \ \min\{F_3(x, v, w, \lambda, \mu, t) \mid (x, v, w, \lambda, \mu) \in \tilde{M}_3(t)\}, t \in [0, 1],$$

where

$$F_3(x, v, w, \lambda, \mu, t) := F_{(A, B, C, D, E, x^0, \lambda^0, \mu^0, v^0, w^0)}(x, v, w, \lambda, \mu) =$$
$$= t[f(x) + \sum_{i \in I} \lambda_i h_i(x, t) + \sum_{j \in J} \mu_j g_j(x, t)] + (1 - t)[(x - x^0)^T A(x - x^0) -$$
$$- (\lambda - \lambda^0)^T B(\lambda - \lambda^0) - (\mu - \mu^0)^T C(\mu - \mu^0)] + (v - v^0)^T D(v - v^0) +$$
$$+ (w - w^0)^T E(w - w^0)$$

(here A is an (n,n) matrix, B and D are (m,m) matrices, C and E are (s,s) matrices)

$$\tilde{M}_3(t) := M_{(x^0,v^0,w^0,w^1,\lambda^0,\mu^0,b,p,q)}(t) = \{(x,\lambda,\mu,v,w) \in {I\!\!R}^{n+2m+2s} \mid$$
$$th_i(x) + (1-t)(v_i - v_i^0) = 0, i \in I,$$
$$tg_j(x) + (1-t)(w_j - w_j^1) \le 0, j \in J,$$
$$-\mu_j \le 0, j \in J$$
$$\|x - x^0\|^2 + b^T x - p \le 0$$
$$\|\lambda - \lambda^0\|^2 + \|\mu - \mu^0\|^2 + \|v - v^0\|^2 + \|w - w^0\|^2 - q \le 0\},$$

$$h_i(x,t) := h_i(x) + (t-1)h_i(x^0), \ i \in I,$$
$$g_j(x,t) := g_j(x) + (t-1)|g_j(x^0)|, \ j \in J$$

(in [3] it is explained why we introduced $h_i(x,t), i \in I$ and $g_j(x,t), j \in J$ in the objective of $\tilde{P}_3(t)$).

We note that the Lagrange multiplier method follows a curve of local $\min_{x,v,w}$-$\max_{\lambda,\mu}$-points (i.e. a local minimizer with respect to x, v, w and a local maximizer with respect to λ, μ) that are saddle points in the set of stationary points for $\tilde{P}_3(t)$.

Then, it is easy to see:

THEOREM 1.3 (cf. [3], too). *$\tilde{P}_3(t)$ has the following properties*

(i) *Choosing $w^0, w^1 \in {I\!\!R}^s$ with $w_j^0 < w_j^1, j \in J$, $\mu^0 \in {I\!\!R}^s$ with $\mu_j^0 > 0, j \in J$, $x^0 \in {I\!\!R}^n$, $v^0, \lambda^0, c \in {I\!\!R}^m$, $\mu^0, d \in {I\!\!R}^s$, the matrices A, B, C, D, E positive definite, $p > 0$, $q > 0$ sufficiently large, then the point $(x^0, v^0, w^0, \lambda^0, \mu^0)$ is a global $\min_{x,v,w}$-$\max_{\lambda,\mu}$-point (i.e. a global minimizer with respect to x, v, w and a global maximizer with respect to λ, μ) for $\tilde{P}_3(0)$. Furthermore, $(x^0, v^0, w^0, \lambda^0, \mu^0)$ is a non-degenerate stationary point.*

(ii) *Analogously to Theorem 1.1(ii).*

(iii) *Analogously to Theorem 1.1(iii).*   □

Now we have the following important

REMARK 1.4 We observe that there is the following strict difference with respect to the starting situation between the penalty and the exact penalty embedding on the one hand, and the Lagrange multiplier embedding on the other hand: For the first two embeddings the known starting points are the only stationary points. For the Lagrange multiplier method at least one global minimizer $(x^1, w^1, \lambda^1, \mu^1)$ will appear, i.e., we have at least two stationary points for $\tilde{P}_i(0), i = 1, 2$. Therefore, we will show that the class of optimization problems for which the penalty and the exact penalty method are surely successful is much larger than the class where the Lagrange multiplier embedding is successful.   □

2 THEORETICAL BACKGROUND

We present a reduced version of Chapter 2 in [2]. We refer to [5,8,9,10,11,12,13,15] for more details.

We consider the one-parametric problem with some additional information that is important for our investigation:

$$P(t) \qquad \min\{f(x,t)|x \in M(t)\}, \tag{2.1}$$

where $t \in \mathbb{R}$, $M(t) = \{x \in \mathbb{R}^n \mid h_i(x,t) = 0, \ i \in I, \ g_j(x,t) \leq 0, \ j \in J\}$, and $f, h_i, g_j \in C^k(\mathbb{R}^n \times \mathbb{R}, \mathbb{R})$, $i \in I, j \in J, k \geq 2$.

Furthermore, we introduce the following notations:

$$\Sigma_{gc} := \{(x,t) \in \mathbb{R}^n \times \mathbb{R} \mid x \text{ is a generalized critical point}^{1)} \text{ of } P(t)\},$$
$$\Sigma_{\text{stat}} := \{(x,t) \in \mathbb{R}^n \times \mathbb{R} \mid x \text{ is a stationary point of } P(t)\},$$
$$\Sigma_{\text{loc}} := \{(x,t) \in \mathbb{R}^n \times \mathbb{R} \mid x \text{ is a local minimizer of } P(t)\},$$
$$H := (h_1, \ldots, h_m)^T, \quad G := (g_1, \ldots, g_s)^T.$$

The Linear Independence Constraint Qualification (briefly LICQ) is satisfied at $\bar{x} \in M(\bar{t})$ if the vectors $D_x h_i(\bar{x}, \bar{t})$, $i \in I$, $D_x g_j(\bar{x}, \bar{t})$, $j \in J_0(\bar{x}, \bar{t})$, are linearly independent $(J_0(x,t) := \{j \in J \mid g_j(x,t) = 0\})$.

The Mangasarian-Fromovitz Constraint Qualification (briefly MFCQ) is satisfied at $\bar{x} \in M(\bar{t})$ if:

(MF1) $D_x h_i(\bar{x}, \bar{t})$, $i \in I$, are linearly independent,

(MF2)   there exists a vector $\xi \in R^n$ with

$$D_x h_i(\bar{x}, \bar{t}) \xi = 0, \quad i \in I, \quad ^{2)}$$
$$D_x g_j(\bar{x}, \bar{t}) \xi < 0, \quad j \in J_0(\bar{x}, \bar{t}).$$

The KKT-system for a given problem $P(t)$ is fulfilled at a point $(\bar{x}, \bar{t})$ if there exists a point $\bar{y} \in \mathbb{R}^{m+s}$ such that $\mathcal{H}(\bar{x}, \bar{y}, \bar{t}) = 0$, where $\mathcal{H} : \mathbb{R}^{n+m+s+1} \to \mathbb{R}^{n+m+s}$ is defined by

$$\mathcal{H}(x,y,t) = \left\{ \begin{array}{l} D_x f(x,t) + \sum\limits_{i \in I} y_i D_x h_i(x,t) + \sum\limits_{j \in J} y_{m+j}^+ D_x g_j(x,t) \\ -h_i(x,t), i \in I \\ y_{m+j}^- - g_j(x,t), j \in J \end{array} \right\} \tag{2.2}$$

(for $\alpha \in \mathbb{R}$ let $\alpha^+ = \max\{\alpha, 0\}$ and $\alpha^- = \min\{\alpha, 0\}$). Obviously, the so-called Kojima-mapping $\mathcal{H}$ in (2.2) is piecewise continuously differentiable. In [13] the classical definition of a regular value of a continuously differentiable function is generalized for piecewise continuously differentiable functions. Furthermore, it is shown that if $0 \in \mathbb{R}^{n+m+s}$ is a regular value of $\mathcal{H}$, then the set $\mathcal{H}^{-1}(0)$ is a piecewise one-dimensional C^1-manifold (briefly PC^1-manifold).

---

[1] For the definition we refer to [12], see [8], too

[2] We consider the gradient $D_x h_i(\bar{x}, \bar{t})$ as a row vector.

Next, we cite our short characterization from [2] of the class $\mathcal{F}$ introduced by Jongen, Jonker and Twilt ([11,12]). In [12] the local structure of Σ_{gc} is completely described if (f, H, G) belongs to a C_s^3-open and dense subset $\mathcal{F}$ of $C^3(\mathbb{R}^n \times \mathbb{R}, \mathbb{R})^{1+m+s}$, where C_s^3 denotes the strong (or Whitney-) C^3-topology (see [8], too).

If $(f, H, G) \in \mathcal{F}$, then Σ_{gc} can be divided into 5 types.

Type 1: A point $\bar{z} = (\bar{x}, \bar{t}) \in \Sigma_{\mathrm{gc}}$ is of Type 1 if the following conditions are satisfied:
There exist $\bar{\lambda}_i, \bar{\mu}_j \in \mathbb{R}$, $i \in I$, $j \in J_0(\bar{z})$ with

$$\Big(D_x f + \sum_{i \in I} \bar{\lambda}_i D_x h_i + \sum_{j \in J_0(\bar{z})} \bar{\mu}_j D_x g_j\Big)\big|_{z=\bar{z}} = 0, \tag{2.3}$$

LICQ is satisfied at $\bar{x} \in \mathbf{M}(\bar{t})$, $\tag{2.4a}$

(therefore $\bar{\lambda}_i$, $\bar{\mu}_j$, $i \in I$, $j \in J_0(\bar{z})$ are uniquely defined)

$$\bar{\mu}_j \neq 0, \quad j \in J_0(\bar{z}), \tag{2.4b}$$

$$D_x^2 L(\bar{x}, \bar{t})|_{T(\bar{z})} \text{ is nonsingular}, \tag{2.4c}$$

where $D_x^2 L$ is the Hessian of the Lagrangian

$$L(x, t) = f(x, t) + \sum_{i \in I} \bar{\lambda}_i h_i(x, t) + \sum_{j \in J_0(\bar{z})} \bar{\mu}_j g_j(x, t),$$

and the uniquely determined numbers λ_i, μ_j are taken from (2.3). Furthermore,

$$T(z) = \{\xi \in \mathbb{R}^n \mid D_x h_i(z)\xi = 0, i \in I, D_x g_j(z)\xi = 0, j \in J_0(z)\}$$

is the tangent space at z. $D_x^2 L(z)|_{T(z)}$ represents $V^T D_x^2 L V$, where V is a matrix whose columns form a basis of $T(z)$.

A point of Type 1 is called a nondegenerate critical point. The set Σ_{gc} is the closure of the set of all points of Type 1, the points of the Types 2–5 constitute a discrete subset of Σ_{gc}. The points of the Types 2–5 represent three basic degeneracies:

Type 2 – violation of (2.4b)

Type 3 – violation of (2.4c)

Type 4 – violation of (2.4a) and $|I| + |J_0(\bar{z})| - 1 < n$

Type 5 – violation of (2.4a) and $|I| + |J_0(\bar{z})| = n + 1$.

For each of these five types Figure 2.1 illustrates the local structure of Σ_{gc} in the neighbourhood of stationary points. Let Σ_{gc}^ν, $\nu \in \{1, \ldots, 5\}$ be the set of g.c. points of Type ν. Figure 2.2 illustrates the local structure of $\mathcal{F}$ in Σ_{loc} and Σ_{stat}. The class $\mathcal{F}$ is defined by

$$\mathcal{F} = \Big\{(f, H, G) \in C^3(\mathbb{R}^n \times \mathbb{R}, \mathbb{R})^{1+m+s} \mid \Sigma_{\mathrm{gc}} \subset \bigcup_{\nu=1}^{5} \Sigma_{\mathrm{gc}}^\nu\Big\}.$$

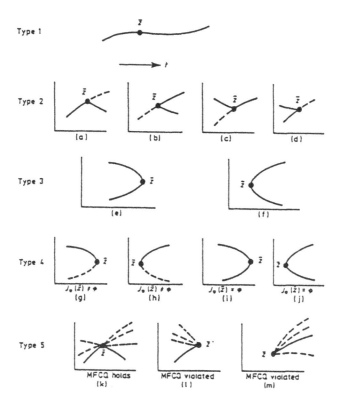

The full curve stands for the curve of stationary points $z = (x, t)$, and the dotted curve represents the curve of g.c. points which are not stationary points.

Figure 2.1

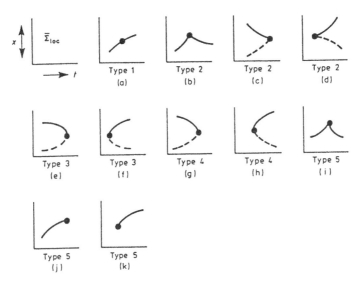

The full curve stands for a curve of local minimizers and the dotted curve in Fig. 2.2(c), (d), (e), (f) represents a curve of stationary points not being local minimizers. The dotted curve in Fig. 2.2(g), (h) stands for a curve of stationary points in case $J_0(\bar{x}, \bar{t}) = \emptyset$.

Figure 2.2

The following theorem provides a special perturbation of (f, H, G) with additional parameters that can be chosen arbitrarily small such that the perturbed function vector belongs to the class $\mathcal{F}$.

THEOREM 2.1 (cf. [15]). *Let* $(f, H, G) \in C^3(\mathbb{R}^n \times R, R^{1+m+s})$. *Then, for almost all* $(b, A, c, D, e, F) \in \mathbb{R}^n \times \mathbb{R}^{n(n+1)/2} \times \mathbb{R}^m \times \mathbb{R}^{mn} \times \mathbb{R}^s \times \mathbb{R}^{sn}$, *we have*

$$(f(x, t) + b^T x + x^T Ax, H(x, t) + c + Dx, G(x, t) + e + Fx) \in \mathcal{F}.$$

Here "almost all" means: each measurable subset of

$$\{(b, A, c, D, e, F) \mid (f(x, t) + b^T x + x^T Ax, H(x, t) + c + Dx, G(x, t) + e + Fx) \notin \mathcal{F}\}$$

has the Lebesgue-measure zero. □

REMARK 2.2 (cf. [15]). Considering Σ_{stat} we note that the condition $(f, H, G) \in \mathcal{F}$ implies that zero is a regular value of the Kojima-mapping $\mathcal{H}$. □

DEFINITION 2.3 Let $K \subseteq \mathbb{R} \cup \{\pm\infty\}$.

(i) The problem $P(t)$ is called regular in the sense of Jongen-Jonker-Twilt – briefly JJT-regular – (with respect to K) if $(f, H, G) \in \mathcal{F}\left((\mathbb{R}^n \times K) \cap \Sigma_{\text{gc}} \subset \bigcup_{\nu=1}^{5} \Sigma_{\text{gc}}^{\nu}\right)$.

(ii) The problem $P(t)$ is called regular in the sense of Kojima-Hirabayashi – briefly KH-regular – (with respect to K) if $0 \in \mathbb{R}^{n+m+s}$ is a regular value of $\mathcal{H}$ $(\mathcal{H}|_{\mathbb{R}^n \times \mathbb{R}^m \times \mathbb{R}^s \times K})$.

THEOREM 2.4 (cf. [13]). *Let* $(f, H, G) \in C^3(\mathbb{R}^n \times \mathbb{R}, \mathbb{R})^{1+m+s}$. *Then, for almost all* $(b, c, d) \in \mathbb{R}^n \times \mathbb{R}^m \times \mathbb{R}^s$, *the problem*

$$P_{(b,c,d)}(t) \quad \min\{f(x,t) + b^T x \left| \begin{array}{l} h_i(x,t) + c_j = 0, \ i \in I, \\ g_j(x,t) + d_j \leq 0, \ j \in J \end{array} \right\}$$

is KH-regular.   □

Now, we present some facts about pathfollowing methods (for more details see [8]). For this, we assume (A1),(A2) and the JJT-regularity of the considered problem $P(t)$.

The algorithm PATH III – which is included in the so-called computer program PAFO – (cf. [8,7,16]) computes a numerical description of a compact connected component in Σ_{gc} and Σ_{stat}, respectively.

In the last part of Chapter 2 we present two theorems that are essential for our analysis.

THEOREM 2.5 (cf. [5]). *We assume*

(C1) $M(t)$ *is non-empty and there exists a compact set C with $M(t) \subseteq C$ for all $t \in [0, 1]$.*

(C2) $P(t)$ *is KH-regular with respect to $[0, 1]$.*

(C3) *There exists a $t_1 > 0$ and a continuous function $x : [0, t_1) \to \mathbb{R}^n$ such that $x(t)$ is the unique stationary point for $P(t)$ for $t \in [0, t_1)$.*

(C4) *MFCQ is satisfied for all $x \in M(t)$ for all $t \in [0, 1]$.*

Then there exists a PC^1-path in Σ_{stat} that connects $(x^0, 0)$ with some point $(x^*, 1)$.   □

Applying Remark 2.2 we obtain

COROLLARY 2.6 *We assume* (C1), (C3), (C4) *and*

(D2) $P(t)$ *is JJT-regular with respect to $[0, 1]$.*

Then there exists a PC^2-path $K(x^0, 0)$ in Σ_{stat} connecting $(x^0, 0)$ with some point $(x^*, 1)$. Furthermore if $(x, t) \in K(x^0, 0)$ than (x, t) belongs to $\bigcup_{\nu \in \{1,2,3,5\}} \Sigma_{\text{gc}}^{\nu}$.   □

REMARK 2.7 Assume (C4). Now we have a look at Fig. 2.2. Since the MFCQ is satisfied, points of Type 5 in (j) and (k) are excluded.   □.

Finally we present a consequence of a general topological stability result given in [9]:

THEOREM 2.8 (cf. [9]). *We assume* (C1) *and* (C4). *Then $M(t_1)$ is homeomorphic with $M(t_2)$ for all $t_1, t_2 \in [0, 1]$.*   □

3 ON CONNECTING CURVES

Considering Theorem 2.5, Corollary 2.6 and Theorem 2.8 we first ask for a condition on the original problem (P) ensuring that the MFCQ is satisfied for all $(x, v, w) \in M_1(t)$, $(x, w) \in M_2(t)$, and $(x, v, w, \lambda, \mu) \in M_3(t)$ for all $t \in [0, 1]$. We discuss the so-called Enlarged Mangasarian-Fromovitz Constraint Qualification (briefly EnMFCQ) introduced in 1985 in [5], and, simultaneously in [14] in a modified version, later on used in [8], [6], [2] and in further papers of di Pillo, Grippo et al.

DEFINITION 3.1 Assume (B1) and $p \geq \tilde{p}$ (cf. (B2)). The EnMFCQ is satisfied if, for each $x \in E(x^0, b, p)$:

(a) $Dh_i(x)$, $i \in I$, are linearly independent,

(b) there exists a vector $\xi \in I\!\!R^n$ with

$$h_i(x) + Dh_i(x)\xi = 0, \quad i \in I,$$
$$g_j(x) + Dg_j(x)\xi < 0, \quad j \in J_+(x) := \{j \in J \mid g_j(x) \geq 0\},$$
$$(2x - 2x^0 + b)^T \xi < 0, \quad \text{if } \|x - x^0\|^2 + b^T x = p. \qquad \square$$

The answer to the question above was published for $\tilde{M}_i(t)$, $i \in \{1, 2\}$ in [6] and [2]

THEOREM 3.2 *Assume* (B1), (B2), *and EnMFCQ. Then we have*

(i) $\tilde{M}_i(t) \neq \emptyset$, $\tilde{M}_i(t) \subseteq C_i$, $i \in \{1, 2, 3\}$ *for all* $t \in [0, 1]$, *where* C_i *are compact sets.*

(ii) *The MFCQ is satisfied for all* $(x, v, w) \in \tilde{M}_1(t)$, $(x, w) \in \tilde{M}_2(t)$ *and* $(x, v, w, \lambda, \mu) \in \tilde{M}_3(t)$ *for all* $t \in [0, 1]$ *if we choose* $\mu_j^0 > 0$, $j \in J$.

Proof. (i) obvious.

(ii) 1) W.r.t. $\tilde{M}_1(t)$: The proof runs analogously the main part of the proof of Theorem 4.2 in [5].

(ii) 2) W.r.t. $\tilde{M}_2(t)$: The proof is given in [2].

(ii) 3) W.r.t. $\tilde{M}_3(t)$: For $y := (x, v, w, \lambda, \mu) \in I\!\!R^{n+2m+2s}$ we consider

$$\tilde{M}_3(t) := \{y \in I\!\!R^{n+2m+2s} \mid H_i(y, t) := th_i(x) + (1 - t)(v_i - v_i^0) = 0, \ i \in I,$$
$$(G_1)_j(y, t) := tg_j(x) + (1 - t)(w_j - w_j^1) \leq 0, \ j \in J,$$
$$(G_2)_j(y, t) := -\mu_j \leq 0, \ j \in J,$$
$$(G_3)(y, t) := \|\lambda - \lambda^0\|^2 + \|\mu - \mu^0\|^2 + \|v - v^0\|^2 + \|w - w^0\|^2 - q \leq 0,$$
$$(G_4)(y, t) := \|x - x^0\|^2 + b^T x - p \leq 0\}, \ t \in [0, 1].$$

We will discuss the three cases of $t = 0$, $t \in (0, 1)$ and $t = 1$ separately. For $y \in \tilde{M}_3(t)$ we define

$$(J_1)_0(y, t) = \{j \in J \mid (G_1)_j(y, t) = 0\},$$
$$(J_2)_0(y, t) = \{j \in J \mid (G_2)_j(y, t) = 0\}.$$

Obviously, $D_y H_i(y, t)$, $i \in I$, are linearly independent for all $t \in [0, 1]$.

Case I: $t = 0$.

Let $y \in \tilde{M}_3(0)$. If $\|x - x^0\|^2 + b^T x = p$ choose $\xi \in \mathbb{R}^n$ from Definition 3.1(b). Otherwise take $\xi \in \mathbb{R}^n$ arbitrarily. We need a vector $(\eta^\lambda, \eta^\mu, \eta^v, \eta^w) \in \mathbb{R}^m \times \mathbb{R}^s \times \mathbb{R}^m \times \mathbb{R}^s$ such that $(\xi, \eta^\lambda, \eta^\mu, \eta^v, \eta^w)$ is a Mangasarian-Fromovitz-vector (briefly MF-vector) for $\tilde{M}_3(0)$. Take $\eta^\lambda = -2(\lambda - \lambda^0)$, $\eta_j^\mu = 1$, $j \in (J_2)_0(y, 0)$, $\eta_j^\mu = 2(\mu_j - \mu_j^0)$, $j \in J \setminus (J_2)_0(y, 0)$, $\eta^v = 0$, $\eta_j^w = -1$, $j \in (J_1)_0(y, 0)$, $\eta_j^w = -2(w_j - w_j^0)$, $j \in J \setminus (J_1)_0(y, 0)$.

Case II: $t \in (0, 1)$.

Let $y \in \tilde{M}_3(t)$ and $(J_1)_0 := (J_1)_0(y, t)$, $(J_2)_0 := (J_2)_0(y, t)$. Choose $\xi \in \mathbb{R}^n$ from Definition 3.1(b). Take $\eta^\lambda = -2(\lambda - \lambda^0)$, $\eta_j^\mu = 1$, $j \in (J_2)_0$, $\eta_j^\mu = -2(\mu_j - \mu_j^0)$, $j \in J \setminus (J_2)_0$,
$\eta_i^v = \frac{t}{t-1} Dh_i(x)\xi$, $i \in I$,

$\eta_j^w \in \left(0, \frac{t}{t-1} Dg_j(x)\xi\right)$, $j \in (J_1)_0 \cap J_+(x)$ and $(w_j - w_j^1) \leq 0$,

$\eta_j^w < 0$, $j \in (J_1)_0 \cap J_+(x)$ and $w_j - w_j^1 > 0$,

$\eta_j^w < 0$, $j \in (J_1)_0 \setminus J_+(x)$ if $Dg_j(x)\xi < 0$,

$\eta_j^w < \frac{t}{t-1} Dg_j(x)\xi$, $j \in (J_1)_0 \setminus J_+(x)$ if $Dg_j(x)\xi \geq 0$,

$\eta_j^w = -2(w_j - w_j^0)$, $j \in J \setminus (J_1)_0$.

Then $(\xi, \eta^\lambda, \eta^\mu, \eta^v, \eta^w)$ is an MF-vector for all $y \in \tilde{M}_3(t)$.

Case III: $t = 1$.

Let $y \in \tilde{M}_3(1)$. Choose $\xi \in \mathbb{R}^n$ from Definition 3.1(b). Take $\eta^\lambda = -2(\lambda - \lambda^0)$, $\eta_j^\mu = 1$, $j \in (J_2)_0(y, 1)$, $\eta_j^\mu = -2(\mu_j - \mu_j^0)$, $j \in J \setminus (J_2)_0(y, 1)$, $\eta^v = -2(v - v^0)$, $\eta^w = -2(w - w^0)$.

Then $(\xi, \eta^\lambda, \eta^\mu, \eta^v, \eta^w)$ is an MF-vector for all $y \in \tilde{M}_3(1)$. $\square$

Using Theorem 2.8 we obtain

COROLLARY 3.3 *Assume* (B1) *and* (B2). *Let the EnMFCQ be fulfilled. Then, for* $i \in \{1, 2, 3\}$, $\tilde{M}_i(t_1)$ *is homeomorphic with* $\tilde{M}_i(t_2)$ *for all* $t_1, t_2 \in [0, 1]$. $\square$.

For $i \in \{1, 2\}$ this result is already included in [6] and [2].

REMARK 3.4 From Corollary 3.3 it follows, in particular, that $\tilde{M}_i(0)$ is homeomorphic with $\tilde{M}_i(1)$. Using the special structure of $\tilde{M}_i(0)$, the set $\tilde{M}_i(0)$ is a convex set for $i \in \{1, 2, 3\}$. This means that $\tilde{M}_i(1)$ is homeomorphic with a convex set and therefore, an unconnected feasible set M of the original problem is excluded for the considered embeddings. This show us how restrictive the EnMFCQ is, but restrictive to the constraint functions only. $\square$

Now, we give an answer to the question whether there exists a "connecting curve between $t = 0$ and $t = 1$" for the three embeddings using Theorem 2.5 and Corollary 2.6. We remind the reader of Remark 1.4 which states that the assumption (C3) is not satisfied for the embedding $\tilde{P}_3(t)$. Hence, Theorem 2.5 and Corollary 2.6 are not applicable.

THEOREM 3.5 *Assume*

(a) $f, h_i, g_j \in C^3(\mathbb{R}^n, \mathbb{R})$, $i \in I$, $j \in J$.

(b) *(B1), (B2), and EnMFCQ.*

(c) $P_1(t)$ *(and* $P_2(t)$, *respectively) is JJT-regular with respect to* $(0, 1]$.

(d) *Choosing* A, B, C (A, B) *positive definite (in such a way that the Hessian* $D^2 F_2(x^0, w^0, 0)$ *is positive definite) and* $w^0, w^1 \in \mathbb{R}^s$ *with* $w_j^0 < w_j^1$, $j \in J$, $(w_j^1 < w_j^0, j \in J)$, $x^0, b \in \mathbb{R}^n$, *with* $b^T x^0 - p < 0$, $v^0 \in \mathbb{R}^m$, $p, q \in \mathbb{R}$ *with* $p > \tilde{p}$ *and* $q > \tilde{q}$ $(x^0, b \in \mathbb{R}^n, \alpha \in \mathbb{R}^s$ *with* $\alpha_j > g_j(x^0)$, $j \in J$, $p, q \in \mathbb{R}$ *with* $p > \tilde{p}$ *and* $q > \tilde{q}$).

Then there exists a P^2-*path in* Σ_{stat} *connecting* $(x^0, v^0, w^0, 0)$ $((x^0, w^0, 0))$ *with some point* $(\hat{x}, \hat{v}, \hat{w}, 1)$ $((\hat{x}, \hat{w}, 1))$, *where* $\hat{x}$ *is a stationary point of* (P) *and* $\Sigma_{\text{stat}}|_{[0,1]} = \bigcup_{\nu=1}^{3} \Sigma_{\text{gc}}^\nu \cap \Sigma_{\text{stat}}$ *for* $n > 1$ *(for* $n = 1$ *a point of Type 5 may appear).*

Remarks to the proof. We check the assumptions of Corollary 2.6: For (C1) see Theorem 1.1 and Theorem 1.2. Furthermore, (x^0, v^0, w^0) $((x^0, w^0))$ is a point of Type 1 and the only stationary point for $\tilde{P}_1(0)$ $(\tilde{P}_2(0))$. Then (C3) is satisfied and (D2) holds, too. Using Theorem 3.2 MFCQ is satisfied for all $(x, v, w) \in \tilde{M}_1(t)$ $((x, w) \in \tilde{M}_2(t))$ for all $t \in [0, 1]$, points of Type 4 cannot appear. Using the special structure of the two-parametric optimization problems, points of Type 5 cannot appear for $n > 1$. $\square$

In the following we discuss the assumption (c) of Theorem 3.5. Let $\mathcal{A} \subset \mathbb{R}^{\frac{1}{2}n(n+1)}$ $(\mathcal{B} \subset \mathbb{R}^{\frac{1}{2}m(m+1)}, \mathcal{C} \subset \mathbb{R}^{\frac{1}{2}s(s+1)})$ be the set of all symmetric nonsingular $(n, n)((m, m), (s, s))$ matrices. Obviously, $\mathcal{A}(\mathcal{B}, \mathcal{C})$ is an open and dense subset of $\mathbb{R}^{\frac{1}{2}n(n+1)}(\mathbb{R}^{\frac{1}{2}m(m+1)}, \mathbb{R}^{\frac{1}{2}s(s+1)})$ and $\mathbb{R}^{\frac{1}{2}n(n+1)} \setminus \mathcal{A}$ $(\mathbb{R}^{\frac{1}{2}m(m+1)} \setminus \mathcal{B}, \mathbb{R}^{\frac{1}{2}s(s+1)} \setminus \mathcal{C})$ has the Lebesgue measure 0.

THEOREM 3.6 *Assume that* $f, h_i, g_j \in C^3(\mathbb{R}^n, \mathbb{R})$, $i \in I$, $j \in J$, *and that* $\tilde{P}_i(1)$ *is JJT-regular (i.e.,* $\tilde{P}_i(t)$ *is JJT-regular with respect to* $K = \{1\}$), $i \in \{1, 2\}$.

(i) *(with respect to* $\tilde{P}_1(t)$). *Then,* $\tilde{P}_1(t)$ *is JJT-regular with respect to* $[0, 1]$ *for almost all* $w^0, w^1 \in \mathbb{R}^s$ *with* $w_j^0 < w_j^1$, $j \in J$, $x^0, b \in \mathbb{R}^n$ *with* $b^T x^0 - p < 0$, $v^0 \in \mathbb{R}^m$, $p, q \in \mathbb{R}$ *with* $p > \tilde{p}$ *and* $q > \tilde{q}$, $A \in \mathcal{A}$, $B \in \mathcal{B}$, $C \in \mathcal{C}$.

(ii) *(with respect to* $\tilde{P}_2(t)$). *Then,* $\tilde{P}_2(t)$ *is JJT-regular with respect to* $[0, 1]$ *for almost* $w^0, w^1 \in \mathbb{R}^s$ *with* $w_j^1 < w_j^0$, $j \in J$, $x^0, b \in \mathbb{R}^n$ *with* $b^T x^0 - p < 0$, $\alpha \in \mathbb{R}^s$ *with* $\alpha_j > g_j(x^0)$, $j \in J$, $p, q \in \mathbb{R}$ *with* $p > \tilde{p}$ *and* $q > \tilde{q}$, $A \in \mathcal{A}$, $B \in \mathcal{B}$.

Remark to the proof. $\tilde{P}_i(t)$, $i \in \{1, 2\}$, is JJT-regular with $t \in [0, 1)$ using the special perturbations and we follow the explanation with respect to the proof of Theorem 5.1 in [6], cf. also [15]. The JJT-regularity for $\tilde{P}_i(1)$ is necessary for the extension to the closed interval $[0, 1]$. $\square$

Now we consider $\tilde{P}_1(t)$ with $A = I^n$, $B = I^m$, $C = I^s$, $b = 0$ and $\tilde{P}_2(t)$ with $A = k \cdot I^n$, $B = k \cdot I^s$, k sufficiently large, $b = 0$ and answer the same questions.

COROLLARY 3.7 (cf. [6] and [2], too). *Assume*

(a) $f, h_i, g_j \in C^2(I\!\!R^n, I\!\!R)$, $i \in I$, $j \in J$.

(b) (B1), (B2), *and EnMFCQ*.

(c) $\tilde{P}_1(t)$ *with* $A = I^n$, $B = I^m$, $C = I^s$, $b = 0$ ($\tilde{P}_2(t)$ *with* $A = k \cdot I^n$, $B = k \cdot I^s$, k *sufficiently large*, $b = 0$, *respectively*) *is KH-regular with respect to* $(0, 1]$.

Choosing $w^0, w^1 \in I\!\!R^s$ *with* $w_j^0 < w_j^1$, $j \in J$ ($w_j^1 < w_j^0$, $j \in J$), $x^0 \in I\!\!R^n$, $v^0 \in I\!\!R^m$, $p, q \in I\!\!R$ *with* $p > \tilde{p}$ *and* $q > \tilde{q}$ ($x^0 \in I\!\!R^n$, $\alpha \in I\!\!R^s$ *with* $\alpha_j > g_j(x^0)$, $j \in J$, $p, q \in I\!\!R$ *with* $p > \tilde{p}$ *and* $q > \tilde{q}$), *there exists a* PC^1-*path in* Σ_{stat} *connecting* $(x^0, v^0, w^0, 0)$ $((x^0, w^0, 0))$ *with some point* $(\hat{x}, \hat{v}, \hat{w}, 1)$ $((\hat{x}, \hat{w}, 1))$, *where* $\hat{x}$ *is a stationary point of* (P). $\square$

COROLLARY 3.8 (cf. [6] and [2], too). *Assume that* $f, h_i, g_j \in C^2(I\!\!R^n, I\!\!R)$, $i \in I$, $j \in J$, *and that* $\tilde{P}_i(1)$ *is KH-regular* (*i.e.*, $\tilde{P}_i(t)$ *is KH-regular with respect to* $K = \{1\}$), $i \in \{1, 2\}$.

(i) (*with respect to* $\tilde{P}_1(t)$ *with* $A = I^n$, $B = I^m$, $C = I^s$, $b = 0$). *Then* $\tilde{P}_1(t)$ *is KH-regular with respect to* $[0, 1]$ *for almost all* $w^0, w^1 \in I\!\!R^s$ *with* $w_j^0 < w_j^1$, $j \in J$, $x^0 \in I\!\!R^n$, $v^0 \in I\!\!R^m$, $p, q \in I\!\!R$ *with* $p > \tilde{p}$ *and* $q > \tilde{q}$.

(ii) (*with respect to* $\tilde{P}_2(t)$ *with* $A = I^n$, $B = I^s$, $b = 0$). *Then* $\tilde{P}_2(t)$ *is KH-regular with respect to* $[0, 1]$ *for almost all* $w^0, w^1 \in I\!\!R^s$ *with* $w_j^1 < w_j^0$, $j \in J$, $x^0 \in I\!\!R^n$, $\alpha \in I\!\!R^s$ *with* $\alpha_j > g_j(x^0)$, $j \in J$, $p, q \in I\!\!R$ *with* $p > \tilde{p}$ *and* $q > \tilde{q}$. $\square$

We observe that the EnMFCQ is a sufficient condition for a "connecting curve between $t = 0$ and $t = 1$". Now we ask for a necessary and sufficient condition. We know that for almost all inputs x^0, w^0, w^1, etc. (cf. Theorem 3.5) the starting point (x^0, v^0, w^0) for $\tilde{P}_1(0)$ and (x^0, w^0) for $\tilde{P}_2(0)$, respectively, lies on a uniquely determined connected component $K(x^0, v^0, w^0, 0)$ and $K(x^0, w^0, 0)$, respectively (briefly $K(y^0, 0)$ for both embeddings). Furthermore, we know that $K(y^0, 0)|_{[0,1]}$ is the only connected component in Σ_{stat} crossing the hyperplane $\{(y, t) \mid t = 0\}$. We introduce the following condition (cf. [2] for $\tilde{P}_2(t)$):

(F1) The MFCQ is satisfied for all $y \in \tilde{M}_i(t)$, $i \in \{1, 2\}$, with $(y, t) \in \text{cl } K(y^0, 0)|_{[0,1]}$ (cl K means the closure of K).

THEOREM 3.9 (with respect to $\tilde{P}_1(t)$ (and $\tilde{P}_2(t)$, respectively)).
Assume (a), (c) *and* (d) *from Theorem 3.5. Then there exists a* PC^2-*path in* Σ_{stat} *connecting* $(x^0, v^0, w^0, 0)$ $((x^0, w^0, 0))$ *with some point* $(\hat{x}, \hat{v}; \hat{w}, 1)$ $((\hat{x}, \hat{w}, 1))$, *where* $\hat{x}$ *is a stationary point of* (P) *if and only if* (F1) *is satisfied.*

Remark to the proof. Use the same concept of proof as in Theorem 2.5 (cf. [5]).

REMARK 3.10 This weakest condition for the existence of a "connecting curve between $t = 0$ and $t = 1$" is of theoretical nature, but we can check the condition for a given problem by PAFO ([7,16]). Once more, the MFCQ plays an essential role. $\square$

4   ILLUSTRATIVE EXAMPLES

(P) $\min\{f(x) \mid g(x) \le 0\}$

EXAMPLE 4.1

$$f(x) = 0.0265073509x^8 - 0.211505207x^7 + 0.25753848x^6 + 1.34579642x^5$$
$$-2.34222067x^4 - 2.65029635x^3 + 3.45664738x^2 + 0.91447716x + 5,$$
$$g(x) = 10(x - 2.5)^2 - 5.$$

Fig. 4.1 shows $f(x)$ and $g(x)$.

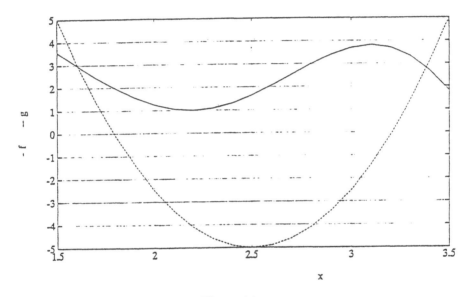

Figure 4.1

Of course, the EnMFCQ is satisfied.

We have chosen the penalty embedding $\tilde{P}_1(t)$ with $A = I^n$, $C = I^s$, $b = 0$, $p = 25$, $q = 36$, $x^0 = -2$, $w^0 = 1$, $w^1 = 0$. Figures 4.2 and 4.3 show the curve of stationary points.

Note that all curves were computed by PAFO ([7,16]). We see that the penalty method (with increasing t) is successful. One point of Type 2 appears.

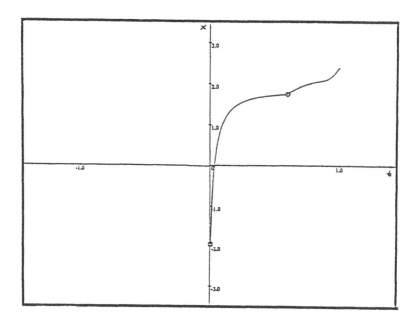

Fig. 4.2

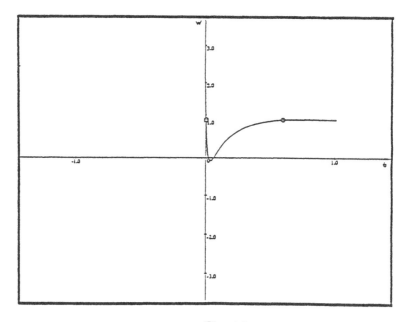

Fig. 4.3

We have chosen the exact penalty embedding $\tilde{P}_2(t)$ with $A = I^n$, $B = I^s$, $\alpha = 200$, $p = 25$, $q = 36$, $x^0 = -2$, $w^0 = 1$, $w^1 = 0$. Fig. 4.4 and 4.5 show the curve of stationary points.

We see that the exact penalty method (with increasing t) is successful. Two points of Type 2 appear.

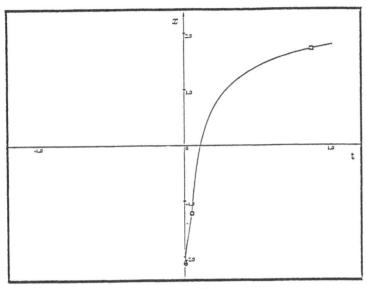

Fig. 4.4

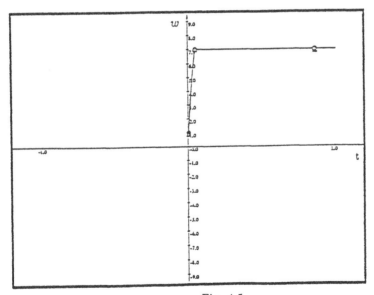

Fig. 4.5

We have chosen the Lagrange multiplier embedding $\tilde{P}_3(t)$ with $A = I^n$, $C = E = I^s$, $p = 25$, $q = 36$, $x^0 = -2$, $w^0 = 1$, $w^1 = 2$, $\mu^0 = 1$. The starting point (x^0, w^0, μ^0) is a global $\min_{x,w}$-$\max_{\mu}$-point for $\tilde{P}_3(0)$. The corresponding curve (cf. Fig. 4.6 – 4.8) does not attain $\{(x, w, \mu, t) \mid t = 1\}$ and returns to the local minimizer $(\bar{x}, \bar{w}, \bar{\mu}) = (-2, 1, 0)$ by using the pathfollowing procedure only. The example shows that the Lagrange multiplier method is not necessarily successful under EnMFCQ. We see that there is another connected component in Σ_{stat} where $(\tilde{x}, \tilde{w}, \tilde{\mu}, 0) = (-2, 1, 7, 0)$ belongs to. $(\tilde{x}, \tilde{w}, \tilde{\mu})$ is the global minimizer for $\tilde{P}_3(0)$. A jump from the first connected component to the second one is possible (cf. Remark 5.3). We observe that the pathfollowing procedure applied to $\tilde{P}_3(t)$ and the second connected component has nothing to do with the classical Lagrange multiplier method (following a curve of $\min_{x,v,w}$-$\max_{\lambda,\mu}$-points of $\tilde{P}_3(t)$).

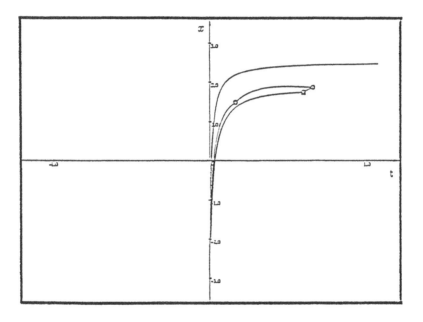

Fig. 4.6

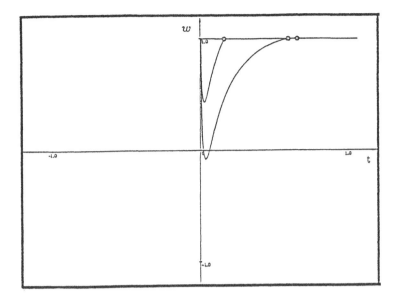

Fig. 4.7

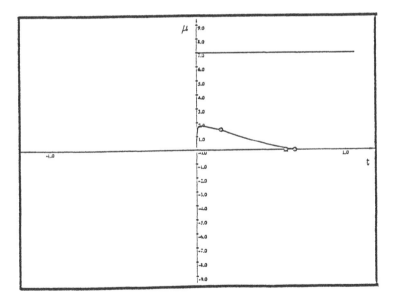

Fig. 4.8

Choose the same embedding $\tilde{P}_3(t)$ and the following example

EXAMPLE 4.2

$$f(x) = 0.0265073509x^8 - 0.211505207x^7 + 0.25753848x^6 + 1.34579642x^5$$
$$-2.34222067x^4 - 2.65029635x^3 + 3.45664738x^2 + 0.91447716x + 5,$$
$$g(x) = 0.1x^6 - 0.3x^5 - 0.25x^4 + x^3 - 2x + 4,$$
$$p = 25, \ q = 36, \ x^0 = -2, \ w^0 = 1, \ w^1 = 2, \ \mu^0 = 1.$$

Fig. 4.9 shows $f(x)$ and $g(x)$.

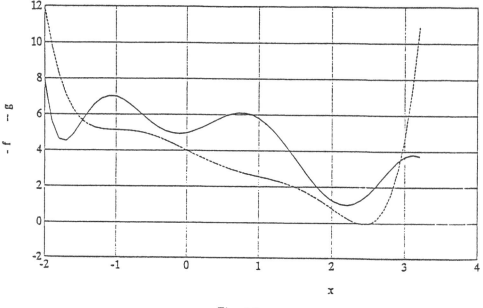

Fig. 4.9

We see that the EnMFCQ is satisfied.

Fig. 4.10 shows the curve of stationary points. We have two points of Type 2 and four points of Type 3, and we attain $\{(x, w, \mu, t) \mid t = 1\}$.

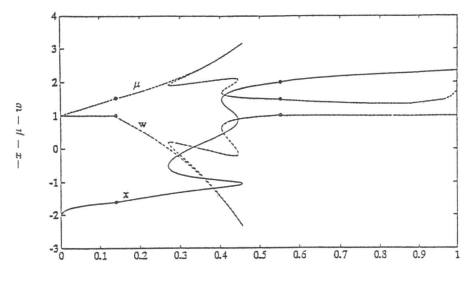

Fig. 4.10

Example 4.2 illustrates that the Lagrange multiplier embedding could also be successful. Since we have four points of Type 3, we have to change the direction of t four times (increasing – decreasing etc.). Of course, the character of the stationary point will be changed at each turning point, i.e., in particular after the first turning point we are outside the classical Lagrange multiplier method from two points of view: t decreases and we follow a curve of stationary points of $\tilde{P}_3(t)$ that are not $\min_{x,w}\text{-}\max_{\mu}$-points.

5 CONCLUDING REMARKS

REMARK 5.1 We present arguments to the following thesis: "Penalty and exact penalty methods are successful for a larger class of optimization problems than the Lagrange multiplier method, more precisely: The Lagrange multiplier method is definitely successful for convex optimization problems only, provided there exists "a connecting curve between $t = 0$ and $t = 1$" for the penalty and exact penalty embedding under EnMFCQ (a condition to the constraint functions only) and certain regularity conditions."

(i) We observe (cf. Remark 1.4) that the starting point (x^0, v^0, w^0) for the penalty embedding and (x^0, w^0) for the exact penalty embedding, respectively, is the only stationary point for $\tilde{P}_1(0)$ and $\tilde{P}_2(0)$, respectively, and it is non-degenerate, i.e., the condition (C3) of Theorem 2.5 is satisfied. For the Lagrange multiplier embedding, the starting point $(x^0, v^0, w^0, \lambda^0, \mu^0)$ is a $\min_{x,v,w}\text{-}\max_{\lambda,\mu}$-point and a stationary point for $\tilde{P}_3(0)$, but not the only one, because there exists at least a global minimizer with respect to (x, v, w, λ, μ) that is a stationary point, too. Considering Example 4.1 we have at least 3 stationary points for $\tilde{P}_3(0)$: The min-max-point $(-2, 1, 1)$ as the known starting point and a local minimizer $(-2, 1, 0)$, both are connected by a curve of stationary points (cf. Figures 4.6 – 4.8). Besides, there exists the global minimizer $(-2, 1, 7)$ for $\tilde{P}_3(0)$, which is a stationary point, too. Example 4.1 satisfies the EnMFCQ because the constraint function $g(x)$ is strictly convex. Figures 4.6 – 4.8 show that the curve connecting $(-2, 1, 1)$ and $(-2, 1, 0)$ returns to $\{(x, v, w, \mu, t) \mid t = 0\}$ at a turning point of Type 2 before the curve attains $\{(x, v, w, \mu, t) \mid t = 1\}$. The penalty and exact penalty method is successful (Example 4.1 illustrates Theorem 3.5).

(ii) The class for which the pathfollowing procedure definitely computes a stationary point for (P) using the penalty embedding $\tilde{P}_1(t)$, $t \in [0, 1]$, and the exact penalty embedding $\tilde{P}_2(t)$, $t \in [0, 1]$, is restricted by the condition (F1) (see Theorem 3.9). The sufficient condition EnMFCQ is a condition to constraint functions only. The objective can be "really non-convex" (including functions with local minimizers and local maximizers that are non-degenerate). Such functions are "only" excluded from the constraint functions.

(iii) Now we give an answer why the Lagrange multiplier method is surely successful for convex problems only. Suppose that (P) is a non-convex optimization problem, then we cannot exclude the appearance of turning points of Type 2 and Type 3. That means, the same situation as in Example 4.1 (Figures 4.6 – 4.8) is possible.

□

REMARK 5.2 The classical penalty, exact penalty and Lagrange multiplier method (with increasing t) are not necessarily successful if there exists a "curve connecting $t = 0$ and $t = 1$". We know that turning points (Type 2 and Type 3) may appear. Then the original methods (for increasing t) stop at such a point. We have to go back. We do not have any problems using PAFO, but we leave the original method. In particular, the character of the stationary point will be changed, for instance: We follow a curve of local minimizers and a point of Type 3 appears, then we have to go back and follow a curve of saddle-points. Here, we refer to Remark 3.1 in [6].
Considering the Lagrange multiplier method which follows a curve of $\min_{\substack{x,v,w \\ \lambda,\mu}}$-max-points (cf. Example 4.2 and Fig. 4.10 where a turning point appears) the character of the stationary point will be changed. We have left the classical Lagrange multiplier method from two points of view: t decreases and we follow a curve of stationary points that are not $\min_{\substack{x,v,w \\ \lambda,\mu}}$-max-points.   □

REMARK 5.3 Assuming that we are not successful with the pathfollowing procedure only, then we refer to our investigation in [2], [3], [6], [8], [10] using pathfollowing methods in the set Σ_{gc} instead of the set Σ_{stat}. In case we are not successful either, we propose pathfollowing procedures with jumps from one connected component to another one in the set Σ_{loc} and Σ_{gc}. We refer to the algorithms JUMP I and JUMP II in [8]. A summary is presented in Chapter 2 in [6] and [3]. JUMP I follows a curve of local minimizers and jumps to another connected component by finding a direction of descent if points of Type 2 or Type 3 appear. If we pass a point $(\bar{x}, \bar{t})$ of Type 4, then the local minimizer turns into a local maximizer. Then we have two cases:

Case I: The value of $f(\cdot, \bar{t})$ decreases.

Case II: The value of $f(\cdot, \bar{t})$ increases.

In Case I it is possible to jump to another branch of local minimizers. In Case II the corresponding connected component of the feasible set becomes empty and we do not have a jump. We note that the MFCQ is violated at a point of Type 4, i.e., under EnMFCQ we are able to jump to another connected component in Σ_{loc} of $\tilde{P}_i(t)$, $i \in \{1, 2\}$, if a point of Type 2 or 3 appears.

This is another possibility in order to reach the level $t = 1$ for increasing t, but the path-following procedure (with increasing and decreasing t) seems to be more efficient. With respect to the Lagrange multiplier embedding $\tilde{P}_3(t)$ we have to use the jumps described in [10] (for a short description cf. Chapter 2 of [3]). There, jumps are presented at points of Type i, $i \in \{2, 3, 4\}$, if we follow a curve of stationary points not being local minimizers. We have jumps for all types, even in points of Type 4. These jumps are applicable to the Lagrange multiplier embedding. We can jump to another curve of local minimizers with respect to (x, v, w, λ, μ). Fig. 5.1 illustrates this situation.

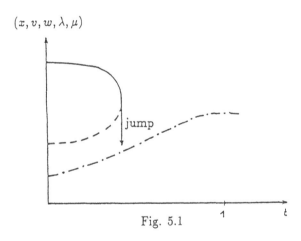

Fig. 5.1

—— local min-max-point
      x,v,w λ,μ

– – stationary point that is not a min-max-point
                                          x,v,w λ,μ

– · – local minimizer with respect to (x, v, w, λ, μ)

In Fig. 5.1 we reach the level $t = 1$. We observe that we are outside the Lagrange multiplier method if we follow a branch of local minimizers. For more details we refer to Remark 3.1 in [3].

Considering $\tilde{P}_3(t)$ with $n > 1$, which is JJT-regular with respect to $[0,1]$, and assuming the EnMFCQ to be satisfied, we attain the level $t = 1$ by pathfollowing procedures with jumps. Under EnMFCQ and $n > 1$ (cf. Theorem 3.5) we have points of Type 2 and 3 only. Therefore, we also have jumps if a point of Type 2 or Type 3 appears on a branch of local minimizers.

REMARK 5.4 Corollary 3.3 and Remark 3.4 show how restrictive the EnMFCQ is with respect to the feasible set M of the original problem (P): An unconnected feasible set M is excluded for the considered embeddings.

Acknowledgement: We thank the anonymous referees for the helpful hints.

References

[1] Bertsekas, D.P. (1982): Constrained optimization and Lagrange multiplier methods. Academic Press New York .

[2] Dentcheva, D., Gollmer, R., Guddat, J., Rückmann, J.-J. (1995): Pathfollowing methods in nonlinear optimization II: Exact penalty methods. In [4], 200-230.

[3] Dentcheva, D., Guddat, J., Rückmann, J.-J., Wendler, K. (1995): Pathfollowing methods in nonlinear optimization III: Lagrange multiplier embedding. ZOR 41, 127-152.

[4] Florenzano, M. et al. (eds.) (1995): Approximation and Optimization in the Caribbean II. In: Ser. approximation and optimization. Verlag Peter Lang, Frankfurt a.M., Berlin, Bern, New York, Paris, Wien.

[5] Gfrerer, H., Guddat, J., Wacker, Hj., Zulehner, W. (1985): Pathfollowing methods for Kuhn-Tucker curves by an active index set strategy. In: Bagchi, A., Jongen, H.Th. (eds.) Systems and optimization. Lecture Notes in Control and Information Sciences 66, Springer-Verlag Berlin, Heidelberg, New York, 111-131.

[6] Gollmer, R., Guddat, J., Guerra, F., Nowack, D., Rückmann, J.-J. (1993): Pathfollowing methods in nonlinear optimization I: Penalty embedding. In: Guddat, J. et al. (eds.) Parametric optimization and related topics III. In: Ser. approximation and optimization. Verlag Peter Lang, Frankfurt a.M., Berlin, Bern, New York, Paris, Wien, 163-214.

[7] Gollmer, R., Kausmann, U., Nowack, D., Wendler, K. (1995): Computerprogramm PAFO, Humboldt-Universität Berlin, Institut für Mathematik.

[8] Guddat, J., Guerra, F., Jongen, H.Th. (1990): Parametric optimization: Singularities, pathfollowing and jumps. BG Teubner, Stuttgart and John Wiley, Chichester.

[9] Guddat, J., Jongen, H.Th., Rückmann, J.-J. (1986): On stability and stationary points in nonlinear optimization. J. Austral. Math. Soc., Ser. B 28, 36-56.

[10] Guddat, J., Rückmann, J.-J. (1994): One-parametric optimization: Jumps in the set of generalized critical points. Control and Cybernetics 23, 1/2, 139-151.

[11] Jongen, H.Th., Jonker, P., Twilt, F. (1986): On one-parametric families of optimization problems. Equality constraints. JOTA 48, 141-161.

[12] Jongen, H.Th., Jonker, P., Twilt, F. (1986): Critical sets in parametric optimization. Math. Programming 34, 333-353.

[13] Kojima, M., Hirabayashi, R. (1984): Continuous deformation of nonlinear programs. Math. Progr. Study 21, 150-198.

[14] Di Pillo, G., Grippo, L. (1986): An exact penalty function method with global convergence properties for nonlinear programming. Math. Programming, 36, 1-18.

[15] Rückmann, J.-J., Tammer, K. (1992): On linear-quadratic perturbations in one-parametric non-linear optimization. Systems Science 18 1, 37-48.

[16] Wendler, K. (1993): Implementation of a pathfollowing procedure for solving non-linear one-parametric optimization problems. In: Brosowski, B. et al. (eds.) Multicriteria decision. In: Ser. approximation and optimization. Verlag Peter Lang, Frankfurt a.M., Berlin, Bern, New York, Paris, Wien, 139-163.

[17] Hasan, M., Poore, A.B.(1996): Bifurcation analysis for singularities on a tangent space for quadratic penalty-barrier and multiplier methods for solving constrained optimization problems. Part I, J. of Math. Anal. and Apll., Vol 197, No.3 658-678.

[18] Hasan, M., Poore, A.B. (1996): Analysis of bifurcation due to loss of linear independence and strict complementarity for penalty methods for solving constrained optimization problems. To appear in: J. of Math. Anal. and Appl.

[19] Lundberg, B.N., Poore, A.B., Yang, B. (1990): Smooth Penalty Functions and Continuation Methods for Constrained Optimization. In E.L. Allgower and K. Georg, eds., *Lectures in Applied Mathematics Series*, Vol. 26, American Mathematical Society, Providence, 389-412.

[20] Poore, A.B., Al-Hassan, Q. (1988): The Expanded Lagrangian System for Constrained Optimization Problems. SIAM J. on Control and Optimization, Vol. 26, No. 2, 417-427.

Hoffman's Error Bound for Systems of Convex Inequalities

DIETHARD KLATTE   Institut für Operations Research, Universität Zürich, CH–8044 Zürich, Switzerland

Abstract: In this paper we point out that Lipschitz properties of convex multifunctions are closely related to a so–called bounded excess property. This applies to convex inequality systems. We show that Slaters' condition and the bounded excess condition imply the existence of a global error bound in Hoffman's sense. In the case of a differentiable convex inequality system for which any subsystem satisfies the bounded excess condition, Abadie's constraint qualification is shown to be equivalent to the existence of Hoffman's error bound.

1 Introduction

Consider the system of convex inequalities

$$f(x) \leq 0, \tag{1.1}$$

where $f(x) = (f_1(x), \ldots, f_m(x))^T$ is a vector–valued function from $\mathbb{R}^n$ to $\mathbb{R}^m$, each of its components is a convex function on $\mathbb{R}^n$. Denote by $\mathcal{M}$ the solution set of (1.1). Let $\| \cdot \|$ and $| \cdot |$ be any two norms in $\mathbb{R}^m$ and $\mathbb{R}^n$, respectively, and let $\text{dist}\,(x, Z) := \inf_{z \in Z} |x - z|$ be the point-to-set distance of $x \in \mathbb{R}^n$ and $Z \subset \mathbb{R}^n$.

In this paper, we study conditions for the existence of a global error bound of the following type:

$$\exists c > 0 \ \forall x \in \mathbb{R}^n : \quad \text{dist}\,(x, \mathcal{M}) \leq c \, \|[f(x)]_+\|, \tag{1.2}$$

where y_+ denotes the vector obtained by replacing each negative component of y by zero. In his famous paper, Hoffman [12] showed that (1.2) holds if f is an affine mapping, and so, the notion *Hoffman's error bound* is often used to refer to (1.2). Various extensions of Hoffman's result were obtained for general convex inequality systems. Robinson [27, 28] showed that (1.2) is true for any convex inequality system which satisfies Slater's condition and has a bounded solution set. The result similarly holds in normed linear spaces.

In order to incorporate also unbounded solution sets of convex inequality systems, one has to require additional assumptions on the asymptotical behavior. For differentiable convex inequalities on $\mathbb{R}^n$, Mangasarian [24] obtained Hoffman's error bound under Slater's condition and some asymptotic constraint qualification (ACQ). Auslender and Crouzeix [2] extended Mangasarian's result to the nonsmooth case and obtained sharp bounds. A generalization of Robinson's [27] result to convex inequality systems in Banach spaces with unbounded solution sets was recently presented by Deng [9, 10]. Luo and Luo [21] considered the finite–dimensional case and avoided to assume any ACQ, but they had to restrict the data to the class of convex quadratic functions. Li [20] recently proved that Abadie's constraint qualification and the existence of Hoffman's error bound are equivalent for convex quadratic systems.

Several interesting papers are devoted to global error bounds being not of Lipschitz type, i.e., a nonlinear function of the residuum appears: Luo and Luo [21] and Luo and Pang [22] study systems with analytic (not necessarily convex) functions, while in Wang and Pang [31] convex quadratic inequalities (without Slater's condition) are considered.

Global error bounds for convex inequalities are highly relevant to theory and methods in optimization and to several applications, for recent surveys of the topic and its applications we refer, e.g., to [15, 20, 21]. For unified approaches and literature studies in the case of *linear* systems, including the discussion of concrete forms of the constant c in (1.2), see, e.g., [8, 16, 17, 19].

The purpose of our paper is to present sufficient conditions for the existence of Hoffman's error bound in the case of an unbounded solution set of the convex inequality system (1.1). Our approach uses the concept of an asymptotic constraint qualification which is related to so–called multifunctions of bounded excess. In particular, many known results are illuminated from this viewpoint. As far as it was possible, we tried to find a self–contained presentation. In Section 2, we shall show that if a convex multifunction F is of bounded excess at some interior point t^0 of its effective domain $\operatorname{dom} F$, then F is lower Lipschitz at t^0. This allows to use a proof technique proposed by Psenichny [25, Thm. III.1.1] in order to derive a global upper Lipschitz bound at t^0. Moreover, under the same assumptions, F turns out to be Lipschitzian around the given point t^0. In Section 3, these results will be applied to the study of Hoffman's error bound. In particular, the following will be shown: if (1.1) is defined by differentiable convex functions then Abadie's constraint qualification is equivalent to the existence of Hoffman's bound, provided that the bounded excess condition holds for any subsystem of (1.1). Here we use recent results obtained by Li [20]. Finally, the bounded excess condition will be discussed for special cases. In particular, systems of quadratic convex inequalities automatically satisfy it. Under the strong form of the Slater condition, the bounded excess condition implies the asymptotic constraint qualification used by Mangasarian [24] and Auslender and Crouzeix [2].

We conclude the introductory section by some notation. Let $B(x, r)$ denote the closed r–neighborhood of $x \in \mathbb{R}^m$ for the underlying norm. According to the standard monograph [29] we use the symbols int D, ri D and aff D for the interior, the relative interior and the affine hull of some convex set D in $\mathbb{R}^m$. We will be particularly interested in multi-valued functions, briefly called *multifunctions*. Given a multifunction F from a convex set $T \subset \mathbb{R}^m$ to $\mathbb{R}^n$, we denote by graph $F :=$ $\{(t, x) \in T \times \mathbb{R}^n \mid x \in F(t)\}$ the *graph* of this multifunction. If graph F is convex (closed), F is called *convex* (*closed*). For nonempty sets $X, Z \subset \mathbb{R}^n$, we define

$$\mathrm{d}(X, Z) \quad := \quad \sup_{x \in X} \mathrm{dist}(x, Z)$$

$$\mathrm{h}(X, Z) \quad := \quad \max\{\mathrm{d}(X, Z), \mathrm{d}(Z, X)\}.$$

For $X \neq \emptyset$ put $\mathrm{d}(X, \emptyset) := +\infty$, and let $\mathrm{d}(\emptyset, Z) := 0$ for any set Z, per definitionem. $\mathrm{d}(X, Z)$ is the *excess* of X relative to Z. In the case of nonempty compact sets X and Z, $\mathrm{h}(X, Z)$ is the usual *Hausdorff distance* between X and Z. A multifunction F as defined above is said to be *Hausdorff upper semicontinuous (H-u.s.c.)* at $t^0 \in T$, if for any sequence $\{t^k\} \subset T$, $t^k \to t^0$ implies $\mathrm{d}(F(t^k), F(t^0)) \to 0$. We shall say that F is *Hausdorff lower semicontinuous (H-l.s.c.)* at $t^0 \in T$, if for any sequence $\{t^k\} \subset T$, $t^k \to t^0$ implies $\mathrm{d}(F(t^0), F(t^k)) \to 0$. If F is both upper and lower semicontinuous at $t^0 \in T$, then F is said to be *H-continuous* at t^0. Note that in the case $F(t^0) \neq \emptyset$, the Hausdorff lower semicontinuity of F includes that the sets $F(t)$ are nonempty for all $t \in T$ near t^0. F is said to be *Lipschitzian* on $D \subset T$, if

$$\exists \varrho > 0 \ \forall t, t' \in D : \quad \mathrm{h}(F(t), F(t')) \leq \varrho \|t - t'\|. \tag{1.3}$$

Note that (1.3) includes $D \subset \mathrm{dom}\, F$. The constant ϱ in (1.3) will be called *Lipschitz modulus* of F. Finally, we shall say that F is *upper Lipschitz(-continuous)* resp. *lower Lipschitz(-continuous)* at $t^0 \in \mathrm{ri}\, T$ if for some constant ϱ and some relative neighborhood U of t^0, $\mathrm{d}(F(t), F(t^0)) \leq \varrho \|t - t^0\|$ resp. $\mathrm{d}(F(t^0), F(t)) \leq \varrho \|t - t^0\|$ hold, whenever $t \in U$.

2 Lipschitz Properties

For convex multifunctions, the following concept will guarantee certain kinds of Lipschitz stability. In Section 3 below it will play the role of an asymptotic constraint qualification.

Definition 2.1 Let F be a multifunction from $T \subset \mathbb{R}^m$ to $\mathbb{R}^n$, and let $F(t^0)$ be nonempty. F is called a *multifunction of bounded excess* at t^0 if

$$\mathrm{d}(F(t^0), F(t)) < +\infty \qquad \forall t \in U \cap \mathrm{dom}\, F$$

holds for some neighborhood U of t^0. $\diamond$

Trivially, if F is bounded–valued then F is of bounded excess at each $t^0 \in \mathrm{dom}\, F$. In Section 3 below, we shall give examples of multifunctions of bounded excess, which may be unbounded–valued, in general.

Theorem 2.2 *Let F be a convex multifunction from $T \subset \mathbb{R}^m$ to $\mathbb{R}^n$, where T is a nonempty convex set. Suppose $F(t) \neq \emptyset$ for all $t \in T$, and let $t^0 \in \operatorname{ri} T$. If F is a multifunction of bounded excess at t^0, then F is lower Lipschitz at t^0.* $\diamond$

Proof. First we note that for some $\varepsilon > 0$, the mapping

$$t \in T \mapsto \varphi(t) := \mathrm{d}\left(F(t^0), F(t)\right) = \sup_{x \in F(t^0)} \operatorname{dist}\left(x, F(t)\right)$$

is finite and convex on $U_\varepsilon := B(t^0, \varepsilon) \cap \operatorname{aff} T$. Indeed, since F is of bounded excess at t^0, where t^0 is a relatively interior point of $T = \operatorname{dom} F$, there exists some positive real number ε such that

$$\varphi(t) = \mathrm{d}\left(F(t^0), F(t)\right) < +\infty \qquad \forall t \in U_\varepsilon.$$

Since graph F is convex, standard arguments from convex analysis easily yield that for any fixed $x \in F(t^0)$, the function

$$t \mapsto \operatorname{dist}\left(x, F(t)\right) = \inf\{|x - y| \mid y \in F(t)\},$$

is convex on U_ε, cf., e.g., Rockafellar [29, §29] or Fiacco and Kyparisis [11]. Therefore, φ is convex on U_ε as the supremum of convex functions.

The function φ is hence continuous at t^0, i.e., F is H–l.s.c. at t^0. This implies that for some $\eta > 0$,

$$\mathrm{d}\left(F(t^0), F(t)\right) < 1 \qquad \forall t \in U_\eta. \tag{2.1}$$

Recall that U_η is the closed (relative) η–neighborhood of t^0. Let $x^0 \in F(t^0)$ and $\bar{t} \in U_\eta \setminus \{t^0\}$ be any two fixed elements. Then

$$t' := t^0 + \eta \|\bar{t} - t^0\|^{-1}(\bar{t} - t^0) \in U_\eta,$$

and, with $\alpha' := \eta^{-1}\|\bar{t} - t^0\|$,

$$\bar{t} = \alpha' t' + (1 - \alpha') t^0.$$

Thus, by (2.1), there is some $y' \in F(t')$ such that

$$|x^0 - y'| < 1.$$

From the convexity of graph F, we then have

$$\bar{y} := \alpha' y' + (1 - \alpha') x^0 \in F(\bar{t}),$$

and so,

$$\operatorname{dist}\left(x^0, F(\bar{t})\right) \le |x^0 - \bar{y}| = \alpha' |x^0 - y'| < \frac{1}{\eta}\|t^0 - \bar{t}\|.$$

Passing to the supremum over all $x^0 \in F(t^0)$, we obtain

$$\mathrm{d}\left(F(t^0), F(\bar{t})\right) \le \frac{1}{\eta}\|t^0 - \bar{t}\|.$$

Since this holds for any $\bar{t} \in B_\eta$, the lower Lipschitz–continuity of F at t^0 is shown. $\square$

Note that the preceding theorem contains the proof that lower Lipschitz continuity at $t^0 \in \mathrm{ri\,dom}\,F$ already follows from Hausdorff lower semicontinuity, this proof is taken from Klatte [14, Satz 3.3.1]. Next, we show that bounded excess of F at $t^0 \in \mathrm{ri\,dom}\,F$ implies a global upper Lipschitz estimate of F at t^0. This result was first presented by Belousov and Andronov in [7, Thm. 3, Section 4.3]. Our proof goes directly back to Theorem 2.2. Note that the property derived in the following theorem is stronger than the standard upper Lipschitz property which has local character.

Theorem 2.3 *Let F be a convex multifunction from $T \subset \mathbb{R}^m$ to $\mathbb{R}^n$, where T is a nonempty convex set. Let t^0 be an element of $\mathrm{ri\,dom}\,F$. If F is of bounded excess at t^0, then there exists a constant $c > 0$ such that*

$$\mathrm{d}\left(F(t), F(t^0)\right) \le c\|t - t^0\| \qquad \forall t \in T. \tag{2.2}$$

$\diamond$

Note. In fact, we shall show that if F is lower Lipschitz at t^0 with modulus μ relative to some η–neighborhood of t^0, i.e., $\mathrm{d}\left(F(t^0), F(t)\right) \le \mu\|t - t^0\|$ for all $t \in U_\eta := B(t^0, \eta) \cap \mathrm{aff}\,T$, then (2.2) holds with the constant $c = \mu$.

Proof. If $t \notin \mathrm{dom}\,F$ then $F(t) = \emptyset$ and hence $\mathrm{d}\left(F(t), F(t^0)\right) = 0$, i.e., the inequality in (2.2) is trivially fulfilled. Therefore, without loss of generality, let $T = \mathrm{dom}\,F$. By the assumptions, T is nonempty and convex, $t^0 \in \mathrm{ri}\,T$ and F is of bounded excess at t^0. Hence Theorem 2.2 applies.

By Theorem 2.2, there exist numbers $c > 0$ and $\eta > 0$ such that one has $F(s) \ne \emptyset$ and

$$\mathrm{d}\left(F(t^0), F(s)\right) \le c\|s - t^0\| \quad \forall s \in U_\eta. \tag{2.3}$$

For $t = t^0$, the inequality in (2.2) is trivially fulfilled. Let $t \in T \setminus \{t^0\}$ and $x \in F(t)$ be arbitrarily chosen. Set

$$\hat{t} := t^0 - \frac{\eta}{\|t - t^0\|}(t - t^0),$$

$$\alpha := \frac{\|t - t^0\|}{\|t - t^0\| + \eta}.$$

Hence, $0 < \alpha < 1$ and

$$t^0 = \alpha\hat{t} + (1 - \alpha)t.$$

Further, by construction, $\hat{t} \in U_\eta$ and so $F(\hat{t}) \ne \emptyset$. Let δ be any positive real number, and let $y \in F(t^0)$ and $z \in F(\hat{t})$ be points such that

$$|x - y| \le \mathrm{dist}\left(x, F(t^0)\right) + \delta \quad \text{and} \quad |y - z| \le \mathrm{dist}\left(y, F(\hat{t})\right) + \delta. \tag{2.4}$$

From the convexity of the multifunction F we obtain

$$z^0 := \alpha z + (1 - \alpha)x \in F(t^0),$$

by using $t^0 = \alpha \hat{t} + (1 - \alpha)t$, $z \in F(\hat{t})$ and $x \in F(t)$. This implies

$$\text{dist}(x, F(t^0)) \le |x - z^0| = \alpha|x - z| \le \alpha(|x - y| + |y - z|).$$

Thus, applying (2.4) and (2.3), together with $\hat{t} \in U_\eta$, we have

$$(1 - \alpha)\text{dist}(x, F(t^0)) \le \alpha(\text{dist}(y, F(\hat{t})) + 2\delta) \le \alpha(c\|t^0 - \hat{t}\| + 2\delta),$$

and so, with $\alpha(t^0 - \hat{t}) = (1 - \alpha)(t - t^0)$,

$$(1 - \alpha)\text{dist}(x, F(t^0)) \le c(1 - \alpha)\|t - t^0\| + 2\alpha\delta.$$

This holds for an arbitrary positive real number $\delta > 0$, which implies

$$\text{dist}(x, F(t^0)) \le c\|t - t^0\|. \tag{2.5}$$

Since c is independent of t and x, and $t \in T \setminus \{t^0\}$ and $x \in F(t)$ were arbitrarily chosen, this completes the proof. $\qquad\qquad\qquad\qquad\qquad\qquad\qquad\qquad\qquad\Box$

The following proposition is an immediate consequence of Theorem 2.2 and 2.3, together with Theorem 3.3.1 in [14]. For completeness, we present the proof.

Theorem 2.4 *Let F be a convex multifunction from $T \subset \mathbb{R}^m$ to $\mathbb{R}^n$, where T is nonempty and convex. Let t^0 be an element of $\mathrm{ri}\,T$. Then the following are equivalent:*

(i) F is H–continuous at t^0.

(ii) For some $\eta > 0$, F is Lipschitzian on $B(t^0, \eta) \cap \mathrm{aff}\,T$.

(iii) $F(t)$ is nonempty for all t in some relative neighborhood of t^0,
*   and F is a multifunction of bounded excess at t^0.* $\qquad\qquad\qquad\Diamond$

Proof.   Obviously, we have (ii) $\Longrightarrow$ (i) $\Longrightarrow$ (iii). It suffices to prove (iii) $\Longrightarrow$ (ii). If (iii) holds, then Theorem 2.2 and Theorem 2.3 particularly imply that F is H–continuous at t^0.

Hence, there is some $\eta > 0$ such that $F(t) \ne \emptyset$ for $t \in U_{3\eta}$,

$$U_{3\eta} \subset T \quad \text{and} \quad \mathrm{h}(F(t), F(t^0)) < \frac{1}{2} \ (\forall t \in U_{3\eta}), \tag{2.6}$$

where again $U_\varepsilon := B(t^0, \varepsilon) \cap \mathrm{aff}\,T$ for $\varepsilon > 0$. Choose any $t, t' \in U_\eta$ with $t \ne t'$. Then

$$\|t - t'\| \le 2\eta$$

and $F(t), F(t') \ne \emptyset$. Let $z \in F(t)$, and define

$$\alpha \quad := \quad \frac{1}{2\eta}\|t - t'\|, \tag{2.7}$$

$$t'' \quad := \quad t - \frac{1}{\alpha}(t - t'). \tag{2.8}$$

Hence, $0 < \alpha \le 1$ and

$$t' = \alpha t'' + (1 - \alpha)t,$$

i.e., t' is a convex combination of t'' and t. Moreover,

$$\|t'' - t^0\| \le \|t'' - t\| + \|t - t^0\| \le \frac{1}{\alpha}\|t' - t\| + \eta = 3\eta.$$

Hence, by (2.6), there exists some $z'' \in F(t'')$ satisfying

$$|z - z''| < 1. \tag{2.9}$$

Since graph F is a convex set, we have

$$\alpha z'' + (1 - \alpha)z \in F(t'),$$

which, along with (2.7) and (2.9), implies

$$\text{dist}\,(z, F(t')) \le \alpha |z - z''| < \frac{1}{2\eta}\|t - t'\|.$$

Since z was arbitrarily taken from $F(t)$,

$$\text{d}\,(F(t), F(t')) \le \frac{1}{2\eta}\|t - t'\|$$

follows. Changing the role of t and t', we obtain the Lipschitz continuity of F on U_η with modulus $\frac{1}{2\eta}$. This completes the proof. $\square$

It is worth noting that the recent book by Belousov and Andronov [7] contains a lot of material concerning Lipschitz and Hausdorff stability of convex multifunctions, in particular, for multifunctions defined by convex polynomial inequality systems. Also the solution set mapping for special convex polynomial programs is studied in this respect.

3 Hoffman's Error Bound

Now we apply the Lipschitz stability results of the previous section in order to obtain conditions for the existence of Hoffman's error bound for the system (1.1) of convex inequalities. Let $f : \mathbb{R}^n \to \mathbb{R}^m$ be the vector-valued function defining (1.1), each of its components is a convex function on $\mathbb{R}^n$. Consider the multifunction given by

$$M(u) := \{x \in \mathbb{R}^n \mid f(x) \le u\}, \ u \in \mathbb{R}^m. \tag{3.1}$$

Definition 3.1 We shall say that an inequality system $f(x) \le 0$ with nonempty solution set satisfies the *bounded excess condition* if the multifunction M is of bounded excess at $u = 0$. $\diamond$

Let the index set $\mathcal{I} := \{1, \ldots, m\}$ be splitted into

$$I_L := \{i \in \mathcal{I} \mid f_i \text{ linear–affine}\}, \qquad I_N := \mathcal{I} \setminus I_L.$$

Then there are vectors a^j and scalars b_j, $j \in I_L$, such that

$$f_j(x) = \left(a^j\right)^T x - b_j \qquad (\forall j \in I_L).$$

Recall that $\mathcal{M}$ denotes the solution set of (1.1). Let I_S be the *index set of singular inequalities* in the system (1.1), i.e.,

$$I_S := \{i \in \mathcal{I} \,|\, f_i(x) = 0 \;\; \forall x \in \mathcal{M}\}.$$

We will allow to handle singular affine–linear inequalities in view of mixed systems of convex inequalities and linear equations.

Definition 3.2 We shall say that the inequality system $f(x) \le 0$ satisfies the *modified Slater condition* if there exists some point $x^0 \in \mathcal{M}$ such that $f_i(x^0) < 0$ for all $i \in I_N$, and $\{a^j, j \in I_S\}$ are linearly independent. $\diamond$

Theorem 3.3 *Consider the convex inequality system (1.1). Suppose that (1.1) satisfies both the modified Slater condition and the bounded excess condition. Then there exists a positive real number c such that*

$$\text{dist}\,(x, \mathcal{M}) \le c \,\|f(x)_+\| \qquad \forall x \in I\!\!R^n \tag{3.2}$$

holds. $\diamond$

Note. If the solution set of (1.1) is nonempty and bounded, then under the modified Slater condition, the bounded excess condition is automatically satisfied. Hence, Theorem 3.3 includes Robinson's classical result [27] on Hoffman's error estimate for the bounded case.

Proof. Let $I := \{i \in I_L \,|\, \exists x \in \mathcal{M} : f_i(x) < 0\}$. By hypothesis, there is some $x^0 \in \mathcal{M}$ such that $f_i(x^0) < 0$ for all $i \in I_N$. Thus, by standard arguments from convex analysis, there exists some $\bar{x} \in \mathcal{M}$ such that

$$f_i(\bar{x}) < 0 \quad \forall i \in J := I \cup I_N, \tag{3.3}$$

and one has $I_S := \{1, ..., m\} \setminus J$, i.e.,

$$\mathcal{M} \subset \{x \in I\!\!R^n \,|\, f_j(x) = 0, j \in I_S\}.$$

By hypothesis, the set $\{a^j \,|\, j \in I_S\}$ is linearly independent. This, together with (3.3) and $f_j(\bar{x}) = 0$, $j \in I_S$, imply that small right-hand side perturbations of the system $[f_i(x) \le 0, i = 1, ..., m]$ preserve feasibility, i.e.,

$$0 \in \text{int dom}\, M.$$

Now, let $z \in I\!\!R^n$ be arbitrarily chosen. Put $u := [f(z)]_+$, hence $z \in M(u)$. Then use Theorem 2.3 by setting $T := \text{dom}\, M$, $F := \mathcal{M}$, $t := u$ and $t^0 := 0$, which yields

$$\text{dist}\,(z, \mathcal{M}) \le \text{d}\,(M(u), M(0)) \le c\|u\| = c\|[f(z)]_+\|.$$

This completes the proof. $\square$

Example 3.4 The constant c in (3.2) is not uniform with respect to variation of the right-hand side u. Consider for varying $u \in I\!\!R$,

$$x^2 - y \le 0, \quad x \le u,$$

hence, for $u \le 0$, the residuum of the origin in $I\!\!R^2$ relative to $M(u)$ is equal to $|u|$, but

$$\text{dist}\,(0, M(u)) = \sqrt{u^2 + u^4} = \sqrt{1 + u^2}\,|u|,$$

with $\sqrt{1 + u^2} \to +\infty$ as $u \to -\infty$. $\diamond$

Now, in the case that no singular inequalities occur, it will be shown that the bounded excess condition implies the asymptotic constraint qualification in the sense of Auslender and Crouzeix [2]. Hence, in this case, Theorem 3.3 is already covered by Theorem 2 in [2].

Let f_i be the components of f in (1.1), and put $\varphi(x) := \max_{1 \leq i \leq m} f_i(x)$. Denote by $\partial\varphi(x)$ the subdifferential of φ at x. Following Auslender and Crouzeix [2, page 250], we shall say that the *asymptotic constraint qualification (A.C.Q.)* holds if for each sequence $\{x_n\}$ satisfying both $\varphi(x_n) = 0 \; \forall n$ and $\lim_n \|x_n\| = +\infty$, the vector zero is not an accumulation point of any sequence $\{c_n\}$, $c_n \in \partial\varphi(x_n) \; (\forall n)$.

Proposition 3.5 *Consider the convex inequality system (1.1). Suppose that there is some point x^0 such that $f_i(x^0) < 0$ for all $i \in \{1, \ldots, m\}$. If (1.1) satisfies the bounded excess condition, then A.C.Q. holds.* ◇

Proof. Let the bounded excess condition be satisfied, i.e., there is some neighborhood U of 0 such that

$$\mathrm{d}\,(\mathcal{M}, M(u)) < +\infty \quad \forall u \in U. \tag{3.4}$$

Assume, on the contrary, that A.C.Q. does not hold. Then there exist sequences $\{x_n\}$ and $\{c_n\}$ such that

$$\varphi(x_n) = 0 \; (\forall n), \quad \|x_n\| \to +\infty, \quad c_n \in \partial\varphi(x_n) \; (\forall n), \quad c_n \to 0.$$

Let $\varepsilon > 0$ be such that

$$\bar{u} = (-\varepsilon, \ldots, -\varepsilon) \in U.$$

By the closedness of $M(\bar{u})$ and by (3.4), there are a sequence $\{y_n\} \subset M(\bar{u})$ and a constant γ such that

$$\|x_n - y_n\| = \mathrm{dist}\,(x_n, M(\bar{u})) \leq \gamma \quad \forall n.$$

Since $\varphi(y_n) \leq -\varepsilon$ for all n, we have

$$-\varepsilon > \varphi(y_n) \geq c_n^T (y_n - x_n) \quad \forall n$$

because of $c_n \in \partial\varphi(x_n)$, and hence

$$0 < \varepsilon < -\varphi(y_n) \leq c_n^T (x_n - y_n) \leq \|c_n\|\|x_n - y_n\| \leq \|c_n\|\gamma.$$

Passing to the limit, we obtain

$$0 < \varepsilon \leq 0,$$

a contradiction. This completes the proof. □

In what follows we shall suppose that the convex components f_i of the vector-valued function f defining (1.1) are differentiable on $\mathbb{R}^n$. Let $\nabla f_i(x)$ denote the gradient of f_i at x. For differentiable nonlinear programs, *Abadie's constraint qualification (Abadie's CQ)* is a well-known concept, see, e.g., Bazaraa and Shetty [4]: Abadie's CQ is satisfied at $x^0 \in \mathcal{M}$ if

$$T_{\mathcal{M}}(x^0) = \{h \in \mathbb{R}^n \mid h^T \nabla f_i(x^0) \leq 0, \; i \in E(x^0)\},$$

where $T_{\mathcal{M}}(x^0)$ denotes the tangent cone of $\mathcal{M}$ at x^0 in the standard sense of convex analysis, and $E(x^0) := \{i \in \{1, ..., m\} \mid f_i(x^0) = 0\}$. Following [20], we shall say that *the system (1.1) satisfies Abadie's CQ* if Abadie's CQ holds at every point in $\mathcal{M}$.

Remark 3.6 It is well–known that the Linear Independence CQ, Cottle's CQ, the Mangasarian–Fromovitz CQ, the (standard) Slater condition and hence the modified Slater condition imply Abadie's CQ, cf., e.g., Bazaraa and Shetty [4, Chapter 5]. Here we mean that (1.1) satisfies the standard *Slater condition* if for some $x^0 \in \mathcal{M}$, $f_i(x^0) < 0$ holds for every i in I_N. In some sense, Abadie's CQ is the weakest CQ for differentiable convex programs to guarantee that the Karush–Kuhn–Tucker (KKT) conditions hold for optimal solutions, for details we refer to Li [20, Lemma 4]. ◇

The following useful lemma was recently proved by Li:

Lemma 3.7 (Li [20, Thm. 11]) *Consider the inequality system (1.1). Suppose that each component f_i of f, $i = 1, \ldots, m$, is differentiable and convex on $\mathbb{R}^n$. Then (1.1) satisfies Abadie's constraint qualification if and only if for any number $r > 0$, there exists a positive constant $\gamma(r)$ such that*

$$\mathrm{dist}\,(x, \mathcal{M}) \le \gamma(r)\,\|f(x)_+\| \qquad \text{when } \|x\| \le r \tag{3.5}$$

holds. ◇

Condition (3.5) is a non–parametric version of metric regularity. Now we are able to characterize the existence of Hoffman's error bound for differentiable convex inequalities by means of Abadie's CQ, provided that each subsystem of (1.1) satisfies the bounded excess condition. The following proof is modeled after the proofs of Theorem 2.2 in [17] and of Theorem 13 in Li [20]. Note that this method of proof via the Karush–Kuhn–Tucker conditions and the reduction to the Linear independence CQ is standard in the context of Hoffman's bound, cf., e.g., [13, 14, 17, 20, 23, 26].

Theorem 3.8 *Consider the inequality system (1.1). Suppose that each component f_i of f, $i = 1, \ldots, m$, is differentiable and convex on $\mathbb{R}^n$ and that each subsystem of (1.1) satisfies the bounded excess condition. Then (1.1) satisfies Abadie's constraint qualification if and only if there exists a positive constant c such that*

$$\mathrm{dist}\,(x, \mathcal{M}) \le c\,\|f(x)_+\| \qquad \forall x \in \mathbb{R}^n \tag{3.6}$$

holds. ◇

Note. Below we shall show that the bounded excess condition is fulfilled if, for example, the functions f_i are convex and quadratic. Thus, Theorem 3.8 recovers Theorem 13 in Li [20].

Proof. *(if)* Suppose that (3.6) holds. Then (3.5) immediately follows, and so, by Lemma 3.7, (1.1) satisfies Abadie's CQ.

(only if) Without loss of generality, let $|\cdot|$ and $\|\cdot\|$ be the Euclidean and the sum norm, respectively. Suppose that Abadie's CQ is satisfied for (1.1). Let x be

any point in $\mathbb{R}^n \setminus \mathcal{M}$, otherwise the inequality (3.6) is trivially fulfilled. $\mathcal{M}$ is a nonempty and closed set, hence, one has for some $y(x) \in \mathcal{M}$,

$$\text{dist}\,(x, \mathcal{M}) = |x - y(x)| = \min\{|x - y| \,|\, y \in \mathcal{M}\}.$$

This holds if and only if $y(x)$ is the solution of the differentiable convex program

$$\min\{(x - y)^T (x - y) \,|\, y \in \mathcal{M}\}.$$

Hence, the Karush–Kuhn–Tucker conditions yield the existence of a vector $\lambda \in \mathbb{R}_+^m$ such that for some index set $I \subset E(y(x)) = \{i | f_i(y(x)) = 0\}$,

$$x - y(x) = \sum_{i \in I} \lambda_i \nabla f_i(y(x)), \qquad \lambda_i > 0,\, i \in I. \tag{3.7}$$

By a version of Caratheodory's theorem, cf., e.g., [29, Corollary 17.1.2], we may assume that $\{\nabla f_i(y(x)),\, i \in I\}$ are linearly independent. This implies, by standard arguments, the existence of some x_I such that

$$f_i(x_I) < 0,\, i \in I,$$

i.e., the system

$$f_i(x) \leq 0,\, i \in I \tag{3.8}$$

satisfies the (modified) Slater condition. Since the functions f_i, $i \in I$, are convex and $y(x)$ belongs to $\mathcal{M}_I$, where $\mathcal{M}_I$ is the solution set of (3.8), the conditions (3.7) are also sufficient for the point $y(x)$ to be a solution of the program

$$\min\{(x - y)^T (x - y) \,|\, y \in \mathcal{M}_I\}.$$

Therefore,

$$\text{dist}\,(x, \mathcal{M}) = \text{dist}\,(x, \mathcal{M}_I).$$

By assumption, (3.8) satisfies the bounded excess condition. This implies the existence of a positive constant c_I such that

$$\text{dist}\,(x, \mathcal{M}) \leq c_I \sum_{i \in I}(f_i(x))_+ \leq c_I \sum_{i=1}^m (f_i(x))_+.$$

Since for varying x, only finitely many index sets I satisfying (3.7) may appear, (3.6) holds with $c := \max_I\{c_I\}$. This completes the proof. $\square$

We conclude the section by discussing in more detail the bounded excess condition. We are looking for special classes of functions or certain common regularity conditions such that the multifunction M defined in (3.1) is a multifunction of bounded excess at $u = 0$.

Since the solution set $M(u)$ of a solvable system of linear inequalities under right–hand side perturbations

$$Ax \leq u$$

may be represented by

$$M(u) = C(u) + M(0),$$

where $C(u)$ is a bounded convex polyhedral set, we immediately have

Proposition 3.9 *Suppose that* $x \mapsto f(x)$ *is an affine mapping from* $\mathbb{R}^n$ *to* $\mathbb{R}^m$. *If the solution set of the system (1.1) is nonempty, then each subsystem of (1.1) satisfies the bounded excess condition.*

Of course, Theorem 3.8 and Proposition 3.9 together recover the classical Hoffman result for linear systems. Note the well-known fact that Abadie's CQ is automatically fulfilled for linear inequality systems. A strong result on global error bounds for convex quadratic inequalities, which was recently proved by Wang and Pang [31], allows to extend Proposition 3.9 to this class. A Slater type condition is not required.

Proposition 3.10 *Suppose that each component* f_i, $i = 1, \ldots, m$, *of the vector-valued function* f *is a convex quadratic function on* $\mathbb{R}^n$. *If the solution set of the system (1.1) is nonempty, then each subsystem of (1.1) satisfies the bounded excess condition.* $\diamond$

Note. Non–trivial examples show that Proposition 3.10 does not carry over to general convex polynomial systems, see [7, §4.3.4], [30, §2] .

Proof. Without loss of generality, $\|\cdot\|$ is the sum norm. Let M be the multifunction according to (3.1). Choose any u in dom M, and let I be any subset of $\{1, \ldots, m\}$. By Theorem 3.1 in Wang and Pang [31], there exist constants $\tau > 0$ and $d > 0$ such that

$$\text{dist}\,(x, M(u)) \leq \tau \sum_{i \in I} (\,f_i(x) - u_i)_+\,)^d.$$

For any $x \in \mathcal{M} = M(0)$ and any $i \in I$, we have $f_i(x) \leq 0$ and

$$(f_i(x) - u_i)_+ = 0 \leq |u_i| \qquad \text{if } f_i(x) \leq u_i,$$
$$(f_i(x) - u_i)_+ = f_i(x) - u_i \leq -u_i = |u_i| \qquad \text{if } u_i < f_i(x).$$

Thus,

$$\text{dist}\,(x, M(u)) \leq \tau \sum_{i \in I} (\,|u_i|\,)^d.$$

The constant on the right–hand side of this inequality is independent of x, hence $\text{d}\,(\mathcal{M}, M(u))$ is finite. $\square$

Next, we mention a concept (Shironin [30]) which allows to extend the previous propositions to a larger class of convex polynomials. In fact, the linear and the convex quadratic case are included in this concept, but we preferred to present both special cases separately, because the publication [30] devoted to the more general class of convex polynomials is not easily available.

Shironin [30] introduces the class of so-called simple convex polynomials. Denote by epi f the epigraph of a function f. Following [3, 30] we say that a convex polynomial function $f : \mathbb{R}^n \to \mathbb{R}$ is *simple* (or *primitive*), if for any two affine subspaces A_1 and A_2 of $\mathbb{R}^n$ such that $\text{h}\,(A_1, A_2) < +\infty$ and $A_i \cap \text{int epi}\, f \neq \emptyset$, $i = 1, 2$, the condition $\text{h}\,(A_1 \cap \text{epi}\, f, A_2 \cap \text{epi}\, f) < +\infty$ is satisfied. In [30, Section V.2, Thm. 2] it is shown that the multifuntion M (3.1) restricted to $T = \text{dom}\, M$ is H-continuous on T, provided that (1.1) is defined by simple convex polynomials f_i. By definition, we immediately obtain in this case that M is a multifunction of bounded excess. This result, of course, carries over to subsystems.

Shironin proves that, for example, the following classes of convex polynomial functions are simple, see [30, Section V.4]: finite sums of simple convex polynomial functions, convex forms, hence quadratic convex functions, convex separable polynomials, convex polynomials being non-linear in at most two variables, convex polynomials of 3 variables up to degree 14.

We conclude the paper by considering sets of the form of a Minkowski sum $\mathcal{M} = C + K$, where C is a (convex) compact set and K is a convex (polyhedral) cone. For convex programs with constraints sets of this type, there are some interesting results concerning the existence of optimal solutions. For an extensive discussion of such results and of representations of the type $\mathcal{M} = C + K$, e.g., for convex polynomial systems, we refer to Belousov [5]. Applications to integer programming and stability theory in optimization can be found, e.g., in [3, 5, 6, 18]. The non-convex case is treated in Auslender [1].

Proposition 3.11 *Suppose that each component f_i, $i = 1, \ldots, m$, of the vector-valued function f is a convex polynomial function on $\mathbb{R}^n$. If the solution set of the system (1.1) is nonempty and has a representation $\mathcal{M} = C_0 + K$, where C_0 is a convex compact set and K is the recession cone of $\mathcal{M}$, then (1.1) satisfies the bounded excess condition.* ◇

Proof. By Theorem 3 in Section 4 of Bank and Mandel [3], the assumptions allow for all $u \in \text{dom } M$, M according to (3.1), a representation

$$M(u) = C(u) + K,$$

where $u \mapsto C(u)$ is H-continuous on $\text{dom } M$. Hence, for each $u \in \text{dom } M$ and each $x \in \mathcal{M} = M(0)$, there are points $z \in C_0$, $h \in K$ and $z(u) \in C(u)$ such that

$$x = z + h, \quad |z - z(u)| = \text{dist}\,(z, C(u)),$$

and so,

$$\text{dist}\,(x, M(u)) \leq |x - (z(u) + h)| = |z - z(u)| = \text{dist}\,(z, C(u)).$$

By the H–continuity of C, there is a neighborhood U of 0 such that $d\,(C_0, C(u))$ is finite if $u \in U \cap \text{dom } M$, hence, $d\,(\mathcal{M}, M(u)) \leq d\,(C_0, C(u))$ implies that the multifunction M is of bounded excess at $u = 0$. This completes the proof. □

Note that the solution set of a convex polynomial system (the same for a quasiconvex polynomial system) has a polyhedral recession cone K. In Proposition (3.11) one could equivalently suppose that

$$\sup\{(d^j)^T x \mid x \in \mathcal{M}\} < +\infty, \quad j = 1, \ldots, k,$$

where $\{d^j, j = 1, \ldots, k\}$ is a set of generators of the dual cone of K. See [3] for the details.

Added note. After this paper had been completed, the author's attention was called to three related preprints. Mangasarian gives a relative error bound for convex inequalities under the Slater condition $\exists x^0 : f_i(x^0) < 0, i = 1, \ldots, m$, and he derives a global error bound under a strengthened Slater condition, cf.

O.L. Mangasarian [Error bounds for nondifferentiable convex inequalities under a strong Slater constraint qualification, *Math. Programming Technical Report 96-04*, Univ. of Wisconsin-Madison, USA (July 1996)]. Lewis and Pang establish a necessary and sufficient condition for a convex set defined by a closed proper convex function to possess a global error bound in Hoffman's sense, and they study many special cases of the main characterization, cf. A.S. Lewis and J.S. Pang [Error bounds for convex inequality systems, *Manuscript*, Univ. of Waterloo, Canada, and Johns Hopkins Univ. Baltimore, Maryland, USA (July 1996)]. Finally, Li and Singer treat global error bounds in a very general setting, namely, for convex multifunctions between normed linear spaces, cf. W. Li and I. Singer [Global error bounds for convex multifunctions and applications, *Manuscript*, Old Dominion Univ. Norfolk, Virginia, USA, and Institute of Mathematics, Romanian Academy, Bucharest, Romania (July 1996)].

Acknowledgement. The author wishes to thank two referees for their careful reading of the manuscript and for several helpful comments.

References

[1] A. Auslender, Non coercive optimization problems. Manuscript, July 1995, *Math. Oper. Res.*, to appear.

[2] A. Auslender and J.-P. Crouzeix, Global regularity theorems. *Math. Oper. Res.* 13: 243–253 (1988).

[3] B. Bank and R. Mandel, *Parametric Integer Optimization*, Akademie-Verlag, Berlin (1988).

[4] S. Bazaraa and C.M. Shetty, *Nonlinear Programming – Theory and Algorithms*, Wiley, New York, Chichester (1979).

[5] E.G. Belousov, *Introduction to Convex Analysis and Integer Programming* (In Russian). Moscow University Publ., Moscow (1977).

[6] E.G. Belousov, On solvability and stability of polynomial optimization problems (In Russian). In: E.G. Belousov and B. Bank, editors, *Mathematical Optimization: Questions of Solvability and Stability* (In Russian), Chapter IV, Moscow University Publ., Moscow (1986).

[7] E.G. Belousov and V.G. Andronov, *Solvability and Stability for Problems of Polynomial Programming* (In Russian), Moscow University Publ., Moscow (1993).

[8] J.V. Burke and P. Tseng, A unified analysis of Hoffman's bound via Fenchel duality. *SIAM J. Optim.* 6: 265–282 (1996).

[9] S. Deng, Computable error bounds for convex inequality systems in reflexive Banach spaces. Manuscript, 1995, *SIAM J. Optim.*, to appear.

[10] S. Deng, Global error bounds for convex inequality systems in Banach spaces. Manuscript, Department of Mathematics, Northern Illinois University, De Kalb (October 1995, revised July 1996).

[11] A.V. Fiacco and J. Kyparisis, Convexity and concavity properties of the optimal value function in parametric nonlinear programming. *J. Optim. Theory Appl.* 48: 95–126 (1986).

[12] A.J. Hoffman, On approximate solutions of systems of linear inequalities. *J. Res. Nat. Bur. Standards* 49: 263–265 (1952).

[13] D. Klatte, Eine Bemerkung zur parametrischen quadratischen Optimierung. *Seminarbericht Nr. 50*, S. 174-185. Sektion Mathematik, Humboldt-Universität Berlin, Berlin (1983).

[14] D. Klatte, *Beiträge zur Stabilitätsanalyse nichtlinearer Optimierungsprobleme*. Dissertation B (Habilitationsschrift), Sektion Mathematik, Humboldt-Universität Berlin, Berlin (1984).

[15] D. Klatte, Lipschitz stability and Hoffman's error bounds for convex inequality systems, Manuscript, October 1995, revised June 1996, to appear in the Proceedings of the Conference *Parametric Optimization and Related Topics IV* (Enschede 1995).

[16] D. Klatte and G. Thiere, Error bounds for solutions of linear equations and inequalities. *ZOR - Mathematical Methods of Operations Research* 41: 191-214 (1995).

[17] D. Klatte and G. Thiere, A note on Lipschitz constants for solutions of linear inequalities and equations. *Linear Algebra Appl.* 244: 365-374 (1996).

[18] B. Kummer, Globale Stabilität quadratischer Optimierungsprobleme. *Wiss. Zeitschrift Humboldt-Univ. Berlin, Math.-Nat.R.* XXVI: Heft 5, 565-569 (1977).

[19] W. Li, The sharp Lipschitz constants for feasible and optimal solutions of a perturbed linear program. *Linear Algebra Appl.* 187: 15-40 (1993).

[20] W. Li, Abadie's constraint qualification, metric regularity, and error bounds for differentiable convex inequalities. Manuscript, 1995, *SIAM J. Optim.*, to appear.

[21] X.D. Luo and Z.Q. Luo, Extension of Hoffman's error bound to polynomial systems. *SIAM J. Optim.* 4: 383-392 (1994).

[22] Z.Q. Luo and J.-S. Pang, Error bounds for analytic systems and their applications. *Math. Programming* 67: 1-28 (1994).

[23] O.L. Mangasarian, A condition number of linear inequalities and equalities. *Methods Oper. Res.* 43: 3-15 (1981).

[24] O.L. Mangasarian, A condition number for differentiable convex inequalities. *Math. Oper. Res.* 10: 175-179 (1985).

[25] B. Psenichny, *Convex Analysis and Extremal Problems* (In Russian), Nauka, Moscow (1980).

[26] S.M. Robinson, Bounds for error in the solution set of a perturbed linear program. *Linear Algebra Appl.* 6: 69-81 (1973).

[27] S.M. Robinson, An application of error bounds for convex programming in a linear space. *SIAM J. Control* 13: 271-273 (1975).

[28] S.M. Robinson, Regularity and stability for convex multivalued functions. *Math. Oper. Res.* 1: 130-143 (1976).

[29] R.T. Rockafellar, *Convex Analysis*, Princeton University Press, Princeton, NJ (1970).

[30] V.M. Shironin, On Hausdorff continuity of convex and convex polynomial mappings (In Russian). In: E.G. Belousov and B. Bank, editors, *Mathematical Optimization: Questions of Solvability and Stability* (In Russian), Chapter V, Moscow University Publ., Moscow (1986).

[31] T. Wang and J.-S. Pang, Global error bounds for convex quadratic inequality systems. *Optimization* 31: 1-12 (1994).

Lipschitzian and Pseudo-Lipschitzian Inverse Functions and Applications to Nonlinear Optimization

BERND KUMMER
Humboldt-Universität zu Berlin
Institut für Mathematik
D-10099 Berlin, Germany

Keywords: inverse functions, pseudo-Lipschitz property, metric regularity, successive approximation, Lipschitzian homeomorphism, local uniqueness, continuous selection, nonlinear optimization, critical points, strong stability, generalized equations, complementarity.

Abstract: As regularity conditions for multifunctions, Lipschitzian and pseudo-Lipschitzian behaviour of inverse mappings are investigated. We verify by a successive approximation scheme, that pseudo-regularity induces the same property with respect to Lipschitzian perturbations. In the case that the original map is a proper function in finite dimension, we characterize the two regularity properties by generalized directional derivatives as well as in terms of an exact penalty function, and show that continuous selections of the inverse map play the crucial role for the equivalence of these regularities.
In particular, KKT-points of nonlinear C^1-optimization problems are investigated. It turns out that pseudo-regularity requires the LICQ constraint qualification, and implies (Lipschitzian) regularity in the convex case. For piecewise C^2-problems, inequalities to zero-Lagrange multipliers are not essential for the problem of equivalence, and the both regularities do not coincide, in general. Finally, we study regularity (strong stability) when the functions involved have only locally Lipschitzian derivatives. The results are extended to critical points of similar problems, in particular to zeros of generalized equations.

1   INTRODUCTION

The basic notion of the present paper is related to a multivalued mapping $S: Y \to X$, where X and Y are normed spaces and $S(y)$ is a subset of X (empty or not). Given x^0 and y^0 satisfying $x^0 \in S(y^0)$, the mapping S is said to be *pseudo-Lipschitz* at (x^0, y^0) if there are neighbourhoods (nbhds) $U \ni x^0$ & $V \ni y^0$ and some constant L satisfying

$$S(y') \cap U \subset S(y'') + L \| y' - y'' \| B_X \qquad \forall y' \& y'' \in V \qquad (1.1)$$

(c.f. notations at the end of the introduction).

The property (1.1) characterizes S around (x^0, y^0) and is of particular interest if $S = F^{-1}$ is the inverse of a multifunction $F: X \to Y$. If F^{-1} fulfils (1.1), then F is said to be *pseudo-regular* at (x^0, y^0). If, additionally, F^{-1} is single-valued on V, we call F *regular* at (x^0, y^0).

Provided that F is a function, the point $y^0 = F(x^0)$ is unique, and we simply say that F is (pseudo-) regular at x^0.

Pseudo-regularity is also called *metric regularity* in several publications. Here, the present terminology seems to be more appropriate because we will permit more general perturbations of F, not covered by metric regularity. For gaining an impression about the value of property (1.1), we refer to J.P. Aubin & I. Ekeland [1], R. Cominetti [3] and B. Mordukhovich [10], where the reader finds also an overview about further references devoted to the multivalued case in general spaces. For historical notes concerning this property as well as for investigations in the context of finite dimensional generalized equations, we refer to the recent paper of A.L. Dontchev & R.T. Rockafellar [4]. There, the notion *Aubin-property* has been used for property (1.1).

The aim of this paper is twofold.

In a general setting, we show that (and how) pseudo-regularity of F permits to apply successive approximation for solving an inclusion $g(x) \in F(x)$, whenever g is a "small Lipschitz function", Lemma 2.3.

On the other hand, we are mainly interested in the particular case of $S = F^{-1}$ being the inverse of a continuous (or Lipschitzian) function F in finite dimension. This case is special enough in order to allow comparisons with the usual notion of regularity, and it is sufficiently general in view of several applications to variational inequalities, generalized equations, complementarity problems and critical point theory in nonlinear optimization.

In § 2, were arbitrary (multi-)functions F are considered, we mainly deal with the question of equivalence of both regularities and with their characterization by means of certain generalized derivatives of F at/near x^0; contingent derivatives and Thibault's (double) limit sets.

To describe solutions of inclusions under small *nonlinear perturbations*, we put, in (1.1), $y^0 = 0$ and replace the set S(y) by the solution set S(g) of the inclusion $g(x) \in F(x)$, $x \in \Omega$.

Now g belongs to a zero-nbhd V in some normed space $(G, |\,.\,|)$ of continuous functions $g: \Omega \to Y$, and Ω is a nbhd of x^0. If the modified condition

$$S(g') \cap U \subset S(g'') + L\,|\,g' - g''\,|\,B \qquad \forall\, g'\, \&\, g'' \in V\,(\subset G) \qquad (1.2)$$

is fulfilled, we call F *pseudo-regular* at (the zero) x^0 *w. r. t. G*.

We will put $G = C^{0.1}(\Omega, Y)$, the space of all bounded functions $g: \Omega \to Y$ having a smallest finite Lipschitz constant L_g on Ω, where $|\,g\,| = \max\{\sup_\Omega \|g(x)\|, L_g\}$. This space is of particular interest since (by Banach's fixed point theorem) regularity implies regularity w. r. t. G if X is complete. Our Th. 2.4 asserts the same concering pseudo-regularity.

For the special case of locally Lipschitzian functions $f: R^n \to R^m$, Th. 2.6 shows that pseudo-regularity can be equivalently characterized by several approaches (in contrast to the general case). Another statement, true for $f \in C(R^n, R^m)$ & $m = n$, but false for multifunctions or for $m \neq n$, is our Lemma 2.1.

In § 3, we study both regularity concepts more specifically in the context of stability of (primal-dual) critical points for optimization problems. There, F is Kojima's function associated to a nonlinear optimization problem, or (for dealing with generalized equations) F is a modification of such a function. For *convex C^1- problems* , the equivalence of both concepts now follows from Lemma 2.1 and E. Michael's selection theorem. For *nonconvex C^1-problems,* we show that pseudo-regularity requires the Linear Independence Constraint Qualification. For special $C^{1.1}$-*optimization problems* (if the derivatives of the functions involved are piecewise differentiable), Th. 3.5 clarifies that both concepts can *only differ if* they already differ with respect to the subproblem without the inequalities having a zero Lagrange multiplier. Hence, for answering the question of equivalence, one may surprisingly delete all the inequalities which violate strict complementarity. Usually, even these constraints create all the difficulties when dealing with stability of critical points. For C^2-*problems,* the remaining system is locally a C^1-equation, thus both concepts are equivalent. This equivalence for C^2-problems (not true for $C^{1.1}$, see Remark 3.3) was a famous result of [4] where generalized equations

$$H(x) + a \in N_C(x), \quad C \text{ polyhedral, } H \in C^1, \quad \text{parameter: a}$$

have been investigated under the same viewpoint. There, the authors applied strong results from B. Mordukhovich [10], S.M. Robinson [14] and S. Scholtes [15] concerning conditions for pseudo-Lipschitzian behaviour of multifunctions, normal maps and connections between open and coherently oriented piecewise linear multifunctions, respectively. Our way is completely different and uses the powerful invariance of domain theorem (hidden in Lemma 2.1). Applied to generalized equations, our approach illustrates the equivalence if an additional parameter b describes the constraint set C (provided the locally defined normal-map is non-degenerated):

$$H(x) + a \in N_{C(b)}(x), \quad C(b) = \{ x \,/\, g(x) \le b \}, \quad H \in C^1, g \in C^2.$$

In § 4, we recall and interpret a regularity criterion, applicable to $C^{1.1}$-optimization problems, and compare it with the condition derived in [4]. Our new topics are the Lemmata 4.2 and 4.3 as well as the extensions to generalized Kojima-functions.

At this point, it seems appropriate to mention the known fact that the concepts of regularity are different for Lipschitz functions and dim X = dim Y = n, indeed. Let us present two examples, perhaps the simplest ones, where n = 2, $f^{-1}(0) = \{0\}$ and $f^{-1}(c) = \{ z(c), -z(c) \}$ for $c \ne 0$.

EXAMPLE 1.   This piecewise linear example has been found by Peter Fusek, Humboldt-University Berlin. Define g(x, y) in the following way:

  g = y if $|y| \le x$,   g = x if $|x| \le y$
  g = -y if $|y| \le -x$,   g = -x if $|x| \le -y$.

Next, take the linear transformation L turning z = (x, y) by $-\pi/4$ in the plane, i.e. $L(z) = \frac{1}{2} \sqrt{2} (y+x, y-x)$, and define $f_1(z) = g(z)$, $f_2(z) = g(L(z))$. Explicitly, f_2 has the form $h / \sqrt{2}$ where

  h = y - x if $x \ge 0$ & $y \ge 0$,   h = -y - x if $x \ge 0$ & $y \le 0$,
  h = x - y if $x \le 0$ & $y \le 0$,   h = y + x if $x \le 0$ & $y \ge 0$.

EXAMPLE 2.   Take the complex function $f(z) = z^2 / |z|$ if $z \ne 0$ & $f(0) = 0$.

Verifying the announced properties is left to the reader.

Notations   Given any normed space G, we denote by B_G and B^o_G the closed and open unit-ball, respectively. If the space is obvious, we also omit the index. Further, we mix operations with sets and points in the sense of the Minkovsky sum of sets. So, x + r B is the closed ball with centre x and radius r. For $a \in R^n$, $r \in R$ & $C \subset R^n$, we simply write a + r C instead of the set { a+rc / c∈ C }, and $r C^T a$ = { $r c^T a$ / c ∈ C }. If the kind of (matrix) multiplication is evident, we will omit the sign of transposition.

The point-to-set distance is denoted by dist(x, A) = inf { ‖ x - a ‖ / a ∈ A } with the usual convention dist(x, $\varnothing$) = ∞ .

By $f \in C^{1.1}$, we indicate that f is a functions having locally Lipschitz first derivatives. An optimization problem defined by $C^{1.1}$-functions is said to be a $C^{1.1}$-problem. "nbhd" abbreviates "neighbourhood".

2   (PSEUDO-) REGULARITY of LOCALLY LIPSCHITZIAN FUNCTIONS

Many known technics and statements of nonsmooth and multivalued analysis are basically of the following form: One considers an inclusion h(x) ∈ H(x) where h is a function, and studies the solution behaviour under certain small perturbations g of the function only: h(x) + g(x) ∈ H(x).

The "input" are assumptions concerning constant perturbations h(x) + y ∈ H(x). Setting F = - h + H , this is the relation between some kind of stability of F and stability of F w.r.t. G. Since x solves the g- inclusion iff x is a fixed point of F^{-1}(g(.)), the properties of F^{-1} are the key of all these statements, and the specific structure of H comes only into the play when the suppositions concerning F^{-1} have to be checked. So, it is not surprising, that the multivalued term H plays often the role of an appendix, and that many statements about equations hold for multifunctions, too (with the same proof). Following this line, we will show that even the basic tool for dealing with functions, successive approximation, can be easily extended to multifunctions and leads to useful results. These are reasons for dealing with pseudo-regular functions particularily. In addition, though the topological properties of a *regular* locally Lipschitz function f: $R^n \to R^n$ are rather simple, there are still several open questions under the weaker pseudo-regularity at a zero x^0. So, it is even not clear whether the zero x^0 is necessarily isolated.

2.1   Basic Observations: Selections and Successive Approximation

In order to guarantee that a multifunction F has a continuous selection on a set U, one could require that there are some set W and some continuous function h such that F(u) ∩ W = { h(u) } for all u ∈ U.

In accordance with Lemma 2.1, this most trivial sufficient condition is even a necessary one if F = f^{-1} and f : $R^n \to R^n$ is continuous.

Selections

LEMMA 2.1.

Let f: $R^n \to R^n$ be continuous, $\varnothing \neq U \subset R^n$ be open and bounded and

h: U → R^n be any continuous selection of f^{-1} on U.
Then, the set W = h(U) is open, $f^{-1}(u) \cap W$ = { h(u) } $\forall$ u ∈ U and h^{-1} = $f|_W$.

Proof:
Since $f^{-1}(u') \cap f^{-1}(u'')$ = ∅ (u' ≠ u''), each selection h: U → W of f^{-1} has the inverse h^{-1}: W → U with h^{-1} = $f|_W$. The function $f|_W$ has the inverse h ; hence it is one-to-one, which tells us that $f^{-1}(u) \cap W$ = { h(u) } for all u ∈ U. Because h and h^{-1} = $f|_W$ are continuous, the sets U and W are homeomorphic. Therefore, W is open due to Brouwer's invariance of domain theorem. ◆

NOTE.
Supposing 0 ∈ U , the Lemma says the following.
The pre-image x^0 := h(0) is an isolated zero of f, and W is a nbhd of x^0 such that the map u → $f^{-1}(u) \cap W$ is single-valued and continuous on U. If such open sets U and W exist, we will also say that *f^{-1} is locally unique and continuous* near $(x^0, 0)$.

LEMMA 2.2.
Let f: R^n → R^n be continuous, ∅ ≠ U ⊂ R^n be open and bounded. Moreover, suppose there is some subset X ⊂ R^n such that the multifunction u → F(u) := $f^{-1}(u) \cap X$ is lower semicontinuous on U and has convex and non-empty images F(u) ($\forall$ u ∈ U). Then, F is single-valued and continuous on U.

Proof: The application of E. Michael's [9] selection theorem yields the existence of a continuous selection h of F on U. By Lemma 2.1, { h(u) } = $f^{-1}(u) \cap h(U)$. Since W = h(U) is open and W ⊂ X, the convexity of $f^{-1}(u) \cap X$ yields that even $f^{-1}(u) \cap X$ is single-valued. ◆

Successive approximation

It is well known that, for solving the equation g(x) = f(x) where g is a small C^1-function, one may study the iterative (fixed point) process of *successive approximation* x^{k+1} = $f^{-1}(g(x^k))$, if f has a well-defined Lipschitzian inverse. For solving an inclusion g(x) ∈ F(x), if the inverse F^{-1} is closed-valued and pseudo-Lipschitzian at $(x^0, 0)$, the situation is completely similar. The same process x^{k+1} ∈ $F^{-1}(g(x^k))$ can be applied with the only difference that now x^{k+1} and x^k should be "sufficiently close".
The main points in this and the next subsection are the facts that g varies in $C^{0.1}$ and F^{-1} is pseudo-Lipschitz and closed valued. Whether F is multivalued or not, does not play any role in this context. With other words, the proofs for multifunctions F are the same as for functions.

LEMMA 2.3.
Let X be a Banach space , Y be a normed space, and let F: X → Y be a multifunction such that both F^{-1}: Y → X is pseudo-Lipschitz at $(\xi^0, 0)$ and the sets $F^{-1}(y)$ are closed. Without loss of generality, let U = ξ^0 + δ B^0_X, V = δ B^0_Y in (1.1), and δ < 1. Put Ω = U and G = $C^{0.1}(\Omega, Y)$. Then, if g ∈ G has sufficiently small norm, | g | ≤ ¼ δ $(L+1)^{-1}$, the mapping Ω ∍ x → $F^{-1}(g(x))$ has a fixed point $x^* ∈ \xi^0$ + 3 (L+1) | g | B.

Moreover, x^* can be computed by the following geometrically convergent process of successive approximation:
Put $x^0 = \xi^0$. Select $x = x^1$ such that
$$x \in F^{-1}(g(x^0)) \quad \& \quad \| x - x^0 \| \le \text{dist}(x^0, F^{-1}(g(x^0))) + \| g(x^0) - 0 \|$$
and, for $k > 0$, take $x = x^{k+1}$ such that
$$x \in F^{-1}(g(x^k)) \quad \& \quad \| x - x^k \| \le \text{dist}(x^k, F^{-1}(g(x^k))) + \| g(x^k) - g(x^{k-1}) \|.$$

Proof: Our formulation for selecting x^1 as well as the proof are made in such a way that the Lemma can directly be extended to the situation of metric spaces, provided that X is complete, $0 \in Y$ and (1.1) means
$$\text{dist}(x', F^{-1}(y'')) \le L\, d(y'', y') \text{ if } x' \in F(y') \cap U \text{ and } y', y'' \in V.$$
Our estimates are not sharp, but keep the proof simple. We fix some $q \in (0, 1)$ such that
$$W := x^0 + [\, 2q + q/(1-q)\,] B_X \subset U, \qquad\qquad \text{e.g} \quad q = \frac{1}{4}\delta$$
and take r such that $0 < r < \delta$ and $r(L+1) \le q$, e.g. $r = q(L+1)^{-1}$.
Now consider any function $g \in G$ with $|g| \le r$.
For $k = 0$, we have $x^0 \in F^{-1}(0)$ by the assumption, and
$$\text{dist}(x^0, F^{-1}(g(x^0))) \le L\, \| g(x^0) - 0 \|.$$
If $g(x^0) = 0$, then $x^1 = x^0$ fulfils the iteration rule.
If $g(x^0) \ne 0$, then, due to pseudo-regularity, the existence of x^1 is also obvious.
In both cases, we see that the iterate $x^1 \in F^{-1}(g(x^0))$ fulfils
$$\| x^1 - x^0 \| \le (L+1)\, \| g(x^0) \| \le (L+1)\, |g| \le q < 1.$$
Clearly, $x^1 \in W$.
Now, beginning with $k = 1$, it holds $x^k \in F^{-1}(g(x^{k-1})) \cap W$.
If $g(x^k) = g(x^{k-1})$, then one may put $x^{k+1} = x^k$.
If $g(x^k) \ne g(x^{k-1})$, then the next iterate x^{k+1} again trivially exists.
We show that $\| x^{k+1} - x^k \| \le q \| x^k - x^{k-1} \|$.
Since $g(x^k)$ & $g(x^{k-1}) \in V$ and $x^k \in F^{-1}(g(x^{k-1}))$, the pseudo-regularity yields
$$\text{dist}(x^k, F^{-1}(g(x^k))) \le L\, \| g(x^k) - g(x^{k-1}) \|.$$
Hence, step k provides us with x^{k+1} satisfying indeed
$$\| x^{k+1} - x^k \| \le (L+1)\, \| g(x^k) - g(x^{k-1}) \| \le (L+1)\, |g|\, \| x^k - x^{k-1} \| \le q \| x^k - x^{k-1} \|.$$
Setting $h = (L+1)|g|$ and summing up the distances
$$\| x^0 - x^1 \| + \ldots + \| x^{k+1} - x^k \| \le h + h(1-h)^{-1} \le q + q(1-q)^{-1},$$
one finds that $x^{k+1} \in W$.
Therefore, the process generates a Cauchy-sequence converging to $x^* \in W$.
Pseudo-regularity and $x^k \in F^{-1}(g(x^{k-1})) \cap W$ ensure that
$$\text{dist}(x^k, F^{-1}(g(x^*))) \le L\, \| g(x^*) - g(x^{k-1}) \| \le L\, |g|\, \| x^* - x^{k-1} \|.$$
So we observe $\text{dist}(x^*, F^{-1}(g(x^*))) = 0$ and, since $F^{-1}(g(x^*))$ is closed,
$$x^* \in F^{-1}(g(x^*)).$$
Taking q and r as above, the *estimate* $x^* \in \xi^0 + 3(L+1)|g|B$ follows from
$h = (L+1)|g| < \frac{1}{4}\delta < \frac{1}{4}$ and $\| x^{k+1} - x^0 \| \le h + h(1-h)^{-1} \le 3h$. ♦

The next statement extends pseudo-regularity to Lipschitzian perturbations.

THEOREM 2.4
Under the hypotheses of Lemma 2.3, the mapping F is pseudo-regular at ξ^0 w.r.t. $G = C^{0.1}(\Omega, Y)$.

Proof.

Let L and δ be as in the Lemma.

First we prove that the multifunction $F_g = g + F$ is pseudo-regular with constant $2L$ at ξ provided that

$g \in G$, $|g| < r := (\delta/3)(L+1)^{-1}$, $0 \in F_g(\xi)$ and $\xi \in \xi^0 + (\delta/3) B^0$.

Possible related nbhds are

$U_g = \xi + (\delta/3) B^0$ and $V_g = \varepsilon B^0$ with $0 < \varepsilon < \delta/3$ and $\varepsilon < (\delta/12)(L+1)^{-1}$.

This first statement generalizes Th. 2.1 of [3] where F has a special structure. Our proof is more direct and constructive.

Let $y' \in F_g(x')$, $x' \in U_g$, $y' \in V_g$.

Given $y'' \in V_g$ we have to find some x'' such that

$\qquad y'' - g(x'') \in F(x'')$ and $\| x'' - x' \| \le 2L \| y'' - y' \|$.

We modify the proof of Lemma 2.3 and exploit that our constants ensure the inequalities

$2L \| y'' - y' \| < 4(L+1)\varepsilon < \delta/3$, $L|g| < Lr < \delta/3 < \frac{1}{2}$ and $\varepsilon + |g| < 2\delta/3$.

Again, we generate a sequence x^k; now by setting $x^0 = x'$ and selecting x^{k+1} near x^k such that $y'' - g(x^k) \in F(x^{k+1})$.

For $k = 0$, it holds $y' - g(x^0) \in F(x^0)$.

Since $x^0 \in U_g \subset U$ and $\max\{\| y' - g(x^0) \|, \| y'' - g(x^0) \|\} < \varepsilon + |g| < 2\delta/3$, we may use that F is pseudo-regular at ξ^0. Hence, there exists x^1 such that

$\qquad \| x^1 - x^0 \| \le L \| y'' - y' \|$ and $y'' - g(x^0) \in F(x^1)$.

So, at least for $k = 1$, we have

$\qquad \| x^k - x^{k-1} \| \le 2^{-(k-1)} L \| y'' - y' \|$ and $y'' - g(x^{k-1}) \in F(x^k)$. $\qquad$ (2.1)

As the induction step, notice that (2.1) implies

x^k & $x^{k-1} \in U \qquad$ due to $\qquad \| x^k - x^0 \| < 2L \| y'' - y' \| < \delta/3$,

whereas

$y'' - g(x^{k-1})$ & $y'' - g(x^k) \in V \qquad$ due to $\| y'' - g(x^k) \| < \varepsilon + |g| < 2\delta/3$.

Using $y'' - g(x^{k-1}) \in F(x^k)$ and pseudo-regularity of F, there exists x^{k+1} such that $y'' - g(x^k) \in F(x^{k+1})$ and

$\| x^{k+1} - x^k \| \le L \| g(x^k) - g(x^{k-1}) \| \le L|g| \; \| x^k - x^{k-1} \| < \frac{1}{2} \| x^k - x^{k-1} \|$.

Thus, $\| x^{k+1} - x^k \| \le 2^{-k} L \| y'' - y' \|$, and (2.1) holds for all k, indeed.

This ensures convergence $x^k \to x^*$ as well as $\| x^* - x^0 \| \le 2L \| y'' - y' \|$.

From $x^k \in F^{-1}(y'' - g(x^{k-1}))$, we derive as in Lemma 2.3

$\qquad \text{dist}(x^k, F^{-1}(y'' - g(x^*))) \le L \| g(x^*) - g(x^{k-1}) \| \le L|g| \; \| x^* - x^{k-1} \|$.

Hence $\text{dist}(x^*, F^{-1}(y'' - g(x^*))) = 0$, and $x'' := x^* \in F^{-1}(y'' - g(x^*))$ has the claimed property.

Next, assume that g' and g'' have small norm in G, $g'(x') \in F(x')$ and x' is close to ξ^0. To find x'' such that $g''(x'') \in F(x'')$ and $\| x'' - x' \|$ is sufficiently small, notice that x' solves $0 \in -g' + F =: F'$.

If $\| x' - \xi^0 \|$ and $|g'|$ are small enough, F' is (uniformly) pseudo-regular at x', according to the first part of the proof.

Considering x'' as a solution to $(g'' - g') \in -g' + F$, Lemma 2.3 applied to x', F' and $(g'' - g')$ yields the assertion with the new constant $L_G = 3(2L+1)$. $\quad \blacklozenge$

An application to Newton-type methods

Knowing both that F is pseudo-regular and that successive approximation can by used to construct solutions, Newton-type methods can be studied like in the smooth case.

To be more precise, let $h \in H$ be an inclusion of interest, and ξ^0 be a solution. Assume that F = -h + H satisfies the assumptions of Lemma 2.3, and put $\Omega = U$, $G = C^{0.1}(\Omega, Y)$. Introducing any function $h_\xi: X \to Y$ for $\xi \in \Omega$ ("linearization" of h near ξ) and setting $g_\xi = h - h_\xi$, $F_\xi = -h_\xi + H$, the inclusions $g_\xi \in F_\xi$ and $h \in H$ coincide. Thus, one may apply Lemma 2.3 to $g_\xi \in F_\xi$: Find $x = x^{k+1}$ such that

$$g_\xi(x^k) \in F_\xi(x) \ \& \ \|x - x^k\| \le \text{dist}(x^k, F_\xi^{-1}(g_\xi(x))) + \|g_\xi(x^k) - g_\xi(x^{k-1})\|. \quad (2.2)$$

Clearly, the starting point x^0 has to be a zero of F_ξ sufficiently close to ξ^0, one could put $\xi = x^k$ to get a "pure" Newton method with new "linearizations" at each step, F_ξ should be simpler than F, and (2.2) can be modelled by minimizing $\| x - x^k \|$ s.t. $g_\xi(x^k) \in F_\xi(x)$.

After collecting the needed straightforward estimates (where Th. 2.4 is useful), it remains, however, a serious problem: We have to ensure that $g_\xi \in G$ and $| g_\xi |$ is sufficiently small on $\Omega = \xi^0 + \delta B^0$ containing ξ. If the constant L is not explicitly known, we need $| g_\xi | = o(\delta)$ to ensure local convergence. This problem concerns the approximations of h only:

> Find a (useful !) approximation h_ξ of h such that, given ξ^0 and ξ, the $C^{0.1}$-norm of h_ξ - h on $\Omega = \xi^0 + \|\xi - \xi^0\| B$ is of type $o(\|\xi - \xi^0\|)$.

If $h \in C^1$, usual linearizations can be applied. For h being locally Lipschitz, among several approaches, we mention the use of (strongly) B-derivatives, c.f. S.M. Robinson [13]. Nevertheless, the claimed approximation is quite strong. Concerning approaches based on weaker assumptions c.f. [8] and [11].

2.2   Lipschitzian and pseudo-Lipschitzian Inverse Functions

Here, we assume that x^0 is zero of a continuos function f in finite dimension.

First observe the following simple *consequences* of the fact that f is pseudo-regular at x^0 w.r.t. G.

(C1) $S(g) \cap U \ne \emptyset \quad \forall g$ sufficiently close to $0 \in G$.

(C2)   The mappings $g \to S_\varepsilon(g) := S(g) \cap U$ and $g \to \text{cl } S_\varepsilon(g)$ are *lower semicontinuous (l.s.c.)* on V.

(C3)   If f is pseudo-regular at x^0 then it fulfils (locally) the so-called *open property*: $f(\Omega) \cap V$ is open for each open subset $\Omega \subset U$.

Projections.

Put $\phi(g) = \inf \{ \| x - x^0 \| / x \in S(g) \}$, and let

$\Psi(g) = \{\xi \in S(g) / \| \xi - x^0 \| = \phi(g)\}$ be the *projection of x^0 onto S(g).*

Taking $r > 0$ sufficiently small and $g, g' \in r B_G$, we observe:

The closed set S(g) is nonempty, and $\emptyset \ne \Psi(g) \subset x^0 + L | g | B$.

If $\xi \in \Psi(g)$, then there is some $x' \in S(g')$ such that $\| x' - \xi \| \le L | g' - g |$, hence $\phi(g') \le \phi(g) + L | g' - g |$. Reversing g' and g, ϕ is Lipschitz on $r B_G$. This yields the well-known observation

(C4)   If card $\Psi(g) = 1$ on r B_G , then Ψ is continuous.

COROLLARY 2.5.

Let $f : R^n \to R^n$ be continuous and pseudo-regular at x^0, $f(x^0) = 0$.

Then the following properties are equivalent.

(i) f^{-1} has a continuous selection h near the origin and $h(0) = x^0$.

(ii)   The projection map Ψ is locally (near 0) single-valued.

(iii)   f is regular at x^0.

Proof: The corollary is a direct consequence of Lemma 2.1 and (C4).   ◆

In particular, the local uniqueness of the projection Ψ does not depend on the norm if $f \in C(R^n, R^n)$ is pseudo-regular. Moreover, if f is pseudo-regular at x^0 without being regular, then x^0 is a bifurcation point of f^{-1} such that, near $(x^0, f(x^0))$, the mapping f^{-1} has not any (single-valued) continuous selection.

Equivalent conditions for pseudo-regularity can be given in terms of Mordukhovich's coderivatives (for general multifunctions in abstract spaces, c.f. [10]) In finite dimension, the key of his definition is the projection of points near (x^0, y^0) on the graph of F.

In our special case, we only need the simpler construction of a *directional contingent derivative* Cf(x; u) for f at x in direction u. Contingent or Bouligand derivatives for multifunctions were essentially used e.g. in [1].

The set Cf(x; u) consists of all limits $v = \lim t^{-1}(f(x + tu) - f(x))$ for $t \downarrow 0$.

Further, given some nbhd $\Omega \ni x^0$, $\alpha > 0$, $x \in \Omega$ and $v \in C^{0.1}(\Omega, R^m)$ we introduce a functional p_α on Ω (an exact penalty function) by

$p_\alpha(y) = \| y - x \| + 2\alpha \| f(x) + v(y) - f(y) \|$ $(y \in \Omega)$,

and put $\Phi_\alpha(x, v) = \arg \min p_\alpha(y)$ s.t. $y \in \Omega$.

NOTE

$\Phi_\alpha(x, v) \neq \emptyset$ whenever both $\| x - x^0 \|$ and $|v|$ are small enough.

Indeed, for $\xi \in \Omega$ and $p_\alpha(\xi) \leq p_\alpha(x)$, it holds

$\qquad \| \xi - x \| \leq p_\alpha(\xi) \leq p_\alpha(x) = 2\alpha \| v(x) \| \leq 2\alpha |v|$ (2.3)

Hence ξ belongs to the compact ball $\| \xi - x^0 \| \leq 2\alpha |v| + \| x - x^0 \|$ which lies entirely in Ω. Here, Ω is not necessarily requested to be open.

THEOREM 2.6.

Let   f: $R^n \to R^m$ be locally Lipschitz, $x^0 \in$ int Ω, $f(x^0) = 0$, and let Ω be bounded. Put G = $C^{0.1}(\Omega, R^m)$. Then the following statements are equivalent.

(i)   f is pseudo-regular at x^0

(ii) $\exists \alpha > 0 :$ $B \subset$ Cf(x ; αB) $\forall x \in x^0 + \alpha^{-1}B$

(iii) $\exists \alpha > 0, \beta > 0 : \xi \in \Phi_\alpha(x, v) \Rightarrow f(\xi) = f(x) + v(\xi)$

$\qquad\qquad\qquad\qquad\qquad\qquad \forall x \in x^0 + \beta B, v \in \beta B_G$

(iv)   f is pseudo-regular at x^0 w.r.t. G.

Proof:

(i) $\Rightarrow$ (ii)   Let U, V and L satisfy (1.1), take some $\beta > 0$ such that

$x \in U \cap f^{-1}(V)$ $\forall x \in x^0 + \beta B^0$ and suppose $v \in$ bd B & $\| x - x^0 \| < \beta$.

For small t > 0,   (1.1) and $x \in f^{-1}(f(x)) \cap U$ ensure the existence of certain

$x' \in f^{-1}(f(x) + t\,v)$ such that $0 < \|x' - x\| \le t\,L$. Obviously,

$$f(x') - f(x) = t\,v \quad \text{and} \quad \|t\,v\| / \|x' - x\| \ge 1/L.$$

Setting $x' = x + \lambda w$, $\|w\| = 1$, $0 < \lambda = \|x' - x\| \le t\,L$, we obtain, with some Lipschitz constant Lip(f) for f near x^0,

$$[f(x + \lambda w) - f(x)] = t\,v \quad, \quad \lambda \ge t\,\text{Lip}(f) \quad \text{and}$$
$$[f(x + t\,(\lambda/t)\,w) - f(x)] / t = v.$$

Clearly, w and λ depend on t.

Passing to the limit $t \to 0$ we get some u satisfying $v = Cf(x; u)$ as a cluster point of the vectors $u(t) = (\lambda/t)\,w$ where $\text{Lip}(f) \le \|u\| \le L$.

This step is justified because of

$$\| [f(x + t\,u(t)) - f(x)] - [f(x + t\,u) - f(x)] \| \le t\,\text{Lip}(f)\,\|u(t) - u\| = o(t).$$

Hence $\text{bd } B \subset Cf(x; \alpha B) \ \forall x \in x^0 + \beta B^0$; with $\alpha = L$. Since $Cf(x; .)$ is positively homogeneous and α may be enlarged, we verified (ii).

(ii) $\Rightarrow$ (iii)   Let α satisfy (ii), and let $\beta \in (0, \frac{1}{2})$ be small, such that

$$\beta + 2\alpha\,\beta < \alpha^{-1} \quad \& \quad x^0 + (\beta + 2\alpha\,\beta)\,B \subset \Omega.$$

Now suppose x and v to be fixed with $\max\{\|x - x^0\|, |v|\} < \beta$.

Using (2.3) we observe $(\varnothing \ne) \ \Phi_\alpha(x, v) \subset x^0 + \alpha^{-1}B$.

Given $\xi \in \Phi_\alpha(x, v)$, put $w = f(\xi) - (v(\xi) + f(x))$, and assume $w \ne 0$.

Setting $h = w / \|w\|$ we obtain

$$Cp_\alpha(\xi;\ r) = r^T(\xi - x) / \|\xi - x\| + 2\alpha\,C(f - v)\,(\xi; r)^T\,h \quad \text{for } \xi \ne x \text{ and}$$
$$Cp_\alpha(\xi;\ r) = \|r\| + 2\alpha\,C(f - v)\,(\xi; r)^T\,h \quad\quad\quad \text{for } \xi = x.$$

Due to (ii), there is some direction $r \in \alpha B$ such that $-h \in Cf(\xi; r)$.

For defining some $h' \in C(f - v)(\xi; r)$, the same sequence $t\downarrow 0$ as for h may be used. Since v is Lipschitz with modulus β, one then finds $h' \in C(f - v)(\xi; r)$ such that $\|h' - h\| \le \beta\,\|r\| \le \beta\,\alpha$.

Thus $Cp_\alpha(\xi;\ r)$ contains an element γ satisfying

$$\gamma \le \|r\| + 2\alpha\,h'^T\,h \le \alpha + 2\alpha\,(-h)^T\,h + 2\beta\,\alpha \le -\alpha + 2\beta\,\alpha < 0.$$

This contradicts $\xi \in \Phi_\alpha(x, v)$, consequently $w = 0$ must be true.

(iii) $\Rightarrow$ (iv)   Let α and β be the constants from (iii), and d be the finite diameter of Ω. After decreasing β, we may suppose that

$$\Phi_\alpha(x, v) \ne \varnothing \quad \text{whenever } |v| \le \beta \text{ and } x \in x^0 + \beta B.$$

For g', g'' $\in \delta B_G$ and $x \in \Omega$, we observe

$$\| g''(y) - g'(x) \| \le \| g''(y) - g''(x) \| + \| g''(x) - g'(x) \| \le d\,\delta + 2\,\delta.$$

Thus, if $\delta > 0$ & $2\,\delta + d\,\delta < \beta$, the function v as $v(y) = g''(y) - g'(x)$ satisfies $|v| \le \beta$.

Now consider any fixed element $x \in S(g') \cap (x^0 + \beta B^0)$.

With $\xi \in \Phi_\alpha(x, v)$, statement (iii) and $f(x) = g'(x)$ yield the equations

$$f(\xi) = f(x) + v(\xi) = f(x) + g''(\xi) - g'(x) = g''(\xi);$$

therefore, $\xi \in S(g'')$.   By (2.3) we obtain

$$\|\xi - x\| \le 2\alpha\,\|v(x)\| = 2\alpha\,\|g''(x) - g'(x)\| \le 2\alpha\,|g'' - g'|.$$

Thus f is pseudo-regular at x^0 w.r.t. G.   ◆

The direction (ii) $\Rightarrow$ (i) of Th.2.6 is already a consequence of the more general theorem 4, § 7.5 in [1]. It is further worth noting that the Lipschitz-property of f is only needed for (i) $\Rightarrow$ (ii). The implications (ii) $\Rightarrow$ (iii) $\Rightarrow$ (iv) remain true for continuous functions f. Clearly, (i) $\Rightarrow$ (iv) follows also from Th. 2.4.

To characterize *regularity* we use, for locally Lipschitz f, the (closed and connected) sets $\Delta f(x; u)$ which, by definition, consist of all possible double limits of the form $\lim t^{-1} (f(x' + t u) - f(x'))$, where $x' \to x$ and $t \downarrow 0$.

These limit sets have been considered already by L.Thibault [16] in order to construct certain subdifferentials for vector-valued functions.

For functionals, they are the key to define F.H. Clarke's [2] directional derivative via $f^C(x ; u) = \sup \Delta f(x ; u)$. The meaning of these sets for regularity and implicit functions has been shown in [6] and [7]. Let us recall a basic statement.

THEOREM 2.7.

Let $f: R^n \to R^n$ be continuous.

f is regular at x^0 iff there exists an $r > 0$ such that
$$\| f(x) - f(y) \| \geq r \| x - y \| \quad \forall x, y \in x^0 + r B.$$
Let $f: R^n \to R^n$ be locally Lipschitz, Ω be some nbhd of x^0, $f(x^0) = 0$.

Then the following statements are equivalent

(i)   f is regular at x^0,
(ii) $0 \in \Delta f(x^0; u)$ implies $u = 0$,
(iii)   f is regular at x^0 w.r.t. $G = C^{0.1}(\Omega, R^n)$.

In the regular case, the locally unique inverse f^{-1} satisfies
$$u \in \Delta f^{-1}(0 ; v) \text{ iff } v \in \Delta f(x^0; u).$$

We refer to [6] for a proof, for relations to F.H. Clarke's [2] inverse function theorem, which gives a sufficient regularity condition in terms of generalized Jacobians, as well as for chain rules concerning Δf for composed functions, in particular for Kojima's function.   ◆

The implication (i) $\Rightarrow$ (iii) follows also with the tools elaborated in [12] or [13]. The Lipschitz-property of f is required for (ii) $\Rightarrow$ (i), the implications (i) $\Rightarrow$ (iii) $\Rightarrow$ (ii) remain true for continuous f.

The theorems 2.6 and 2.7 yield relations between the different directional derivatives $Cf(x; .)$ and $\Delta f(x; .)$ near x^0 by comparing the two regularity concepts for locally Lipschitz functions. $Cf(x; .)$ has to be surjective in a specified uniform manner, $\Delta f(x; .)$ should be injective.

COROLLARY 2.8.

If $f: R^n \to R^n$ is locally Lipschitz, then $0 \notin \Delta f(x^0; bd B)$ implies the surjectivity property (ii) of Th.2.6.

COROLLARY 2.9.

For $f \in C^1(R^n, R^n)$, *pseudo-regular* implies *regular*.

Proof: We have $\Delta f(x^0; u) = Cf(x^0; u) = \{ Df(x^0)u \}$. Hence, f is pseudo-regular at $x^0 \Rightarrow Df(x^0)$ is regular $\Rightarrow$ f is regular at x^0.   ◆

2.3   Continuous Multivalued Selections of pseudo-Lipschitzian Maps

In this section, we assume that $f: R^n \to R^n$ is continuous, $f(x^0) = 0$.

REMARK 2.1.

Let f be pseudo-regular w.r.t. G at an isolated zero x^0.

Then there are some nbhd U' of x^0 and some $r > 0$ such that the multifunction $g \to H(g) := S(g) \cap U'$ has compact images $H(g)$, H is continuous (upper & lower semicontinuous) on $r B_G$, and $H(0) = \{ x^0 \}$.

Indeed, setting $H(g) = S(g) \cap (x^0 + 2 L \mid g \mid B)$, (with L from (1.2)), it is evident that H has the claimed properties.   ♦

Generally, the existence of a nontrivial *continuous compact-valued selection* $H \subset S$, seems to be an open problem. Nevertheless, the existence can easily be shown for piecewise C^1-functions (briefly PC^1).

REMARK 2.2.

Let $f: R^n \to R^n$ be a PC^1 function, pseudo-regular at x^0.

Then x^0 is an isolated zero. Thus, near the origin, f^{-1} has a continuous multivalued selection H with compact images and $H(0) = \{ x^0 \}$.

Indeed, by definition, f is continuous, and $f(x) = h_k(x)$ is always true for some function h_k, where $k = k(x)$ and $h_k \in \{ h_i \mid i \in I \}$, a finite family of C^1-functions. We may assume that the sets $X_i = \{ x \mid f(x) = h_i (x) \}$ fulfil $x^0 \in cl (int X_i)$, otherwise the local representation of f would need less functions h_i. Applying Th.2.6 (property ii) to $x \in int X_i \cap (x^0 + \alpha^{-1}B)$ we obtain $\| Dh_i (x)^{-1} \| \le 1/\alpha$. Hence, f^{-1} is locally defined by the regular, inverse functions h_i^{-1}: $\exists \gamma > 0 : f^{-1}(a) \cap (x^0 + \gamma B) \subset \{ h_i^{-1}(a) \mid i \in I \}$ for sufficiently small $\|a\|$. Thus x^0 is an isolated zero, and H exists as required.   ♦

The next proof shows how Lemma 2.1 and the selection H can be used. For Lemma 3.4, we need a more skilful construction.

COROLLARY 2.10.

Let $f \in C(R, R)$. If f is pseudo-regular at x^0, then f is regular at x^0.

Proof: The zero must be isolated. Otherwise, one finds intervals $[x^k , x^{k+1}]$ where $f(x^k) = 0$ and $x^k \to x^0$. Let ξ^k be a global extremal point of f in the open interval (x^k , x^{k+1}), say a maximizer. Since $f(x) = a + f(\xi^k)$, $a > 0$ has no solution in the interval, we get a contradiction. So, x^0 is isolated. Using the selection $H \subset f^{-1}$ and taking the smallest element h(c) of H(c) , h becomes continuous. By Lemma 2.1, the assertion follows.   ♦

3   OPTIMIZATION PROBLEMS IN FINITE DIMENSION, the C^1-CASE

Given an optimization problem

(P)   min $\{ f(x) \mid g(x) \le 0 \}$ with $f \in C^1(R^n, R)$, $g \in C^1(R^n, R^m)$,

we define a function $F = (F_1, F_2)$ depending on the primal-dual vector $z = (x, y)$

as

$$F_1(z) = Df(x) + \Sigma_i \ y_i^+ \ Dg_i(x) \ \ (= D_x L(x, y))$$
$$F_2(z) = \ g(x) - y^-,$$

where $y_i^+ = \max\{0, y_i\}$, $y_i^- = \min\{0, y_i\}$.
This function was used (perhaps for the first time) by M. Kojima [5]. Let us briefly summarize some well-known facts.

3.1   Preliminaries; Kojima's system, KKT-points and Generalized Equations

The equation $F(z) = 0$ is closely related to the *generalized equation*

$$H(z) \in N_C(z), \quad z \in C \tag{3.1}$$

where

$$H_1(z) = Df(x) + y \ Dg(x) = Df(x) + \Sigma \ y_i \ Dg_i(x)$$
$$H_2(z) = g(x), \qquad\qquad\qquad C = R^n \times R_+{}^m$$

and $N_C(z) = \{ (0, \eta) \in R^{n+m} \ / \ \eta_i = 0 \text{ if } y_i > 0 \ \& \ \eta_i \le 0 \text{ if } y_i = 0 \}$.
The solutions of (3.1) are the *Karush-Kuhn-Tucker points* of (P).

Zeros of F (also called *critical points* of P) can be Lipschitzian mapped into Karush-Kuhn-Tucker points and voice versa. More precisely, we have

$$\begin{array}{llll} & (x, y) \ \text{KKT- point} & \Rightarrow \ (x, y + g(x)) & \text{critical} \\ \text{and} & (x, y) \ \text{critical} & \Rightarrow \ (x, y^+) & \text{KKT- point.} \end{array} \tag{3.2}$$

The set $F^{-1}(a, b)$ describes the critical points of the linearly perturbed problem
P(a, b) $\min \{ f(x) - a^T x \ / \ g(x) \le b \}$.
The natural parametrization of (3.1) leads to

$$H(z) + p \in N_C(z), \quad z \in C. \tag{3.1.p}$$

As long as $H = (H_1, H_2)$ has the above meaning, the parameter p coincides with (a, b), and (3.1.p) describes again $F^{-1}(a, b)$ up to the transformations (3.2).

If f, $g_i \in C^2$, then, after the linearization of H_1 and H_2 at some solution z^0, one obtains with $Q = D^2_x L(x^0, y^0)$ and $A = Dg(x^0)$
$$\begin{array}{llll} L_1(z) = DH_1(z^0)(z - z^0) & = & Df(x^0) + Q(x - x^0) + yA \\ L_2(z) = H_2(z^0) + DH_2(z^0)(z - z^0) & = & g(x^0) + A(x - x^0). \end{array}$$

The linearized system

$$L(z) \in N_C(z), \quad z \in C, \tag{3.1}'$$

(again with solution z^0) coincides with (3.1) applied to the parametric quadratic problem
(PQ)(a, b) $\min \{ q(x) - a^T x \ / \ g(x^0) + A(x - \underline{x}^0) \le b \}$
             with $q(x) = Df(x^0)^T x + \frac{1}{2}(x-x^0)^T Q(x-x^0)$
for $(a, b) = (0, 0)$.

Problem (P) may also include *equations* h(x) = 0 as constraints.
Then the dual vector includes the related Lagrange multipliers, say μ, and Kojima's system becomes larger due to $z = (x, y, \mu)$,
$$F_1(z) = Df(x) + \Sigma_i \ y_i^+ \ Dg_i(x) + \Sigma_k \ \mu_k \ Dh_k(x)$$
and because of the additional requirement $F_3(z) := h(x) = 0$.

Similarly, the generalized equation must be changed. However, these formal modifications would make the only difference in the following. Concerning the subsequent statements, equations play the same role as inequalities with positive Lagrange variables y^0_i, so we omit them. Of course, h must be affine-linear in convex problems.

Moreover, several of the following statements remain true if the first component of the function F is only affine-linear in y^+, that is to say
$$F_1(z) = \Phi(x) + \Sigma_i \ y_i^+ \ \Psi_i(x)$$
$$F_2(z) = g(x) - y^-,$$
and Φ and Ψ have the size and smoothness of Df and Dg.
To indicate that F has this structure, we call F a *generalized Kojima-function*.
Since F is still the product of the matrix
$$\begin{array}{ccc} \Phi(x) & \Psi(x) & 0 \\ g(x) & 0 & -E \end{array}$$
with the vector $(1, y^+, y^-)$, one may compute the generalized derivatives ΔF, for locally Lipschitzian Φ, Ψ and g, as shown in [6], § 5. Then the usual chain rules hold in set-valued form of equation-type.

As an example, let H in (3.1) be an *arbitrary* function and assume some description of C, say $C = \{ z \ / \ A \ z \le \alpha \}$. Then (3.1) is equivalent to $H(z) = y^+ A$ & $A \ z - \alpha = y^-$. This is a generalized Kojima system with $\Phi = -H$ and $g(z) = A \ z - \alpha$. The equation $F(z,y) = (a, b)$ now means $- H(z) - a + y^+ A = 0$ & $A \ z - \alpha - y^- = b$, and describes the solutions z of $H(z) + a \in N_{C(b)}(z)$ with $z \in C(b) = \{ z \ / \ A \ z - \alpha \le b \}$.
Disregarding the interpretation of the perturbed systems, we will, indeed, never use that Df appearing in F is a derivative of some function.
Hence, after replacing Df by -H, the stability of the generalized equation (3.p) and of Kojima's system differ in the description and parametrization of the feasible set only. Therefore, our Theorem 3.5 will also say:
For $H \in C^1$, the solution map S(a,b) to (3.1.a, b) is pseudo-Lipschitz at a solution $z^0 \in S(0,0)$ only if S(a,b) is locally unique.
Compared with the parametrization (3.1.p) used in [4] (there $H \in C^1$ and C is polyhedral) we need a stronger assumption (because of the variation of b) and obtain a stronger assertion (for the same reason). Additionally, we may allow that C(b) has a description of the form $C(b) = \{ z \ / \ g(z) \le b \}$, $g \in C^2$ as long as $N_{C(b)}(z)$ denotes the cone generated by the "active" gradients Dg_i.

Instead of calling F regular at some zero (x^0, y^0), one also says that (P) is *strongly stable* or *strongly regular* at this point. With slightly different meanings, this notions were introduced and investigated by M. Kojima [5] and S.M. Robinson [12] for C^2-optimization problems. Finally, let us make the following conventions.
Since inactive constraints, $g_i(x^0) < 0$, do not play any role for the local analysis of our problems, we will assume throughout the rest of this paper that *all* constraints are active at x^0, i.e., $g(x^0) = 0$. Then the set $I = \{1, ...,m\}$ will be covered by the sets $I^+ = \{ i \in I \ / \ y^0_i > 0 \}$ and $I^0 = \{ i \in I \ / \ y^0_i = 0 \}$. To simplify (PQ), one may further assume that $x^0 = 0$.

3.2   Relations between the Regularity Conditions

LEMMA 3.1.   (Necessity of LICQ)
Let F be a generalized Kojima function, where Φ and $\Psi \in C$, $g \in C^1$,
and let F be pseudo-regular at some zero $z^0 = (x^0, y^0)$. Then the gradients of
active constraints { $Dg_i(x^0)$ / $g_i(x^0) = 0$ } are linearly independent.

Proof: Recall that we may assume $g(x^0) = 0$.
Let $\mathbf{1} = (1...1) \in R^m$ and $z(\varepsilon) = (x^0, y^0 + \varepsilon\,\mathbf{1})$, $\varepsilon > 0$.
Setting $a(\varepsilon) = \varepsilon \sum \Psi_i(x^0)$, the point $z(\varepsilon)$ solves $F_1 = a(\varepsilon)$, $F_2 = 0$.
Let α satisfy $\sum \alpha_i Dg_i(x^0) = 0$.
With $\delta > 0$, we consider solutions (x, y) to $F_1 = a(\varepsilon)$, $F_2 = \delta\,\alpha$.
For small fixed ε, and for δ tending to zero, now there are solutions (x, y)
satisfying the pseudo-Lipschitz inequality
$$\| (x, y) - (x^0, y^0 + \varepsilon\mathbf{1}) \| \leq L\,\delta \| \alpha \| . \qquad (3.3)$$
The components of $y^0 + \varepsilon\mathbf{1}$ are not smaller than ε. Hence, $y_i > 0$ for small δ.
This implies $\delta\alpha_i = g_i(x) \; \forall i$.
By the mean-value theorem and $g(x^0) = 0$, we derive the identities
$g_i(x) = Dg_i(\theta_i) (x - x^0)$, where $\theta_i \to x^0_i$ (as $\delta \to 0$).
Considering $\alpha^T g(x)$ the equation $\delta \|\alpha\|^2 = \sum \alpha_i Dg_i(\theta_i) (x - x^0)$
is evident.
Since Dg is continuous and $\delta \to 0$, it holds that
$$\sum \alpha_i Dg_i(\theta_i) \to \sum \alpha_i Dg_i(x^0) = 0.$$
Recalling (3.3) one thus observes that $\| x - x^0 \| \leq L\,\delta \| \alpha \|$ and
$$\delta \|\alpha\|^2 \leq \| \sum \alpha_i Dg_i(\theta_i) \| \; \| x - x^0 \| = o(\delta)$$
for arbitrarily small δ. This inequality implies $\alpha = 0$.   ◆

In the context of optimization (with the original function F) , Lemma 3.1 ensures
the uniqueness of the Lagrange-multipliers $y = y(x)$ for critical points
$(x, y) \in F^{-1}(a, b)$ near (x^0, y^0) and small $\|(a, b)\|$, in the pseudo-Lipschitz-case.
The same is true for the related KKT-points. Our proof directly indicates which
perturbations disturb the pseudo-Lipschitz property, provided LICQ is violated.
The next statement particularly says that pseudo-regularity of F and regularity
coincide for *convex* C^1-problems.

THEOREM 3.2.
Let f and g_i be *convex* C^1- functions on R^n, and $z^0 = (x^0, y^0)$ be a zero of F.
Assume that there exist positive ε and δ such that the mapping
$(a,b) \to F^{-1}(a,b) \cap (z^0 + \varepsilon B^0)$ is lower semicontinuous for $\| (a, b) \| < \delta$.
Then, F^{-1} is single-valued and continuous on some ball $\| (a, b) \| < \delta'$.

Proof: Again, we assume that $g(x^0) = 0$, hence z^0 is also a KKT-point.
Because of (3.2), our suppositions are also true for the KKT-map T^{-1}. Setting
$V = z^0 + \varepsilon B^0$, the mapping $(a, b) \to T^{-1}(a,b) \cap V$ has nonempty, convex
images on some ball U defined by $\| (a,b) \| < \delta'$.
"nonempty" follows from $\emptyset \neq T^{-1}(0,0) \cap V$ and the lower semicontinuity of F^{-1}.
"convex"   is ensured by convexity of f and g_i .

Setting $R(a,b) = T^{-1}(a,b) \cap V$ for $(a, b) \neq 0$ and $R(0,0) = \{ z^0 \}$, the mapping $R(.)$ is l.s.c. and convex-valued, too. Hence, $R(.)$ has a continuous selection σ on U. Because of (3.2), we also get a continuous selection h for F^{-1}, and $h(0,0) = \sigma(0,0) = z^0$. Lemma 2.1 yields

$$F^{-1}(a,b) \cap h(U) = \{ h(u) \} \text{ on U} . \tag{3.4}$$

Each proper line-segment s in the convex set $T^{-1}(a,b)$ will by transformed, by (3.2), into an arc $s' \subset F^{-1}(a,b)$. Since $h(U)$ is open, (3.4) indicates that $T^{-1}(a,b)$ and $F^{-1}(a,b)$ are singletons.   ◆

REMARK 3.1.

Because of the transformations (3.2) (which preserve continuous selections), Th. 3.2 remains true if F^{-1} is replaced by the map T^{-1} assigning, to P(a,b), the set of KKT- points.

REMARK 3.2

From the topological point of view, the theorem tells us that $T^{-1}(a,b)$ cannot be some proper polyhedron which converges to z^0 for $(a,b) \to 0$.

Without convexity, we would need the uniqueness of the primal solutions $x = x(a,b)$ in order to see that the statement is still valid.

Zero- Langrange multipliers

The next two statements show that zero-Lagrange-multipliers do not play an essential role for the relation between both kinds of regularity. In the context of generalized Kojima-functions, the functions Φ and g have to be continuous, only.

LEMMA 3.3.   (deleting constraints with zero LM , *pseudo-regular*)
Let F be a generalized Kojima function with continuous Φ, Ψ and g, and let F be pseudo-regular at some zero $z^0 = (x^0, y^0)$ with $y^0{}_m = g_m(x^0) = 0$. Define $F^{(m)}$ by removing both, the m-th component of F_2 and the product $y^+{}_m \Psi_m$ in F_1. Then $F^{(m)}$ is again pseudo-regular at z^0, considered in R^{n+m-1}.

Proof: We suppose that $(a, \beta), (x, \eta) \in R^{n+m}$ & $(a, b), (x, y) \in R^{n+m-1}$, specify the norm to be the maximum-norm, and put $S = (F^{(m)})^{-1}$.
In (1.1), let V be the open ball $\delta B^0 \subset R^{n+m}$, and
$U = z^0 + \varepsilon B^0 \subset R^{n+m}$, where $\varepsilon > 0$ is already small enough such that
$$\| x' - x^0 \| \leq \varepsilon \text{ implies } \qquad g_m(x') < \delta/3.$$
Again by continuity, there is some $r \in (0 , \delta)$ such that
$$\| x'' - x^0 \| \leq \varepsilon + L\,r \text{ implies } \qquad g_m(x'') < \tfrac{1}{2}\, \delta.$$
Now, let (a', b') & (a'', b'') $\in \tfrac{1}{2}\, r\, B^0$ and $(x', y') \in (z^0 + \varepsilon B^0) \cap S(a', b')$ be arbitrarily fixed in R^{n+m-1}.
We have to show that there is some point $(x'', y'') \in S(a'', b'')$ such that
$$\| (x', y') - (x'', y'') \| \leq L \| (a', b') - (a'', b'') \|. \tag{3.5}$$
For this reason, we define the vector $(a', \beta') \in \delta B^0$, which differs from (a', b') by the additional component $\beta'{}_m = \tfrac{1}{2}\, \delta$ only. Similarly, put $\beta''{}_m = \tfrac{1}{2}\, \delta$ for defining (a'', β'') by using (a'', b'').
Because of $g_m(x') < \delta/3 < \beta'{}_m$,   the m-th inequality is not active at x'.
Hence the point $(x', \eta') := (x', y', g_m(x') - \beta'{}_m)$ belongs to $F^{-1}(a', \beta')$.
The pseudo-regularity of F provides us with some $(x'', \eta'') \in F^{-1}(a'', \beta'')$ satisfying

$\| (x', \eta') - (x'', \eta'') \| \le L \| (a', \beta') - (a'', \beta'') \| = L \| (a', b') - (a'', b'') \| < L\,r.$
The choice of r yields $g_m(x'') < \frac{1}{2}\,\delta = \beta''_m$, which gives $\eta''_m = g_m(x'') - \beta''_m < 0$.
Therefore, the point (x'', y'') defined by deleting η''_m in (x'', η'') belongs to
$S(a'', b'')$ and satisfies (3.5). ◆

LEMMA 3.4. (deleting constraints with zero LM , *not regular*)
Let F be a generalized Kojima function with continuous Φ, g and locally Lipschitz
Ψ, and let F be pseudo-regular at some zero $z^0 = (x^0, y^0)$.
Suppose that F^{-1} has a closed-valued, continuous selection H such that
$H(0) = \{ z^0 \}$, c.f. § 2.3. Finally, let $y^0_m = g_m(x^0) = 0$, and define $F^{(m)}$ as in
Lemma 3.3. Then, if F is not regular at z^0, so is $F^{(m)}$.

Proof: Consider the mapping $M(a, b) = \arg\min \{ y_m / (x, y) \in H(a, b) \}$ and
assume first that M is single-valued on some ball r B.
Then, by the properties of H, M is a continuous selection of F^{-1}, $M(0, 0) = \{ z^0 \}$
and, in accordance with Lemma 2.1, F is regular at z^0.
Hence, $M(a,b)$ *is not single-valued* for some sequence $(a,b) \to (0,0)$.
We thus obtain the existence of certain elements
(x', y') & $(x'', y'') \in M(a, b)$ satisfying
$$(u, v) := (x'', y'') - (x', y') \ne 0 \quad \text{and} \quad y''_m = y'_m . \tag{3.6}$$
Our assumptions concerning H ensure that $(u, v) \to (0, 0)$ & $y'_m \to y^0_m = 0$.
Next, consider the inverse map S of $F^{(m)}$ at the parameter points
$p' = (a - (y'_m)^+ \Psi_m(x') , b_{redu})$ and $p'' = (a - (y''_m)^+ \Psi_m(x'') , b_{redu})$,
where b_{redu} is the projection of b onto R^{m-1}.
Deleting the y_m-coordinates of the points in (3.6), we define points
(x', η') and (x'', η'') satisfying $(x', \eta') \in S(p')$ and $(x'', \eta'') \in S(p'')$.
Due to $v_m = 0$, the difference of the parameters becomes
$p' - p'' = (y''_m)^+ \Psi_m(x'') - (y'_m)^+ \Psi_m(x') = (y'_m{}^+) [\Psi_m(x'') - \Psi_m(x')]$.
To show that $F^{(m)}$ is not regular at z^0 , we assume the contrary.
With some Lipschitz-constant K for Ψ_m near x^0, and
$(u, v)_m := (u, v_1,...,v_{m-1})$, one then obtains
$\|(u, v)_m\| \le L \|p' - p''\| = L (y'_m)^+ \| \Psi_m(x'') - \Psi_m(x') \| \le L(y'_m)^+ K \| u \|.$
Since $y'_m \to 0$ and $u \to 0$, the estimate implies $(u, v)_m = 0$ and, due to $v_m = 0$,
even $(u, v) = 0$. Recalling (3.6), $F^{(m)}$ cannot be regular. ◆

Both lemmata permit a simple reduction procedure.

THEOREM 3.5. (Reduction for PC^2 -functions)
Let $f, g_i \in C^1$, and let Df, Dg_i be PC^1- functions.
Let $z^0 = (x^0, y^0)$ be a zero of F and $I^+ = \{ i / y^0_i > 0 \}$.
Define the reduced problem (P^r) of (P) by deleting all constraints belonging to
$i \notin I^+$. If F is pseudo-regular but not regular at z^0, then the same is true for
Kojima's function F^r related to (P^r).

Proof: The function F is PC^1, hence we may apply remark 2.2; the mapping
F^{-1} has the required continuous multivalued selection H as long as F is pseudo-
regular. Due to the lemmas 3.3 and 3.4, one may successively remove all
constraints with $y^0_i = 0$. After deleting the related components, z^0 is still a zero
of F^r , which remains pseudo-regular but not regular. ◆

The (reduced) point z^0 fulfils the strict complementarity condition for (P^r). So, in the C^2- case, system $F^r = 0$ is (locally) a C^1-equation. Hence, both regularities coincide. However, the general situation is less simple.

REMARK 3.3.

There is a function $h \in C^1(R^2, R)$ such that Dh is *piecewise linear* and the critical-point map $S = (Dh)^{-1}$ is pseudo-Lipschitz without being Lipschitz.

EXAMPLE 3.

Take the function f of Example 1 (P. Fusek) in the introduction and put $h(z) = \sqrt{2}\ f_1(z)\ f_2(z)$; $z = (x, y)$. To describe h as well as the partial derivatives $D_x h$, $D_y h$ explicitly, we write z in polar-coordinates r (cos ω, sin ω) and regard the closed cones $C(k) = \{ z / \omega \in [(k-1)\ \pi/4 ,\ k\ \pi/4] \}$, $1 \le k \le 8$. Then one obtains:

	$D_x h$	$D_y h$	h
C(1)	- y	2y - x	y (y - x)
C(2)	-2x + y	x	x (y - x)
C(3)	2x + y	x	x (y + x)
C(4)	- y	-2y - x	- y (y + x) .

On the remaining cones C(k+4), $(1 \le k \le 4)$, h is defined as in C(k). Studying the Dh-image of the sphere, it is not difficult to see (but needs some effort) that Dh is continuous and $(Dh)^{-1}$ is pseudo-Lipschitz at 0.

Moreover, for $a \in R^2 \setminus \{ 0 \}$, there are exactly 3 solutions of $Dh(z) = a$. ◆

We finally notice that the reduction of Th. 3.5 is possible even for $C^{1.1}$-optimization problems and for the generalized Kojima function with Φ, Ψ and g as in Lemma 3.4, provided that the multifunction H required there exists in each step. Does it always exist ?

4   STRONG STABILITY OF $C^{1.1}$-OPTIMIZATION PROBLEMS

To characterize regularity of Kojima's function F (strong stability) for $C^{1.1}$-optimization problems, a direct approach based on Th. 2.7 has been developed in [6]. There, one finds our Th. 4.1, and system (4.2) is formulated (like in [4]) as a second order condition in conjunction with LICQ. Moreover, in [7], applications to the critical values and to derivatives of critical curves have been elaborated, based on an implicit-function theorem for Lipschitz functions.

THEOREM 4.1.

Let (x^0, y^0) be a zero of F & $g(x^0) = 0$.   Put $A_i = Dg_i(x^0)$, $I^+ = \{ i / y^0_i > 0 \}$, $I^0 = \{ i / y^0_i = 0 \}$ and, in case of C^2, $Q = D^2_{xx}L(x^0, y^0)$. If $f, g_i \in C^2$, then F is regular at (x^0, y^0) if and only if the system

$$Q u + \Sigma_{k \in I^+}\ v_k A_k + \Sigma_{i \in I^0}\ r_i v_i A_i = 0 \qquad (4.1.1)$$
$$A_k u \qquad\qquad\qquad\quad = 0 \quad (k \in I^+) \qquad (4.1.2)$$
$$A_i u\ - (1 - r_i)\ v_i \quad = 0 \quad (i \in I^0) ,\ r_i \in [0, 1] \qquad (4.1.3)$$

is solvable for $(u, v) = 0$ only.

If $f, g_i \in C^{1.1}$, then system (4.1) similarly characterizes regularity:
Only equation (4.1.1) must be written as an inclusion
$$Q(u) + \Sigma_{k \in I^+} v_k A_k + \Sigma_{i \in I^0} r_i v_i A_i \ni 0 \qquad (4.1.1)'$$
where the set $Q(u) = \Delta D_x L(\,.\,, y^0)(x^0 ; u) = \Delta F_1(\,.\,, y^0)(x^0 ; u)$ is formed by all possible limits
$$q_u = \lim t^{-1} [D_x L(x' + t u , y^0) - D_x L(x', y^0)] \quad \text{for } t \downarrow 0 \text{ and } x' \to x^0.$$

Proof: in [6].   ◆

For C^2-problems, system (4.1) does not change after replacing problem (P) by its quadratic approximation (PQ)(0, 0). Hence, strong stability of (P) and (PQ)(0, 0) at the common critical point (x^0, y^0) coincide. Recall that the latter fact was an important (particular) result of [12].

In the $C^{1.1}$-case, problem (PQ) is not defined. Nevertheless, system (4.1) has a special meaning.
Using chain rules for the limit sets ΔF, one can verify that the left-hand sides of system (4.1), for fixed (u, v) and varying r, just form the set $\Delta F((x^0, y^0); (u, v))$ appearing in Th. 2.7.
This fact is of importance if some nonsmooth Newton method, based on generalized derivatives, should be applied for computing a zero of F, and it makes clear that Th. 4.1 is nothing but an application of Th. 2.7 to the specified locally Lipschitz function F.

The same application shows how Th. 4.1 can be directly extended to generalized Kojima-functions
$$F_1 (z) = \Phi(x) + \Sigma_i y_i^+ \Psi_i (x)$$
$$F_2 (z) = g(x) - y^-,$$
provided that Φ, Ψ and Dg are locally Lipschitz.
Now, $Q(u) = \Delta F_1(\,.\,, y^0)(x^0 ; u)$ is again defined by the limits of
$t^{-1} [F_1(x' + t u , y^0) - F_1(x', y^0)]$; $t \downarrow 0$ & $x' \to x^0$; and A_v in (4.1.1)' must be replaced by $\Psi_v (x^0)$. The equations (4.1.2) and (4.1.3) remain the same.

System (4.1) can be reformulated by the transformations
$$\alpha_i = r_i v_i , \qquad \beta_i = (1-r_i) v_i , \qquad \text{and conversely by}$$
$$v_i = \alpha_i + \beta_i , \qquad r_i = \alpha_i / v_i \text{ if } v_i \neq 0 \ \& \ r_i = 1 \text{ if } v_i = 0.$$
Then, (4.1) becomes equivalent to the more convenient system
$$Q u + \alpha A = 0 \qquad\qquad\qquad (4.2.1)$$
$$A u - \beta = 0, \qquad \beta_i = 0 \ (i \in I^+), \quad \alpha_i \beta_i \geq 0 \ (i \in I^0). \qquad (4.2.2)$$
where *regularity means that* $(u, \alpha, \beta) = 0$ *is the only solution of (4.2)*.

This condition can be equivalently written in several languages.
However, the present characterization has a certain advantage in comparison with equivalent other conditions for strong stability of C^2-problems, formulated e.g. in [4], [5] and [15] by means of subdeterminants of the matrices in (4.2) or in terms of normal and tangent cones:
To characterize regularity of $C^{1.1}$- problems, equation (4.2.1) must be replaced, once again, by the inclusion $Q(u) + \alpha A \ni 0$ only.

Let us compare our condition with the related condition in [4].

REMARK 4.1
After applying the Farkas lemma, the critical face theorem 5 in [4] asserts for
C^2-problems:
F is regular at $(x^0, y^0) \Leftrightarrow (u, \alpha, \beta) = 0$ for each solution of (4.2) satisfying $\beta_i \leq 0$
if $\alpha_i \neq 0$, $i \in I^0$.

So, u is even a feasible direction for some reduced linearized problem $PQ_r(0, 0)$
at x^0 and, since $(\alpha_i > 0$ & $\beta_i > 0)$ is excluded, now more problems are regular ...

Really, the different descriptions of the same thing are induced by the different
approaches.
Our condition is a direct characterization of regularity (based on Th. 2.7),
whereas the critical face condition characterizes (based on [10]), the formally
weaker pseudo-regularity (so, the class of unstable problems looks smaller).
The equivalence of the regularities in question (for C^2) now tells us that, to
each nontrivial solution of (4.2), there is a (possibly) second one satisfying
$\beta_i \leq 0$, if $\alpha_i \neq 0$, $i \in I^0$. This can be seen without any algebra: If (4.2) has a
nontrivial solution, then regularity is violated (Lemma 4.3. will even indicate
where double solutions exist). Hence, by [4], such a special solution exist.

Let us add a second observation concerning feasible directions. Denote, by [P],
the class of problems (P') which can be generated from problem (P) by
reversing arbitrary inequalities $g_i(x) \leq 0$, $i \in I^0$ into $g_i(x) \geq 0$.

LEMMA 4.2
Regularity of F at a critical point (x^0, y^0) for a $C^{1.1}$-problem (P) is an invariant
property of all problems in [P].

Proof: After reversing the constraints, (x^0, y^0) is still critical for (P') due to
$y^0_i = 0$. For the same reason, the Hessian Q (or the map $u \to Q(u)$) remains the
same; the only difference between the systems (4.2) and (4.2)' related to (P) and
(P'), respectively, is the sign of $Dg_i(x^0) = A_i$.
Having a solution (u, α, β) of (4.2) one obtains a solution (u, α', β') of (4.2)' by
changing the signs of α_i and β_i, and vice versa. Hence system (4.2) describes
even strong stability of (x^0, y^0) in problem (P)'.   ♦

Given any solution of (4.2), there is always some problem (P') in the class [P]
such that u is a feasible direction at x^0 for the linearized constraints of (P').
In conjunction with the lemma, this explains why additional requirements like
$A_i u \leq 0$ for certain $i \in I^0$ can be avoided.

Finally, a further nice meaning of system (4.2) should be mentioned. It indicates
that and where, near (x^0, y^0) and for certain small (a, b), there are two different
KKT-points to the quadratic problem (PQ)(a, b), provided regularity fails to hold.

LEMMA 4.3.   (two close critical points, quadratic problems).
Let (x^0, y^0) be a KKT-point for (PQ)(0, 0), $g(x^0) = 0$.

(i)   If (x, y) and (x', y') are KKT-points for (PQ)(a,b), sufficiently close to (x^0, y^0), then $u = x' - x$, $\alpha = y' - y$ and $\beta = A u$ solve (4.2).

(ii)   If (u, α, β) solves (4.2) and has small norm, then, setting $a = -\alpha^- A$ & $b = \beta^+$, the points $(x, y) = (x^0, y^0 - \alpha^-)$ and $(x', y') = (x, y) + (u, \alpha)$ are KKT-points for (PQ)(a, b).

Proof:

(i)   If (x, y) and (x', y') are KKT-points, then $u = x' - x$ and $\alpha = y' - y$ fulfil $Qu + \alpha A = 0$. If $\beta_i = A_i u > 0$, then x cannot be active for the constraint i because $x' = x + u$ is feasible. Hence $y_i = 0$ and $\alpha_i = (y' - y)_i \geq 0$. Considering x', one shows similarly that $\beta_i < 0$ implies $\alpha_i \leq 0$. If $i \in I^+$, then y_i' and y_i (near y^0_i) are positive. Thus $\beta_i = A_i u = 0$. Summarizing, (4.2) is satisfied.

(ii)   Consider now any solution of (4.2).

With our specified $b = \beta^+$, we see that x^0 and $x^0 + u$ are feasible.

The identity $Dq(x^0) + y^0 A - a - \alpha^- A = 0$ shows that $(x^0, y^0 - \alpha^-)$ is a KKT point for the problem (PQ)(a, b) *provided* that the complementarity condition holds.

Using (4.2.1), the equation $Dq(x^0) + y^0 A - a - \alpha^- A + Qu + \alpha A = 0$ ensures the same for $(x', y') = (x^0 + u, y^0 + \alpha^+)$.

The needed *complementarity* follows from (4.2.2); let $i \in I^0$.

Compl. at x^0 :   If $-\alpha_i^- > 0$, then $\alpha_i < 0$, $\beta_i \leq 0$ & $A_i x^0 = 0 = \beta_i^+$.

Compl. at $x^0 + u$:   If $\alpha_i^+ > 0$, then $\alpha_i > 0$, $\beta_i \geq 0$ & $A_i (x^0 + u) = \beta_i^+$. ◆

Summarizing, we notice that pseudo-regularity (like regularity) *induces* certain important local properties under closedness of the inverse images:

1. Successive approximation is applicable for solving $g \in F$ if $|g|_C 0.1$ is small.
2. Pseudo- regularity w.r.t. small $C^{0.1}$-variations is true.

Concerning *equivalence* of the both regularity notions we obtained:

3. For $f \in C(R^n, R^n)$: equivalence holds if f^{-1} has a continuous selection; in particular, if the preimages $f^{-1}(y)$ are convex.
4. For critical points of optimization problems: equivalence holds for C^2-problems and for convex C^1- problems; generally, it fails to hold if PC^1-derivatives appear; for this class of problems, the investigation of equivalence is reducible to the case of strict complementarity under LICQ, i.e. to an usual equation.

Acknowledgement

I am indebted to Jiri Outrata, who started the discussion about the present subject during his stay at the Humboldt-University in April 1995.
Furthermore, I would like to thank D.Klatte, B.Mordukhovich, K.Tammer and S.Scholtes for their permanent interest and helpful discussions and remarks.

REFERENCES

1. Aubin, J.P. & Ekeland, I.   Applied Nonlinear Analysis. Wiley, New York 1984

2. Clarke, F.H.   Optimization and Nonsmooth Analysis. Wiley, New York, 1983

3. Cominetti, R. Metric Regularity, Tangent Sets, and Second-Order Optimality Conditions. Appl. Math. Optim. 21 (1990), 265-287

4. Dontchev A.L., Rockafellar R.T.   Characterizations of Strong Regularity for Variational Inequalities over Polyhedral Convex Set.   Preprint 1995

5. Kojima, M. Strongly stable stationary solutions in nonlinear programs. In: Analysis and Computation of Fixed Points, S.M. Robinson ed., Academic Press, New York, 1980, 93-138

6. Kummer, B.   Lipschitzian Inverse Functions, Directional Derivatives and Application in $C^{1,1}$-Optimization". Journal of Optimization Theory & Appl. 70, No.3, (1991), 559-580

7. Kummer, B. An Implicit Function Theorem for $C^{0,1}$- Equations and Parametric $C^{1,1}$-Optimization. Journal of Mathematical Analysis & Appl. 158, No 1, (1991) 35-46

8. Kummer, B. Newton's Method based on generalized derivatives for nonsmooth functions : Convergence Analysis. in : Lecture Notes in Economics and Mathematical Systems 382; Advances in Optimization ,   W. Oettli , D. Pallaschke (Eds.) , *Proceedings 6th French-German Colloquium on Optimization, Lambrecht , FRG, 1991.* Springer, Berlin 1992,   171-194

9.   Michael, E. Continuous selections I.   Ann. of Math. 63 (1956), 361-382

10. Mordukhovich, B. Stability theory for parametric generalized equations and variational inequalities via nonsmooth analysis. Trans. Amer. Math. Soc. 343 (1994), 609-657

11. Qi, L. & J. Sun   A nonsmooth version of Newton's method. Math. Programming 58 (1993), 353- 367

12. Robinson, S.M. Strongly regular generalized equations. Mathematics of OR, 5 (1980), 43-62

13. Robinson, S.M. An implicit-function theorem for B-differentiable functions. Mathematics of OR, 16, No. 2, (1991), 292-309

14. Robinson, S.M. Normal maps induced by linear transformations. Mathematics of OR, 17 (1992), 691-714

15. Scholtes, S. Introduction to piecewise differentiable equations. Preprint 53/1994, Institut für Statistik und Mathematische Wirtschaftstheorie, Univ. Karlsruhe, 1994

16. Thibault, L. Subdifferentials of compactly Lipschitzian vector- valued functions. Ann. Mat. Pura Appl. 4 , 125 (1980), 157-192

On Well-Posedness and Stability Analysis in Optimization

R. Lucchetti
Dipartimento di Matematica, Università di Milano,
Via Saldini 50, 20133 Milano, Italy

T. Zolezzi
Dipartimento di Matematica, Università di Genova,
Via Dodecaneso 35, 16146 Genova, Italy

Abstract

We review the basic concepts of well-posedness in scalar optimization: Hadamard, Tikhonov, Levitin-Polyak and strong well-posedness. We discuss in some detail a more recent approach, well-posedness by perturbations. We provide several examples, paying special attention to mathematical and convex programming.

1 INTRODUCTION.

Notions of well-posedness in scalar optimization, as documented in the mathematical literature, can be formulated by following two different approaches. One can mimick the classical concept firstly defined by Hadamard in the context of problems with partial differential equations. The minimization problem (X, f) of globally minimizing the proper extended real valued function f over the space X is called *Hadamard well-posed* if there exists a unique minimizer which depends continuously upon problem's data. Of course this concept requires a topological framework to be more precisely defined. The other approach, introduced by Tikhonov, is instead motivated by consideration of the performance of numerical methods in order to get approximate solutions of (X, f). More precisely, the problem (X, f) is called *Tikhonov well-posed* if there exists a unique minimizer, toward which every minimizing sequence converges. Optimization problems abund in which the uniqueness of the minimizer is a too strong requirement and a weakening is needed. Accordingly, the problem (X, f) is called *Tikhonov well-posed in the generalized sense* if every minimizing sequence has some subsequence converging to an optimal solution. Corresponding notions of Hadamard well-posedness exist for problems with many minimizers.

This subject is still under fast development. In the last years two books were devoted to this topic (see [12, 20]), not to mention [4, 13] among surveys dealing in part with such issues within mathematical programming problems.

A survey about the various well-posedness notions, their relationships and several applications can be found in [12]. For connections among them, see also [22].

The aim of this paper is to focus mainly on a new approach to well-posedness, that in some sense contains ideas of both the previous ones. Firstly we shall review some topological aspects of Hadamard and Tikhonov well-posedness, focusing especially on examples. Then we introduce the notion of *well-posedness by perturbations*[1], which takes into account the effects of parametric perturbations of a given optimization problem. In this sense, we go back to the Hadamard idea of well-posedness. Tikhonov approach is instead reflected by the requirement of convergence of approximate solutions of nearby problems to the minimizer of the original one. Here we provide examples of applications as well as some characterizations, generalizing former ones known for Tikhonov well-posedness. The final sections are devoted to some aspects of convex optimization, and to Levitin-Polyak and strong well-posedness for problems with explicit constraints.

2   HADAMARD WELL-POSEDNESS.

To motivate the notion of Hadamard well-posedness, let us consider the problem of globally minimizing a given real valued function over some constraint set: think to mathematical programming as an example. Quite often, it is impossible to solve the problem as it is. Trying to figure out a solution numerically, one is led to introduce approximation schemes of various

---

[1]As pointed out by a referee, the name of well-posedness by perturbations maybe is not very suitable for this notion: however it is already used in several papers, and to avoid possible confusion we like better not to change it.

types. Even more, at the moment of defining constraint sets and objective functions, it is possible that our model ought to be simplified, intoducing again a new approximation process. But, how these changes in the data reflect in changing the solutions? Differently stated, under which assumptions some kind of approximation does not affect too much the solution? It is quite clear that this subject can produce a huge amount of situations to consider: here we shall treat the abstract case of an extended real valued function, defined on a topological space X. This allows including (in an abstract fashion) constrained problems in the theory, in the standard way: by this we mean to substitute the problem of minimizing f over the subset $A \subset X$, with the equivalent one of minimizing $f + I_A$ over X, where I_A is the indicator function of the set A: $I_A(x) = 0$ if $x \in A$, $I_A(x) = \infty$ if $x \notin A$.

In this paper, when referring to Hadamard well-posedness, we shall mean continuity of the solution mapping and of the value function; of course this is just one of the aspects one could consider. To just mention a very important issue we do not consider here, consider Lipschitz continuity of the solution set, that provides also upper bounds for the errors.

So, let us be given a global minimization problem (X, f), where X is some topological (or convergence) space, and $f : X \to (-\infty, \infty]$ is a given extended real valued function. Let us suppose for the moment that the function is bounded from below, and that the problem has exactly one solution, i.e. the set

$$\arg \min f \overset{\text{def}}{=} \{x \in X : f(x) = \inf f(X)\}$$

is a singleton. Suppose we approximate f with a sequence of functions (f_n) converging to f, and suppose the problem of minimizing f_n has solution, say x_n. Does the sequence (x_n) converge to the exact solution of (X, f)? This is one of the naive schemes of well-posedness in this area. Actually, some generalizations are quite natural, because the uniqueness requirement can be too severe in many cases, and also one should not require a priori existence of minimizers for the problems (X, f_n), but rely only on approximate solutions. The first point to be addressed is thus to provide a suitable convergence scheme for functions, in order to ensure convergence of the optimal solutions. The first attempt, of course, is with pointwise convergence. But the following example shows that this is not a good idea.

Example 2.1 Let X be the interval $[0,1]$ and the sequence (f_n) so defined:

$$f_n(x) = \begin{cases} 0 & \text{if} & 0 \leq x \leq 1 - \frac{2}{n}, \\ -nx + n - 2 & \text{if} & 1 - \frac{2}{n} \leq x \leq 1 - \frac{1}{n}, \\ 2nx + 1 - 2n & \text{otherwise.} \end{cases}$$

The pointwise limit function f is 0 everywhere but at 1, where it is valued 1. Observe the following unpleasant facts: the limit function f is not lower semicontinuous, even if all the approximating functions were continuous: a bad feature when dealing with minimization processes. Moreover,

$$-1 = \lim_{n \to \infty} \inf f_n(X) \neq 0 = \inf f(X),$$

even if the sequence of minima (x_n) of f_n converges; furthermore, the limit point of x_n does not minimize f. We have here a total lack of well-posedness. The point is, easy but crucial observation, that the pointwise limit function f is not the right one in a variational sense.

This led to define another kind of limits, in the variational setting. The idea is to appeal to convergence of sets. Taking into account that for most of these convergences the closedness and convexity properties are preserved in the limit, one identifies a given function with its epigraph:

$$\text{epi} f \overset{\text{def}}{=} \{(x, a) \in X \times \mathbf{R} : f(x) \leq a\}.$$

This provides the following advantages:

- As a function f is lower semicontinuous if and only if its epigraph is closed, the limit of a sequence of lower semicontinuous functions will automatically be lower semicontinuous.

- In a linear context, as the epigraph is convex if and only if f is convex, the same property holds for convex functions.

This at least will allow working in natural classes of functions that will be closed for the topologies considered. But much more is true. To illustrate the situation, let us consider the classical set convergence known as *Kuratowski convergence*, see [5]. In our previous example this provides the limit function g defined as 0 for all x but 1, where it values -1. So that, we have in this case convergence of both the minimizers to a minimum point, and of the values to the value of the limit problem. This behaviour can be detected in a much more general setting. Namely, the following theorem holds true:

Theorem 2.1 *Let X be a first countable topological space. Suppose the sequence $f_n : X \to (-\infty, \infty]$ converges to f in Kuratowski sense. Then*

$$\lim_{n \to \infty} \sup \inf f_n(X) \leq \inf f(X),$$

and if there is x_n minimizing f_n and the sequence (x_n) has a limit point x, then

$$\lim_{n \to \infty} \inf f_n(X) = \inf f(X),$$

and x minimizes f.

Proof. Kuratowski convergence of the sequence f_n to f implies the following, see [5], p. 159:

$$\text{i)} \quad \forall x \in X \; \exists x_n \to x \; : \; \limsup_{n \to \infty} f_n(x_n) \le f(x),$$

and

$$\text{ii)} \quad x_n \to x \; \text{implies} \quad \liminf_{n \to \infty} f_n(x_n) \ge f(x).$$

Thus given $\epsilon > 0$ and x such that $f(x) \le \inf f(X) + \epsilon$ (or $f(x) \le -\frac{1}{\epsilon}$ if $\inf f(X) = -\infty$), there is x_n as in i), so that

$$\limsup_{n \to \infty} \inf f_n(X) \le \limsup_{n \to \infty} f_n(x_n) \le f(x) \le \inf f(X) + \epsilon,$$

proving the first claim. Let us now consider x_n as in the statement of the theorem. Then

$$\liminf_{n \to \infty} \inf f_n(X) = \liminf_{n \to \infty} f_n(x_n) \ge f(x)$$

$$\ge \inf f(X) \ge \limsup_{n \to \infty} \inf f_n(X). \quad \blacksquare$$

Theorem 2.1 is in any case a quite abstract result, but it can be considered as a suitable framework to work with. For instance, when we want to handle a problem of the form $f + I_A$, to explicitly take into account the role of the constraints, it is not clear what kind of conditions (for instance on the constraint sets) will guarantee Kuratowski convergence of the sum. This is due to the fact that usually set convergence of two sequences of functions does not necessarily imply set convergence of their sum. However, the following result in convex programming can be proved.

Let $f_n : \mathbf{R}^k \to (-\infty, \infty]$ be a sequence of convex lower semicontinuous functions converging in Kuratowski sense to f. Let be given h sequences $g_{ni} : \mathbf{R}^k \to \mathbf{R}$ of convex functions converging again in Kuratowski sense to g_i, $i = 1, \cdots, h$. Suppose also that a Slater condition is fulfilled, i.e. there exists a point $x \in \text{int dom } f$ such that $g_i(x) < 0$, for $i = 1, \cdots, h$ (As usual, for $f : X \to (-\infty, \infty]$, dom $f \stackrel{\text{def}}{=} \{x \in X : f(x) < \infty\})$. Consider the problem of minimizing f over the constraint set

$$A \stackrel{\text{def}}{=} \{x \in \mathbf{R}^k : g_i(x) \leq 0, i = 1, \cdots, h\},$$

and the associated perturbed problems. Then the following theorem holds:

Theorem 2.2 *Under the previous assumptions we have:*

$$\lim_{n \to \infty} \sup \inf f_n(A_n) \leq \inf f(A).$$

Moreover, if there is a sequence (x_n) of solutions of the approximating problems that has a limit point x, then

$$\lim_{n \to \infty} \inf f_n(A_n) = \inf f(A),$$

and x is a solution of the original problem.

Theorem 2.2 does not have an immediate extension to the infinite dimensional setting. This depends on the fact that the Slater condition allows to get a result of Kuratowski convergence of the constraint sets only in euclidean spaces: the classical extension to reflexive (infinite dimensional) Banach spaces of Kuratowski convergence of functions and sets, i.e. Mosco convergence (see [5]), does not enjoy the sum property, unless we require a much more severe constraint qualification condition ([2]). However the same result as in Theorem 2.2 holds in infinite dimensional spaces, if we require a finer convergence on the functions: the so called bounded Hausdorff topology, see [6, 7].

Remark 2.1 Theorem 2.2 can be found in [19], and its generalization to the infinite dimensional setting in [7].

The two previous results deal with upper semicontinuity of the function $f \to \inf f(X)$ and of the closedness of the graph of the multifunction $f \to$ arg min f. Several other aspects of Hadamard well-posedness can be considered as well. To cite one of them, we could consider different types of semicontinuity of $f \to$ arg min f. We refer to [12] and [20] for more about this.

3 TIKHONOV WELL-POSEDNESS.

This section is dedicated to Tikhonov well-posedness, in the previous abstract setting. This means that we consider in particular constraints of the minimum problem absorbed in the function f to be minimized, redefined ∞ outside the constraint set. Of course, this approach does not take into account the possibility of violating the constraints to some extent, when trying to find out a solution. For this type of approach other kind of well-posedness are more suitable. We shall give a brief description of them in our last section. For more, see [12].

We shall focus here more on examples, rather than illustrating the theory. Some aspects of Tikhonov well-posedness will be described in the subsequent sections, when we generalize it to well-posedness by perturbations.

Suppose X is a convergence space. Then the minimization problem (X, f) is called *Tikhonov well-posed* if:

(1) there exists a unique minimizer $u^* = \arg\min(X, f)$,

i.e. $f(u^*) \leq f(x)$ for every $x \in X$;

(2) every minimizing sequence converges toward u^*,

i.e. $x_n \in X$ and $f(x_n) \to \inf f(X) = f(u^*)$, imply $x_n \to u^*$.

This means that x_n, for sufficiently large n, can be considered as an approximate solution of (X, f). Sometimes u^* is called *strongly unique*.

As we already observed, the uniqueness requirement could be weakened, giving rise to Tykhonov well-posedness in the generalized sense:

(3) $\arg\min(X, f)$ is nonempty;

(4) every minimizing sequence has a subsequence converging to some solution.

Let us see a few examples.

Example 3.1 Uniqueness of the minimizer does not imply well-posedness as $f(x) = x^2 e^{-x}$, $x \in \mathbf{R}$, shows.

Example 3.2 Let K be a compact metric space, and let $f : K \to \mathbf{R}$ be lower semicontinuous. Then the problem (K, f) is Tykhonov well-posed in the generalized sense. So that, whenever the direct method for finding a minimizer is applicable, then the problem is actually well-posed (in the generalized sense).

Example 3.3 Let $f : \mathbf{R}^n \to (-\infty, \infty]$ be convex, lower semicontinuous and with a unique minimizer. Then the problem $(\mathbf{R}^n, f)$ is Tikhonov well-posed. To see this, without loss of generality we can suppose that 0 minimizes f. Any bounded minimizing sequence converges to 0. On the other hand, if we suppose the existence of an unbounded minimizing sequence (x_n), then also the sequence $y_n \overset{\text{def}}{=} \frac{x_n}{\|x_n\|}$ is minimizing, by convexity. But in such a case y_n would have a norm one cluster point minimizing f, which is impossible. An analogous argument shows that if the set of the minimizers is compact nonempty, then the problem is well-posed in the generalized sense. The result fails in infinite dimensions. Consider a separable Hilbert space with orthonormal basis $\{e_n : n \in \mathbf{N}\}$ and the function

$$f(x) = \sum_{n=1}^{\infty} \frac{(x, e_n)^2}{n^2}.$$

In general, well-posedness of a convex function can be related to smoothness of its Fenchel conjugate. See [1].

Example 3.4 Let X be a Banach space with norm $\| \cdot \|$ and consider the *best approximation problem*: given $z \in X$ and a closed convex set $K \subset X$, let us minimize

$$f(x) = \|x - z\|$$

over K. In this case well-posedness is related to the geometry of the Banach space X: The problem is well-posed for each choice of z and K, if and only if X is an E-space, which means that X is a reflexive, strictly convex space such that the norm and weak topologies agree on the boundary of the unit ball.

This result generalizes to the case when we have a linear bounded operator $L : X \to Y$, where X and Y are E-spaces, and we minimize

$$f(x) = \|x - z\| + \|Lx - u\|$$

on a given closed convex subset $K \subset X$, for given $z \in X$ and $u \in Y$ [16, 18]. This in turn allows to get several well-posedness results, as for instance for the linear regulator problem [24].

For more examples, see [12, 15].

4 WELL-POSEDNESS BY PERTURBATIONS.

A further notion of well-posedness, introduced in [26], takes into account the effect of parametric modifications of a given minimization problem. Given the convergence space X and the proper objective function

$$f : X \to (-\infty, \infty],$$

one considers a metric space P of parameters, a fixed point $p^* \in P$ and a ball L in P of center p^* and positive radius. Then one introduces a new function

$$F : X \times L \to (-\infty, \infty],$$

such that for the fixed point $p^* \in P$ one has:

(5) $F(x, p^*) = f(x), x \in X.$

Thus F models perturbations, or deformations, of the original problem (X, f) depending on the parameter p, and can be considered as an embedding of (X, f) in a parametrized family of modified minimization problems $(X, F(\cdot, p))$, $p \in L$; when $p = p^*$ we get back (X, f).

Problem (X, f) is called *well-posed by perturbations* (with respect to the embedding F) if the following hold:

(1) there exists a unique minimizer $u^* = \arg \min(X, f)$;

(6) the value function $V(p) = \inf\{F(x, p) : x \in X\}$ is finite for every $p \in L$;

(7) $x_n \in X, p_n \to p^*$ in $P, F(x_n, p_n) - V(p_n) \to 0$ imply $x_n \to u^* = \arg \min(X, f).$

In the following, we shall write $\arg \min(p) = \arg \min(X, F(\cdot, p))$ and problem (p) for the associated minimum problem $(X, F(\cdot, p))$. Moreover, we shall call sequences (x_n) as in (7) *asymptotically minimizing* corresponding to the sequence (p_n).

The meaning of well-posedness by perturbations is the following: Every approximate minimization method performed on small deformations $(X, F(\cdot, p))$ of the original problem (X, f) corresponding to parameters p sufficiently close to p^* produces good approximations for the solution of (X, f).

By taking $p_n = p^*$ we see that well-posedness by perturbations implies Tikhonov well-posedness. More is true: denote by

$$\epsilon - \arg\min(p) = \epsilon - \arg\min(X, F(\cdot, p)), \quad p \in L, \quad \epsilon > 0,$$

the set of those points $x \in X$ such that

$$F(x, p) \leq V(p) + \epsilon.$$

(We are assuming that $V(p) > -\infty$). If the convergence in X is induced by a topology, then well-posedness by perturbations is equivalent to (1), (6) and upper semicontinuity of the multifunction

$$(\epsilon, p) \to \epsilon - \arg\min(p),$$

at $\epsilon = 0, p = p^*$. Hence the optimal solution $u^* = \arg\min(p^*)$ depends continuously on p, encompassing a form of Hadamard well-posedness. In particular, suppose that every nearby problem $(p), p \in L$, has a (not necessarily unique) optimal solution $y(p)$. Then $y(p) \to u^*$ as $p \to p^*$.

By considering the trivial embedding $F(x, p) = f(x)$ for all $p \in L$ and $x \in X$, we see that well-posedness by perturbations is a strengthening of Tikhonov well-posedness. On the other hand, the difference with Hadamard well-posedness lies on the fact that, besides considering approximated solutions of nearby problems, in this new approach usually we focus on the particular choice of the embedding F.

It is quite clear that well-posedness by perturbations strongly depends on the choice of F. For instance, consider the following example:

Example 4.1 Let $X = P = \mathbf{R}$, $f(x) = x^2$, $F_1(x, p) = x^2 + px$, $F_2(x, p) = x^2 e^{-px}$. Then the original problem is well-posed by perturbations with respect to F_1, but fails to be so with respect to F_2.

We provide now some important examples of perturbation schemes: in each one of them we give also the references where to find conditions for well-posedness by perturbations.

Example 4.2 Mathematical programming. The original problem of minimizing the real-valued function $h(x)$ subject to the inequality constraint

$$g(x) \leq 0,$$

can be embedded in the following parametrized problem: minimize $h(x)$ subject to the constraint

$$g(x) \leq p.$$

Observe that here g can be assumed to be $\mathbf{R}^k$ valued, to take into account a finite number of constraints. In this case, inequalities are intended componentwise. See [10] for this example in a local setting, and the following Theorem 6.2.

Example 4.3 Calculus of variations. Given a bounded open set $\Omega \subset R^N$, a continuous function $g : R \times R^N \rightarrow R$, and a boundary datum $\varphi : \partial\Omega \rightarrow R$, the minimization (on the appropriate Sobolev space) of

$$J(u) = \int_\Omega g(u, Du)dx,$$

under the constraint $u = \varphi$ on $\partial\Omega$, can be embedded in the parametrized problem of minimizing J under the constraint $u = p$ on $\partial\Omega$. See [26, 27, 30] for results about this problem.

Example 4.4 Optimal control of ordinary differential equations. Let the original problem be that of minimizing the performance functional

$$J(u) = \int_0^T g[y(s), u(s)]ds,$$

subject to the dynamic constraint (state equations)

(8) $\dot{y}(s) = h[y(s), u(s)]$

almost everywhere in $(0, T)$, $y(0) = x^*$, and

(9) $y(T) = y^*, u(s) \in K$ almost everywhere in $[0,T]$,

for given points x^*, y^*. Setting $p = (t, x)$ with $0 \leq t \leq T$ and x close to x^*, we embed the above problem in the following parametrized family of optimal control problems: minimize

$$\int_t^T g[y(s), u(s)]ds,$$

subject to (8) almost everywhere in $(t, T), y(t) = x$, and (9). This approach is intimately related to the dynamic programming method and leads to significant connections with the Hamilton-Jacobi-Bellman equation: see [25, 26, 28].

5   WELL-POSEDNESS BY PERTURBATIONS (CONTINUED).

A number of characterizations of well-posedness by perturbations can be found in [9, 26, 29]. We present in this section a partial survey of these criteria and some new results. All the characterizations discussed here have their counterparts in basic criteria about Tikhonov well-posedness, which are thereby extended.

In the following we assume that the problem (X, f) and the embedding $F = F(x, p)$ are fixed such that (5) holds. Moreover X is a metric space with metric d, and θ denotes the metric of P. We shall write for short *well-posed* instead of well-posed by perturbations with respect to the embedding F.

The following proposition yields a sufficient condition under which the two concepts of well-posedness agree, i.e. Tikhonov well-posedness implies well-posedness by perturbations.

Proposition 5.1 *Let X be a compact space, let*

$$F : X \times L \to (-\infty, +\infty]$$

be lower semicontinuous at every point of $X \times \{p^*\}$, V finite on L and upper semicontinuous at p^*. If problem (p^*) is Tikhonov well-posed, then it is so by perturbations (with respect to F).

Proof.   Let $p_n \to p^*$ and let x_n be asymptotically minimizing corresponding to p_n. We have

$$V(p_n) \le F(x_n, p_n) \le V(p_n) + \epsilon_n,$$

where $\epsilon_n \to 0$ and $y_n \to y$, for some subsequence y_n of x_n. Hence

$$F(y, p^*) \le \lim_{n \to \infty} \inf F(y_n, p_n) \le \lim_{n \to \infty} \sup V(p_n) \le V(p^*).$$

Tikhonov well-posedness provides $y = \arg \min (p^*)$, uniqueness of arg min (p^*) yields convergence of the whole sequence (x_n) towards y.   ∎

A consequence (due to Čobanov) follows from Berge theorems (see [12], prop.2 p.335).

Corollary 5.1 *Let X be a compact space. If F is continuous on $X \times L$, then the conclusion of Proposition 5.1 holds true.*

Without compactness of the space X, in general Tikhonov well-posedness does not imply well-posedness by perturbations, as the embedding F_2 of Example 4.1 shows.

Theorem 5.1 *If (X, f) is well-posed then*

(10) diam $\cup\{\epsilon-\arg \min(p) : \theta(p, p^*) < \epsilon\} \to 0$ as $\epsilon \to 0$.

Conversely, assuming that V is finite on L, (10) implies well-posedness of (X, f) provided X is complete and F is lower semicontinuous on $X \times \{p^*\}$.

Proof. Let us outline here the proof. (10) is an easy consequence of the definition of well-posedness. Conversely, let (p_n) be a sequence converging to p^*, and let x_n be such that $F(x_n, p_n) \leq V(p_n) + \epsilon_n$, for some sequence (ϵ_n) converging to 0. Condition (10) shows that (x_n) is a Cauchy sequence, and by completeness there exists the limit, say x. We have to show that x minimizes $F(\cdot, p^*)$. Suppose not, and take $y_n \in X$ such that $F(y_n, p^*) \leq V(p^*) + \epsilon_n$. Then there must exist $a > 0$ such that $d(y_n, x) \geq 2a$ for all large n, otherwise, by lower semicontinuity of F, x would minimize $F(\cdot, p^*)$. This implies that, for all large n, $d(x_n, y_n) \geq a$. But for each $\epsilon > 0$, it happens that x_n and y_n both belong to $\cup\{\epsilon-\arg \min(p) : \theta(p, p^*) < \epsilon\}$ for all large n, and this is a contradiction. ■

Theorem 5.1 strenghtens a criterion proved in [9].

An immediate consequence of Theorem 5.1 is the following basic criterion for Tikhonov well-posedness, due to Furi-Vignoli (see [12], Theorem 11 p.5).

Corollary 5.2 *If (X, f) is Tikhonov well-posed, then*

(11) diam$[\epsilon-\arg \min(p^*)] \to 0$ as $\epsilon \to 0$.

Conversely, assuming that f is bounded from below, (11) implies well-posedness of (X, f) provided X is complete and f is lower semicontinuous.

Observe that, if F is lower semicontinuous at every point of $X \times \{p^*\}$, then well-posedness implies lower semicontinuity of V at p^* (a well known fact independent of well-posedness provided X is compact).

The following example shows that the assumption (10), i.e.

$$\text{diam} \cup \{\epsilon - \text{argmin}(p) : \theta(p, p^*) < \epsilon\} \to 0 \text{ as } \epsilon \to 0,$$

cannot be weakened to:

$$\text{diam}[\epsilon - \text{argmin}(p)] \to 0 \text{ as } (\epsilon, p) \to (0, p^*).$$

Example 5.1 In the plane of points $(p, x) \in R^2$, let T denote the union of the two closed disks of center $(0, 1)$, radius 1 and center $(0, -1)$, radius $1/2$. Let $X = R$, $L = P = [0, 1]$ and $p^* = 1/2$. Consider

$$F(x, p) = \begin{cases} x & \text{if } (x, p) \in T \\ \infty & \text{otherwise.} \end{cases}$$

For $\epsilon > 0$ and sufficiently small we have

$$\text{diam}[\epsilon - \text{argmin}(p)] \leq \epsilon,$$

for every p. However, problem (p^*) is ill-posed as we see by choosing

$$p_n = 1/2 + 1/n, x_n = 1 - \sqrt{1 - p_n^2} = V(p_n),$$

so that x_n is asymptotically minimizing corresponding to p_n but $x_n \to 1 - \sqrt{3}/2 \neq -1 = \text{argmin}(p^*)$.

A little more sophisticated example shows that even with continuity of the value function, condition (10) cannot be weakened in the sense described above:

Example 5.2 Let X be an infinite dimensional separable Hilbert space, equipped with the norm topology, and with Hilbertian basis (e_n) and let P be the closed unit ball of X, equipped with a distance θ compatible with the weak topology. Finally, let $F(x, p) = \|x - p\|$ and $p^* = 0$. It is clear that all the assumptions in the second half of Theorem 5.1 are fulfilled, but condition (10). Moreover, V is continuous, and $\text{diam}[\epsilon\text{-arg min}(p)] = 2\epsilon$ for all p. However, if we take $p_n = x_n = e_n$ (and $\epsilon_n = 0$), we have that $p_n \to 0$ in P, but the sequence (x_n) does not converge to 0 in X for the strong topology, showing that there is no well-posedness by perturbations.

The following definition will be required in stating the next result. Let D be a subset of $\mathbf{R}_+^2$ containing $(0,0)$. Then the function

$$c : D \to [0, +\infty),$$

will be called *forcing* if $c(0,0) = 0$ and if for every sequence $(t_n, s_n) \in D$, $s_n \to 0$ and $c(t_n, s_n) \to 0$ imply $t_n \to 0$.

Theorem 5.2 *If problem (p^*) is well-posed, then there exist a forcing function c and a point $u \in X$ such that:*

(12) $F(x,p) \geq V(p) + c[d(x,u), \theta(p,p^*)]$ *for every $x \in X$ and $p \in P$.*

Conversely, (12) implies well-posedness provided F is lower semicontinuous on $X \times \{p^*\}$, and V is finite on L.

In the special case where no perturbations are present, i.e. $F(x,p) = f(x)$ for every x and p, a significant example of (generalized Tykhonov) well-posed problem arises when $c = c(t)$ is a linear function of t. One obtains in this way the notion of a sharp minimum, see [11] and the references therein.

Again, Theorem 5.2 extends a similar result for Tikhonov well-posedness (see [12], Corollary 13 p.7). Moreover, as in the case of Tikhonov well-posedness, condition (12), which in a sense is necessary and sufficient for well-posedness, can be formulated in an equivalent way that is better suited for obtaining quantitative estimates. If a forcing function c is known fulfilling (12), then an estimate of the convergence rate of the minimizing sequences toward the minimizer is available, according to the next result.

Theorem 5.3 *If problem (p^*) is well-posed, then there exist a point $u \in X$ and a function $q = q(s, \epsilon) \geq 0, s \geq 0, \epsilon \geq 0$ such that:*

(13) $q(s, \cdot)$ *is increasing for each $s, q(s, \epsilon) \to 0$ as $(s, \epsilon) \to (0,0)$,*

(14) $d(x,u) \leq q[\theta(p,p^*), F(x,p) - V(p)]$ *for every $x \in X$ and $p \in L$.*

Conversely if for some u and some q fulfilling (13) we have (14), then problem (p^*) is well-posed provided F is lower semicontinuous on $X \times \{p^*\}$, bounded from below, and V is finite on L.

Proof. Let problem (p^*) be well-posed and consider $u = \arg\min(p^*)$. Fix $T > 0$, and for every $s \geq 0, \epsilon \geq 0$ define

$$q(s, \epsilon) = \sup\{t \in [0, T] : c(t, s) \leq \epsilon\},$$

where c is the forcing function given by Theorem 5.2. Of course $q(s, \epsilon) \geq 0$ and $q(s, \cdot)$ is increasing. Let $s_n > 0, \epsilon_n > 0$ be such that $s_n \to 0, \epsilon_n \to 0$. Then for each n we can find $t_n \geq 0$ such that

$$q(s_n, \epsilon_n) - 1/n \leq t_n, \quad c(t_n, s_n) \leq \epsilon_n.$$

The forcing character of c yields $t_n \to 0$, hence $q(s_n, \epsilon_n) \to 0$. Then $q(s, \epsilon) \to 0$ as $(s, \epsilon) \to 0$, and (14) follows by definition of q. Conversely, let (14) hold. Define

$$c(t, s) = \inf\{\epsilon \geq 0 : q(s, \epsilon) \geq t\}.$$

Of course $0 \leq c(t, s) \leq +\infty, c(0, 0) = 0$ and $0 \leq t \leq q(s, \epsilon)$ implies $c(t, s) \leq \epsilon$. Thus by (14)

$$F(x, p) \geq V(p) + c[d(x, u), \theta(p, p^*)], x \in X, \quad p \in L.$$

The proof will be complete once we show that c is forcing, by an application of Theorem 5.2. Let $t_n \geq 0$ and $s_n \to 0$ be such that $c(t_n, s_n) \to 0$. For every n there exists $\epsilon_n \geq 0$ such that

$$c(t_n, s_n) + 1/n \geq \epsilon_n, \quad q(s_n, \epsilon_n) \geq t_n,$$

hence $\epsilon_n \to 0$, whence $q(s_n, \epsilon_n) \to 0$. This yields $t_n \to 0$ and proves that c is forcing. ∎

A Corollary of Theorem 5.3 is the following. If problem (p^*) is well-posed, then given $\epsilon > 0$ and $p \in L$ pick $x, y \in \epsilon - \arg\min(p)$. By (14) and $q(s, \cdot)$ increasing we have

$$\text{diam}[\epsilon - \arg\min(p)] \leq 2q(\theta(p, p^*), \epsilon),$$

which may be useful in estimating the shrinking to $\arg\min(p^*)$ of the sets $\epsilon - \arg\min(p)$ as $(\epsilon, p) \to (0, p^*)$ (compare with (10) of Theorem 5.1).

Remark 5.1 Theorems 5.1 and 5.3 are chosen from [9, 29] where also their proofs and some other results may be found.

6 EXTENDED WELL-POSEDNESS AND STABILITY BY PERTURBATIONS.

In this section we consider natural extensions of the definition of well-posedness by perturbations, by relaxing the uniqueness requirement for the solution of the unperturbed problem. The first definition is reminiscent of the analogous generalization of Tykhonov well-posedness.

Problem (p^*) is called *well-posed in the extended sense* if the following hold:

(3) arg min(p^*) is nonempty;

(6) the value function $V(p) = \inf\{F(x,p) : x \in X\}$ is finite for every $p \in L$;

(15) for every sequence $p_n \to p^*$, every asymptotically minimizing sequence corresponding to p_n has a subsequence converging to some optimal solution of problem (p^*).

As an immediate consequence of the definition, we have the following:

Proposition 6.1 *Let* $V(p) > -\infty, p \in L$. *If problem* (p^*) *is well-posed in the extended sense, then the multifunction*

$$(\epsilon, p) \to \epsilon - \arg\min(p)$$

is upper semicontinuous at $(0, p^*)$. *The converse holds provided* arg min (p^*) *is compact nonempty.*

We consider now an extension of the concept of *stability* of a given (unperturbed) minimization problem (see [12], p.26). Set, for a point x and a set A, $d(x, A) \overset{\text{def}}{=} \inf\{d(x, a) : a \in A\}$.

Problem (p^*) is called *stable by perturbations* if the following hold:

(3) arg min(p^*) is closed nonempty;

(6) the value function $V(p) = \inf\{F(x,p) : x \in X\}$ is finite for every $p \in L$;

(16) $p_n \to p^*$ in P, x_n asymptotically minimizing corresponding to p_n, imply $d[x_n, \arg\min(p^*)] \to 0$.

This concept is a weak version of extended well-posedness. Roughly speaking, stability by perturbations means that if p is sufficiently close to p^*, every approximate minimization method applied to problems (p) produces points which are close to some minimizer of problem (p^*). However there is no guarantee that these points cluster to some solution of the unperturbed problem, as in the case of extended well-posedness.

To formalize:

Proposition 6.2 *If problem (p^*) is well-posed in the extended sense, then it is stable by perturbations. The converse is true provided arg $min(p^*)$ is compact.*

We characterize now stability by perturbations by using forcing functions, as follows.

Proposition 6.3 *If problem (p^*) is stable by perturbations, then there exists a forcing function c such that:*

(17) $F(x,p) \geq V(p) + c[\mathrm{d}(x, \arg \min(p^*)], \theta(p, p^*)]$ *for every $p \in L$ and $x \in X$.*

Conversely, if V is finite on L, $F(\cdot, p^*)$ lower semicontinuous, arg $\min(p^*)$ nonempty and (17) holds, then problem (p^*) is stable by perturbations.

Proof. Let problem (p^*) be stable by perturbations. Consider

$$c(t, s) = \inf\{[F(x,p) - V(p) : p \in L, \theta(p, p^*) = s, \mathrm{d}[x, \arg\min(p^*)] = t\},$$

$s \geq 0, t \geq 0$. Then $c(t, s) \geq 0, c(0,0) = 0$ since arg min (p^*) is nonempty, and of course (17) holds. Let $t_n \geq 0, s_n \geq 0, s_n \to 0, c(t_n, s_n) \to 0$. Then a sequence (ϵ_n) can be found such that $\epsilon_n > 0$, $\epsilon_n \to 0$ and $c(t_n, s_n) < \epsilon_n$ for every n. It follows that there exist sequences (p_n), (x_n) such that $\theta(p_n, p^*) = s_n, \mathrm{d}[x_n, \mathrm{argmin}(p^*)] = t_n, F(x_n, p_n) - V(p_n) \leq \epsilon_n$. Thus x_n is asymptotically minimizing corresponding to p_n. Hence, by stability, $t_n \to 0$, so that c is forcing. Conversely, (17) implies stability, as required. ∎

By combining the previous results, we get the following:

Proposition 6.4 *If problem (p^*) is well-posed in the extended sense, then arg $\min(p^*)$ is a compact set and there exists a forcing function c such that (17) holds. Conversely, let V be finite on L, $F(\cdot, p^*)$ lower semicontinuous and arg $\min(p^*)$ nonempty and compact. If there exists a forcing function c such that (17) holds, then problem (p^*) is well-posed in the extended sense.*

The next result deals with a characterization making use of the Kuratowski measure of noncompactness of a set $A \subset X$, defined by

$$\alpha(A) = \inf\{k > 0: A \text{ has a finite cover by sets of diameter} < k\}.$$

Theorem 6.1 *Let $V(p)$ be finite for every $p \in L$, X complete, let F be lower semicontinuous on $X \times \{p^*\}$ and V upper semicontinuous at p^*. Then problem (p^*) is well-posed in the extended sense if and only if*

$$\alpha(\cup\{\epsilon - \arg\min(p) : \theta(p, p^*) < \epsilon\}) \to 0,$$

as $\epsilon \to 0$.

In the last result of this section we consider finite-dimensional convex Mathematical Programming, as in Theorem 2.2 and Example 4.2 above. Let

$$g_1, g_2, \cdots, g_k, h : \mathbf{R}^N \to \mathbf{R}$$

be given convex functions and consider, for any $p \in \mathbf{R}^k$, the sets

$$S(p) = \{x \in \mathbf{R}^N \; g(x) \leq p\},$$

where $g(x)$ is the vector of components $g_i(x)$, $i = 1, \cdots, k$ and the inequality between vectors is intended componentwise. Now define (using the indicator function)

$$F(x, p) = h(x) + I_{S(p)}(x), \quad x \in \mathbf{R}^N, p \in \mathbf{R}^k.$$

We consider extended well-posedness of the minimization of h on $S(0)$ (with respect to the embedding F).

Theorem 6.2 *Problem (0) is well-posed in the extended sense provided $S(0) \neq \emptyset$, the function V is upper semicontinuous at 0 and there exist a vector $a \in \mathbf{R}^N$ such that $a_i > 0$ for $i = 1, \cdots, N$ and $b \in \mathbf{R}$ such that the set*

$$\{x \in \mathbf{R}^N : g(x) \leq a, \quad h(x) \leq b\}$$

is nonempty and bounded.

Proof. We apply Proposition 6.1. The boundedness condition implies that arg min(0) is compact nonempty. We need to show upper semicontinuity of the $\epsilon - $ arg min mapping. Arguing by contradiction, assume that there exist an open set T containing arg min(0), sequences $\epsilon_n \to 0$, $p_n \to 0$ and $x_n \in \mathbf{R}^N$ such that

$$g(x_n) \leq p_n, \quad h(x_n) \leq V(p_n) + \epsilon_n, \quad x_n \notin T,$$

for every n. By the boundedness condition some subsequence (y_n) of (x_n) converges to some point y. Continuity of g and h and upper semicontinuity of the value function imply that $y \in$ arg min(0), hence $y_n \in T$ for all large n, which shows the required contradiction. ∎

Remark 6.1 The Slater condition implies (upper semi) continuity of the value function V at $p = p^*$. In presence of this condition, Theorem 6.2 is a consequence of Theorem 2.2.

Remark 6.2 All the above criteria, except Theorem 6.2, are proved in [29], where further results may be found.

7 ASYMPTOTICALLY STATIONARY SEQUENCES IN THE CONVEX CASE.

Here we consider again the notion of stability by perturbations of problem (p^*), mainly in the convex setting, for the following reasons. This notion is appropriate for handling optimization problems with many solutions. In the convex setting, i.e. when for every $p \in L$ we assume that $F(\cdot, p)$ is convex on the Banach space X, there exist minimization methods which generate a stationary sequence, i.e. a sequence of points along with an associate sequence of subgradients converging to 0. Quite often in applications, such methods are not applied to the unperturbed problem (p^*), but rather to given approximations or modifications of it, modeled by problem (p) as $p \to p^*$. Sometimes one combines a basic sequential optimization procedure, like a subgradient method, with a sequential approximation of problem (p^*), described by the sequence of problems (p_n) where $p_n \to p^*$, by changing the approximation at each step of the procedure. See [3] and the references therein.

In the convex case there is a link between these procedures and stability by perturbations of problem (p^*), as we shall show below.

The following property (18) is related to the desired behavior of approximate optimization procedures combined with perturbations of problem (p^*), yielding stationary sequences in the convex setting, as mentioned above.

Suppose X is a real Banach space and $F(\cdot, p)$ is proper, lower semicontinuous and convex on X for each $p \in L$. For any fixed p denote by $\partial F(x, p)$ the subgradient at x of $F(\cdot, p)$ (perhaps empty). Consider the following property:

(3) arg min(p^*) is nonempty;

(18) $p_n \to p^*, x_n \in X$, d$[0, \partial F(x_n, p_n)] \to 0$ imply d$[x_n, \text{arg min}(p^*)] \to 0$.

Sequences (x_n) such that $u_n \to 0$ for some $u_n \in \partial F(x_n, p_n)$ will be referred to as *asymptotically stationary* corresponding to p_n.

Theorem 7.1 *If* (18) *holds, then problem* (p^*) *is stable by perturbations. Conversely, every bounded asymptotically stationary sequence* (x_n) *fulfils* (18) *provided problem* (p^*) *is stable by perturbations, the function* $p \to F(x, p)$ *is upper semicontinuous at* $p = p^*$ *for all* x, *and* V *is continuous at* p^*.

Proof. Assume (18). If $p_n \to p^*$ and (x_n) is an asymptotically minimizing sequence corresponding to p_n, then there exists a sequence of positive numbers $\epsilon_n \to 0$ such that

$$F(x_n, p_n) \le V(p_n) + \epsilon_n^2,$$

for every n. Then by the Brondsted-Rockafellar theorem ([21] p.51), for every n there exists $y_n \in X, u_n \in \partial F(y_n, p_n)$ such that

(19) $\| u_n \| \le \epsilon_n, \| y_n - x_n \| \le \epsilon_n$.

Hence y_n is asymptotically stationary corresponding to p_n. By (18) and (19) we get d$[x_n, \text{arg min}(p^*)] \to 0$, whence stability. Conversely let (x_n) be any bounded asymptotically stationary sequence corresponding to (p_n). Given $\epsilon > 0, p \in L, x \in X, z \in \epsilon-\text{arg min}(p)$ and $u \in \partial F(x, p)$ we have

$$F(x, p) - V(p) \le \epsilon+ < u, x - z > \le \epsilon+ \| u \| \| x - z \|,$$

hence

$$F(x, p) - V(p) \le \epsilon + d(0, \partial F(x, p)) d(x, \epsilon - \text{arg min}(p)).$$

Thus, for every $\epsilon > 0$, if $u_n \in \partial F(x_n, p_n)$ and $u_n \to 0$,

(20) $F(x_n, p_n) - V(p_n) \leq \epsilon + \| u_n \| \, d[x_n, \epsilon\text{-arg min}(p_n)]$.

We want to show now that there is $c > 0$ such that:

$$d(x_n, \epsilon - \text{arg min}(p_n)) \leq c.$$

For, given $z \in X$ such that $F(z, p^*) \leq V(p^*) + \frac{\epsilon}{2}$, by upper semicontinuity of $p \to F(z, p)$ and continuity of V, for all large n we have:

$$F(z, p_n) \leq F(z, p^*) + \frac{\epsilon}{4} \leq V(p^*) + \frac{3\epsilon}{4} \leq V(p_n) + \epsilon.$$

As the sequence (u_n) converges to 0, we then have, from (20):

$$\limsup[F(x_n, p_n) - V(p_n)] \leq \epsilon.$$

It follows that x_n is asymptotically minimizing corresponding to p_n, thus by stability $d[x_n, \text{arg min}(p^*)] \to 0$, whence (18). ∎

Remark 7.1 Related results can be found in [3] and [14]. The asymptotically stationary sequences are called diagonally stationary in [3]. In the same paper, notions of equi-well-posed problems are introduced and characterized, which differ from the notion of stability by perturbations: there, every problem (p_n) is assumed to be solvable, and the asymptotic behavior of x_n is meant in the sense that $d[x_n, \text{arg min } (p_n)] \to 0$, without reference to any unperturbed problem.

8   WELL-POSEDNESS AND CONSTRAINTS.

Our last section is devoted to other types of Tikhonov-like well-posedness, when in the minimum problem some explicit constraints are present. As we already remarked, minimizing f over a constraint set A is in principle the same problem as minimizing $f + I_A$ over the whole space X, where I_A is the indicator function of the set A. However it is clear that the new formulation of the problem does not take into account the fact that many numerical methods (as for instance exterior penalization) usually allow violating the constraints to some (small) extent. In order to better handle this, other notions of well-posedness were proposed, and here we focus on *Levitin-Polyak* and *strong* well-posedness. The idea underlying the two definitions is the same, and goes back to Tikhonov well-posedness, i. e. convergence of minimizing sequences. The point is in the definition of a minimizing sequence.

Suppose we are given a topological space X, an extended real valued function f defined on X, and a subset $A \subset X$.

The problem (A, f) is called *Levitin-Polyak* well-posed if:

(21) there exists a unique minimizer $A \ni u^* = \arg \min(A, f)$,

i.e. $f(u^*) \leq f(x)$ for every $x \in A$;

(22) every sequence $(x_n) \subset X$ such that $f(x_n) \to f(u^*)$ and $d(x_n, A) \to 0$ converges to u^*.

The problem (A, f) is called *strongly* well-posed if:

(21) there exists a unique minimizer $A \ni u^* = \arg \min(A, f)$,

i.e. $f(u^*) \leq f(x)$ for every $x \in A$;

(23) every sequence $(x_n) \subset X$ such that $\limsup_{n \to \infty} f(x_n) \leq f(u^*)$ and $d(x_n, A) \to 0$ converges to u^*.

In other words, we consider as a possible approximation of the solution, sequences that approach the constraint set A, and in the first case their value must approximate the true value of the problem, while in the second one they can also have values going below the value of the problem.

Of course it is possible and natural to define the corresponding notions of generalized well-posedness, as we did for Tykhonov's.

It is quite clear that strong well-posedness implies Levitin-Polyak well-posedness, and this in turn is more demanding than Tikhonov well-posedness. Rather artificial examples show that the two notions differ when dealing with general lower semicontinuous functions on linear spaces, or with continuous functions in general metric spaces. But below we give a theorem showing that they agree in many interesting situations.

The main reason to consider strong well-posedness is because of the basic criterion provided by Corollary 5.2, that has a natural extension in this setting.

Suppose $\inf f(A) \in \mathbf{R}$ and let us set:

$$\epsilon - \arg \min(A, f) \stackrel{\text{def}}{=} \{x \in X : f(x) \leq \inf f(A) + \epsilon \quad \text{and} \quad d(x, A) \leq \epsilon\}.$$

The following result is proved in [8]:

Theorem 8.1 *Let* X *be a complete metric space,* $A \subset X$ *a closed set and* $f : X \to (-\infty, \infty]$ *lower semicontinuous and bounded from below. Then the following are equivalent:*

(i) *The minimum problem* (A, f) *is strongly well-posed;*

(ii) $\inf_{\epsilon > 0} \operatorname{diam} \epsilon - \arg \min(A, f) = 0.$

Thus the useful characterization via sublevel sets applies in natural classes to strong well-posedness rather than Levitin-Polyak well-posedness.

Before providing some examples, let us state two theorems giving sufficient conditions in order to have coincidence between the two notions. We suppose that X is a metric space, $f : X \to (-\infty, \infty]$ is lower semicontinuous and $A \subset X$ is a closed set.

Theorem 8.2 *Either one of the following conditions guarantees that Levitin-Polyak well-posedness implies strong well-posedness:*

(a) f *is uniformly continuous (or uniformly continuous on bounded sets and* A *is bounded);*

(b) f *is continuous and* X *is a normed space;*

(c) X *is a reflexive Banach space,* f *is proper convex and* A *is a convex set.*

Proof. (Outline). Under condition (a) the theorem is obvious. Now, suppose (b) holds. Take a sequence (x_n) such that $\limsup_{n \to \infty} f(x_n) \leq \inf f(A)$ and $d(x_n, A) \to 0$. There is $z_n \in A$ such that $d(z_n, x_n) \to 0$. Define, for all n, $X \ni t_n$ in this way: if n is such that $f(x_n) \geq \inf f(A)$, let $t_n = x_n$. Otherwise, select in the segment $[x_n, z_n]$ a point t_n such that $f(t_n) = \inf f(A)$. The sequence (t_n) converges to the unique solution u^* of the problem (A, f), by Levitin-Polyak well-posedness. Thus also (x_n) converges to u^*. As far as point (c) is concerned, suppose (x_n) is a sequence such that $d(x_n, A) \to 0$ and $\limsup_{n \to \infty} f(x_n) \leq \inf f(A)$. If (x_n) is bounded, there is $\hat{x} \in A$ such that $x_n \rightharpoonup \hat{x}$ (maybe up to a subsequence). Thus $\liminf_{n \to \infty} f(x_n) \geq f(\hat{x}) \geq \inf f(A)$ and so $\lim_{n \to \infty} f(x_n) = \inf f(A)$. By Levitin-Polyak well-posedness we have $x_n \to u^*$, the solution of the problem (A, f) (uniqueness of u^* guarantees convergence of the whole sequence). Supposing the sequence (x_n) has an unbounded subsequence leads to a contradiction. For, let us call

$$z_n = \left(1 - \frac{a}{\|x_n - u^*\|}\right)u^* + \frac{a}{\|x_n - u^*\|}x_n,$$

where $0 < a < \|x_n - u^*\|$. It is routine to show that (z_n) is bounded, $\limsup_{n \to \infty} f(z_n) \leq \inf f(A)$ and that $d(z_n, A) \to 0$. Thus, as before, $z_n \to u^*$. But this is impossible, as $\|z_n - u^*\| = a$. ∎

Remark 8.1 Theorem 8.2 can be improved, especially as far as condition (b) is concerned. See [23].

The next theorem gives equivalence between Tikhonov and strong (hence also Levitin-Polyak) well-posedness in an interesting case. Its proof can be found in [7].

Theorem 8.3 *Let X be a Banach space, $A \subset X$ a closed convex set. Let $f : X \to (-\infty, \infty]$ be convex, lower semicontinuous and proper. If f is continuous at some point of A, or else if dom $f \cap$ int $A \neq \emptyset$, then strong well-posedness of (A, f) is equivalent to Tikhonov well-posedness of $f + I_A$.*

Remark 8.2 Examples show that, without a constraint qualification condition, the conclusion of Theorem 8.3 need not hold.

For more connections between these notions, and especially between these kinds of well-posedness and Hadamard well-posedness, we remind the paper [22].

Let us now see some examples.

Example 8.1 Let $X = \mathbf{R}^2$, $A = \mathbf{R} \times \{0\}$, $f(x, y) = x^2 - x^4 y^2$. Then (A, f) is Tikhonov well-posed. The sequence $((n, \frac{1}{n}))$ fulfils (22) and does not converge. Thus (A, f) is not Levitin-Polyak well-posed.

Example 8.2 Let $A \subset \mathbf{R}^n$ be a closed convex set, let $f : \mathbf{R}^n \to \mathbf{R}$ be convex, lower semicontinuous and with unique minimizer on A. Then the problem (A, f) is Levitin-Polyak (and strongly) well-posed. This follows from Example 3.3 and Theorem 8.3.

Example 8.3 The best approximation problem is Levitin-Polyak well-posed if and only if the space X is an E-space. This readily follows from Example 3.4 and Theorem 8.3.

Example 8.4 Suppose X is a normed linear space, $f, g : X \to \mathbf{R}$ such that g is convex, lower semicontinuous and proper, and $\lim_{\|x\| \to \infty} f(x) = \infty$.

Suppose moreover f is bounded from below, and there exists a point $\bar{x} \in X$ such that $g(\bar{x}) < 0$. Set

$$A \stackrel{\text{def}}{=} \{x \in X \ : g(x) \leq 0\},$$

and suppose the problem (A, f) strongly well-posed. If a sequence (x_n) is such that, for some sequence $\epsilon_n \to 0$,

$$f(x_n) + n \max\{g(x_n), 0\} \leq \inf_{x \in X} \{f(x) + n \max\{g(x), 0\}\} + \epsilon_n,$$

then (x_n) converges to the solution of the problem. Thus, to find out approximate solutions of the constrained problem (A, f), one is led in this case to (approximatively) solve the unconstrained problems of minimizing $f(x) + n \max\{g(x), 0\}$. The mixing of penalization technique and well-posedness guarantee that, for large n, x_n is close to arg min (A, f).

Let us show the claimed convergence. It is clear that boundedness from below of f guarantees that $\limsup_{n \to \infty} g(x_n) \leq 0$. Moreover, convexity of g and existence of $\bar{x}$ such that $g(\bar{x}) < 0$ together imply that $d(x_n, A) \to 0$. To see this, define a sequence $(y_n) \in A$ as follows: if $g(x_n) \leq 0$, set $y_n \stackrel{\text{def}}{=} x_n$. Otherwise, let

$$y_n \stackrel{\text{def}}{=} \frac{g(x_n)}{g(x_n) - g(\bar{x})} \bar{x} + (1 - \frac{g(x_n)}{g(x_n) - g(\bar{x})}) x_n.$$

Then $y_n \in A$ for all n, and $d(x_n, y_n) \to 0$ (as the sequence (x_n) is bounded). Finally, it is $\limsup_{n \to \infty} f(x_n) \leq \inf f(A)$. To conclude the proof, use strong wellposedness.

Variants of this example can be provided for instance by imposing convexity on f instead of the growth condition used here, or by using strong rather Levitin-Polyak well-posedness.

Acknowledgement. Work partially supported by MURST, fondi 40%. We also thank the referees for helpful comments.

References

[1] Asplund, E. and Rockafellar, R.T.: *Gradients of convex functions*, Trans. Amer. Math. Soc. **139** (1969), 443-467.

[2] Attouch, H., Azé and D. Wets, R. J-B.: *On continuity properties of the partial Legendre transform: convergence of sequences of augmented Lagrangian functions, Moreau-Yosida approximates, and subdifferential operators*, in J.B Hirriart-Urruty (ed), Mathematics of Optimization, Elsevier, 1986.

[3] Bahraoui, M.A. and Lemaire, B.: *Convergence of diagonally stationary sequences in convex optimization*, Set-Valued Anal. **2** (1994), 49-61.

[4] Bank, B, Guddat, J., Klatte, D., Kummer, B. and Tammer, K.: Nonlinear parametric optimization, *Academie Verlag*, 1993.

[5] Beer, G.: Topologies on closed and closed convex sets, *Kluwer Acad. Publishers* **268**, 1993.

[6] Beer, G. and Lucchetti, R.: *Convex optimization and the epidistance topology*, Trans. Amer. Math. Soc. **327** (1991), 795-813.

[7] Beer, G. and Lucchetti, R.: *The epidistance topology: continuity and stability results with applications to convex minimization problems*, Math. Oper. Res. **17** (1992), 715-726.

[8] Beer, G. and Lucchetti, R.: *Solvability for constrained problems*, Dip. Matem. Milano **3** (1991).

[9] Bennati, M.L.: *Well-posedness by perturbation in optimization problems and metric characterizations*, to appear on Rend. Mat.

[10] Bennati, M.L.: *Local well-posedness of constrained problems*, to appear on Optimization.

[11] Burke, J.V. and Ferris, M.C.: *Weak sharp minima in mathematical programming*, SIAM J. Control Optim. **31** (1993), 1340-1359.

[12] Dontchev, A. and Zolezzi, T.: Well-posed optimization problems, *Lecture Notes in Math.* **1543**, Springer, 1993.

[13] Fiacco, A.V.: Introduction to sensitivity and stability analysis in nonlinear programming, *Academic Press*, 1983.

[14] Lemaire, B.: *Bounded diagonally stationary sequences in convex optimization*, J. Convex Anal. **1** (1994), 75-86.

[15] Lucchetti, R.: *Hadamard and Tikhonov well-posedness in optimal control*, Methods Oper. Res. **45** (1983), 113-126.

[16] Lucchetti, R.: *Convergence of sets and projections*, Boll. UMI Serie VI, **1** (1985), 477-483.

[17] Lucchetti, R.: *Hypertopologies and applications*, in Recent Developments in Well-posed Variational Problems, Lucchetti, R. and Revalski, J. (eds) *Kluwer Acad. Publishers* **331**, 1995, 193-209.

[18] Lucchetti, R. and Patrone, F.: *Hadamard and Tikhonov well-posedness of a certain class of convex functions*, J. Math. Anal. Appl. **88** (1982), 204-215.

[19] Lucchetti, R. and Patrone, F.: *Closure and upper semicontinuity results in mathematical programming, Nash and economic equilibria*, Optimization **17** (1986), 619-628.

[20] Lucchetti, R. and Revalski, J.(eds): Recent developments in well-posed variational problems, *Kluwer Acad. Publishers* **331**, 1995.

[21] Phelps, R.: Convex functions, monotone operators and differentiability, *Lecture Notes in Math.* **1364**, Springer, 1989.

[22] Revalski, J.: *Various aspects of well-posedness of optimization problems*, in Recent developments in well-posed variational problems, Lucchetti, R. and Revalski, J. (eds) *Kluwer Acad. Publishers* **331**, 1995, 229-256.

[23] Revalski, J. and Zhivkov, N.: *Well-posed optimization problems in metric spaces*, J. Optim. Theory Appl. **76** (1993), 145-163.

[24] Zolezzi, T.: *Characterizations of some variational perturbations of the abstract linear-quadratic problem*, SIAM J. Control Optim **16** (1978), 106-121.

[25] Zolezzi, T.: *Well-posedness of optimal control problems*, Control Cybern, **3** (1994), 289-301.

[26] Zolezzi, T.: *Well-posed Criteria in Optimization with Application to the Calculus of Variations*, Nonlinear Anal. Theory Methods Appl. **25** (1995), 437-453.

[27] Zolezzi, T.: *Well-posed problems of the calculus of variations for nonconvex integrals*, J. Convex Analysis **2** (1995), 375-383.

[28] Zolezzi, T.: *Extended well-posedness of optimal control problems*, J. Discrete Continuous Dynamical Systems **1** (1995), 547-553.

[29] Zolezzi, T.: *Extended well-posedness of optimization problems*, to appear in J. Optim. Theory Appl.

[30] Zolezzi, T.: *Well-posedness of multiple integrals in the calculus of variations*, to appear on Ricerche Mat.

Convergence of Approximations to Nonlinear Optimal Control Problems

KAZIMIERZ MALANOWSKI [1]   Systems Research Institute, Polish Academy of Sciences, ul.Newelska 6, 01-447 Warszawa, Poland

CHRISTOF BÜSKENS   Westfälische Wilhelms-Universität Münster, Institute für Numerische und instrumentelle Mathematik, Einsteinstrasse 62, 48149 Münster, Germany

HELMUT MAURER   Westfälische Wilhelms-Universität Münster, Institute für Numerische und instrumentelle Mathematik, Einsteinstrasse 62, 48149 Münster, Germany

Dedicated to Professor Anthony V.Fiacco

Abstract: An approach of studying convergence of approximations to cone constrained optimization problems in Banach spaces is proposed. In this approach stability results for parametric optimization problems, based on Robinson's implicit function theorem for generalized equations, are exploited. An estimate of the rate of convergence is obtained, provided that the problem is uniformly strongly regular. The abstract results are applied to estimate the rate of convergence of Euler's approximation to nonlinear optimal control problems subject to mixed control-state constraints. Numerical examples are provided.

AMS subject classification: 49M25, 49K30, 49K40, 58C15, 65K10.

Key words: convergence of approximation, stability analysis, Robinson's implicit function theorem, uniform strong regularity, Euler's approximation, nonlinear optimal control, control-state constraints.

1   Introduction

Systematical studies of the properties of solutions to parametric optimization problems were initiated in middle seventies by pioneering papers of Fiacco [9], Levitin [16] and Robinson [29]. Since that time the stability and sensitivity analysis for such problems has been developed into a well advanced thery.

[1]Supported by grant No.3 P403 002 05 from the State Committee for Scientific Research (Komitet Badań Naukowych)

The results are the most complete for finite dimensional parametric mathematical programs. However, starting with eighties stability and sensitivity analysis for infinite dimensional problems has ben also substencially developed (see e.g. [1, 6, 7, 17, 18, 31]). In this theory the crucial role is played by the analysis of constraint qualifications and sufficient optimality conditions.

One of the important class of infinite dimensional optimization problems is optimal control. Stability and sensitivity of solutions to parametric optimal control problems has been recently studied (see [6, 7, 13, 17, 19, 21]), where in particular constraints qualifications and second order sufficient optimality conditions have been developed for this class of problems [21, 25, 35].

The results of stability analysis find applications in those areas, where the dependence of solutions to optimization problems on parameters has to be studied. A typical application is in the convergence analysis of some iterative algorithms of optimization (see e.g. [29, 1]).

Another prospective field of applications is in covergence analysis of approximations to optimal control problems. The approximating problems can be treated as perturbations of the original one, depending on a parameter of approximation destined to tend to zero. Convergence properties for small values of the parameter are studied.

This class of results has a direct applications. Namely, in order to solve numerically optimal control problems for continuous systems, at some stage of computation we have to approximate an infinite dimensional problem by a finite dimensional one and to know that, for sufficiently refined approximation the approximate solution is arbitrarily close to the original one.

Not very much has been done in that area. One should mention here some general results concerning the convergence of approximations in the sense of the value of the cost functional [2, 8, 27].

The approach based on stability analysis allows to obtain close estimates of the rate of convergence of the approximate solutions. For linear systems, this approach was exploited in [4] (see also [5]), and for nonlinear systems in [6]. A modification of Robinson's [30] implicit function theorem for multivalued mappings was used there. It was shown that, under the assumptions virtually the same as in stability analysis, the error of Euler's approximation to nonlinear optimal control problems with convex control constraints is of the same order as the mesh spacing.

The approach presented in this paper is very similar to that in [6]. However, we do not use the theorem from that paper, but the original Robinson's implicit function theorem. In application to stability analysis of parametric optimization and optimal control problems, this theorem allows to show local Lipschitz continuity of the Karush-Kuhn-Tucker (KKT) points, provided that the so called *strong regularity condition* is satisfied at the reference point.

Strong regularity means local Lipschitz continuity of the linearized (KKT) points with respect to the additive perturbations.

Exactly the same technique is used here in studying convergence of approximations. In Section 2 an estimate of the error of the approximate (KKT) points is derived, under the assumption that the *strong regularity is*, in a sense *uniform with respect to the parameter of approximation.*

This result is quite general and it can be applied to different schemes of approximation. In each case we have to verify the uniform strong regularity.

We apply this approach to Euler's approximation of nonlinear optimal control problems subject to mixed control-state constraints. The crucial role is played here by constraint qualifications and strong second order sufficient optimality conditions recently developed for this class of optimal control problems in connection with stability and sensitivity analysis (see [21, 20]).

It is shown that these conditions are stable under small perturbations. That allows to prove uniform strong regularity and to estimate the error of Euler's approximation.

In Section 6 numerical examples of controlled Rayleigh equation are presented. All assumptions needed in theoretical part are carefully checked and the numerical computations fully confirm theoretical convergence results.

The theoretical part of this paper is due to the first author, whereas the numerical results are provided by the second and third author.

Throughout the paper by $\langle \cdot, \cdot \rangle$ and $(\cdot, \cdot)$ we denote the inner products in $\mathbb{R}^n$ and in $L^2(0, T)$, respectively. The norms in $\mathbb{R}^n$ are denoted by $|\cdot|$ and in Hilbert or Banach spaces by $\|\cdot\|$, if necessary with appropriate subscripts. $L(X; Y)$ denotes the space of linear continuous operators from Banach space X into Y, supplied with the usual norm. $D_x f(x, u), D_{xu}^2 f(x, u)$ etc. denote Fréchet derivatives of appropriate order, with respect to the given variable.

2   Abstract Approximation Problem

In this section we are going to use Robinson's implicit function theorem to obtain the estimate of the rate of convergence of solutions to approximations of nonlinear generalized equations. This result will be applied to estimate the rate of convergence of (KKT) points for approximations of optimal control problems.

Let X be a Banach space and $C \subset X$ a closed convex set. By $\partial \psi_C$ we denote the normal cone operator for C :

$$\partial \psi_C(w) := \begin{cases} \{y \in X^* \mid (y, c - w) \leq 0 \text{ for } c \in C\} & \text{if } w \in C, \\ \emptyset & \text{if } w \notin C. \end{cases} \quad (2.1)$$

Let $\Omega \subset X$ be an open subset and $\varphi : \Omega \mapsto X^*$ be a Fréchet differentiable function.

Assume that the generalized equation in X

$$0 \in \varphi(w) + \partial \psi_C(w) \quad (2.2)$$

has w_0 as a solution. Let H be a Banach space of parameters and $G \subset H$ a neighborhood of the origin θ. For each value $h \in G \setminus \theta$ let X_h denote a subspace of X supplied with the same norm $\|\cdot\|_X$ and let $C_h \subset X_h$ be a closed convex set. Let $\varphi_h : X_h \mapsto X_h^*$ be a Fréchet differentiable function such that

(I.1) $\|D_x \varphi_h(x') - D_x \varphi_h(x'')\|_{X^*} \leq c \|x' - x''\|_X$ for all $x', x'' \in X_h,$ (2.3)
where $c > 0$ is independent of h.

Consider a family of the following generalized equations in X_h for $h \in G \setminus \theta$:

$$0 \in \varphi_h(x) + \partial \psi_{C_h}(x), \quad (2.4)$$

where $\partial \psi_{C_h}$ is the normal cone operator for C_h.

We are interested in conditions under which, for h sufficiently small there exists a locally unique solution x_h of (2.4) such that

$$\|w_0 - x_h\|_X \to 0 \quad \text{for } \|h\|_H \to 0. \tag{2.5}$$

Moreover we will estimate the rate of the above convergence. We assume

(I.2) (Approximation condition)
For each $h \in G \setminus \theta$ there exist elements $s_h \in X_h^*$ and $w_h \in X_h$ such that

$$\begin{aligned}
&0 \in \varphi_h(w_h) + s_h + \partial\psi_{C_h}(w_h), \quad \text{and} \\
&\|w_h - w_0\|_X \to 0, \ \|s_h\|_{X^*} \to 0 \quad \text{for } \|h\|_H \to 0.
\end{aligned} \tag{2.6}$$

Let us introduce the following linearizations of the generalized equations (2.6) perturbed by $\delta \in X_h^*$:

$$\delta \in \varphi_h(w_h) + s_h + D_w\varphi_h(w_h)(w - w_h) + \partial\psi_{C_h}(w). \tag{2.7}$$

(I.3) (Uniform strong regularity)
There exist $\eta > 0, \chi > 0$ and $\varsigma > 0$ such that for each $h \in G_\eta := \{h \in G \mid 0 < \|h\|_H \leq \eta\}$ and each $\delta \in \Delta_{h,\chi} := \{\delta \in X_h^* \mid \|\delta\|_{X^*} \leq \chi\}$ there is a unique in $\mathcal{X}_{h,\varsigma} := \{w \in X_h \mid \|w - w_h\|_X \leq \varsigma\}$ solution $w_h(\delta)$ of (2.7) and

$$\|w_h(\delta') - w_h(\delta'')\|_X \leq \ell\|\delta' - \delta''\|_{X^*} \quad \text{for all } \delta', \delta'' \in \Delta_{h,\chi}, \tag{2.8}$$

where $\ell > 0$ is independent of h.

Note that for fixed h, (I.3) corresponds to the condition of *strong regularity* of w_h introduced in [30]. In (I.3) we require that this strong regularity is uniform with respect to h.

We will show that if (I.1)-(I.3) are satisfied, then (2.5) holds. Let us start with the following

LEMMA 2.1 *If (I.1)-(I.3) hold, then for each $\epsilon > 0$ there are constants $\xi > 0$ and $\zeta > 0$ independent of h, such that for each fixed $h \in G_\eta$ and each $s \in \mathcal{X}_{h,\xi}^* := \{s \in X_h^* \mid \|s - s_h\|_{X^*} \leq \xi\}$ the generalized equation*

$$0 \in \varphi_h(x) + s + \partial\psi_{C_h}(x) \tag{2.9}$$

has a solution $x_h(s)$ unique in $\mathcal{X}_{h,\zeta}$ and

$$\|x_h(s') - x_h(s'')\|_X \leq (\ell + \epsilon)\|s' - s''\|_{X^*}. \tag{2.10}$$

$\diamond$

Proof For fixed h, the generalized equation (2.9) depends on the parameter s. Hence, if (I.2) holds then the existence of $\xi(h)$ and $\zeta(h)$ such that (2.10) is satisfied, follows immediately from Therem 2.1 and Corollary 2.2 in [30]. Thus, to prove the assertion of the lemma it remains to show that $\xi(h)$ and $\zeta(h)$ can be chosen independent of $h \in G_\eta$. To this end we have to reexamine briefly the proof of Theorem 2.1 in [30], where $h \in G_\eta$ will be treated as a fixed parameter.

In this proof there is defined the mapping

$$r_h : X_h^* \times X_h \mapsto X_h^*,$$
$$r_h(s, x) := (\varphi_h(w_h) + s_h) + D_x \varphi_h(w_h)(x - w_h) - (\varphi_h(x) + s) \tag{2.11}$$

and for fixed $\epsilon > 0$ we choose $\rho > 0$ such that $\ell \rho < \epsilon/(\ell + \epsilon)$.

By (2.3) we can choose $\xi > 0$ and $\zeta \in [0, \varsigma)$, independent of h such that

$$r_h(\sigma, x) \in \Delta_{h,\chi} \quad \text{and} \tag{2.12}$$
$$\|D_x r_h(\sigma, x)\|_{L(X;X^*)} = \|D_x \varphi_h(w_h) - D_x \varphi_h(x)\|_{L(X;X^*)} \le \rho \tag{2.13}$$

for all $x \in \mathcal{X}_{h,\zeta}$ and $s \in \mathcal{X}_{h,\xi}^*$.

By (I.2) and (2.12), for any $s \in \mathcal{X}_{h,\xi}^*$ the mapping

$$\Phi_{h,s} : \mathcal{X}_{h,\zeta} \to X_h, \quad \Phi_{h,s}(x) = y_h(r_h(s, x))$$

is well defined. It can be checked that $x \in \mathcal{X}_{h,\zeta} \cap \Phi_{h,s}$ if and only if $x \in \mathcal{X}_{h,\zeta}$ and x is a solution to (2.9).

It was shown in [30] that by (2.8) and (2.13), $\Phi_{h,s}$ is a contractive self-map on $\mathcal{X}_{h,\zeta}$. Hence, by the contraction principle there exists a unique in $\mathcal{X}_{h,\zeta}$ solution x_h of (2.9), and (2.10) follows from the contraction estimates. $\square$

THEOREM 2.2 *If assumptions* (I.1)-(I.3) *hold then there exists* $\eta > 0$ *such that for each* $h \in G_\eta$ *there exists a locally unique solution* x_h *of* (2.4) *and a constant* $\ell' > 0$, *independent of* h *such that*

$$\|x_h - w_0\|_X \le \ell'(\|s_h\|_{X^*} + \|w_h - w_0\|_X). \tag{2.14}$$

$\diamond$

Proof By (2.6) we can choose $\eta > 0$ so small that $\|s_h\|_{X^*} \le \xi/2$ for all $h \in G_\eta$. Hence, putting $s' = 0$ we find that there exists a unique in $\mathcal{X}_{h,\xi}^*$ solution x_h of (2.4). Choosing $s'' = s_h$ we get from (2.10)

$$\|x_h - w_h\|_X \le (\ell + \epsilon)\|s_h\|_{X^*}.$$

Hence

$$\|x_h - w_0\|_X \le (\|x_h - w_h\|_X + \|w_h - w_0\|_X) \le \ell'(\|s_h\|_{X^*} + \|w_h - w_0\|_X),$$

where $\ell' = \max\{1, \ell + \epsilon\}$. $\square$

3 Optimal Control Problem

In this section we introduce an optimal control problem subject to mixed control-state constraints. The convergence of Euler's approximation to this problem will be investigated in next sections.

The problem is the same as that considered in [21] and in [20]. The assumptions will be identical with those in [20].

(O) Find $(x_0, u_0) \in W^{1,\infty}(0, T; \mathbb{R}^n) \times L^\infty(0, T; \mathbb{R}^m)$ such that

$$F(x_0, u_0) = \min\{F(x, u) := \int_0^T f^0(x(t), u(t))dt + g(x(T))\}$$

subject to

$$\dot{x}(t) - f(x(t), u(t)) = 0, \qquad \text{for a.a. } t \in [0, T],$$
$$x(0) - \eta_0 = 0, \quad \xi(x(T)) = 0, \tag{3.1}$$
$$\theta(x(t), u(t)) \leq 0, \qquad \text{for a.a. } t \in [0, T],$$

where $\xi : I\!\!R^n \mapsto I\!\!R^d$, $\theta : I\!\!R^n \times I\!\!R^m \mapsto I\!\!R^k$.
We assume:

(II.1) The functions $f^0(\cdot, \cdot), g(\cdot), f(\cdot, \cdot), \xi(\cdot)$ and $\theta(\cdot, \cdot)$ are twice Fréchet differentiable in all arguments, and the respective derivatives are locally Lipschitz continuous in x, u.

(II.2) There exists a possibly local solution

$$(x_0, u_0) \in C^1(0, T; I\!\!R^n) \times C(0, T; I\!\!R^m).$$

We will introduce constraint qualifications and second order sufficient optimality conditions that assure local uniqueness and regularity of (x_0, u_0).

For the sake of simplicity we denote

$$A(t) := D_x f(x_0(t), u_0(t)), \qquad B(t) := D_u f(x_0(t), u_0(t)),$$
$$\Xi := D_x \xi(x_0(T)),$$
$$\Theta_x(t) := D_x \theta(x_0(t), u_0(t)), \qquad \Theta_u(t) := D_u \theta(x_0(t), u_0(t)).$$

Let us denote by
$Z = W^{1,\infty}(0, T; I\!\!R^n) \times L^\infty(0, T; I\!\!R^m)$ and
$Y = L^\infty(0, T; I\!\!R^n) \times I\!\!R^n \times I\!\!R^d \times L^\infty(0, T; I\!\!R^k)$
the spaces of arguments and constraints, respectively. Moreover, introduce Hilbert spaces $\widehat{Z}$ and $\widehat{Y}$ defined in the same way as Z and Y, but with ∞ substituted by 2 and define $X = Z \times Y$, $\widehat{X} = \widehat{Z} \times \widehat{Y}$. We will put $\widehat{Y}^* = \widehat{Y}$.

Let us denote $I := \{1, 2, ..., k\}$ and for $\sigma \geq 0$ define

$$I_\sigma(t) = \{i \in I \mid \theta^i(x_0(t), u_0(t)) \geq -\sigma\}, \quad \imath_\sigma(t) = \text{ card } I_\sigma(t) \text{ and}$$
$$\Theta_x^\sigma(t) := [D_x \theta^i(x_0(t), u_0(t))], \quad \Theta_u^\sigma(t) := [D_u \theta^i(x_0(t), u_0(t))] \tag{3.2}$$
$$\text{for } i \in I_\sigma(t).$$

Introduce the sets

$$\Omega_\sigma^i = \{t \in [0, T] \mid i \in I_\sigma(t)\}, \qquad \Omega_\sigma = \prod_i \Omega_\sigma^i, \tag{3.3}$$

and define the space

$$Y^\sigma = L^\infty(0, T; I\!\!R^n) \times I\!\!R^n \times I\!\!R^d \times L^\infty(\Omega_\sigma; I\!\!R^k), \tag{3.4}$$

where $L^\infty(\Omega_\sigma; I\!\!R^k) := \prod_i L^\infty(\Omega_\sigma^i; I\!\!R^1)$. The space $\widehat{Y}^\sigma$ is defined in the same way but with ∞ substituted by 2. Assume:

(II.3) There exist positive constants $\sigma > 0$ and $\beta > 0$ such that

$$|[\Theta_u^\sigma(t)]^* z| \geq \beta |z| \quad \text{for all } z \in I\!\!R^{l_\sigma(t)} \text{ and a.a. } t \in [0, T].$$

(II.4) For any $e \in I\!\!R^d$ the following boundary value system has a solution (is completely output controllable)

$$\begin{aligned}
&\dot{y}(t) - \widetilde{A}(t)y(t) - \widetilde{B}v(t) = 0, \\
&y(0) = 0, \quad \Xi y(T) = e,
\end{aligned} \tag{3.5}$$

where
$$\begin{aligned}
\widetilde{A}(t) &= A(t) - B(t)\Theta_u^\sigma(t)^*(\Theta_u^\sigma(t)\Theta_u^\sigma(t)^*)^{-1}\Theta_x^\sigma(t), \\
\widetilde{B}(t) &= B(t)(J - \Theta_u^\sigma(t)^*(\Theta_u^\sigma(t)\Theta_u^\sigma(t)^*)^{-1}\Theta_x^\sigma(t))
\end{aligned}$$

and J denotes the unit matrix.

REMARK 3.1 It can be shown (see Lemma 5.1 in [17]) that in (II.3) it is enough to assume $\sigma = 0$, since this condition satisfied at $\sigma = 0$ implies that it is also satisfied for certain $\sigma > 0$. The same refers to (II.4).

Condition (II.4) can be expressed in terms of the full rank of a controllability matrix (see [21]). $\diamond$

LEMMA 3.2 *If* (II.3) *and* (II.4) *hold, then there exists a constant* $\bar{\beta} > 0$ *such that the mapping* $\mathcal{C}: L(Z; Y^\sigma) \cup L(\widehat{Z}; \widehat{Y}^\sigma)$

$$\mathcal{C}(y, v) = \begin{bmatrix} \dot{y} - Ay - Bv \\ y(0) \\ \Xi y(T) \\ \Theta_x^\sigma y + \Theta_u^\sigma v \end{bmatrix} \tag{3.6}$$

satisfies the following surjectivity conditions

$$\begin{aligned}
&\|\mathcal{C}(y, v)\|_{Y^\sigma} \geq \bar{\beta}\|(y, v)\|_Z \quad \text{for all } (y, v) \in Z, \\
&\|\mathcal{C}(y, v)\|_{\widehat{Y}^\sigma} \geq \bar{\beta}\|(y, v)\|_{\widehat{Z}} \quad \text{for all } (y, v) \in \widehat{Z}.
\end{aligned} \tag{3.7}$$

$\diamond$

The proof of the lemma is very similar to that of Lemma 1 in [21]. Since the elements of this proof will be needed later on in the proof of Lemma 5.2, it is presented in Appendix.

Let us introduce the Lagrangian

$$\begin{aligned}
&\mathcal{L}: Z \times Y \mapsto I\!\!R^1, \\
&\mathcal{L}(x, u, p, \pi, \rho, \lambda) = F(x, u) - (p, \dot{x} - f(x, u)) + \langle \mu, x(0) - \eta_0 \rangle + \\
&+ \langle \pi, \xi(x(T)) \rangle + (\lambda, \theta(x, u)).
\end{aligned} \tag{3.8}$$

By Lemma 3.2 there exist unique Lagrange multipliers $(p_0, \mu_0, \pi_0, \lambda_0) \in Y^*$ associated with (x_0, u_0) such that (KKT) conditions are satisfied at $w_0 := (x_0, u_0, p_0, \mu_0, \pi_0, \lambda_0)$. Moreover, it follows from (II.3) and (II.4) that the Lagrange multipliers are more regular. Namely $(p_0, \mu_0, \pi_0, \lambda_0) \in Y$.

Still, let us introduce the following augmented Hamiltonian:

$$\mathcal{H} : \mathbb{R}^n \times \mathbb{R}^m \times \mathbb{R}^n \times \mathbb{R}^k \mapsto \mathbb{R}^1,$$
$$\mathcal{H}(x, u, p, \lambda) = f^0(x, u) + \langle p, f(x, u) \rangle + \langle \lambda, \theta(x, u) \rangle. \tag{3.9}$$

The stationarity condition of the Lagrangian takes on the form of the adjoint equation

$$\dot{p}_0 + D_x \mathcal{H}(x_0, u_0, p_0, \lambda_0) = 0,$$
$$p_0(0) + \mu_0 = 0, \quad p_0(T) - D_x(g(x_0(T)) + \pi_0^* \xi(x_0(T)))^* = 0, \tag{3.10}$$

and of the condition

$$D_u \mathcal{H}(x_0, u_0, p_0, \lambda_0) = 0. \tag{3.11}$$

Let us define the space

$$\widetilde{X} = L^\infty(0, T; \mathbb{R}^n) \times \mathbb{R}^n \times \mathbb{R}^n \times L^\infty(0, T; \mathbb{R}^m) \times$$
$$\times L^\infty(0, T; \mathbb{R}^n) \times \mathbb{R}^n \times \mathbb{R}^d \times L^\infty(0, T; \mathbb{R}^k),$$

which can be considered as a subspace of X^*.

In the space $L^\infty(0, T; \mathbb{R}^k)$ we introduce the cone of non-negative functions

$$K = \{\lambda \in L^\infty(0, T; \mathbb{R}^k) \mid \lambda(t) \geq 0 \ \text{ for a.a. } t \in [0, T]\},$$

and define the following closed convex cone

$$C = L^\infty(0, T; \mathbb{R}^n) \times \mathbb{R}^n \times \mathbb{R}^n \times L^\infty(0, T; \mathbb{R}^m) \times$$
$$\times L^\infty(0, T; \mathbb{R}^n) \times \mathbb{R}^n \times \mathbb{R}^d \times K. \tag{3.12}$$

The (KKT) conditions for (O) can be expressed in the form of the following generalized equation in X :

$$0 \in \mathcal{F}(w_0) + \partial \psi_C(w_0), \tag{3.13}$$

where $w_0 := (x_0, u_0, p_0, \mu_0, \pi_0, \lambda_0)$,

$$\mathcal{F}(w_0) = \begin{bmatrix} \dot{p}_0 + D_x \mathcal{H}(x_0, u_0, p_0, \lambda_0) \\ p_0(0) + \mu_0 \\ p_0(T) - D_x(g(x_0(T)) + \pi_0^* \xi(x_0(T)))^* \\ D_u \mathcal{H}(x_0, u_0, p_0, \lambda_0) \\ \dot{x}_0 - f(x_0, u_0) \\ x_0(0) - \eta_0 \\ \xi(x_0(T)) \\ -\theta(x_0, u_0) \end{bmatrix}, \quad \partial \psi_C(w_0) = \begin{bmatrix} 0 \\ 0 \\ 0 \\ 0 \\ 0 \\ 0 \\ 0 \\ \partial \psi_K(\lambda_0) \end{bmatrix}, \tag{3.14}$$

and $\partial \psi_C, \partial \psi_K$ are the normal cone operators for C and K respectively, given by (2.1).

Still we will need coercivity conditions in the same form as in [21].

For a given $\alpha \geq 0$ denote $I_+^\alpha(T) = \{i \in I_0(t) \mid \lambda_0^i(t) > \alpha\}$, and define

$$\widehat{\Theta}_x^\alpha(t) := [D_x \theta^i(x_0(t), u_0(t))], \quad \widehat{\Theta}_u^\alpha(t) := [D_u \theta^i(x_0(t), u_0(t))],$$
$$\text{for } i \in I_+^\alpha(t).$$

Moreover, we introduce the sets

$$\Upsilon_\alpha^i = \{t \in [0, T] \mid i \in I_+^\alpha(t)\}. \tag{3.15}$$

In addition to (II.1)-(II.4) we assume the following two conditions:

(II.5) There exists $\gamma > 0$ such that

$$v^* D_{uu}^2 \mathcal{H}_0(t)v \geq \gamma |v|^2 \text{ for all } v \in I\!\!R^m \text{ satisfying } \widehat{\Theta}_u^\alpha(t)v = 0.$$

(II.6) The following Riccati equation

$$\dot{Q}(t) = -Q(t)A(t) - A(t)^*Q(t) - D_{xx}^2 \mathcal{H}_0(t) +$$

$$+ \left\{ \left[\begin{array}{c} D_{ux}^2 \mathcal{H}_0(t) \\ \widehat{\Theta}_x^\alpha(t) \end{array} \right]^* + Q(t) \left[\begin{array}{c} B(t)^* \\ 0 \end{array} \right]^* \right\} \times$$

$$\times \mathcal{K}^\alpha(t)^{-1} \left\{ \left[\begin{array}{c} B(t)^* \\ 0 \end{array} \right] Q(t) + \left[\begin{array}{c} D_{ux}^2 \mathcal{H}_0(t) \\ \widehat{\Theta}_x^\alpha(t) \end{array} \right] \right\},$$

has a solution Q bounded on $[0, T]$, that satisfies the following boundary condition:

$$y^*(\Gamma - Q(T))y \geq 0 \text{ for all } y \in I\!\!R^n \text{ such that } \Xi y = 0,$$

where $\mathcal{H}_0(t) := \mathcal{H}(x_0(t), u_0(t), p_0(t), \lambda_0(t))$,

$$\Gamma := D_{xx}^2(g(x_0(T)) + \pi_0^*\xi(x_0(T))) \tag{3.16}$$

and the matrix

$$\mathcal{K}^\alpha(t) = \left[\begin{array}{cc} D_{uu}^2 \mathcal{H}_0(t) & \widehat{\Theta}_u^\alpha(t)^* \\ \widehat{\Theta}_u^\alpha(t) & 0 \end{array} \right],$$

is non-singular by (II.3) and (II.5).

The following result is proved in [21] (Lemma 2):

LEMMA **3.3** *If* (II.5) *and* (II.6) *are satisfied then there exist* $\bar{\gamma} > 0$ *such that*

$$((y, v), \mathcal{Q}(y, v)) := ((y, v), \left(\begin{array}{cc} D_{xx}^2 \mathcal{L}(w_0) & D_{xu}^2 \mathcal{L}(w_0) \\ D_{ux}^2 \mathcal{L}(w_0) & D_{uu}^2 \mathcal{L}(w_0) \end{array} \right)(y, v)) =$$

$$= \int_0^T \left[\begin{array}{c} y(t) \\ v(t) \end{array} \right]^* \left[\begin{array}{cc} D_{xx}^2 \mathcal{H}_0(t) & D_{xu}^2 \mathcal{H}_0(t) \\ D_{ux}^2 \mathcal{H}_0(t) & D_{uu}^2 \mathcal{H}_0(t) \end{array} \right] \left[\begin{array}{c} y(t) \\ v(t) \end{array} \right] dt +$$

$$+ y(T)^*\Gamma y(T) \geq \bar{\gamma}(\|y\|_{1,2}^2 + \|v\|_2^2)$$

for all $(y, v) \in W^{1,2}(0, T; I\!\!R^n) \times L^2(0, T; I\!\!R^m)$ *satisfying*

$$\dot{y}(t) - A(t)y(t) - B(t)v(t) = 0,$$
$$y(0) = 0, \qquad \Xi y(T) = 0,$$
$$\widehat{\Theta}_x^\alpha(t)y(t) + \widehat{\Theta}_u^\alpha(t)v(t) = 0.$$

◇

REMARK 3.4 Conditions (II.5) and (II.6) satisfied for $\alpha = 0$ are also satisfied for some $\alpha > 0$ (see Lemma 5 in [7]). Hence it is enough to put in (II.5) and (II.6), $\alpha = 0$.   ◇

As a by-product of our assumptions, using the same technique as in [11] we obtain the following regularity of primal and dual optimal variables:

LEMMA 3.5 *If assumptions* (II.1)-(II.6) *are satisfied, then* $(\dot{x}_0, u_0, \dot{p}_0, \lambda_0)$ *are Lipschitz continuous on* $[0, T]$.   ◇

4   Euler's Approximation to Optimal Control Problem

In this section we will consider Euler's approximation of optimal control problem (O) introduced in Section 3 and formulate (KKT) conditions for approximating problems.

Let N be a natural number and $h = \frac{T}{N}$ be the mesh spacing, which will be treated as the parameter of approximation. Denote $t_j = jh$, $j = 0, 1, ..., N$.

For given h we introduce the space $L_h^\infty(0, T; I\!\!R^n)$ of piecewise constant functions v defined on $[0, T]$ by

$$v(t) = v(t_j) \text{ for } t \in [t_j, t_{j+1}), \quad j = 0, 1, ..., N - 1, \tag{4.1}$$

with the norm

$$\|v\|_\infty = \max\{|v(t_j)| \mid j = 0, 1, ...N - 1\}.$$

Similarly, by $L_h^2(0, T; I\!\!R^n)$ we denote the space of elements (4.1) supplied with the norm

$$\|v\|_2 = \left[h\Sigma_{j=0}^{N-1}|y(t_j)|^2\right]^{\frac{1}{2}}.$$

Moreover we define the spaces $W_h^{1,\infty}(0, T; I\!\!R^n)$ and $W_h^{1,2}(0, T; I\!\!R^n)$ of piecewise linear elements

$$y(t) = y(t_j) + (t - t_j)\nabla y(t_j) \text{ for } t \in [t_j, t_{j+1}], j = 0, 1, ..., N - 1, \tag{4.2}$$

where $\nabla y(t_j) := \frac{1}{h}(y(t_{j+1}) - y(t_j))$, supplied with the following norms:

$$\|y\|_{1,\infty} = \max\{|y(t_0)|, |\nabla y(t_j)| \mid j = 0, 1, ..., N - 1\},$$
$$\|y\|_{1,2} = \left\{|y(t_0)|^2 + h\Sigma_{j=0}^{N-1}|\nabla y(t_j)|^2\right\}^{\frac{1}{2}}.$$

Consider the following approximation of (O):

(O$_h$)   Find $(x_h, u_h) \in W_h^{1,\infty}(0, T; I\!\!R^n) \times L_h^\infty(0, T; I\!\!R^m)$ such that
$$F_h(x_h, u_h) = \min\{F_h(x, u) := h\Sigma_{j=0}^{N-1}f^0(x(t_j), u(t_j)) + g(x(t_N))\}$$
subject to
$$\nabla x(t_j) - f(x(t_j), u(t_j)) = 0, \qquad \text{for } j = 0, 1, ..., N - 1,,$$
$$x(t_0) - \eta_0 = 0, \quad \xi(x(t_N)) = 0, \tag{4.3}$$
$$\theta(x(t_j), u(t_j)) \leq 0, \qquad \text{for } j = 0, 1, ..., N - 1.$$

Define the spaces
$Z_h = W_h^{1,\infty}(0,T;I\!\!R^n) \times L_h^\infty(0,T;I\!\!R^m)$,
$Y_h = L_h^\infty(0,T;I\!\!R^n) \times I\!\!R^n \times I\!\!R^d \times L_h^\infty(0,T;I\!\!R^k)$, and $X_h = Z_h \times Y_h$.
The spaces $\widehat{Z}_h, \widehat{Y}_h, \widehat{X}_h$ are defined in the same way with ∞ substituted by 2. We will put $\widehat{Y}_h^* = \widehat{Y}_h$.

Let us introduce the Lagrangian for (O_h) in the form analogous to (3.8)

$$\mathcal{L}_h : Z_h \times Y_h \mapsto I\!\!R^1,$$

$$\mathcal{L}_h(x,u,p,\mu,\pi,\lambda) = h\Sigma_{j=0}^{N-1} f^0(x(t_j),u(t_j)) + g(x(t_N)) -$$
$$- h\Sigma_{j=0}^{N-1} \langle p(t_{j+1}), \nabla x(t_j) - f(x(t_j),u(t_j)) \rangle + \langle \mu, x(t_0) - \eta_0 \rangle + \tag{4.4}$$
$$+ \langle \pi, \xi(x(t_N)) \rangle + h\Sigma_{j=0}^{N-1} \langle \lambda(t_j), \theta(x(t_j),u(t_j)) \rangle.$$

As in (3.10) and (3.11) the stationarity conditions of the Lagrangian take on the form:

$$\nabla p_h(t_j) + D_x\mathcal{H}(x_h(t_j),u_h(t_j),p_h(t_{j+1}),\lambda_h(t_j)) = 0, \tag{4.5}$$
$$j = 0,1,...,N-1,$$
$$p_h(t_0) + \mu_h = 0, \quad p_h(t_N) - D_x(g(x_h(t_N) + \pi_h^*\xi(x_h(t_N)))^* = 0$$
$$D_u\mathcal{H}(x_h(t_j),u_h(t_j),p_h(t_{j+1}),\lambda_h(t_j)) = 0, \; j = 0,1,...,N-1, \tag{4.6}$$

where the augmented Hamiltonian $\mathcal{H}$ is defined in (3.9).

As in Section 3, to define the (KKT) conditions for (O_h) we introduce the space

$$\widetilde{X}_h = L_h^\infty(0,T;I\!\!R^n) \times I\!\!R^n \times I\!\!R^n \times L_h^\infty(0,T;I\!\!R^m) \times$$
$$\times L_h^\infty(0,T;I\!\!R^n) \times I\!\!R^n \times I\!\!R^d \times L_h^\infty(0,T;I\!\!R^k),$$

and the cone

$$C_h = L_h^\infty(0,T;I\!\!R^n) \times I\!\!R^n \times I\!\!R^n \times L_h^\infty(0,T;I\!\!R^m) \times$$
$$\times L_h^\infty(0,T;I\!\!R^n) \times I\!\!R^n \times I\!\!R^d \times K_h, \tag{4.7}$$

where $K_h = \{\lambda \in L_h^\infty(0,T;I\!\!R^k) \mid \lambda(t_j) \geq 0, \; j = 0,1,...,N-1\}$.

The (KKT) conditions for (O_h) can be expressed in the form of the generalized equation analogous to (3.13):

$$0 \in \mathcal{F}_h(w_h) + \partial\psi_{C_h}(w_h), \tag{4.8}$$

where $w_h := (x_h,u_h,p_h,\mu_h,\pi_h,\lambda_h) \in X_h$ and

$$\mathcal{F}_h(w_h) = \begin{bmatrix} \nabla p_h(t_j) + D_x\mathcal{H}(x_h(t_j),u_h(t_j),p_h(t_{j+1}),\lambda_h(t_j)) \\ p_h(t_0) + \mu_h \\ p_h(t_N) - D_x(g(x_h(t_N)) + \pi_h^*\xi(x_h(t_N)))^* \\ D_u\mathcal{H}(x_h(t_j),u_h(t_j),p_h(t_{j+1}),\lambda_h(t_j)) \\ \nabla x_h(t_j) - f(x_h(t_j),u_h(t_j)) \\ x_h(t_0) - \eta_0 \\ \xi(x_h(t_N)) \\ -\theta(x_h(t_j),u_h(t_j)) \end{bmatrix}, \tag{4.9}$$

$$
\partial\psi_{C_h}(w_h) = \begin{bmatrix} 0 \\ 0 \\ 0 \\ 0 \\ 0 \\ 0 \\ 0 \\ \partial\psi_{K_h}(\lambda_h(t_j)) \end{bmatrix}.
$$

5 Convergence of Approximation

We are going to apply the general approach developed in Section 2 to estimate the rate of convergence of the solutions of (O_h) to that of (O). First of all note that, in view of (II.1), there exists a constant $c > 0$, independent of h such that

$$
\|D_w\mathcal{F}_h(w') - D_w\mathcal{F}_h(w'')\|_{L(X;\widetilde{X})} \le c\|w' - w''\|_X \tag{5.1}
$$

i.e., the abstract assumption (I.1) holds. Now, we have to define the element w_h needed in (I.2). We put $w_h = (\widehat{x}_h, \widehat{u}_h, \widehat{p}_h, \widehat{\mu}_h, \widehat{\pi}_h, \widehat{\lambda}_h)$, where

$$
\begin{array}{lll}
\widehat{x}_h \in W_h^{1,\infty}(0,T;\mathbb{R}^n), & \widehat{x}_h(t_j) = x_0(t_j) & j = 0,1,...N, \\
\widehat{u}_h \in L_h^{\infty}(0,T;\mathbb{R}^m), & \widehat{u}_h(t_j) = u_0(t_j) & j = 0,1,...N-1, \\
\widehat{p}_h \in W_h^{1,\infty}(0,T;\mathbb{R}^n), & \widehat{p}_h(t_j) = p_0(t_j) & j = 0,1,...N, \\
\widehat{\lambda}_h \in L_h^{\infty}(0,T;\mathbb{R}^k), & \widehat{\lambda}_h(t_j) = \lambda_0(t_j) & j = 0,1,...N-1, \\
\widehat{\mu}_h = \mu_0, \quad \widehat{\pi}_h = \pi_0.
\end{array} \tag{5.2}
$$

Since by Lemma 3.5, $(\dot{x}_0, u_0, \dot{p}_0, \lambda_0)$ are Lipschitz continuous functions of time, we find that

$$
\|w_h - w_0\|_X \le ch, \tag{5.3}
$$

where w_0 is the (KKT) point of (O) and $c > 0$ is independent of h.

REMARK 5.1 In [6] the concept of the so called *averaged modulus of smoothness* is used, which allows to get some estimates of $\|w_h - w_0\|_X$ in case, where w_0 is not necessarily Lipschitz continuous. ◇

Using definition (5.2) and comparing (3.13) and (4.8), we find that w_h satisfies the following generalized equation

$$
0 \in \mathcal{F}_h(w_h) + s_h + \psi_{C_h}(w_h), \tag{5.4}
$$

where

$$
s_h(t_j) := \begin{bmatrix} s_h^1(t_j) \\ s_h^2(t_j) \\ s_h^3(t_j) \\ s_h^4(t_j) \\ s_h^5(t_j) \\ s_h^6(t_j) \\ s_h^7(t_j) \\ s_h^8(t_j) \end{bmatrix} = \begin{bmatrix} (\dot{p}_0(t_j) + D_x\mathcal{H}(x_0(t_j), u_0(t_j), p_0(t_j), \lambda_0(t_j))) - \\ -(\nabla p_0(t_j) + D_x\mathcal{H}(x_0(t_j), u_0(t_j), p_0(t_{j+1}), \lambda_0(t_j))) \\ 0 \\ 0 \\ D_u\mathcal{H}(x_0(t_j), u_0(t_j), p_0(t_j), \lambda_0(t_j)) - \\ -D_u\mathcal{H}(x_0(t_j), u_0(t_j), p_0(t_{j+1}), \lambda_0(t_j)) \\ \dot{x}_0(t_j) - \nabla x_0(t_j) \\ 0 \\ 0 \\ 0 \end{bmatrix}.
$$

In view of (II.1) and Lemma 3.5, we find that a constant $c > 0$, independent of h exists such that

$$\|s_h\|_{\widetilde{X}} \le ch. \tag{5.5}$$

Conditions (5.3) and (5.5) show that the abstract assumption (I.2) holds.

The main difficulty in verifying assumptions of Theorem 2.1 are connected with the *uniform strong regularity* (I.3). This verification will be performed in the following steps:

1) We show that the generalized equation (2.7) constitutes (KKT) conditions for accessory linear-quadratic optimal control problems (LO_δ).

2) We modify (LO_δ) in such a way to get problems ($\widetilde{\text{LO}}_\delta$) for which conditions of coercivity and of linear independence of gradients of all constraints are satisfied.

3) We introduce new variables, in which ($\widetilde{\text{LO}}_\delta$) yield problems ($\widetilde{\text{LO}}'_\delta$) with constraints independent of the perturbations δ. For these problems we obtain Lipschitz continuity of the (KKT) points with respect to perturbations.

4) Finally we prove that, for sufficiently small perturbations δ, the (KKT) points of ($\widetilde{\text{LO}}'_\delta$) are the solutions of the initial generalized equations. Thus the condition (I.3) is verified.

Let us start with introducing the following linear generalized equation corresponding to (2.7):

$$\delta \in \mathcal{F}_h(w_h) + s_h + D_w \mathcal{F}_h(w_h)(w - w_h) + \partial \psi_{C_h}(w),$$

where $\delta = (a^1, a^2, a^3, a^4, b^1, b^2, b^3, b^4) \in \widetilde{X}_h$ and $w = (y, v, q, \nu, \tau, \kappa) \in X_h$ i.e., we have

$$\nabla q(t_j) + A(t_j)^* q(t_{j+1}) + (D_x \theta(x_0(t_j), u_0(t_j))^* \kappa(t_j) +$$
$$+ D^2_{xx} \mathcal{H}(t_j) y(t_j) + D^2_{xu} \mathcal{H}(t_j) v(t_j) + \widehat{s}^1_h(t_j) - a^1(t_j) = 0,$$
$$q(t_0) + \nu - a^2 = 0,$$
$$q(t_N) - D^2_{xx}(g(x_0(t_N)) + \pi^*_0 \xi(x_0(t_N))) y(t_N) +$$
$$+ D_x \xi(x_0(t_N))^* \tau + \widehat{s}^3_h - a^3 = 0,$$
$$D^2_{ux} \mathcal{H}(t_j) y(t_j) + D^2_{uu} \mathcal{H}(t_j) v(t_j) + B(t_j) q(t_{j+1}) + \tag{5.6}$$
$$+ D_u \theta(x_0(t_j), u_0(t_j))^* \kappa(t_j) + \widehat{s}^4_h(t_j) - a^4(t_j) = 0,$$
$$\nabla y(t_j) - A(t_j) y(t_j) - B(t_j) v(t_j) + \widehat{s}^5_h(t_j) - b^1(t_j) = 0,$$
$$y(t_0) + \widehat{s}^6_h - b^2 = 0,$$
$$\Xi y(t_j) + \widehat{s}^7_h - b^3 = 0,$$
$$-\Theta_x(t_j) y(t_j) - \Theta_u(t_j) v(t_j) + \widehat{s}^8_h(t_j) - b^4(t_j) \in \partial \psi_{K_h}(\kappa(t_j)),$$

where $\mathcal{H}(t_j) := \mathcal{H}(x_0(t_j), u_0(t_j), p_0(t_{j+1}), \lambda_0(t_j))$ and

$$
\begin{aligned}
\widehat{s}_h^1(t_j) &= -D_{xx}^2\mathcal{H}(t_j)x_0(t_j) - D_{xu}^2\mathcal{H}(t_j)u_0(t_j) + D_x f^0(x_0(t_j), u_0(t_j)), \\
\widehat{s}_h^3 &= -D_{xx}^2(g(x_0(t_N)) + \pi_0^*\xi(x_0(t_N)))x_0(t_N) + D_x g(x(t_j)), \\
\widehat{s}_h^4(t_j) &= -D_{ux}^2\mathcal{H}(t_j)x_0(t_j) - D_{uu}^2\mathcal{H}(t_j)u_0(t_j) + D_u f^0(x_0(t_j), u_0(t_j)), \\
\widehat{s}_h^5(t_j) &= -\nabla x_0(t_j) + A(t_j)x_0(t_j) + B(t_j)u_0(t_j), \\
\widehat{s}_h^6 &= \eta_0, \quad \widehat{s}_h^7 = -\Xi x_0(t_N) \\
\widehat{s}_h^8(t_j) &= -\theta(x_0(t_j), u_0(t_j)) + \Theta_x(t_j)x_0(t_j) + \Theta_u(t_j)u_0(t_j).
\end{aligned}
\tag{5.7}
$$

An inspection reveals that (5.6) constitutes the (KKT) conditions for the following linear-quadratic discrete optimal control problems depending on the parameter δ :

(LO_δ) Find $(y_\delta, v_\delta) \in W_h^{1,\infty}(0, T; \mathbb{R}^n) \times L_h^\infty(0, T; \mathbb{R}^m)$ such that

$$\mathcal{I}_h(y_\delta, v_\delta, \delta) = \min\{\mathcal{I}_h(y, v, \delta) :=$$

$$
:= h\Sigma_{j=0}^{N-1}\left(\frac{1}{2}\left[\begin{array}{c} y(t_j) \\ v(t_j) \end{array}\right]^*\left[\begin{array}{cc} D_{xx}^2\mathcal{H}(t_j) & D_{xu}^2\mathcal{H}(t_j) \\ D_{ux}^2\mathcal{H}(t_j) & D_{uu}^2\mathcal{H}(t_j) \end{array}\right]\left[\begin{array}{c} y(t_j) \\ v(t_j) \end{array}\right] + \right.
$$

$$+ [\widehat{s}_h^1(t_j) - a^1(t_j)]^* y(t_j) + [\widehat{s}_h^4(t_j) - a^4(t_j)]^* v(t_j)) +$$

$$\left. + \frac{1}{2}y(t_N)^*\Gamma y(t_N) + [\widehat{s}_h^3 - a^3]^* y(t_N)\right\}$$

subject to

$$
\begin{aligned}
&\nabla y(t_j) - A(t_j)y(t_j) - B(t_j)v(t_j) + \widehat{s}_h^5(t_j) - b^1(t_j) = 0, \\
&y(t_0) + \widehat{s}_h^6 - b^2 = 0, \quad \Xi y(t_N) + \widehat{s}_h^7 - b^3 = 0, \\
&\Theta_x(t_j)y(t_j) + \Theta_u(t_j)v(t_j) + \widehat{s}_h^8(t_j) - b^4(t_j) \leq 0,
\end{aligned}
\tag{5.8}
$$

where Γ is defined in (3.16).

We have to show that for δ sufficiently small there is a unique (KKT) point of (LO_δ), which is a Lipschitz continuous function of δ.

It will be convenient to modify (LO_δ) as follows:

$(\widetilde{\text{LO}}_\delta)$ Find $(\widetilde{y}_\delta, \widetilde{v}_\delta) \in W_h^{1,2}(0, T; \mathbb{R}^n) \times L_h^2(0, T; \mathbb{R}^m)$ such that

$$\mathcal{I}_h(\widetilde{y}_\delta, \widetilde{v}_\delta, \delta) = \min \mathcal{I}_h(y, v, \delta)$$

subject to

$$
\begin{aligned}
&\nabla y(t_j) - A(t_j)y(t_j) - B(t_j)v(t_j) + \widehat{s}_h^5(t_j) - b^1(t_j) = 0, \\
&y(t_0) + \widehat{s}_h^6 - b^2 = 0, \quad \Xi y(t_N) + \widehat{s}_h^7 - b^3 = 0, \\
&\langle \Theta_x^i(t_j), y(t_j)\rangle + \langle \Theta_u^i(t_j), v(t_j)\rangle + \\
&+ (\widehat{s}_h^8)^i(t_j) - (b^4)^i(t_j) \begin{cases} = 0 & \text{if } t_j \in \Upsilon_\alpha^i, \\ \leq 0 & \text{if } t_j \in \Omega_\sigma^i \setminus \Upsilon_\alpha^i, \end{cases}
\end{aligned}
\tag{5.9}
$$

where Ω_σ^i and Υ_α^i are given in (3.3) and (3.15), respectively.

Note that $(\widetilde{\text{LO}}_\delta)$ is considered in $\widehat{Z}_h$ rather than in Z_h. Since these spaces are finite dimensional, they are isomorphic.

Lagrange multipliers associated with $(\widetilde{y}_\delta, \widetilde{v}_\delta)$ are denoted by $(\widetilde{q}_\delta, \widetilde{\nu}_\delta, \widetilde{\tau}_\delta, \widetilde{\kappa}_\delta)$. Note that for $\delta = 0$ we have

$$(\widetilde{y}_0, \widetilde{v}_0, \widetilde{q}_0, \widetilde{\nu}_0, \widetilde{\tau}_0, \widetilde{\kappa}_0) = (y_0, v_0, q_0, \nu_0, \tau_0, \kappa_0) = (\widehat{x}_h, \widehat{u}_h, \widehat{p}_h, \widehat{\mu}_h, \widehat{\pi}_h, \widehat{\lambda}_h). \quad (5.10)$$

Moreover

$$(\widetilde{y}_\delta, \widetilde{v}_\delta, \widetilde{q}_\delta, \widetilde{\nu}_\delta, \widetilde{\tau}_\delta, \widetilde{\kappa}_\delta) = (y_\delta, v_\delta, q_\delta, \nu_\delta, \tau_\delta, \kappa_\delta), \quad (5.11)$$

provided that

$$\Theta_x^i(t_j)\widetilde{y}_\delta(t_j) + \Theta_u^i(t_j)\widetilde{v}_\delta(t_j) + (\widehat{s}_h^8)^i(t_j) - (\widehat{b}^4)^i(t_j) < 0 \quad \text{for } t_j \notin \Omega_\sigma^i,$$
and
$$\widetilde{\kappa}_\delta^i(t_j) > 0 \qquad\qquad\qquad\qquad\qquad\quad \text{for } t_j \in \Upsilon_\alpha^i. \quad (5.12)$$

In view of (5.10) and of the definitions of Ω_σ^i and Υ_α^i, the conditions (5.12) will be satisfied if

$$|\Theta_x^i(t_j)(\widetilde{y}_\delta(t_j) - \widetilde{y}_0(t_j)) + \Theta_x^i(t_j)(\widetilde{y}_\delta(t_j) - \widetilde{y}_0(t_j)) - (\widetilde{b}^4(t_j))^i| \le \sigma, \text{ and}$$
$$|\widetilde{\kappa}_\delta^i(t_j) - \widetilde{\kappa}_0^i(t_j)| \le \alpha, \quad (5.13)$$

for all $i \in I$ and $j = 0, 1, ..., N - 1$.

In a similar way as in (3.4) we define the space

$$Y_h^\sigma = L_h^\infty(0, T; \mathbb{R}^n) \times \mathbb{R}^n \times \mathbb{R}^d \times L_h^\infty(\Omega_\sigma; \mathbb{R}^k), \quad (5.14)$$

where $L_h^\infty(\Omega_\sigma; \mathbb{R}^k) := \prod_i L_h^\infty(\Omega_\sigma^i; \mathbb{R}^1)$, and $L_h^\infty(\Omega_\sigma^i; \mathbb{R}^1)$ is the space of piece-wise constant functions such that $t_j \in \Omega_\sigma^i$.

The following two lemmas are proved in Appendix. They show that surjectivity and coercivity conditions given in Lemma 3.2 and Lemma 3.3 are preserved for approximations, provided that the mesh spacing h is sufficiently small.

LEMMA 5.2 *If* (II.3) *and* (II.4) *hold, then there exists* $h_1 > 0$ *such that for all* $h < h_1$ *the mapping* $\mathcal{C}_h : L(Z_h; Y_h^\sigma) \cup L(\widehat{Z}_h; \widehat{Y}_h^\sigma)$

$$\mathcal{C}_h(y, v) = \begin{bmatrix} \nabla y - Ay - Bv \\ y(t_0) \\ \Xi y(t_N) \\ \Theta_x^\sigma y + \Theta_u^\sigma v \end{bmatrix} \quad (5.15)$$

satisfies the following surjectivity conditions

$$\|\mathcal{C}_h(y, v)\|_{Y_h^\sigma} \ge \tfrac{\bar{\beta}}{2}\|(y, v)\|_Z \quad \text{for all } (y, v) \in Z_h,$$
$$\|\mathcal{C}_h(y, v)\|_{\widehat{Y}_h^\sigma} \ge \tfrac{\bar{\beta}}{2}\|(y, v)\|_{\widehat{Z}} \quad \text{for all } (y, v) \in \widehat{Z}_h, \quad (5.16)$$

where $\bar{\beta} > 0$ *is given in* (3.7). $\diamond$

LEMMA 5.3 *If* (II.5) *and* (II.6) *hold, then there exists* $h_2 > 0$ *such that for all* $h < h_2$ *we have*

$$((y, v), \mathcal{Q}_h(y, v)) :=$$

$$:= h\Sigma_{j=0}^{N-1} \begin{bmatrix} y(t_j) \\ v(t_j) \end{bmatrix}^* \begin{bmatrix} D_{xx}^2\mathcal{H}(t_j) & D_{xu}^2\mathcal{H}(t_j) \\ D_{ux}^2\mathcal{H}(t_j) & D_{uu}^2\mathcal{H}(t_j) \end{bmatrix} \begin{bmatrix} y(t_j) \\ v(t_j) \end{bmatrix} + \quad (5.17)$$

$$+ y(t_N)^*\Gamma y(t_N) \ge \tfrac{\bar{\gamma}}{2}(\|y\|_{1,2}^2 + \|v\|_2^2)$$

for all $(y, v) \in W_h^{1,2}(0, T; \mathbb{R}^n) \times L_h^2(0, T; \mathbb{R}^m)$ *such that*

$$
\begin{aligned}
&\nabla y(t_j) - A(t_j) y(t_j) - B(t_j) v(t_j) = 0, \\
&y(t_0) = 0, \qquad \Xi y(t_N) = 0, \\
&\widehat{\Theta}_x^\alpha(t_j) y(t_j) + \widehat{\Theta}_u^\alpha(t_j) v(t_j) = 0,
\end{aligned}
\tag{5.18}
$$

where $\bar{\gamma} > 0$ *is given in Lemma 3.3.*   ◇

It is easy to see that by (5.16) and (5.17), for any $\delta \in \widetilde{X}_h$ and any $h < \min\{h_1, h_2\}$, problem $(\widetilde{\mathrm{LO}}_\delta)$ has a unique solution $(\widetilde{y}_\delta, \widetilde{v}_\delta)$ and unique Lagrange multipliers $(\widetilde{q}_\delta, \widetilde{\nu}_\delta, \widetilde{\tau}_\delta, \widetilde{\kappa}_\delta)$.

Let us prove the following

LEMMA 5.4 *If assumptions* (II.1)-(II.6) *hold then there exists* $\ell_1 > 0$ *such that for all* $h < \min\{h_1, h_2\}$

$$
\begin{aligned}
&\|\widetilde{y}_{\delta'} - \widetilde{y}_{\delta''}\|_{1,2}, \|\widetilde{v}_{\delta'} - \widetilde{v}_{\delta''}\|_2 \leq \ell_1 \|\delta' - \delta''\|_{\widetilde{X}}, \\
&\|\widetilde{q}_{\delta'} - \widetilde{q}_{\delta''}\|_{1,2}, |\widetilde{\nu}_{\delta'} - \widetilde{\nu}_{\delta''}|, |\widetilde{\tau}_{\delta'} - \widetilde{\tau}_{\delta''}|, \|\widetilde{\kappa}_{\delta'} - \widetilde{\kappa}_{\delta''}\|_2 \leq \ell_1 \|\delta' - \delta''\|_{\widetilde{X}}.
\end{aligned}
\tag{5.19}
$$

◇

Proof Let $\mathcal{J}_h$ denotes the canonical isomorphism between $\widehat{Z}_h$ and $\widehat{Z}_h^*$. By (5.16) the operator $\mathcal{C}_h \mathcal{J}_h^{-1} \mathcal{C}_h^* \in L(\widehat{Y}^\sigma, \widehat{Y}^\sigma)$ is invertible and

$$
\|(\mathcal{C}_h \mathcal{J}_h^{-1} \mathcal{C}_h^*)^{-1}\|_{L(\widehat{Y}^\sigma, \widehat{Y}^\sigma)} \leq \frac{\bar{\beta}^2}{4}.
\tag{5.20}
$$

In $(\widetilde{\mathrm{LO}}_\delta)$ we introduce new variables (z, w) putting

$$
\begin{bmatrix} z \\ w \end{bmatrix} = \begin{bmatrix} y \\ v \end{bmatrix} + \mathcal{J}_h^{-1} \mathcal{C}_h^* (\mathcal{C}_h \mathcal{J}_h^{-1} \mathcal{C}_h^*)^{-1} k_\delta,
\tag{5.21}
$$

where

$$
k_\delta(t_j) = \begin{bmatrix} \widehat{s}_h^5(t_j) - b^1(t_j) \\ -b^2 \\ -b^3 \\ (\widehat{s}_h^8(t_j) - \widehat{b}^4(t_j))^\sigma \end{bmatrix}
$$

and for any $d(t_j) \in \mathbb{R}^k$, $(d(t_j))^\sigma := [d^i(t_j)]$ for $i \in I_\sigma(t_j)$.

In new variables (z, w) problem $(\widetilde{\mathrm{LO}}_\delta)$ takes on the form

$(\widetilde{\mathrm{LO}}_\delta')$ Find $(z_\delta, w_\delta) \in W_h^{1,2}(0, T; \mathbb{R}^n) \times L_h^2(0, T; \mathbb{R}^m)$ such that

$\qquad \widehat{\mathcal{I}}_h(z_\delta, w_\delta, \delta) = \min \widehat{\mathcal{I}}_h(z, w, \delta)$

$\qquad$ subject to

$$
\begin{aligned}
&\nabla y(t_j) - A(t_j) y(t_j) - B(t_j) v(t_j) = 0, \\
&y(t_0) = 0, \quad \Xi y(t_N) = 0, \\
&\langle \Theta_x^i(t_j), y(t_j) \rangle + \langle \Theta_u^i(t_j), v(t_j) \rangle \begin{cases} = 0 & \text{if } t_j \in \Upsilon_\alpha^i, \\ \leq 0 & \text{if } t_j \in \Omega_\sigma^i \setminus \Upsilon_\alpha^i, \end{cases}
\end{aligned}
\tag{5.22}
$$

where

$$\widehat{\mathcal{I}}_h(z, w, \delta) = \tfrac{1}{2}((z, w), \mathcal{Q}_h(z, w))_{\widehat{Z}} + ((n_\delta^1, n_\delta^2), (z, w))_{\widehat{Z}}, \quad \text{and}$$

$$\begin{bmatrix} n_\delta^1 \\ n_\delta^2 \end{bmatrix} = -\mathcal{Q}_h \mathcal{J}_h^{-1} \mathcal{C}_h^* (\mathcal{C}_h \mathcal{J}_h^{-1} \mathcal{C}_h^*)^{-1} k_\delta + \begin{bmatrix} m_\delta^1 \\ m_\delta^2 \end{bmatrix}, \tag{5.23}$$

$$m_\delta^1 = \begin{bmatrix} \widehat{s}_h^1 - a^1 \\ \widehat{s}_h^3 - a^3 \end{bmatrix}, \qquad m_\delta^2 = \widehat{s}_h^4 - a^4.$$

The well known condition of optimality yields

$$(\mathcal{Q}_h(\widetilde{z}_\delta, \widetilde{w}_\delta) + (n_\delta^1, n_\delta^2), (z, w) - (\widetilde{z}_\delta, \widetilde{w}_\delta))_{\widehat{Z}} \geq 0$$

for all feasible (z, w). Let us choose arbitrary $\delta', \delta'' \in \widetilde{Z}_h$. Since the constraints in (5.22) are independent of δ, $(\widetilde{z}_{\delta''}, \widetilde{w}_{\delta''})$ is feasible for $(\widetilde{\mathrm{LO}}_{\delta'}')$, and vice versa $(\widetilde{z}_{\delta'}, \widetilde{w}_{\delta'})$ is feasible for $(\widetilde{\mathrm{LO}}_{\delta''}')$. Hence

$$(\mathcal{Q}_h(\widetilde{z}_{\delta'}, \widetilde{w}_{\delta'}) + (n_{\delta'}^1, n_{\delta'}^2), (\widetilde{z}_{\delta''}, \widetilde{w}_{\delta''}) - (\widetilde{z}_{\delta'}, \widetilde{w}_{\delta'}))_{\widehat{Z}} \geq 0,$$
$$(\mathcal{Q}_h(\widetilde{z}_{\delta''}, \widetilde{w}_{\delta''}) + (n_{\delta''}^1, n_{\delta''}^2), (\widetilde{z}_{\delta'}, \widetilde{w}_{\delta'}) - (\widetilde{z}_{\delta''}, \widetilde{w}_{\delta''}))_{\widehat{Z}} \geq 0.$$

Adding these inequalities and using (5.17) we obtain

$$((n_{\delta'}^1, n_{\delta'}^2) - (n_{\delta''}^1, n_{\delta''}^2), (\widetilde{z}_{\delta''}, \widetilde{w}_{\delta''}) - (\widetilde{z}_{\delta'}, \widetilde{w}_{\delta'}))_{\widehat{Z}} \geq$$
$$\geq (\mathcal{Q}_h((\widetilde{z}_{\delta''}, \widetilde{w}_{\delta''}) - (\widetilde{z}_{\delta'}, \widetilde{w}_{\delta'})), (\widetilde{z}_{\delta''}, \widetilde{w}_{\delta''}) - (\widetilde{z}_{\delta'}, \widetilde{w}_{\delta'}))_{\widehat{Z}} \geq$$
$$\geq \tfrac{\gamma}{2}\|(\widetilde{z}_{\delta''}, \widetilde{w}_{\delta''}) - (\widetilde{z}_{\delta'}, \widetilde{w}_{\delta'})\|_{\widehat{Z}}^2.$$

Hence

$$\|(n_{\delta'}^1, n_{\delta'}^2) - (n_{\delta''}^1, n_{\delta''}^2)\|_{\widehat{Z}*} \geq \frac{\bar{\gamma}}{2}\|(\widetilde{z}_{\delta''}, \widetilde{w}_{\delta''}) - (\widetilde{z}_{\delta'}, \widetilde{w}_{\delta'})\|_{\widehat{Z}}. \tag{5.24}$$

Note that $(n_{\delta'}^1, n_{\delta'}^2) - (n_{\delta''}^1, n_{\delta''}^2)$ is a linear function of $(\delta' - \delta'')$. Moreover, in view of (5.20) there exist constants $c_1 > 0$ and $c_2 > 0$, independent of h such that

$$\|\mathcal{J}_h^{-1}\mathcal{C}_h^*(\mathcal{C}_h \mathcal{J}_h^{-1}\mathcal{C}_h^*)^{-1}\|_{L(\widehat{Y};\widehat{Z})} \leq c_1 \quad \text{and}$$
$$\|\mathcal{Q}_h \mathcal{J}_h^{-1}\mathcal{C}_h^*(\mathcal{C}_h \mathcal{J}_h^{-1}\mathcal{C}_h^*)^{-1}\|_{L(\widehat{Y};\widehat{Z}*)} \leq c_2. \tag{5.25}$$

Hence, by (5.21) and (5.23) we find that

$$\|(n_{\delta'}^1, n_{\delta'}^2) - (n_{\delta''}^1, n_{\delta''}^2)\|_{\widehat{Z}*} \leq c\|\delta' - \delta''\|_{\widetilde{X}}. \tag{5.26}$$

Substituting this estimate to (5.24) and using (5.21) we obtain the first inequality in (5.19).

To obtain the second inequality in (5.19), let us note that by the stationarity of the Lagrangian of $(\widetilde{\mathrm{LO}}_\delta')$

$$\mathcal{Q}_h(\widetilde{z}_\delta, \widetilde{w}_\delta) + (n_\delta^1, n_\delta^2) + \mathcal{C}_h^*(\widetilde{q}_\delta, \widetilde{\nu}_\delta, \widetilde{\tau}_\delta, \widetilde{\kappa}_\delta) = 0.$$

Hence

$$\mathcal{Q}_h((\widetilde{z}_{\delta'}, \widetilde{w}_{\delta'}) - (\widetilde{z}_{\delta''}, \widetilde{w}_{\delta''})) + ((n_{\delta'}^1, n_{\delta'}^2) - (n_{\delta''}^1, n_{\delta''}^2)) +$$
$$+\mathcal{C}_h^*((\widetilde{q}_{\delta'}, \widetilde{\nu}_{\delta'}, \widetilde{\tau}_{\delta'}, \widetilde{\kappa}_{\delta'}) - (\widetilde{q}_{\delta''}, \widetilde{\nu}_{\delta''}, \widetilde{\tau}_{\delta''}, \widetilde{\kappa}_{\delta''})) = 0.$$

That, in view of (5.20),(5.26) and of the first estimate in (5.19) yields the second one.   □

Note that (5.19) gives us Lipschitz continuity of the (KKT) points for $(\widetilde{\text{LO}}'_\delta)$ only in the *weaker norm* of the Hilbert space $\widehat{X}$. This result is not strong enough to assure (5.13) and to allow to use (5.11). We will take advantage of the structure of problem $(\widetilde{\text{LO}}'_\delta)$ to get Lipschitz continuity in the stronger norm of X.

LEMMA 5.5 *If assumptions* (II.1)-(II.6) *hold then there exist* $\ell > 0$ *and* $\bar{h} > 0$ *such that for all* $h < \bar{h}$

$$\|\widetilde{y}_{\delta'} - \widetilde{y}_{\delta''}\|_{1,\infty}, \|\widetilde{v}_{\delta'} - \widetilde{v}_{\delta''}\|_\infty \le \ell\|\delta' - \delta''\|_{\widetilde{X}},$$
$$\|\widetilde{q}_{\delta'} - \widetilde{q}_{\delta''}\|_{1,\infty}, |\widetilde{\nu}_{\delta'} - \widetilde{\nu}_{\delta''}|, |\widetilde{\tau}_{\delta'} - \widetilde{\tau}_{\delta''}|, \|\widetilde{\kappa}_{\delta'} - \widetilde{\kappa}_{\delta''}\|_\infty \le \ell\|\delta' - \delta''\|_{\widetilde{X}}. \tag{5.27}$$

◇

Proof Note that by the maximum principle for $(\widetilde{\text{LO}}'_\delta)$ (see [12] p.280) $\widetilde{v}_\delta(t_j)$ and $\widetilde{\kappa}_\delta(t_j)$ can be treated as the solution and the Lagrange multiplier of the following quadratic mathematical program in $I\!\!R^m$, depending on the vector parameter $\zeta(t_j) = (\delta(t_j), \widetilde{y}_\delta(t_j), \widetilde{q}_\delta(t_j))$.

$$(\text{MP}_{\zeta(t_j)}) \quad \min_{v \in I\!\!R^m} \{\tfrac{1}{2}v^* D^2_{uu}\mathcal{H}(t_j)v +$$
$$+ [\widetilde{y}_\delta(t_j)^* D^2_{xu}\mathcal{H}(t_j) + \widetilde{q}_\delta(t_j)^* B(t_j) + \widehat{s}^3_h(t_j)^* - a^3(t_j)^*]v\}$$

subject to

$$\langle \Theta^i_x(t_j), y(t_j)\rangle + \langle \Theta^i_u(t_j), v(t_j)\rangle +$$
$$+ (\widehat{s}^8_h(t_j))^i - (b^4(t_j))^i \begin{cases} = 0 & \text{if } t_j \in \Upsilon^i_\alpha, \\ \le 0 & \text{if } t_j \in \Omega^i_\sigma \setminus \Upsilon^i_\alpha. \end{cases}$$

Recall that $\mathcal{H}(t_j) := \mathcal{H}(x_h(t_j), u_h(t_j), p_h(t_{j+1}), \lambda_h(t_j))$. Hence, by virtue of (II.5) and of Lipschitz continuity of $p_h(\cdot)$ there exists $\bar{h} \le \min\{h_1, h_2\}$ such that, for all $h < \bar{h}$ we have

$$v^* D^2_{uu}\mathcal{H}(t_j)v \ge \tfrac{\gamma}{2}|v|^2$$

for all $v \in I\!\!R^m$ satisfying $\widehat{\Theta}^\alpha_u(t_j)v = 0$ and all $j = 0, 1, ..., N - 1$. $\tag{5.28}$

In view of (I.3) and (5.28), by the well known stability result for solutions to parametric mathematical programs (see [30]), we find that there exists a constant $c > 0$, dependent only on β and γ such that

$$|\widetilde{v}_{\delta'}(t_j) - \widetilde{v}_{\delta''}(t_j)|, |\widetilde{\kappa}_{\delta'}(t_j) - \widetilde{\kappa}_{\delta''}(t_j)| \le$$
$$\le c[|\delta'(t_j) - \delta''(t_j)| + |\widetilde{y}_{\delta'}(t_j) - \widetilde{y}_{\delta''}(t_j)| + \widetilde{q}_{\delta'}(t_j) - \widetilde{q}_{\delta''}(t_j)|].$$

On the other hand, it follows from (5.19) that

$$|\widetilde{y}_{\delta'}(t_j) - \widetilde{y}_{\delta''}(t_j)|, |\widetilde{q}_{\delta'}(t_j) - \widetilde{q}_{\delta''}(t_j)| \le c\|\delta' - \delta''\|_{\widetilde{X}}.$$

Hence

$$\|\widetilde{v}_{\delta'}(t_j) - \widetilde{v}_{\delta''}(t_j)\|_\infty, \|\widetilde{\kappa}_{\delta'}(t_j) - \widetilde{\kappa}_{\delta''}(t_j)\|_\infty \le \widetilde{\ell}\|\delta' - \delta''\|_{\widetilde{X}}, \tag{5.29}$$

where $\widetilde{\ell}$ is independent of h.

Using (5.29) and the state and adjoint equations, we finally arrive at (5.27).

□

PROPOSITION **5.6** *If assumptions* (II.1)-(II.6) *hold then there exist a constant* $\bar{h} > 0$ *and constants* $\ell > 0, \chi > 0, \varsigma > 0$ *such that for each* $h < \bar{h}$ *and each* $\delta \in \Delta_{h,\chi} :=$ $\{\delta \in \widetilde{X}_h \mid \|\delta\|_{X^*} \leq \chi\}$ *there exist a* (KKT) *point* $w_\delta = (y_\delta, v_\delta, q_\delta, \nu_\delta, \tau_\delta, \kappa_\delta)$, *unique in* $\mathcal{X}_{h,\varsigma} := \{w \in X_h \mid \|w - w_h\|_X \leq \varsigma\}$ *and*

$$\|y_{\delta'} - y_{\delta''}\|_{1,\infty}, \|v_{\delta'} - v_{\delta''}\|_\infty \leq \ell\|\delta' - \delta''\|_{\widetilde{X}},$$

$$\|q_{\delta'} - q_{\delta''}\|_{1,\infty}, |\nu_{\delta'} - \nu_{\delta''}|, |\tau_{\delta'} - \tau_{\delta''}|, \|\kappa_{\delta'} - \kappa_{\delta''}\|_\infty \leq \ell\|\delta' - \delta''\|_{\widetilde{X}}. \tag{5.30}$$

for all $\delta', \delta'' \in \Delta_{h,\chi}$. ◇

Proof Let $\bar{h}$ be as in Lemma 5.5. Choose

$$\varsigma' = \frac{\sigma}{2}[\max_i(\|\Theta_x^i\|_\infty + \|\Theta_u^i\|_\infty)],$$

and put $\varsigma = \min\{\varsigma', \alpha\}$, where σ and α are given in (II.3) and (II.5), respectively.
By (5.27) we can find $\chi' > 0$, independent of h such that

$$\|\widetilde{y}_\delta - \widetilde{y}_0\|_\infty, \|\widetilde{v}_\delta - \widetilde{v}_0\|_\infty, \|\widetilde{\kappa}_\delta - \widetilde{\kappa}_0\|_\infty \leq \varsigma \quad \text{for all } \delta \in \Delta_{h,\chi'},$$

and put $\chi = \min\{\frac{\sigma}{2}, \chi'\}$.
With such a choice of χ conditions (5.13) are satisfied for all $\delta \in \Delta_{h,\chi}$. Hence (5.11) holds and (5.30) follows from (5.27).
The uniqueness of w_δ in $\mathcal{X}_{h,\varsigma}$ follows from the uniqueness of $\widetilde{w}_\delta$. □

By Proposition 5.6 the abstract condition (I.3) is satisfied, so all assumptions of Theorem 2.2 hold. In view of (5.3) and (5.5), we obtain from that theorem our main convergence result:

THEOREM **5.7** *If assumptions* (II.1)-(II.6) *hold then there exist* $\eta > 0$ *such that for each* $h < \eta$ *there exists a locally unique* (KKT) *point* $(x_h, u_h, p_h, \mu_h, \pi_h, \lambda_h)$ *of* (O_h) *and*

$$\|x_h - x_0\|_{1,\infty}, \|u_h - u_0\|_\infty \leq \ell'|h|,$$

$$\|p_h - p_0\|_{1,\infty}, |\mu_h - \mu_0|, |\pi_h - \pi_0|, \|\lambda_h - \lambda_0\|_\infty \leq \ell'|h|, \tag{5.31}$$

where $\ell' > 0$ *is independent of* h. ◇

Using (5.17), by the argument virtually the same as in the proof of Lemma 5.3 we obtain

LEMMA **5.8** *If assumptions* (II.1)-(II.6) *hold then there exists* $h_3 > 0$ *such that for all* $h < h_3$

$$((y,v), \begin{pmatrix} D_{xx}^2\mathcal{L}_h(w_h) & D_{xu}^2\mathcal{L}_h(w_h) \\ D_{ux}^2\mathcal{L}_h(w_h) & D_{uu}^2\mathcal{L}_h(w_h) \end{pmatrix} (y,v))_{\widehat{Z}} \geq \frac{\bar{\gamma}}{4}(\|y\|_{1,2}^2 + \|v\|_2^2) \tag{5.32}$$

for all $(y,v) \in W_h^{1,2}(0,T;\mathbb{R}^n) \times L_h^2(0,T;\mathbb{R}^m)$ *such that*

$$\nabla y(t) - D_x f(x_h(t_j), u_h(t_j))y(t_j) - D_u f(x_h(t_j), u_h(t_j))v(t_j) = 0,$$

$$y(t_0) = 0, \qquad \Xi y(t_N) = 0,$$

$$D_x\theta^i(x_h(t_j), u_h(t_j))y(t_j) + D_u\theta^i(x_h(t_j), u_h(t_j))v(t_j) = 0 \quad \text{for } i \in I_+^\alpha(t_j).$$

◇

Since (5.32) constitutes a sufficient optimality condition for (O_h) we get

COROLLARY **5.9** *For* $\eta > 0$ *sufficiently small* (x_h, u_h) *in* (5.31) *is a solution to* (O_h) *and* $(p_h, \mu_h, \pi_h, \lambda_h)$ *the associated Lagrange multipliers.* ◇

6 Convergence Analysis for the Controlled Rayleigh Equation

The purpose of this section is to illustrate the convergence result in Theorem 5.7 by a numerical example for which Assumptions (II.1)-(II.6) are carefully checked. We consider a modification of the Rayleigh problem that has been studied in [14, 15, 34] via differential dynamic programming. The modified Rayleigh problem has also been treated in [22] from the point of view of sensitivity analysis. We have mentioned that the assumptions for sensitivity analysis (see [21, 23, 24]) comprise Assumptions (II.1)-(II.6) in Theorem 5.7. Hence, we can rely on parts of the analysis in [22].

Consider the electric circuit (tunneldiode oscillator) shown in Figure 1 where L denotes inductivity, C capacity, R resistance, I electric current, V, and where D is the diode.

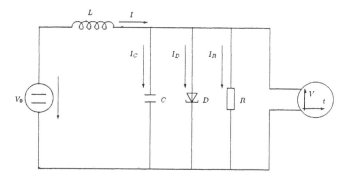

Figure 1 : Tunneldiode oscillator

The state variable $x(t)$ represents the electric current I at time t. The voltage $v_0(t)$ at the generator is treated as a control function. After a suitable transformation of $v_0(t)$ we arrive at the following specific Rayleigh equation with a scalar control $u(t)$ (see [14, 15, 34]):

$$\ddot{x}(t) = -x(t) + \dot{x}(t)\,(\,1.4 - 0.14\,\dot{x}(t)^2\,) + 4\,u(t). \tag{6.1}$$

It can be verified numerically that the Rayleigh equation (6.1) with zero control $u(t) \equiv 0$ has a *limit cycle* in the $(x, \dot{x})$–plane. The aim is to control oscillations of the electric current in the tunneldiode oscillator shown in
Figure 1. Introducing the state variables $x_1 = x$ and $x_2 = \dot{x}$, the controlled Rayleigh equation can be stated as the following control problem: minimize the functional

$$J(u) = \int_0^T (\,u(t)^2 + x_1(t)^2\,)\,dt \tag{6.2}$$

subject to

$$\dot{x}_1(t) = x_2(t)\,,\;\; \dot{x}_2(t) = -x_1(t) + x_2(t)\,(\,1.4 - 0.14\,x_2(t)^2\,) + 4\,u(t), \tag{6.3}$$

$$x_1(0) = x_2(0) = -5, \tag{6.4}$$

$$|u(t)| \leq 1 \quad \text{for } t \in [0, T]. \tag{6.5}$$

In the following numerical analysis, we consider three cases:

Case (a): Final time $T = 4.5$ and final conditions $x_1(T) = x_2(T) = 0$;

Case (b): Final time $T = 2.5$;

Case (c): Final time $T = 2.5$ and *mixed* constraint $u(t) + x_1(t)/6 \leq 0$ instead of the pure control constraint (6.5).

Cases (b) has been studied in [14, 15, 34] although junction points with the boundary have not been computed exactly by these authors. We shall see that Case (a) with a larger final time leads to a more interesting control structure.

The optimal solution of the Rayleigh problem is computed in *two steps*. The *first step* consists in applying Euler's method (4.3) with a sufficiently small stepsize h. This step yields a fairly accurate idea of the optimal control structure (see Figures 2, 3, 4) that will be refined in the *second step*.

First step: There are two ways of implementing Euler's method (4.3). Both methods are equivalent from a theoretical point of view but not from a numerical one. The *first* method treats both the control and state variables as optimization variables. Here, the Euler discretization of the ODE is considered as an equality constraint in the form $\nabla x(t_j) - f(x(t_j), u(t_j)) = 0$. This approach leads to a large nonlinear programming problem with a sparse structure of the Jacobian and Hessian matrix. The *second* method treats only the control variables $u(t_j)$ as decision variables while the state is computed recursively as a function of the control using Euler's approximation. This approach has been studied extensively in [10], [33]. An efficient and robust implementation of this approach has been developed in [3] using the NLP-code E04UCF of the NAG-library. The numerical results at the end of this section were obtained by the implementation of the second method in [3].

Second step: The approximate solution obtained via Euler's method is used as initial estimate for solving the complete boundary value problem (BVP) associated with Pontryagin's minimum principle. One reliable method for solving (BVP) is the shooting method. The following computations are performed with the shooting code BNDSCO in [28] yielding state and costate variables that are correct with ten decimals. This numerical solution allows us to verify the Assumptions (II.1)–(II.6) with high precision.

Case (a): Final time $T = 4.5$ and final conditions $x_1(T) = x_2(T) = 0$.

The augmented Hamiltonian (3.9) for problem (6.2)–(6.5) becomes

$$\begin{aligned}
\mathcal{H}(x, u, p, \lambda, h) &= u^2 + x_1^2 + p_1 x_2 + p_2 \left(-x_1 + x_2 (1.4 - 0.14 x_2^2) + 4u\right) \\
&+ \lambda_1(u - 1) + \lambda_2(-u - 1),
\end{aligned} \tag{6.6}$$

where $p = (p_1, p_2)$ denote the adjoint variables associated with x_1, x_2 and $\lambda = (\lambda_1, \lambda_2)$ is the multiplier corresponding to the two inequality constraints in (6.5). The multipliers satisfy $\lambda_i(t) \geq 0$ for $0 \leq t \leq T$ and $\lambda_1(t) = 0$ when $u(t) < 1$ resp. $\lambda_2(t) = 0$ when $-1 < u(t)$.

On time intervals with inactive inequality constraints (6.5), the optimal control is determined via the minimum condition

$$\mathcal{H}_u = 2u + 4p_2 = 0, \quad \text{i.e.} \quad u(t) = -2p_2(t). \tag{6.7}$$

The adjoint equations (3.10) are

$$\dot{p}_1 = p_2 - 2x_1, \quad \dot{p}_2 = 0.42\, p_2\, x_2^2 - 1.4\, p_2 - p_1. \tag{6.8}$$

The approximate solution computed via Euler's method suggests the following structure of the optimal control; see Figure 2 :

$$u(t) = \begin{cases} 1 & \text{for} \quad 0 \le t \le \tau_1, \\ -2p_2(t) & \text{for} \quad \tau_1 \le t \le \tau_2, \\ -1 & \text{for} \quad \tau_2 \le t \le \tau_3, \\ -2p_2(t) & \text{for} \quad \tau_3 \le t \le 4.5. \end{cases} \tag{6.9}$$

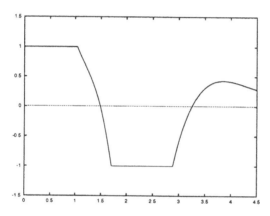

Figure 2 : Optimal control for $T = 4.5$, $|u(t)| \le 1$, $x_1(T) = x_2(T) = 0$

Here, the *junction points* τ_i, $i = 1, 2, 3$, with the boundary satisfy $0 < \tau_1 < \tau_2 < \tau_3 < T = 4.5$. The control with $|u(t)| < 1$ is substituted from (6.7). The Hamiltonian (6.6) is continuous along the optimal solution and is regular in the sense that it admits a unique minimum with respect to u. From these properties it follows that the control is *continuous* at the junction points τ_i, $i = 1, 2, 3$, which yields the following three junction conditions in view of (6.7) :

$$-2p_2(\tau_1) = 1, \quad -2p_2(\tau_2) = -1, \quad -2p_2(\tau_3) = -1. \tag{6.10}$$

In summary, we have to solve the boundary value problem (BVP) consisting of 1. the ODE's (6.3) and (6.8) with control law (6.9); 2. the boundary conditions (6.4) and $x_1(T) = x_2(T) = 0$; 3. three junction conditions (6.10) with three unknowns τ_1, τ_2, τ_3.

The shooting code BNDSCO in [28] yields the following initial values of adjoint variables, the junction points and the optimal functional value:

$$
\begin{aligned}
p_1(0) &= -12.70997710, & p_2(0) &= -4.596316827, \\
p_1(4.5) &= 0.05196024178, & p_2(4.5) &= -0.1399040781, \\
\tau_1 &= 1.041982827, & \tau_2 &= 1.706631091, \\
\tau_3 &= 2.884944106, & J^* &= 43.92548861.
\end{aligned}
\tag{6.11}
$$

It remains to check the sign conditions $\lambda_i(t) \geq 0$ for the multipliers. Explicit expressions for the multipliers are obtained from the minimum condition $\mathcal{H}_u(t) = 0$. One can verify numerically that the following *strict complementarity conditions* hold

$$\lambda_1(t) = -2 - 4p_2(t) > 0 \quad \text{for } 0 \leq t < \tau_1,$$
$$\lambda_2(t) = -2 + 4p_2(t) > 0 \quad \text{for } \tau_2 < t < \tau_3.$$

The next step is to verify Assumptions (II.1)–(II.6) for this optimal solution candidate. Assumptions (II.1)–(II.3) and (II.5) are trivially satisfied. Complete controllability in Assumption (II.4) can not be checked directly. However, the proof of Theorem 2.1 in [21] reveals that controllability is a byproduct of the regularity of the shooting matrix associated with the above (BVP). The code BNDSCO in [28] provides the information that the shooting matrix is regular.

Finally, we show the existence of a bounded solution of the Riccati equation defined in Assumption (II.6). Since the Rayleigh problem has dimension $n = 2$, we set up the symmetric 2×2 matrix $Q(t)$ in the form

$$Q(t) = \begin{pmatrix} q_1(t) & q_2(t) \\ q_2(t) & q_4(t) \end{pmatrix}.$$

In time intervals $[t_1, t_2]$ and $[t_3, T]$ where the inequality constraints (6.5) are inactive, the Riccati equation in Assumption (II.6) is given by

$$\dot{Q} = -Qf_x - f_x^* Q - \mathcal{H}_{xx} + (\mathcal{H}_{xu} + Qf_u)(\mathcal{H}_{uu})^{-1}(\mathcal{H}_{ux} + f_u^* Q). \tag{6.12}$$

This leads to the equations

$$\begin{aligned}
\dot{q}_1 &= 2q_2 + 8q_2^2 - 2, \\
\dot{q}_2 &= -q_1 + (0.42\,x_2^2 - 1.4)\,q_2 + q_4 + 8\,q_2\,q_4, \\
\dot{q}_4 &= -2\,q_2 + (0.84\,x_2^2 - 2.8)\,q_4 + 8\,q_4^2 + 0.84\,p_2\,x_2.
\end{aligned} \tag{6.13}$$

For active time intervals, it is difficult to write down explicitly the Riccati equation in Assumption (II.6). However, for a *scalar* control with $m = 1$, the Riccati equation degenerates into a *linear* ODE that has been evaluated in [24] as:

$$\begin{aligned}
\dot{Q} &= -Q(f_x - f_u\theta_u^{-1}\theta_x) - (f_x - f_u\theta_u^{-1}\theta_x)^* Q + W, \tag{6.14} \\
W &= -\mathcal{H}_{xx} + \mathcal{H}_{xu}\theta_u^{-1}\theta_x + \theta_x^*(\theta_u^{-1})(\mathcal{H}_{ux} - \mathcal{H}_{uu}\theta_u^{-1}\theta_x).
\end{aligned}$$

Note that for state-independent constraints the Riccati equation (6.14) reduces to the *linear* equation

$$\dot{Q} = -Qf_x - f_x^* Q - \mathcal{H}_{xx}. \tag{6.15}$$

Hence, on active time intervals $[0, t_1]$ and $[t_2, t_3]$ we have the following linear ODE's:

$$\begin{aligned}
\dot{q}_1 &= 2q_2 - 2, \\
\dot{q}_2 &= -q_1 + (0.42\,x_2^2 - 1.4)\,q_2 + q_4, \\
\dot{q}_4 &= -2q_2 + (0.84\,x_2^2 - 2.8)\,q_4 + 0.84\,p_2 x_2.
\end{aligned} \tag{6.16}$$

Since the final state is fixed, the boundary condition for $Q(T)$ in Assumption (II.6) is automatically satisfied. We choose the special boundary condition $Q(T) = 0$ and wish to determine a bounded solution of the Riccati equations composed by (6.13) in the time intervals $[\tau_1, \tau_2]$ and $[\tau_3, T]$ and by (6.15) in the time intervals $[0, \tau_1]$ and $[\tau_2, \tau_3]$. Numerical integration along the nominal solution (6.11) shows indeed that there exists a bounded solution on $[0, T]$ with

$$q_1(0) = 2.3985819, \quad q_2(0) = 0.89023233, \quad q_4(0) = -1.2662091,$$

such that $|q_i(t)| \leq 2.5$ for $t \in [0, T]$. It is interesting to note that the full Riccati equation (6.12) with $Q(T) = 0$ does *not* possess a bounded solution in the *whole* interval $[0, T]$. Hence, this solution can not be used to check the *strong* second order sufficient conditions developed in [25], Theorem 5.1.

Hereby, we have checked all Assumptions (II.1)–(II.6) for Theorem 5.7. In particular, we have verified that the control shown in Figure 2 is a local minimum for the Rayleigh problem (6.2)–(6.5) in Case (a).

Now, let us determine the *numerical* order α of convergence, i.e. an estimate of the form

$$\|u_h - u_0\|_\infty \leq \ell' \, |h|^\alpha. \tag{6.17}$$

We compare the *theoretical* value $\alpha = 1$ with the *numerical* value obtained for different stepsizes. Using stepsizes h and $h/2$, the order α and the constant ℓ' are computed approximately by the expressions

$$\alpha_h = \log_2\left(\frac{\|u_h - u_0\|_\infty}{\|u_{h/2} - u_0\|_\infty}\right), \quad \ell'_h = \frac{\|u_h - u_0\|_\infty}{h^\alpha}. \tag{6.18}$$

The convergence results for the optimal control in Case (a) are given in Table 6.1. The convergence analysis of the two state variables and the adjoint variables displays the same behaviour.

$h = T/2^i$	$\| u_h - u_0 \|_\infty$	α_h in (6.18)	ℓ'_h in (6.18)
$i = 3$	$1.9690 \cdot 10^{\pm 0}$	1.18439	2.12782
$i = 4$	$8.6638 \cdot 10^{-1}$	0.16632	3.08044
$i = 5$	$7.7204 \cdot 10^{-1}$	0.90796	5.49007
$i = 6$	$4.1145 \cdot 10^{-1}$	1.10307	4.93740
$i = 7$	$1.9154 \cdot 10^{-1}$	0.93853	5.44824
$i = 8$	$9.9939 \cdot 10^{-2}$	1.01234	5.68542
$i = 9$	$4.9544 \cdot 10^{-2}$	–	5.63700

Table 6.1: Euler discretisation : $T = 4.5$, $|u(t)| \leq 1$ and $x_1(T) = x_2(T) = 0$

The numerical values for α in Table 6.1 reproduce quite accurately the predicted value $\alpha = 1$. A better coincidence is obtained by considering

Case (b): $T = 2.5$ and $|u(t)| \leq 1$.

The optimal control has the structure

$$u(t) = \begin{cases} 1 & \text{for} & 0 \le t \le \tau_1, \\ -2p_2(t) & \text{for} & \tau_1 \le t \le 2.5. \end{cases} \tag{6.19}$$

with only one boundary arc in $[0, \tau_1]$. Here we have to solve the (BVP) with 1. ODE's (6.3), (6.8) with control law (6.19); 2. boundary conditions (6.4) and $p_1(T) = p_2(T) = 0$, and 3. junction condition $-2p_1(\tau_1) = 1$. The code BNDSCO in [28] yields the solution

$$\begin{aligned} p_1(0) &= -12.57716212, & p_2(0) &= -4.504636178, \\ \tau_1 &= 1.021086841, & J^* &= 42.667114891. \end{aligned}$$

The optimal control is shown in Figure 3. Assumptions (II.1)-(II.6) can be checked easily for this solution. Table 6.2 presents the convergence analysis.

$h = T/2^i$	$\| u_h - u_0 \|_\infty$	α_h in (6.18)	l'_h in (6.18)
$i = 3$	$3.8614 \cdot 10^{-1}$	0.74530	1.23564
$i = 4$	$2.3035 \cdot 10^{-1}$	0.96382	1.47424
$i = 5$	$1.1810 \cdot 10^{-1}$	1.19024	1.51168
$i = 6$	$5.1755 \cdot 10^{-2}$	0.85629	1.32492
$i = 7$	$2.8588 \cdot 10^{-2}$	0.92143	1.46372
$i = 8$	$1.5094 \cdot 10^{-2}$	1.00231	1.54563
$i = 9$	$7.5349 \cdot 10^{-3}$	–	1.54316

Table 6.2 : Euler discretisation: $T = 2.5$, $|u(t)| \le 1$

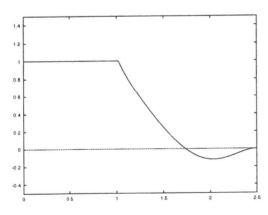

Figure 3 : Optimal control for $T = 2.5$, $|u(t)| \le 1$

Case (c): We consider the final time $T = 2.5$ and replace the pure control constraint $|u(t)| \le 1$ in (6.5) by the *mixed* constraint

$$u(t) + x_1(t)/6 \le 0. \tag{6.20}$$

Here, the adjoint equations (6.8) are modified as follows

$$\dot{p}_1 = p_2 - 2x_1, \quad \dot{p}_2 = 0.42 \, p_2 \, x_2^2 - 1.4 \, p_2 - p_1 - \lambda/6. \tag{6.21}$$

where λ is the multiplier associated with the constraint (6.20). On the boundary of (6.20) we have the control law $u = -x_1/6$. The multiplier λ is determined from $\mathcal{H}_u = 2u + 4p_2 + \lambda = 0$ which gives

$$\lambda = -2u - 4p_2 = x_1/3 - 4p_2. \tag{6.22}$$

Numerical analysis shows that the inequality constraint (6.20) is active in intervals $[0, \tau_1]$ and $[\tau_2, T]$. The junction points τ_1 and τ_2 are determined by two junction conditions

$$\lambda(\tau_i) = x_1(\tau_i)/3 - 4p_2(\tau_i) = 0, \quad i = 1, 2.$$

The code BNDSCO in [28] yields the following solution:

$$
\begin{aligned}
p_1(0) &= -10.62540221, & p_2(0) &= -4.132910052, \\
\tau_1 &= 1.239072135, & \tau_2 &= 2.293615900, \\
J^* &= 43.13227211.
\end{aligned}
$$

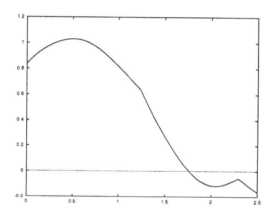

Figure 4 : Optimal control for $T = 2.5$, $u(t) + x_1(t)/6 \le 0$

The Assumptions (II.1)-(II.6) can be checked in the same way as for the other examples. Table 6.3 shows the convergence analysis.

$h = T/2^i$	$\| u_h - u_0 \|_\infty$	α_h in (6.18)	l'_h in (6.18)
$i = 3$	$3.6729 \cdot 10^{-1}$	0.94561	1.17533
$i = 4$	$1.9070 \cdot 10^{-1}$	1.52900	1.22048
$i = 5$	$6.6081 \cdot 10^{-2}$	1.14999	0.84584
$i = 6$	$2.9778 \cdot 10^{-2}$	1.03670	0.76232
$i = 7$	$1.4515 \cdot 10^{-2}$	0.96427	0.74317
$i = 8$	$7.4395 \cdot 10^{-3}$	1.00274	0.76180
$i = 9$	$3.7127 \cdot 10^{-3}$	–	0.76036

Table 6.3 : Euler discretisation: $T = 2.5$, $u(t) + \frac{x_1(t)}{6} \le 0$

A Proofs of Auxiliary Lemmas

Proof of Lemma 3.2

We have to show that, for any right-hand side the equation

$$
\begin{aligned}
&\dot{y} - Ay - Bv = p, \\
&y(0) = q, \\
&\Xi y(T) = r, \\
&\Theta_x^\sigma y + \Theta_u^\sigma v = s,
\end{aligned}
\tag{A.1}
$$

has a solution, which is a bounded function of (p, q, r, s).

Let us denote $M(t) = \Theta_u^\sigma(t)^*(\Theta_u^\sigma(t)\Theta_u^\sigma(t)^*)^{-1}\Theta_u^\sigma(t)$. From the last equation in (A.1) we get

$$
v(t) = \Theta_u^\sigma(t)^*(\Theta_u^\sigma(t)\Theta_u^\sigma(t)^*)^{-1}[s(t) - \Theta_x^\sigma(t)y(t)] + (J - M(t))\vartheta(t),
\tag{A.2}
$$

where $\vartheta(t) \in \mathbb{R}^m$ is arbitrary. Substituting (A.2) into the first equation in (A.1) we obtain

$$
\dot{y} - \widetilde{A}y - \widetilde{B}\vartheta = p,
\tag{A.3}
$$

$$
y(0) = q,
$$

$$
\Xi y(T) = r.
\tag{A.4}
$$

Let $\Phi(t)$ be the fundamental solution of the matrix equation

$$
\dot{\Phi}(t) - \widetilde{A}(t)\Phi(t) = 0, \quad \Phi(0) = J,
\tag{A.5}
$$

then the solution to (A.3) is given by

$$
y(t) = \Phi(t)q + \int_0^t \Phi(t)\Phi(\tau)^{-1}[\widetilde{B}(\tau)\vartheta(\tau) + p(\tau)]d\tau.
\tag{A.6}
$$

Hence (A.4) is satisfied if

$$
\int_0^T P(t)\vartheta(t)dt = \widetilde{r},
\tag{A.7}
$$

where $P(t) = \Xi\Phi(T)\Phi(t)^{-1}\widetilde{B}(t)$ and $\widetilde{r} = r - \Xi[\Phi(T)q - \int_0^T \Phi(T)\Phi(t)^{-1}p(t)]dt$.

By controllability assumption (II.4), the matrix $E := \int_0^T P(t)P(t)^*dt$ is non-singular i.e., there exists $\rho < \infty$ such that

$$
\|E^{-1}\| = \rho.
\tag{A.8}
$$

We can choose the solution of (A.7) in the form

$$
\vartheta(t) = P(t)^*E^{-1}\widetilde{r}.
\tag{A.9}
$$

Substituting (A.9) to (A.6) and (A.2) and using (A.7) we obtain

$$
\begin{bmatrix} y \\ v \end{bmatrix} = \begin{bmatrix} \mathcal{E}^{11} & \mathcal{E}^{12} & \mathcal{E}^{13} & 0 \\ \mathcal{E}^{21} & \mathcal{E}^{22} & \mathcal{E}^{23} & \mathcal{E}^{24} \end{bmatrix} \begin{bmatrix} p \\ q \\ r \\ s \end{bmatrix},
\tag{A.10}
$$

where

$$
\begin{aligned}
(\mathcal{E}^{11}p)(t) &= \int_0^t \Phi(t)\Phi(\tau)^{-1}\{p(\tau) - \widetilde{B}(\tau)P(\tau)^* E^{-1} \times \\
&\quad \times \int_0^T \Phi(T)\Phi(\alpha)^{-1}p(\alpha)d\alpha\}d\tau, \\
(\mathcal{E}^{12}q)(t) &= [\Phi(t) - \int_0^t \Phi(t)\Phi(\tau)^{-1}\widetilde{B}(\tau)P(\tau)^* E^{-1}\Xi\Phi(T)d\tau]q, \\
(\mathcal{E}^{13}r)(t) &= \int_0^t \Phi(t)\Phi(\tau)^{-1}\widetilde{B}(\tau)P(\tau)^* E^{-1}d\tau r, \\
(\mathcal{E}^{21}p)(t) &= -\Theta_u^\sigma(t)^*(\Theta_u^\sigma(t)\Theta_u^\sigma(t)^*)^{-1}\Theta_x^\sigma(t)(\mathcal{E}^{11}p)(t) - \\
&\quad -(J - M(t))P(t)^* E^{-1}\int_0^T \Phi(T)\Phi(t)^{-1}p(t)dt, \\
(\mathcal{E}^{22}q)(t) &= -\Theta_u^\sigma(t)^*(\Theta_u^\sigma(t)\Theta_u^\sigma(t)^*)^{-1}\Theta_x^\sigma(t)(\mathcal{E}^{12}q)(t) - \\
&\quad -(J - M(t))P(t)^* E^{-1}\Xi\Phi(T)q, \\
(\mathcal{E}^{23}r)(t) &= -\Theta_u^\sigma(t)^*(\Theta_u^\sigma(t)\Theta_u^\sigma(t)^*)^{-1}\Theta_x^\sigma(t)(\mathcal{E}^{13}r)(t) + \\
&\quad +(J - M(t))P(t)^* E^{-1}r, \\
(\mathcal{E}^{24}s)(t) &= \Theta_u^\sigma(t)^*(\Theta_u^\sigma(t)\Theta_u^\sigma(t)^*)^{-1}s(t).
\end{aligned}
\tag{A.11}
$$

Certainly, the linear operator

$$
\mathcal{E} = \begin{bmatrix} \mathcal{E}^{11} & \mathcal{E}^{12} & \mathcal{E}^{13} & 0 \\ \mathcal{E}^{21} & \mathcal{E}^{22} & \mathcal{E}^{23} & \mathcal{E}^{24} \end{bmatrix}
$$

is bounded from Y^σ into Z and from $\widehat{Y}^\sigma$ into $\widehat{Z}$. $\qquad\square$

Proof of Lemma 5.2

As in the proof of Lemma 3.1 we have to show that, for $h < h_1$ and for any $(p, q, r, s) \in Y_h^\sigma$, the equation

$$
\begin{aligned}
&\nabla y(t_j) - A(t_j)y(t_j) - B(t_j)v(t_j) = p(t_j), \\
&y(t_0) = q, \\
&\Xi y(t_N) = r, \\
&\Theta_x^\sigma(t_j)y(t_j) + \Theta_u^\sigma(t_j)v(t_j) = s(t_j),
\end{aligned}
\tag{A.12}
$$

has a solution which is a function of (p, q, r, s) uniformly bounded with respect to h.

In the same way as in (A.2) we choose

$$
v(t_j) = \Theta_u^\sigma(t_j)^*(\Theta_u^\sigma(t_j)\Theta_u^\sigma(t_j)^*)^{-1}[s(t_j) - \Theta_x^\sigma(t_j)y(t_j)] + (J - M(t_j))\vartheta(t_j), \tag{A.13}
$$

and obtain from (A.12)

$$
\begin{aligned}
&\nabla y(t_j) - \widetilde{A}(t_j)y(t_j) - \widetilde{B}(t_j)\vartheta(t_j) = p(t_j), \\
&y(t_0) = q, \\
&\Xi y(t_N) = r.
\end{aligned}
$$

Denote by $\Phi_h(t) \in W_h^{1,\infty}(0, T; \mathbb{R}^{n \times n})$ the fundamental solution of the matrix equation

$$
\nabla \Phi_h(t_j) - \widetilde{A}(t_j)\Phi_h(t_j) = 0, \qquad \Phi(t_0) = J.
$$

By convergence of Euler's approximation to linear ordinary differential equations (see e.g., [32])

$$\lim_{h \to 0} \|\Phi - \Phi_h\|_\infty = 0, \tag{A.14}$$

where Φ is given in (A.5).

As in (A.7), let us define

$$P_h(t_j) = \Xi \Phi_h(t_N)\Phi_h(t_j)^{-1}\widetilde{B}(t_j), \quad E_h := \int_0^T P_h(t)P_h(t)^* dt.$$

By (A.8) and (A.14), for any $\bar{\rho} < \rho$ there exists $\bar{h} > 0$ such that for any $h < \bar{h}$

$$\|E_h^{-1}\| \le \bar{\rho}. \tag{A.15}$$

Using the same procedure as in the proof of Lemma 3.1 we construct a solution (y, v) to (A.12) given by

$$\begin{bmatrix} y \\ v \end{bmatrix} = \begin{bmatrix} \mathcal{E}_h^{11} & \mathcal{E}_h^{12} & \mathcal{E}_h^{13} & 0 \\ \mathcal{E}_h^{21} & \mathcal{E}_h^{22} & \mathcal{E}_h^{23} & \mathcal{E}_h^{24} \end{bmatrix} \begin{bmatrix} p \\ q \\ r \\ s \end{bmatrix},$$

where $\mathcal{E}_h^{ij}$ are given by (A.11) with Φ substituted by Φ_h.

In view of (3.7), (A.14) and (A.15), we find that if $h_1 > 0$ is sufficiently small, then (5.16) holds for any $h < h_1$. □

Proof of Lemma 5.3

Similarly to (3.6) and (5.15), let us denote

$$\widehat{\mathcal{C}}(y, v) = \begin{bmatrix} \dot{y} - Ay - Bv \\ y(0) \\ \Xi y(T) \\ \widehat{\Theta}_x^\alpha y + \widehat{\Theta}_u^\alpha v \end{bmatrix}, \quad \widehat{\mathcal{C}}_h(y, v) = \begin{bmatrix} \nabla y - Ay - Bv \\ y(t_0) \\ \Xi y(t_N) \\ \widehat{\Theta}_x^\alpha y + \widehat{\Theta}_u^\alpha v \end{bmatrix}.$$

We have to show that for all $h < h_2$, where h_2 is sufficiently small

$$((y, v), \mathcal{Q}_h(y, v))_{\widehat{Z}} \ge \frac{\bar{\gamma}}{2}(\|y\|_{1,2}^2 + \|v\|_2^2) \tag{A.16}$$

for all $(v, y) \in \widehat{Z}_h$ such that $\widehat{\mathcal{C}}_h(y, v) = 0$. \tag{A.17}

Note that $\lim_{h \to 0} \|\mathcal{Q} - \mathcal{Q}_h\|_{\widehat{Z} \to \widehat{Z}} = 0$. Hence in view of Lemma 3.3, $h_2 > 0$ exists such that for all $h < h_2$ we have

$$((y, v), \mathcal{Q}_h(y, v))_{\widehat{Z}} \ge \frac{3}{4}\bar{\gamma}(\|y\|_{1,2}^2 + \|v\|_2^2) \tag{A.18}$$

for all $(v, y) \in \widehat{Z}$ such that $\widehat{\mathcal{C}}(y, v) = 0$. \tag{A.19}

Let us note that for any $(y, v) \in \widehat{Z}_h$

$$\widehat{\mathcal{C}}(y, v) = \widehat{\mathcal{C}}_h(y, v) + a_h, \tag{A.20}$$

where

$$a_h(t) = \widehat{\mathcal{C}}(y,v)(t) - \widehat{\mathcal{C}}_h(y,v)(t) = \begin{bmatrix} A(t)y(t) - A(t_j)y(t_j) \\ 0 \\ 0 \\ \Theta_x(t)y(t) - \Theta_x(t_j)y(t_j) \end{bmatrix} \quad \text{for } t \in [t_j, t_{j+1}).$$

Hence

$$\lim_{h \to 0} \frac{\|a_h\|_{\widehat{Y}}}{\|(y,v)\|_{\widehat{Z}}} = 0 \tag{A.21}$$

Let $\mathcal{J}_h$ denote the canonical isomorphism between $\widehat{Z}_h$ and $\widehat{Z}_h^*$. By (5.16) the mapping $(\widehat{\mathcal{C}}_h \mathcal{J}_h^{-1} \widehat{\mathcal{C}}_h^*)^{-1}$ is bounded uniformly with respect to $h < h_1$.

Let $(y,v) \in \widehat{Z}_h$ satisfy (A.17), then by (A.20) the element

$$(z,w) = (y,v) - \mathcal{J}_h^{-1} \widehat{\mathcal{C}}_h^* (\widehat{\mathcal{C}}_h \mathcal{J}_h^{-1} \widehat{\mathcal{C}}_h^*)^{-1} a_h \in \widehat{Z}_h$$

satisfies (A.19), so for (z,w) (A.18) holds. On the other hand, in view of (A.21)

$$\lim_{h \to 0} \frac{\|\mathcal{J}_h^{-1} \widehat{\mathcal{C}}_h^* (\widehat{\mathcal{C}}_h \mathcal{J}_h^{-1} \widehat{\mathcal{C}}_h^*)^{-1} a_h\|_{\widehat{Z}}}{\|(y,v)\|_{\widehat{Z}}} = 0.$$

Hence, for any $\kappa > 0$ we can find $h_2 > 0$ such that

$$\|\mathcal{J}_h^{-1} \widehat{\mathcal{C}}_h^* (\widehat{\mathcal{C}}_h \mathcal{J}_h^{-1} \widehat{\mathcal{C}}_h^*)^{-1} a_h\|_{\widehat{Z}} \leq \kappa \|(y,v)\|_{\widehat{Z}}$$

for all $h < h_2$. Therefore, by Lemma 5.5 in [26] we find that (A.16) holds. $\qquad\square$

References

[1] W.Alt, " Stability of Solutions and the Lagrange-Newton Method for Nonlinear Optimization and Optimal Control Problems" (Habilitationsschrift) Universität Bayreuth, Bayreuth (1990).

[2] V.M.Budak and F.P.Vasil'ev, *Some Numerical Aspects of Optimal Control Problems* Izdatielstvo Moskovskogo Universiteta, Moscow (1975) (in Russian).

[3] Ch. Büskens, "Direkte Optimierungsmethoden zur numerischen Berechnung optimaler Steuerungen", Diploma thesis, Institut für Numerische und instrumentelle Mathematik, Universität Münster, Münster, Germany (1993).

[4] A.L.Dontchev, Error estimates for a discrete approximations to ·constrained control problems, *SIAM J. Numer. Anal.*, 13: 500-514 (1981).

[5] A.L.Dontchev, *Perturbations, approximations and sensitivity analysis of optimal control problems*, Lecture Notes in Control and Information Sciences, Vol. 52, Springer-Verlag, New York (1983).

[6] A.L.Dontchev and W.W.Hager, Lipschitz stability in nonlinear control and optimization, *SIAM J. Contr. Optim.*, 31: 569-603 (1993).

[7] A.L.Dontchev, W.W.Hager, A.B.Poore and B.Yang, Optimality, stability and covergence in nonlinear control, *Appl. Math. Optim.*, 31: 297-326 (1995).

[8] Yu.M.Ermol'ev, V.P.Gulenko and T.I.Tsarenko, *Finite-Difference Method in Optimal Control Problems*, Naukova Dumka, Kiyev (1978) (in Russian).

[9] A.V.Fiacco, Sensitivity analysis for nonlinear programming using penalty methods, *Math.Programming*, 10: 287-311 (1976).

[10] I. I. Grachev and Yu. G. Evtushenko, A library of programs for solving optimal control problems, *U.S.S.R. Comput. Maths. Math. Phys.*, 19: 99-119 (1979).

[11] W.W.Hager, Lipschitz continuity for constrained processes, *SIAM J. Contr. Optim.*, 17: 321-338 (1979).

[12] A.D.Ioffe and V.M.Tihomirov, *Theory of extremal problems*, North Holland, Amsterdam (1979).

[13] K.Ito and K.Kunisch, Sensitivity analysis of solutions to optimization problems in Hilbert spaces with applications to optimal control and estimation, *J. Diff. Equations*, 99: 1-40 (1992).

[14] D. H. Jacobson, New Second-Order and Firts-Order Algorithms for Determining Optimal Control: A Differential Dynamic Programming Approach, *J. Optim. Theory Appl.*, 6: 411-440 (1968).

[15] D. H. Jacobson and D. Q. Mayne, *Differential Dynamic Programming*, American Elsevier Publishing Company Inc., New York (1970).

[16] E.S.Levitin, On the local perturbation theory of a problem of mathematical programming in Banach spaces, *Soviet Math.Dokl.*, 16: 1354-1358 (1975).

[17] K.Malanowski, Second order conditions and constraint qualifications in stability and sensitivity analysis of solutions in optimization problems in Hilbert spaces, *Appl. Math. Optim.*, 25: 51-79 (1992).

[18] K.Malanowski, Two-norm approach in stability and sensitivity analysis of optimization and optimal control problems, *Advances in Math.Sciences Appl.*, 2: 397-443 (1993).

[19] K.Malanowski, Stability and sensitivity of solutions to nonlinear optimal control problems, *Appl. Math. Optim.*, 32: 111-141 (1995).

[20] K.Malanowski, "Stability analysis of solutions to parametric optimal control problems", to appear in *Proceedings of the 4th Conference on Parametric Optimization and Related Topics*, Enschede, June 6-9, 1995.

[21] K.Malanowski and H.Maurer, Sensitivity analysis for parametric optimal control problems with control-state constraints, *Comp. Optim. Appl.*, 5: 253-283 (1996).

[22] H. Maurer and D. Augustin, "Second Order Sufficient Conditions and Sensitivity Analysis for the Controlled Rayleigh Equation", to appear in *Proceedings of the 4th Conference on Parametric Optimization and Related Topics*, Enschede, June 6-9, 1995.

[23] H. Maurer and H. J. Pesch, Solution Differentiability for Parametric Nonlinear Control Problems, *SIAM J. Contr. Optim.*, 32: 1542-1554 (1994).

[24] H. Maurer and H. J. Pesch, Solution Differentiability for Parametric Nonlinear Control Problems with Control–State Constraints, *Control and Cybernetics*, 23: 201-227 (1994).

[25] H. Maurer and S. Pickenhain, Second Order Sufficient Conditions for Optimal Control Problems with Mixed Control–State Constraints, *J. Optim. Theory Appl.*, 86: 649-667 (1995).

[26] H.Maurer and J.Zowe, First- and second-order conditions in infinite-dimensional programming problems, *Math. Programming,* 16: 98-110 (1979).

[27] B.S.Mordukhovich, On difference approximation of optimal control systems, Appl. Math. Mech., 42: 452-461 (1978).

[28] H. J. Oberle and W. Grimm, "BNDSCO - A Program for the Numerical Solution of Optimal Control Problems, " Institute for Flight Systems Dynamics, DLR, Oberpfaffenhofen, Germany, Internal Report No. 515-89/22 (1989).

[29] S.M.Robinson, Perturbed Kuhn-Tucker points and rate of convergence for a class of nonlinear programming algorithms, *Math. Programming,* 7: 1-16 (1974).

[30] S.M.Robinson, Strongly regular generalized equations, *Math. Oper. Research,* 5: 43-62 (1980).

[31] A.Shapiro and J.F.Bonnans, Sensitivity analysis of parametrized programs under cone constraints, *SIAM J. Contr. Optim.*, 30: 97-116 (1992).

[32] J.Stoer and R.Bulirsch, *Introduction to Numerical Analysis*, Springer-Verlag, New York (1980).

[33] K. L. Teo, C. J. Goh and K. H. Wong, *A Unified Computational Approach to Optimal Control Problems*, Longman Scientific and Technical, New York (1991).

[34] T. Tun and T. S. Dillon, Extensions of the Differential Dynamic Programming Method to include Systems with State Dependent Control Constraints and State Variable Inequality Constraints, *J. Appl. Science Engineering, A, 3*: 171-192 (1978).

[35] V. Zeidan, The Riccati Equation for Optimal Control Problems with Mixed State-Control Constraints: Necessity and Sufficiency, *SIAM J. Contr. Optim.*, 32: 1297-1321 (1994).

A Perturbation-Based Duality Classification for Max-Flow Min-Cut Problems of Strang and Iri

RYÔHEI NOZAWA* Department of Mathematics, School of Medicine, Sapporo Medical University, Chuou-ku, Sapporo, Japan
email : *nozawa@sapmed.ac.jp*

K. O. KORTANEK Department of Management Sciences, College of Business Administration, The University of Iowa, Iowa City, IA 52242
email : *kort@krona.biz.uiowa.edu*

Abstract

A definitive classification is established for all conceivable and permissible duality states of two slightly different versions (Strang and Iri) of infinite dimensional max–flow min–cut dual optimization problems. The approach builds on three elements: (1) previous work of the first author on duality gaps in max–flow problems under appropriate choices of function spaces, (2) construction of a class of perturbations in infinite convex optimizations germane to max–flow, and (3) a previously published classification theory of duality states defined by combinations of (extended) extremal values of closed convex bifunction dual families and their homogeneous derivant bifunctions.

1   INTRODUCTION: MAX–FLOW UNDER COMPACT CONVEX INCLUSIVE MAPPINGS

Strang [Str83] formulated a pair of optimization problems over a Euclidean domain which are closely related to some problems in mechanics principally in plastic limit analysis, see Kohn and Temam [KT83] and Christiansen [Chr86]. These infinite optimizations are extensions of the classical max–flow and min–cut problems developed on discrete networks. They are termed continuous versions of max–flow and min–cut problems of Strang or Iri type, in contrast to other types of continuous flow problems some of which are more measure theoretic in construction, see [Daf80, FL81, Iri79, JS83, Neu84]. One can refer to Ford and Fulkerson [FF62] or Iri [Iri69] for statements of the classical problem. In the present paper, we follow a classification scheme for general convex programming problems appearing in the

---

Dedicated to Richard J. Duffin, mathematician, (1909–1996).

*Research supported in part from Hokkaido prefecture providing for a two–week visit to The University of Iowa in Summer, 1995

literature, [Duf56, BICK71, KW71, Rom70, Kor74, Kor75], to use a perturbation-based approach for investigating all types of feasibility occurring in max–flow min–cut infinite dimensional problems of this type.

The first author studied max–flow and min–cut problems of Strang and Iri type in [Noz90, Noz94]. These problems are formulated in terms of classes of essentially bounded vector fields and functions of bounded variation. For the case where the constraint set is determined by a bounded and continuous set-valued function to be reviewed in this paper they are in a *perfect duality*, [Kor77], that is, the maximal flow value equals the minimal cut capacity (i.e., no duality gap). On the other hand, an example having a duality gap was given in [Noz94] for the general case where the capacity constraint is unbounded.

In this paper we investigate the properties of asymptotic duality that arise from the closure operation applied to a class of perturbations in general convex programming, typically developed in convex bifunction form, [Roc70]. The perturbation approach is applicable to a wide class of convex programming problems, and different problems can give rise to their own classes of perturbations. The 1956 seminal paper by Richard J. Duffin introduced perturbations in infinite linear programming which generated *subconsistent* solutions, which actually were asymptotically feasible solutions for inconsistent infinite linear programs. With this foundation generalized convex programs were formulated with implicit perturbations, see [Gal67, Gol72], and then with much more general classes of perturbations, see [Roc70, Kor74] for generalized convex bifunction optimizations.

The optimization problems in this paper are infinite dimensional convex, and hence the perturbations and resultant classification of studied in [Kor74] are pursued within the problem framework developed in function space setting of [Noz90, Noz94]. The main result for classifying the duality of the max–flow and min–cut problems is that there are only four duality states for problems of Strang's type. For a literature on convex classification theory for linear spaces having Hahn Banach extension property, see [BICK69, GKR70, BICK71, KR71, KW71, Kor72, KS72, KRS73, Soy73, Kor74, Kor75, GKS76].

Now we recall the precise formulation of the continuous version of max–flow and min–cut problems of Strang's type, which are defined in [Noz90]. We note that α, α' and Λ in [Noz90] should be taken as α, $\alpha' = 0$ and $\Lambda = \partial\Omega$. Let Ω be a bounded domain in the n-dimensional Euclidean space R^n with Lipschitz boundary $\partial\Omega$. Furthermore let $F \in L^n(\Omega)$ and $f \in L^\infty(\partial\Omega)$, where $L^\infty(\partial\Omega)$ is the set of all essentially bounded measurable functions on $\partial\Omega$ with respect to the $(n-1)$-dimensional Hausdorff measure H_{n-1} .

We assume that

$$\int_\Omega F dx + \int_{\partial\Omega} f dH_{n-1} = 0.$$

This is the conservation law. We term a set–valued mapping Γ from Ω to R^n compact convex inclusive if it satisfies the following two conditions:

$(H1)$ $\Gamma(x)$ is a compact convex set containing a ball with center at 0
          for each $x \in \Omega$.

$(H2)$ For any $\varepsilon > 0$ and any compact subset Ω_0 of Ω,
          there is $\delta > 0$ such that $\Gamma(x) \subset \Gamma(y) + B(0, \varepsilon)$

if $x, y \in \Omega_0$ and $|x - y| < \delta$.

We assume throughout this paper that Γ is CCI. Let ν be the unit outer normal to $\partial\Omega$. According to Kohn and Temam [KT83], if $\sigma = (\sigma_1, \ldots, \sigma_n) \in L^\infty(\Omega; R^n)$ with $\operatorname{div} \sigma \in L^n(\Omega)$, then $\sigma \cdot \nu$ is defined as a function in $L^\infty(\partial\Omega)$ in a weak sense, where $\operatorname{div} \sigma = \sum_{i=1}^n \partial\sigma_i/\partial x_i$ is understood in the sense of distribution. (See also [Noz90, Theorem 2.3].) Using this fact, we can define the max–flow problem (MF) of Strang's type corresponding with (Ω, Γ, F, f) as follows:

$$(MF) \qquad \sup \lambda$$

$$\text{subject to } \lambda \in R, \ \sigma \in L^\infty(\Omega; R^n),$$

$$\sigma(x) \in \Gamma(x) \quad \text{for a.e. } x \in \Omega,$$

$$\operatorname{div} \sigma = -\lambda F \quad \text{a.e. on } \Omega \text{ and}$$

$$\sigma \cdot \nu = \lambda f \quad H_{n-1}-\text{a.e. on } \partial\Omega.$$

REMARK 1.1 *Throughout this paper various minimization or maximization problems will be presented bearing names such as (MF) above. We refer to these as Programs and denote their optimal values, including $\pm\infty$, by the appropriate max/min or sup/inf followed by the program name in parantheses. Thus, the optimal value of Program (MF) is simply denoted $\sup(MF)$.*

PROPOSITION 1.1 (MF) *is consistent and* $\sup(MF) \geq 0$.

Proof. It follows from $(H1)$, that $(\lambda, \sigma) = (0, 0)$ is a feasible solution. In particular, the value of (MF) is not less than 0.

The choice of function spaces are quite technical. In fact, $L^n(\Omega)$ and $L^\infty(\partial\Omega)$ are closely related with the class of cuts. However, even in a simple example of (MF), a non-smooth vector field in $L^\infty(\Omega; R^n)$ appears as an optimal flow so that $L^\infty(\Omega; R^n)$ seems to fit for the class of flows.

Before stating the min–cut problem, we introduce the space $BV(\Omega)$ of functions of bounded variation on Ω:

$$BV(\Omega) = \{u \in L^1(\Omega); \ \nabla u \text{ is a Radon measure of bounded variation on } \Omega\},$$

where $\nabla u = (\partial u/\partial x_1, \cdots, \partial u/\partial x_n)$ is also understood in the sense of distribution. We denote the characteristic function of a subset S of Ω by χ_S and set

$$Q = \{S \subset \Omega; \ \chi_S \in BV(\Omega)\}$$

We note that $S \in Q$ can be regarded as a cut. Let $S \in Q$ and define the cut capacity as follows. We set $\beta(v, x) = \sup_{w \in \Gamma(x)} v \cdot w$ for $v \in R^n$. Then $\beta(\cdot, x)$ is the support functional of $\Gamma(x)$. Let $\nabla u/|\nabla u|$ be the Radon-Nikodym derivative of ∇u with respect to $|\nabla u|$ and set

$$\psi(u) = \int_\Omega \beta(\nabla u/|\nabla u|, x) d|\nabla u|(x)$$

for $u \in BV(\Omega)$. Then the cut capacity $C(S)$ of S is defined by $C(S) = \psi(\chi_S)$.

Since $\nabla \chi_S$ is concentrated on ∂S, $C(S)$ is determined by Γ on $\partial S \cap \Omega$. If ∂S is smooth, then $-\nabla \chi_S / |\nabla \chi_S|$ is equal to the unit outer normal to S. Hence in the case where $\Gamma(x) = \{w \in R^n; \ |w| \leq c(x)\}$ for some continuous and positive function $c(x)$,

$$C(S) = \int_{\Omega \cap \partial S} c(x) dH_{n-1}(x).$$

Using a notion of the reduced boundary, we can generalize this relation. The reduced boundary $\partial^* S$ of S is the set of all $x \in \partial S$ where Federer's normal $\nu^S = \nu^S(x)$ to S exists. It is known that $\partial^* S$ is a measurable set with respect to both H_{n-1} and the measure $|\nabla \chi_S|$ of total variation of $\nabla \chi_S$, $|\nabla \chi_S|(R^n - \partial^* S) = 0$ and $|\nabla \chi_S|(E) = H_{n-1}(E)$ for each $|\nabla \chi_S|$-measurable subset E of $\partial^* S$. Furthermore the trace γu of $u \in BV(\Omega)$ belongs to $L^1(\partial \Omega)$. For these facts, one can refer to [Giu84, Maz85]. Then Maz'ja [Maz85, Theorem 6.6.2] implies that $\gamma \chi_S = \chi_{\partial^* S \cap \partial \Omega}$ H_{n-1}–a.e. on $\partial \Omega$. Accordingly, the cut capacity of S is written as follows:

$$C(S) = \int_{\Omega \cap \partial^* S} \beta(-\nu^S(x), x) dH_{n-1}.$$

Since β is continuous and nonnegative by $(H1)$ and $(H2)$, $C(S)$ is defined and nonnegative. Furthermore we set

$$L(u) = \int_{\Omega} F u dx + \int_{\partial \Omega} f \gamma u dH_{n-1}$$

for $u \in BV(\Omega)$. Since $BV(\Omega) \subset L^{n/(n-1)}(\Omega)$, $L(u)$ is finite.

Now the min–cut problem (MC) of Strang's type is defined as follows:

(MC) inf $C(S)/L(\chi_S)$

     subject to $S \in Q$ and $L(\chi_S) > 0$.

If S is a feasible solution of (MC), then S is called a feasible cut of (MC).

One can obtain (MC) by applying an equality of coarea formula type to the following (MF^*), which is defined as a dual problem of (MF): Taking $X = R$, $P = R^+$, $K = K_\Gamma$, $Y = X_0$, $Z = BV(\Omega)$, $L_X(\xi, u) = \xi L(u)$ and $L_Y(\sigma, u) = (\sigma \nabla u)(\Omega)$ for $\xi \in R$, $\sigma \in X_0$ and $u \in BV(\Omega)$ in [Noz90, §3] or [Noz94, Appendix A], we get

(MF^*) inf $\psi(u)$

     subject to $u \in BV(\Omega)$ and $L(u) = 1$.

In contrast to usual Sobolev spaces, $BV(\Omega)$ contains sufficiently many characteristic functions and any bounded set of $BV(\Omega)$ has a compactness property, which may be useful to deal with existence of optimal cut. However, there is a simple example of (MC) with no optimal cut. To deal with the existence of optimal cut, we need a relaxation of the min–cut problem.

We note that (MF), (MF^*), (MC) correspond to $(M\Phi_1)$, $(M\Phi_1^*)$, $(M\Gamma_1)$ defined in [Noz90] respectively. Hence [Noz90, Theorems 4.4 and 4.9] implies the following max–flow min–cut theorem:

THEOREM 1.2 *Let Γ be CCI. Then $sup(MF) \leq \inf(MC) = \inf(MF^*)$. Furthermore if Γ is bounded, that is, $\cup_{x \in \Omega} \Gamma(x)$ is bounded, then $\sup(MF) = \inf(MC) = \inf(MF^*)$.*

As proved in [Noz94], $\sup(MF) \neq \inf(MC)$, both finite, may occur if $\cup_{x \in \Omega} \Gamma(x)$ is not bounded.

In the Section 2, we reformulate (MF) and (MF^*) using a linear mapping and define the notions of consistency. Furthermore we show that only three cases in a classification table given in [Kor74] can actually occur for our problems if $(F, f) \neq (0, 0)$. In Section 3, we establish existence of these states in our context by giving specific examples. Finally, in Section 4, after recalling a bifunction formulation of optimization problems due to [Roc70], we refer to max–flow and min–cut problems of Iri's type.

2   ASYMPTOTIC PROBLEMS AND CONSISTENCY

Let F, f be as in Section 1 and X, Y be real linear spaces defined by

$$
\begin{aligned}
X &= X_0 \times R = \{\sigma \; ; \; \sigma \in L^\infty(\Omega; R^n), \; \operatorname{div}\sigma \in L^n(\Omega)\} \times R, \\
Y &= \{(y_1, y_2) \; ; \; y_1 \in L^n(\Omega), \; y_2 \in L^\infty(\partial\Omega)\}.
\end{aligned}
$$

We consider the norm $\|\sigma\|$ defined by

$$
\|\sigma\| = \|\sigma\|_\infty + \|\operatorname{div}\sigma\|_n
$$

on X_0, where $\|\sigma\|_\infty$ denotes the essential supremum of $|\sigma|$ over Ω and $\|\cdot\|_n$ is the canonical norm on $L^n(\Omega)$. Then X_0 is a Banach space. Let X_0^* be the topological dual space of X_0. On X, we consider the product topology and let X^* be the topological dual space of X. To define a topology on Y, let

$$
W^{1,1}(\Omega) = \{u \in L^1(\Omega); \; \nabla u \in L^1(\Omega; R^n)\}.
$$

Then $W^{1,1}(\Omega) \subset BV(\Omega)$ so that $W^{1,1}(\Omega) \subset L^{n/(n-1)}(\Omega)$. Furthermore

$$
\{\gamma u; \; u \in W^{1,1}(\Omega)\} = L^1(\partial\Omega)
$$

by Gagliardo [Gag57]. Now we consider the weak topology on Y with respect to the following bilinear form on $Y \times W^{1,1}(\Omega)$:

$$
\langle (y_1, y_2), u \rangle = \int_\Omega y_1 u\, dx + \int_\Omega y_2 \gamma u\, dH_{n-1}.
$$

Then the topological dual space Y^* of Y is identified with $W^{1,1}(\Omega)$. Let T be a linear mapping from X to Y such that

$$
T(\sigma, \lambda) = (\operatorname{div}\sigma + \lambda F, -\sigma \cdot \nu + \lambda f),
$$

where $\sigma \cdot \nu$ is understood to be a function defined in Section 1. Since Green's formula

$$
\int_\Omega \sigma \cdot \nabla u\, dx + \int_\Omega u \operatorname{div}\sigma\, dx = \int_{\partial\Omega} \gamma u \sigma \cdot \nu\, dH_{n-1} \tag{1}
$$

holds for $\sigma \in L^\infty(\Omega; R^n)$ with $\operatorname{div}\sigma \in L^n(\Omega)$ and $u \in W^{1,1}(\Omega)$, by our definition of topologies on X and Y, T is continuous and the adjoint T^* of T is defined as a linear mapping from $W^{1,1}(\Omega)$ to X^*. We note that the Green's formula is extended such as

$$(\sigma \nabla u)(\Omega) + \int_\Omega u \operatorname{div} \sigma dx = \int_{\partial\Omega} \gamma u \sigma \cdot \nu dH_{n-1} \tag{2}$$

for $u \in BV(\Omega)$ by [KT83], where $(\sigma \nabla u)$ is a bounded measure on Ω in the sense of [KT83].

Next let $0_{X_0^*}$ be the zero element in X_0^* and set $e = (0_{X_0^*}, 1) \in X^*$. Then denoting $x^*(x)$ by $\langle x, x^* \rangle$ for $x \in X$ and $x^* \in X^*$, we have

$$\langle (\sigma, \lambda), e \rangle = \lambda.$$

Furthermore let 0_Y be the origin of Y and set $K = (K_\Gamma \times R) \cap X$, a convex set, where

$$K_\Gamma = \{ \sigma \in L^\infty(\Omega; R^n);\ \sigma(x) \in \Gamma(x) \text{ for a.e. } x \in \Omega \}.$$

Since the two equalities $\operatorname{div} \sigma = -\lambda F$ and $\sigma \cdot \nu = \lambda f$ are equivalent to $T(\sigma, \lambda) = 0_Y$, using the sextuplet $(X, Y, T, e, 0_Y, K)$, we rewrite (MF) as

$$(P) \qquad \sup \langle (\sigma, \lambda), e \rangle$$
$$\text{subject to } (\sigma, \lambda) \in K, \quad T(\sigma, \lambda) = 0_Y$$

and the dual problem is written as

$$(P^*) \qquad \min \eta$$
$$\text{subject to } u \in W^{1,1}(\Omega),\ \eta \in R^+ \text{ and}$$
$$\langle (\sigma, \lambda), T^*u - e \rangle \geq -\eta \text{ for all } (\sigma, \lambda) \in K,$$

where R^+ is the set of all nonnegative numbers. (P^*) is constructed formally from the finite dimensional program (I), [BICK71, page 678].

If there is a feasible solution (u, η) to (P^*), then we say that (P^*) is consistent, CONS, for short. Otherwise it is inconsistent, INC. The goal is to construct a dual problem to (MF) which is in perfect duality, namely having program value equal to $\sup(MF)$. The fundamental construction of *subconsistent solutions* due to Richard J. Duffin, appeared in 1956 in infinite linear programming see [Duf56]. An extension to convex optimization based on general perturbations appeared in [Roc70, BICK71, KW71, Kor74]. Following this extension we construct an asymptotic problem corresponding to (P) as follows.

$$(AP) \qquad \operatorname{Sup}_{\{(\sigma_i, \lambda_i)_i \in AFP\}} \{ \limsup_i \lambda_i \},$$

where AFP, the set of **asymptotically feasible solutions** to (P) is defined by :

$$\{ (\sigma_i, \lambda_i)_i \subset K;\ \lim_i (\operatorname{div} \sigma_i + \lambda_i F, -\sigma_i \cdot \nu + \lambda_i f) = (0, 0) \}$$

with respect to the topology of Y. Program (AP) is of interest even though (P) itself is consistent and so trivially asymptotically consistent.

If (AP) is finite, then (P) is termed **asymptotically bounded**. By the definition of the topology on Y, if $\operatorname{div}\sigma_i + \lambda_i F \to 0$ and $-\sigma_i \cdot \nu + \lambda_i f \to 0$ with respect to the canonical weak*-topologies on $L^n(\Omega)$ and $L^\infty(\partial\Omega)$ respectively, then the generalized sequence $(\sigma_i, \lambda_i)_i \subset K$ is an asymptotically feasible solution. (P) is termed **asymptotically consistent**, AC, if and only if $AFP \neq \emptyset$; otherwise it is **strongly inconsistent**, SINC and one may define $\sup(\text{AP})=-\infty$. This is the standard way of assigning values to inconsistent programs, see [Roc70] and [Kor74, Table IV].

PROPOSITION 2.1 *It follows that* $\sup(AP) = \inf(P^*) = \inf(MC)$. *Furthermore,* (P^*) *is consistent if and only if there is a feasible cut $S \in Q$ such that $C(S)$ is finite.*

Proof. It is known by [Duf56] and [Roc70, Corollary 30.2.2] that the maximal value of (AP) equals $\inf(P^*)$. We review the infinite convex programming duality from the theory of convex bifunction perturbations in Section 4.

We turn now to the case of consistency. Let (u, η) be a feasible solution to (P^*). Then by Green's formula (1),

$$
\begin{aligned}
\langle (\sigma, \lambda), T^*u - e \rangle &= \langle T(\sigma, \lambda), u \rangle - \langle (\sigma, \lambda), e \rangle \\
&= \int_\Omega (\operatorname{div}\sigma + \lambda F)u\, dx + \int_{\partial\Omega} (-\sigma \cdot \nu + \lambda f)\gamma u\, dH_{n-1} - \lambda \\
&= \int_\Omega \operatorname{div}\sigma \cdot u\, dx + \int_{\partial\Omega} -\sigma \cdot \nu\gamma u\, dH_{n-1} + \lambda\Big(\int_\Omega Fu\, dx + \int_{\partial\Omega} f\gamma u\, dH_{n-1} - 1\Big) \\
&= -\int_\Omega \sigma \cdot \nabla u\, dx + \lambda(L(u) - 1) \geq -\eta
\end{aligned}
$$

for all $(\sigma, \lambda) \in K$.

It follows that $L(u) - 1 = 0$ and

$$
\int_\Omega \sigma \cdot \nabla u\, dx \leq \eta
$$

for all $\sigma \in K_\Gamma \cap X_0$. Hence by $(H1)$ and [Noz90, Lemma 2.6] we obtain

$$
\psi(u) = \sup_{\sigma \in K_\Gamma} \int_\Omega \sigma \cdot \nabla u\, dx = \sup_{\sigma \in K_\Gamma \cap X_0} \int_\Omega \sigma \cdot \nabla u\, dx \leq \eta.
$$

Since $u \in W^{1,1}(\Omega) \subset BV(\Omega)$, u is a feasible solution to (MF^*) and $\inf(MC) = \inf(MF^*) \leq \inf(P^*)$.

In view of Green's formula (2), it is easy to check that $\inf(MC) \geq \sup(AP)$ holds so that the converse inequality is also proved, because $\sup(AP) = \inf(P^*)$.

Finally, we can directly prove the last assertion by $\inf(MC) = \inf(P^*)$. This completes the proof.

By virtue of this proposition, we say that (MC) is consistent if there is a feasible cut S of (MC) such that $C(S)$ is finite.

We next develop an asymptotic program (AP^*) corresponding to (P^*):

$$
(AP^*) \qquad \inf_{\{(u_i, \sigma_i^*, \eta_i)_i \in AFAP^*\}}\{\liminf_i \eta_i\},
$$

where $AFAP^*$, the set of **asymptotically feasible solutions** to (AP^*) is defined by :

$$\{ (u_i, \sigma_i^*, \eta_i)_i \subset W^{1,1}(\Omega) \times X_0^* \times R; \lim_i \left(\int_\Omega \sigma \cdot \nabla u_i dx + \langle \sigma, \sigma_i^* \rangle \right) = 0 \text{ for each } \sigma \in X_0,$$

$$\text{and } L(u_i) \to 1, \ \langle \sigma, \sigma_i^* \rangle \geq -\eta_i \text{ for each } \sigma \in K_\Gamma \cap X_0 \}.$$

If (AP^*) is finite, then program (P^*) is termed **asymptotically bounded**.

However, we say that (P^*) is **asymptotically consistent** (AC) if $AFAP^* \neq \emptyset$. It is **strongly inconsistent** (SINC) if $AFAP^* = \emptyset$, and we formally assign $\inf(AP^*) = +\infty$, as usual.

Let $AFAP_{BV}^*$ be the modification of $AFAP^*$ obtained by replacing $W^{1,1}(\Omega)$ by $BV(\Omega)$, and

$$\int_\Omega \sigma \cdot \nabla u_i dx \quad \text{by} \quad (\sigma \nabla u_i)(\Omega).$$

We can formulate a slightly different asymptotic problem from (AP^*):

$$(AP_{BV}^*) \qquad \inf_{\{(u_i, \sigma_i^*, \eta_i)_i \in AFAP_{BV}^*\}} \{\liminf_i \eta_i\},$$

where $AFAP_{BV}^*$, the set of **asymptotically feasible solutions** to AP_{BV}^* is defined by :

$$\{ (u_i, \sigma_i^*, \eta_i)_i \subset BV(\Omega) \times X_0^* \times R; \lim_i \left((\sigma \nabla u_i)(\Omega) + \langle \sigma, \sigma_i^* \rangle \right) = 0 \text{ for each } \sigma \in X_0,$$

$$L(u_i) \to 1, \ \langle \sigma, \sigma_i^* \rangle \geq -\eta_i \text{ for each } \sigma \in K_\Gamma \cap X_0 \}.$$

According to the Green's formula (2), we easily see that $\sup(MF) \leq \inf(AP_{BV}^*)$. Thus, the following result follows immediately, in view of $\sup(P) = \inf(AP^*)$ again by [Duf56] and [Roc70, Corollary 30.2.2].

PROPOSITION 2.2 *It follows that* $\sup(MF) = \inf(AP_{BV}^*) = \inf(AP^*)$.

When (P^*) is (AC), we say that (P^*) is **properly asymptotically consistent**, (PAC), if there is a feasible solution $(u_i, \sigma_i^*, \eta_i)_i$ such that $\liminf \eta_i < \infty$. It is termed **improperly asymptotically consistent**, (IAC), if it is AC but not PAC, implying $\inf(AP^*) = +\infty$ (as an infimum).

We review the definition of *homogemeous consistency*, a property that will be used in the proof of Theorem 2.3.

DEFINITION Let O^+K be the asymptotic (or recession) cone of K which is defined by

$$O^+K = \{(\sigma, \lambda); \ t(\sigma, \lambda) + (\sigma_0, \lambda_0) \in K \text{ for all } t \in R^+\}$$

for any $(\sigma_0, \lambda_0) \in K$, see [Roc70]. If there is a $(\sigma, \lambda) \in O^+K$ such that $T(\sigma, \lambda) = (0,0)$ and $\langle (\sigma, \lambda), e \rangle > 0$, then (P) is termed *homogeneous consistent*. Otherwise, (P) is termed *homogeneous inconsistent*.

THEOREM 2.3 *There are 4 mutually exclusive and collectively exhaustive cases for the status of* (P^*) *consistency listed in the leftmost column of Table 1. These cases can occur only with the associated given in middle column and the corresponding program values given in the rightmost column of the table. Moreover,* P^* *is in case (4) if and only if* $(F, f) = (0, 0)$.

Table 1: Possible Duality States for (P) & (P^*), [19, Table II]

Case & P^* Status	Duality State, #	Program Values
(1) CONS	1	$\sup(MF) < +\infty$ and $\sup(AP) < +\infty$
(2) INC and PAC	5	$\sup(MF) < +\infty$ and $\sup(AP) = +\infty$
(3) IAC	7	$\sup(MF) = \sup(AP) = +\infty$
(4) SINC	8	$\inf(AP^*) = -\infty$

Proof Logically, cases (1)–(4) are mutually exclusive and collectively exhaustive with respect to (P^*). Since (P) is consistent, case (1) simply corresponds to duality state (DS) 1. Now the only way (P^*) can be INC and PAC, namely case (2) with (P) indentified with (MF), is to be in DS 5 associated with $\sup(MF) < +\infty$ and $\sup(AP) = +\infty$. In case(3), (P^*) is IAC, and this can occur only in DS 7 with $\sup(MF) = \sup(AP) = +\infty$.

By property $(H1)$, $O^+K = \{0\} \times R$. Hence, (P) is homogeneous consistent if and only if $(F, f) = (0, 0)$. Now the remaining case (4) has (P^*) SINC, and this can only occur in DS 8, which has (P) homogeneous consistent. Hence, (P^*) is SINC if and only if $(F, f) = (0, 0)$.

REMARK 2.1 *We reproduce the relevant portion of the classification table at the end of Section 3 because it is in that section that we establish existence of these four states by specific examples. Actually, homogeneous consistency was introduced by R. J. Duffin [Duf56] to characterize subconsistent (asymptotically consistent) solutions in infinite linear programming. It provides a certain uniqueness in defining , which we have used in the above proof.*

3   EXISTENCE OF DUALITY STATES IN MAX–FLOW MIN–CUT PROBLEMS

By Theorem 2.3, Case (4) if and only if $(F, f) = (0, 0)$. Hence we assume $(F, f) \neq (0, 0)$. Then we have three possible cases for consistency by Theorem 2.3. In this section, we give some examples corresponding with the three cases.

One can easily give an example of (MF) such that $\sup(MF) = \sup(P) = \inf(MC) = \inf(P^*) < \infty$, which belongs to Case (1). Furthermore one can find an example of (MF) such that $\sup(MF) = \sup(P) < \inf(MC) \leq \inf(P^*) < \infty$ in [Noz94]. Such an example also belongs to Case (1).

To give an example in Case (2), we modify another example in [Noz94], which was originally constructed as an example of max–flow min–cut problems of Iri's type. we assume that $n = 2$, $\Omega = (0, 1) \times (0, 1)$, $F = 0$ on Ω and f is defined by $f = 1$ on $A = \{0\} \times (0, 1)$, $f = -1$ on $B = \{1\} \times (0, 1)$ and $f = 0$ on $\partial\Omega - (A \cup B)$.

EXAMPLE 1 for Case (2).

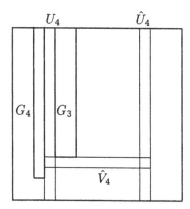

Figure 1: Illustration of sets used in Example 1 and Theorem 3.1

To define Γ, we need sequences of subsets of Ω. Let

$$
\begin{aligned}
F_k &= [2 \cdot 2^{-k} + 2^{-k-2}, 3 \cdot 2^{-k} - 2^{-k-2}] \times [2 \cdot 2^{-k} + 2^{-k-2}, 1), \\
G_k &= [2 \cdot 2^{-k}, 3 \cdot 2^{-k}] \times [2 \cdot 2^{-k}, 1)
\end{aligned}
$$

for each integer $k \geq 3$. See Figure 1 for G_k. ($F_k \subset G_k$ and we do not consider F_2, G_2.) Let $h(x)$ be a continuous function such that $h(x) = 0$ on $\cup_{k=3}^{\infty} F_k$, $h(x) = 1$ on $\Omega - \cup_{k=3}^{\infty} G_k$ and $0 \leq h \leq 1$ on Ω. We set

$$
c(x_1, x_2) = 1 + h(x_1, x_2)(1/x_1^2 + 1/(1 - x_1)^2)(1/x_2) \tag{3}
$$

and $\Gamma(x) = \{w \in R^2; \ |w| \leq c(x)\}$ for every $x = (x_1, x_2) \in \Omega$. Then (MF) corresponding with these data is an example such that $\sup(MF) = \sup(P) = 1 < \inf(MC) = \infty = \inf(P^*) = \sup(AP)$ by the following theorem and thus belongs to case (2).

THEOREM 3.1 *Let a_1, b_1 be the values of $(MF), (MC)$ corresponding with the above data. Then $a_1 = 1$ and $b_1 = \infty$.*

Proof. Since $\sigma = (-1, 0)$ is a feasible flow, $a_1 \geq 1$. In terms of $c_n(x) = \min(c(x), n)$ for a positive integer n, we can prove that the converse inequality in a way similar to that of [Noz94, Lemma 3.4] so that $a_1 = 1$. We prove $b_1 = \infty$. It is sufficient to show that

$$
C(S) = \int_{\Omega \cap \partial^* S} c(x) dH_1 = \infty
$$

for any subset $S \in Q$ satisfying

$$
\int_{\partial \Omega \cap \partial^* S} f(x) dH_1(x) > 0. \tag{4}
$$

As in [Noz94], we set $U_k = (3 \cdot 2^{-k}, 4 \cdot 2^{-k}) \times (0, 1)$ for $k \geq 3$. We note that $c(x_1, x_2) = 1 + (1/x_1^2 + 1/(1 - x_1)^2)(1/x_2) \geq 1/(x_1^2 x_2)$ for all $(x_1, x_2) \in U_k$ and $k \geq 3$.

Let $S \in Q$ satisfy (4). If there exists $b > 0$ such that $H_1(\partial^* S \cap U_k) \geq 2^{-2k}b$ for infinitely many k, then as shown in the proof of [Noz94, Lemma 3.6], $C(S) = \infty$.

Thus for any $b > 0$, we may assume that $H_1(\partial^* S \cap U_k) < 2^{-2k}b$ for all k except finitely many k. By [Noz94, Lemma 3.5],

$$\min(m_2(U_k \cap S), m_2(U_k - S)) \leq 2^k \cdot \kappa \cdot 2^{-4k}b^2 = 2^{-3k}\kappa b^2$$

for such k, where m_2 denotes the 2-dimensional Lebesgue measure and κ is a constant independent of k. However, in the case where, for a fixed $b > 0$, $m_2(U_k \cap S) \leq 2^{-3k}\kappa b^2$ for infinitely many k, as in the proof of [Noz94, Lemma 3.6] one can see that $H_1(A \cap \partial^* S)$ must be 0. This is a contradiction since $\int_{A \cap \partial^* S} f(x)dH_1(x) > 0$ by (4). Now we may assume that

$$m_2(U_k - S) \leq 2^{-3k}\kappa b^2 \tag{5}$$

for all k except finitely many k.

Symmetrically let $\hat{U}_k = (1 - 4 \cdot 2^{-k}, 1 - 3 \cdot 2^{-k}) \times (0, 1)$. Then we may assume that

$$\min(m_2(\hat{U}_k \cap S), m_2(\hat{U}_k - S)) \leq 2^{-3k}\kappa b^2.$$

Furthermore if $m_2(\hat{U}_k - S) \leq 2^{-3k}\kappa b^2$ for infinitely many k, then we can prove that

$$H_1(B \cap \partial^*(\Omega - S)) = H_1(B - \partial^* S) = 0$$

and this is a contradiction. Thus we may assume that

$$m_2(\hat{U}_k \cap S) \leq 2^{-3k}\kappa b^2 \tag{6}$$

hold for all k except finitely many k.

Now we prove that $C(S) = \infty$ for $S \in Q$ satisfying (5) and (6). Let

$$\hat{V}_k = (3 \cdot 2^{-k}, 1 - 3 \cdot 2^{-k}) \times (3 \cdot 2^{-k}, 4 \cdot 2^{-k})$$

for $k \geq 3$ as in Figure 1. Then

$$\begin{aligned} m_2(U_k \cap \hat{V}_k \cap S) &= m_2(U_k \cap \hat{V}_k) - m_2(U_k \cap \hat{V}_k - S) \\ &\geq 2^{-2k}(1 - 2^{-k}\kappa b^2) \end{aligned}$$

for all sufficiently large k. On the other hand

$$m_2((\hat{U}_k \cap \hat{V}_k) \cap S) \leq m_2((\hat{U}_k \cap S) \leq 2^{-3k}\kappa b^2$$

for all sufficiently large k. By Fubini's theorem and [Noz94, Lemma 3.7], there are lines $l_k : x_1 = t_k$ and $\hat{l}_k : x_1 = \hat{t}_k$ such that

$$\begin{aligned} H_1(l_k \cap S) &= H_1(l_k \cap \partial^* \hat{S}_k) \geq 2^{-k}(1 - 2^{-k}\kappa b^2), \\ H_1(\hat{l}_k \cap S) &= H_1(\hat{l}_k \cap \partial^* \hat{S}_k) \leq 2^{-2k}\kappa b^2 \end{aligned}$$

and

$$3 \cdot 2^{-k} < t_k < 4 \cdot 2^{-k}, \quad 1 - 4 \cdot 2^{-k} < \hat{t}_k < 1 - 3 \cdot 2^{-k},$$

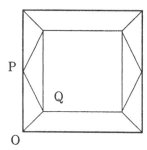

Figure 2: Three points O,P,Q are as indicated in Example 2 for Case (3)

where $\hat{S}_k = \{x = (x_1, x_2) \in \hat{V}_k \cap S; \; t_k < x_1 < \hat{t}_k\}$. Let

$$\hat{\gamma}_k = \{x = (x_1, x_2) \in \partial^* \hat{S}_k \cap \hat{V}_k; \; t_k < x_1 < \hat{t}_k\}.$$

Then $\hat{\gamma}_k \subset \partial^* S$. In this paragraph, we denote Federer's normal to $\hat{S}_k$ by $\nu = (\nu_1, \nu_2)$. By Green's formula,

$$\int_{\hat{V}_k \cap \partial^* \hat{S}_k} -\nu_1 dH_1 = \int_{\partial^* \hat{S}_k} -\nu_1 dH_1 = 0.$$

Hence

$$
\begin{aligned}
H_1(\hat{\gamma}_k) &\geq \int_{\hat{\gamma}_k} \nu_1 dH_1 = -\int_{l_k \cap \partial^* \hat{S}_k} \nu_1 dH_1 - \int_{\hat{l}_k \cap \partial^* \hat{S}_k} \nu_1 dH_1 \\
&= H_1(l_k \cap \partial^* \hat{S}_k) - H_1(\hat{l}_k \cap \partial^* \hat{S}_k) \\
&\geq 2^{-k}(1 - 2^{-k}\kappa b^2) - 2^{-2k}\kappa b^2 = 2^{-k}.
\end{aligned}
$$

Accordingly

$$C(S) = \int_{\Omega \cap \partial^* S} c(x) dH_1(x) \geq \sum_k \int_{\hat{\gamma}_k} cdH_1 \geq \sum_k (4 \cdot 2^{-k})^{-1} \cdot 2^{-k} = \infty,$$

where the summation is taken over all sufficiently large k. This completes the proof.

REMARK 3.1 *In the terminology of [Kor74] (P) is consistent and bounded, asymptotically unbounded, homogeneous inconsistent, but homogeneous asymptotically consistent. This determines duality state 5.*

Finally we give an example which belongs to Case (3).

EXAMPLE 2 for Case(3) Let Ω, F and f be as in Example 1. Let $\hat{c}(x_1, x_2) = 1/x_1^2 + 1/x_2^2 + 1/(1 - x_1)^2 + 1/(1 - x_2)^2$ and set $\Gamma(x) = \{w \in R^2; \; |w| \leq \hat{c}(x)\}$. Then $\sup(MF) = \sup(P) = \inf(MC) = \inf(P^*) = \infty$. In particular, (P) is consistent and unbounded and (P^*) is improperly asymptotically consistent. To prove this, let $0 < \epsilon < 1/2$ and let E be the triangular domain with the three points $O(0,0)$, $P(0, 1/2)$, $Q(\epsilon, \epsilon)$ as vertices. (See Figure 2.) We set $\sigma(x_1, x_2) = (-1, -1 + 1/(2\epsilon))$ in

Table 2: Duality States for Program (P^*) with 0 Designating an Impossible State, with HS Denoting HSINC

INC							CONS			
SINC			AC							
			IAC	PAC						
HAC		HS		AUBD		ABD	AUBD		ABD	
			HS	HAC			HCONS	HINC	HS	
HCONS	HINC			HCONS	HINC	HS	UBD		BD	
0	0	8	7	0	0	5	0	0	0	1

E. Then $\sigma \cdot \nu = 1$ on $A \cap \partial E$, $\sigma \cdot \nu = (2\sqrt{2}\epsilon)^{-1}$ on $\overline{OQ} \cap \partial E$ and $\sigma \cdot \nu = 0$ on $\overline{PQ} \cap \partial E$. In the trapezoid with $O, Q, (1-\epsilon, \epsilon), (1, 0)$ as vertices, we set $\sigma(x) = ((2\sqrt{2}\epsilon)^{-1}, 0)$. Let $\Omega_\epsilon = (\epsilon, 1-\epsilon) \times (\epsilon, 1-\epsilon)$. Extending this σ to a vector field on $\Omega - \Omega_\epsilon$ symmetrically and setting $\sigma = 0$ on Ω_ϵ, we get a feasible flow such that $|\sigma| \le (2\sqrt{2}\epsilon)^{-1}$. We denote it by σ again. Then for $\lambda_\epsilon = 2\sqrt{2}\epsilon \inf_{\Omega-\Omega_\epsilon} \hat{c}$, we can see that $\lambda_\epsilon \sigma$ is a feasible flow and $\lambda_\epsilon \sigma \cdot \nu = \lambda_\epsilon f$ H_1-a.e. on $\partial\Omega$. Since $\lim_{\epsilon \to 0} \lambda_\epsilon = \infty$, we conclude that $\sup(MF) = \infty$.

REMARK 3.2 *(P) is consistent, unbounded, and homogeneous asymptotically consistent but homogeneous inconsistent. (P^*) is inconsistent and improperly asymptotically consistent. This determines duality state 7. The four possible for (P^*) are listed in the tabular classification taxonomy of [Kor74, page 744].*

4   CONVEX BIFUNCTION PERTURBATIONS FOR THE STRANG AND IRI DUALITY

In the previous sections, we are mainly concerned with Strang's problems, in which constraints $\text{div}\,\sigma = -\lambda F$ and $\sigma \cdot \nu = \lambda f$ are of one dimensional type imposed on Ω and $\partial\Omega$ respectively. Then taking

$$S_1 = \{x \in \Omega;\ F(x) > 0\} \quad \text{and} \quad S_2 = \{x \in \partial\Omega;\ f(x) > 0\}$$

as the sources, we can regard

$$\lambda\left(\int_{S_1} F dx + \int_{S_2} f dH_{n-1}\right) = \int_{S_1} -\text{div}\,\sigma\, dx + \int_{S_2} \sigma \cdot \nu dH_{n-1}$$

as the flow value, which is proportional to the optimizing value λ.

On the other hand, in Iri's problems, no constraints of $\sigma \cdot \nu$ are imposed on a source A and a sink B which are given subsets of $\partial\Omega$ and the flow value $\int_A \sigma \cdot \nu dH_{n-1}$ itself is optimized, although $\text{div}\,\sigma = 0$ on Ω and $\sigma \cdot \nu = 0$ are assumed outside of the source and the sink. One can recognize that Iri's Problems are more direct extensions of classical max–flow and min–cut problems.

In this section, we state a correspondence between the setting in Section 2 and a bifunction construction in [Roc70, Part VI] and apply it to the problems of Iri's type as well as Strang's type.

Let us recall formulations of $(P), (P^*)$ in Section 2. Although K is not a cone, we can rewrite these problems as usual linear programming problems as follows: We set

$$
\begin{aligned}
P_S &= \{(t(\sigma,\lambda),t);\ t \geq 0,\ (\sigma,\lambda) \in K\} \subset X \times R, \\
Q_S &= \{0\} \times \{0\} \subset Y \times R, \\
c_0 &= (e,0) \in X \times R, \\
d_0 &= (0_Y,1) \in Y \times R, \\
A_S((\sigma,\lambda),t) &= (T(\sigma,\lambda),t).
\end{aligned}
$$

Then A_S is a continuous linear mapping from $X \times R$ to $Y \times R$ and the adjoint A_S^* of A_S is given by $A_S^*(w,r) = (T^*(w),r)$ for $(w,r) \in Y^* \times R$. Let P_S^+, Q_S^+ be nonnegative cones of P_S, Q_S defined by

$$
\begin{aligned}
P_S^+ &= \{v \in X^* \times R;\ \langle u,v \rangle \geq 0 \text{ for all } u \in P_S\}, \\
Q_S^+ &= \{w \in Y^* \times R;\ \langle z,w \rangle \geq 0 \text{ for all } z \in Q_S\}.
\end{aligned}
$$

It follows that

$$
\begin{aligned}
\sup(P) &= \sup\{\langle u,c_0 \rangle;\ u \in P_S,\ d_0 - A_S u \in Q_S\}, \\
\inf(P^*) &= \inf\{\langle d_0,w \rangle;\ w \in Q_S^+,\ A_S^* w - c_0 \in P_S^+\}.
\end{aligned}
$$

Define the following convex bifunction based on the structure of (P), see [Roc70, Section 29]. Let

$$
(Gz)(p,q) = \langle -(c_0,0), (p,q) \rangle \tag{7}
$$
$$
+\delta(\ (p,q) \mid (p,q) \in P_S \times Q_S,\ z \in Y \times R,\ [A_S, I] \begin{pmatrix} p \\ q \end{pmatrix} = d_0 + z),
$$

where δ is the indicator function. Hence, in this case the adjoint bifunction is:

$$
(G^*(p^*,q^*))(w) \tag{8}
$$
$$
= \inf_{z \in Y \times R,\ p \in P_S,\ q \in Q_S} \{(Gz)(p,q) - \langle p,p^* \rangle - \langle q,q^* \rangle - \langle z,w \rangle\}
$$
$$
= \langle -d_0,w \rangle + \delta(\ w \mid A_S^T w - c_0 - p^* \in P_S^+, w - q^* \in Q_S^+\}
$$

where $p^* \in X^* \times R$, q^* *and* $w \in Y^* \times R$. Two perturbation functions, $(\inf G)$ and $(\sup G^*)$, convex and concave respectively, are used to state two duality equalities, (9) and (10) below :

$$
(\inf G)(z) = \inf_{p,q}(Gz)(p,q) \quad \text{and} \quad (\sup G^*)(w) = \sup_w (G^*(p^*,q^*))(w).
$$

They and their lower semi–continuous hulls (closures) enter into the following key duality equalities which are valid for general convex–bifunctions over linear topologies having the Hahn–Banach extension property, see [Roc70, Corollary 30.2.2] and [Kor74].

$$
(\inf G)(0) = (cl(\sup G^*))(0), \text{ or equivalently, } \sup(P) = \inf(AP^*) \tag{9}
$$

$$(\sup G^*)(0) = (cl(\inf G))(0), \text{ or equivalently, } \inf(P^*) = \sup(AP) \qquad (10)$$

Finally we are address the max–flow and min–cut problems of Iri's type which are originally formulated in Iri [Iri79] as an approximation of dense networks. Let Ω, Γ be as in the previous sections and let A, B be disjoint Borel subsets of $\partial\Omega$. Then the max–flow problem and min–cut problems of Iri's type corresponding with (Ω, Γ, A, B) are formulated as follows:

(MFI) $\qquad \sup \displaystyle\int_A \sigma \cdot \nu dH_{n-1}$

$\qquad\qquad$ subject to $\sigma(x) \in \Gamma(x)$ for a.e. $x \in \Omega$,

$\qquad\qquad$ div $\sigma = 0$ a.e. on Ω, $\sigma \cdot \nu = 0$ H_{n-1}−a.e. on $\partial\Omega - (A \cup B)$.

(MCI) $\qquad \inf C(S)$

$\qquad\qquad$ subject to $S \in Q$,

$\qquad\qquad H_{n-1}(A - \partial^* S) = H_{n-1}(B \cap \partial^* S) = 0$,

where $Q, C(S)$ are as defined in Section 2. The dual problem corresponding with (MFI) is written as

(MFI^*) $\qquad \inf \psi(u)$

$\qquad\qquad$ subject to $u \in W^{1,1}(\Omega)$,

$\qquad\qquad \gamma u = 1$ H_{n-1}−a.e. on A and $\gamma u = 0$ H_{n-1}−a.e. on B.

Then $(MFI), (MFI^*)$ can be reformulated as linear programming problems using a similar technique as for Strang's type. For the completeness, we state the formulations. Let X_0, X, Y be real linear spaces and K_Γ be the convex subset of $L^\infty(\Omega; R^n)$ as defined in Section 2. Instead of A_S, P_S, Q_S, we set

$$\begin{aligned} A_I(\sigma, t) &= (\text{div } \sigma, -\sigma \cdot \nu, t), \\ P_I &= \{(t\sigma, t) \in X; \ \sigma \in K_\Gamma \cap X_0, \ t \geq 0\}, \\ Q_I &= \{(0, y_2, 0) \in Y \times R; \ y_2 = 0 \ H_{n-1}-\text{a.e. on } \partial\Omega - (A \cup B)\}. \end{aligned}$$

Then A_I is a continuous linear mapping from X to $Y \times R$ and the adjoint A_I^* is given by

$$A_I^*(u, r) = (l_u, r),$$

where l_u is an element in X_0^* such that

$$\langle l_u, \sigma \rangle = \int_\Omega \sigma \cdot \nabla u dx$$

for all $\sigma \in X_0$. Furthermore we fix $u_0 \in W^{1,1}(\Omega)$ such that $\gamma u_0 = 1$ H_{n-1}−a.e. on A and $\gamma u_0 = 0$ H_{n-1}−a.e. on $\partial\Omega - A$, and set $c_I = (l_{u_0}, 0) \in X^*$, $d_I = (0_Y, 1) \in Y \times R$. Then we have

$$\begin{aligned} \sup(MFI) &= \sup\{\langle(\sigma, t), c_I\rangle; \ (\sigma, t) \in P_I, \ d_I - A_I(\sigma, t) \in Q_I\}, \\ \inf(MFI^*) &= \inf\{\langle d_I, (u, r)\rangle; \ (u, r) \in Q_I^+, \ A_I^*(u, r) - c_I \in P_I^+\}. \end{aligned}$$

Max-flow Min–cut Dualities

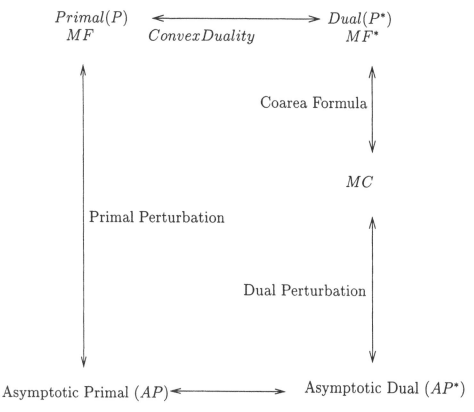

Figure 3: Max–flow min–cut dualities for Γ being compact convex inclusive. The program names below denote program values. It follows that $\inf(AP^*_{BV}) =$

$$\inf(AP^*) = \sup(P) = \sup(MF) \leq \inf(MC) = \inf(MF^*) = \inf(P^*) = \sup(AP)$$

Thus the same analysis applied to problems of Strang's type can be applied to problems of Iri's type. Since (MFI) is always homogeneous inconsistent, we get three possible cases of consistency. One can show examples for the three cases by building on an example in Section 3 of [Noz94] together with a slight modification of Example 2. Figure 3 provides broad relationships between the various dualities associated with max–flow min–cut problems.

5   CONCLUSIONS

One of the contributions of this paper is a synthesis of various known results about choices of function spaces and conditions for a max–flow min–cut duality free of duality–gaps. However, even with appropriate function space choices, other duality states can occur when boundedness conditions are relaxed on the underlying compact convex inclusive mappings. In this paper we define and establish existence of all duality states for the function–theoretic construction of infinite max–flow min–cut problems developed by Gilbert Strang and developed by Masao Iri.

References

[BICK69] A. Ben-Israel, A. Charnes, and K. O. Kortanek. Duality and asymptotic solvability over cones. *Bull. Amer. Math. Soc.*, 75:318–324, 1969. Erratum, Bull. Amer. Math. Soc. 76 (1970) 428.

[BICK71] A. Ben-Israel, A. Charnes, and K. O. Kortanek. Asymptotic duality over closed convex sets. *J. Math. Anal. Appl.*, 35:677–691, 1971. Erratum, Bull. Amer. Math. Soc. 38 (1972).

[Chr86] E. Christiansen. On the collapse solution in limit analysis. *Arch. Rational Mech. Anal.*, 91:119–135, 1986.

[Daf80] S. Dafermos. Continuum modelling of transportation networks. *Transportation Res.*, 14B:295–301, 1980.

[Duf56] R. J. Duffin. Infinite programs. In H.W. Kuhn and A.W. Tucker, editors, *Linear Inequalities and Related Systems*, pages 157 – 170. Princeton University Press Princeton, 1956.

[FF62] L. R. Ford and D. R. Fulkerson. *Flows in Networks*. Princeton University Press, Princeton, New Jersey, 1962.

[FL81] B. Fuchssteiner and W. Lusky. *Convex Cones*. Mathematical Studies 56. North-Holland, Amsterdam, The Netherlands, 1981.

[Gag57] E. Gagliardo. Caratterizzazioni delle tracce sulla frontiera relative ad alcune classi di funzioni in n variabili. *Rend. Sem. Mat. Univ. Padova*, 27:284 – 305, 1957.

[Gal67] D. Gale. A geometric duality theorem with economic applications. *Rev. Econ. Studies*, 34:19–24, 1967.

[Giu84] E. Giusti. *Minimal surfaces and functions of bounded variation*. Birkhäuser-Verlag, Basel-Boston, Massachusetts, 1984.

[GKR70] S. -Å. Gustafson, K. O. Kortanek, and W. O. Rom. Non-chebysevian moment problems. *SIAM J. Numer. Anal.*, 7:335–342, 1970.

[GKS76] W. Gochet, K. O. Kortanek, and Y. Smeers. On a classification scheme for geometric programming and complementarity theorems. *Applicable Analysis*, 6:47–59, 1976.

[Gol72] E. G. Golstein. *Theory of Convex Programming*. Trans. Math. Mono. Amer. Math. Soc., Providence, RI, 1972.

[Iri69] M. Iri. *Network flow, transportation and scheduling - theory and algorithms*. Academic Press, New York, New York, 1969.

[Iri79]   M. Iri. Theory of flows in continua as approximation to flows in net-
           works. In A. Prekopa, editor, *Survey of Math. Programming*, pages 263
           – 278, Amsterdam, The Netherlands, 1979. Mathematical Programming
           Society, North-Holland.

[JS83]   K. Jacobs and G. Seiffert. A measure-theoretical max–flow problem, part
           i. *Bull. Inst. Math. Acad. Sini.*, 11:261 – 280, 1983.

[Kor72]   K. O. Kortanek. On a compound duality classification scheme with homo-
           geneous derivants. *Rendiconti Di Matematica(Rome)*, 5:349–356, 1972.

[Kor74]   K. O. Kortanek. Classifying convex extremum problems over linear
           topologies having separation properties. *J. Math. Anal. Appl.*, 46:725–
           755, 1974.

[Kor75]   K. O. Kortanek. Extended abstract of classifying convex extremum prob-
           lems. *Zentralblatt fur Mathematik*, 283:491–496, 1975.

[Kor77]   K. O. Kortanek. Constructing a perfect duality in infinite programming.
           Applied Mathematics and Optimization, 3:357–372, 1977.

[KR71]   K. O. Kortanek and W. O. Rom. Classification schemes for the strong
           duality of linear programming over cones. *Opns. Res.*, 19:1571–1585,
           1971.

[KRS73]   K. O. Kortanek, W. O. Rom, and A. L. Soyster. On classification schemes
           for solution sets of chemical equilibrium problems. *Opns. Res.*, 21:240–
           246, 1973.

[KS72]   K. O. Kortanek and A. L. Soyster. On refinements of some duality the-
           orems in linear programming over cones. *Opns. Res.*, 20:137–142, 1972.

[KT83]   R. Kohn and R. Temam. Dual spaces of stresses and strains with appli-
           cations to hencky plasticity. *Appl. Math. Optim.*, 10:1 – 35, 1983.

[KW71]   C. Kallina and A. C. Williams. Linear programming in reflexive spaces.
           SIAM Rev., 13:350 – 376, 1971.

[Maz85]   W. Maz'ja. *Sobolev spaces*. Springer-Verlag, Berlin-New York, 1985.

[Neu84]   M. M. Neumann. A Ford-Fulkerson type theorem concerning vector-
           valued flows in infinite networks. *Czechoslovak Math. J.*, 34:156–162,
           1984.

[Noz90]   R. Nozawa. A max–flow min–cut theorem in an anisotropic network.
           Osaka J. Math., 27:805 – 842, 1990.

[Noz94]   R. Nozawa. Examples of max–flow and min–cut problems with duality
           gaps in continuous networks. *Mathematical Programming*, 63:213 – 234,
           1994.

[Roc70]   R. T. Rockafellar. *Convex Analysis*. Princeton University Press, Prince-
           ton, New Jersey, 1970.

[Rom70]   Walter O. Rom. *Classification Theory in Mathematical Programming and Applications.* PhD thesis, Cornell University, Industrial Engineering and Operations Research, Ithaca, New York, June 1970.

[Soy73]   Allen L. Soyster. *Classification Methods in Convex Programming Problems and Duality Over Convex Sets with Applications.* PhD thesis, Carnegie Mellon University, Graduate School of Industrial Adminiatration, Pittsburgh, PA, May 1973.

[Str83]   G. Strang. Maximal flow through a domain. *Mathematical Programming*, 26:123 – 143, 1983.

Central and Peripheral Results in the Study of Marginal and Performance Functions

Jean-Paul Penot
Laboratoire de Mathématiques Appliquées CNRS UPRES A 5033 Faculté des Sciences
Av. de l'Université 64000 Pau, France

1 INTRODUCTION

It is our purpose here to study some differentiability properties of the *performance function*

$$p(w) := \inf_{x \in F(w)} f(w, x)$$

of a parametrized minimization problem

$$(\mathcal{P}_w^F) \qquad \text{Minimize} \quad f(w, x) : x \in F(w),$$

where $f : W \times X \to \mathbb{R}$ is a differentiable function on the product of two Banach spaces W (parameter space) and X (decision space) and $F : W \rightrightarrows X$ is a feasible multimapping. We are especially interested in the case the feasible set $F(w)$ of $(\mathcal{P}_w)$ is defined by an implicit system of equalities and inequalities as in mathematical programming problems. Then, $(\mathcal{P}_w^F)$ takes the form

$$(\mathcal{P}_w) \qquad \text{Minimize} \quad f(w, x) : x \in X, \quad g(w, x) \in C,$$

where C is a closed convex cone of a Banach space Z and $g : W \times X \to Z$ is a continuously differentiable mapping, so that $F(w) = g_w^{-1}(C)$. In the sequel f_w and g_w stand for the partial maps $x \mapsto f(w, x)$ and $x \mapsto g(w, x)$ respectively. The cone C is not supposed to be polyhedral because this property is not satisfied in the case the spaces are usual functional spaces which serve for the modelization of a number of concrete problems. Even in the finite dimensional case, a number of situations such as semi-definite programming (see [98] and its references) deal with non polyhedral cones. Moreover, such a framework allows to use simple notations and has a more geometric nature.

In the case of problem $(\mathcal{P}_w)$, under appropriate assumptions, one can relate the set of multipliers of the problem to the subdifferential of the performance function p. In general the performance function is not differentiable everywhere and one has to replace the derivative of p by a convenient notion of generalized derivative. A similar situation occurs for the *marginal function*

$$m(w) := \sup_{x \in F(w)} f(w, x)$$

which is defined, according to the use in mathematical economics, as the value function of a parametrized maximization problem. However, one has to observe that due to the unilateral (i.e. one-sided) character of the subdifferentials considered here, one cannot transpose immediately the results concerning p into results concerning m or conversely.

Besides these results which can be considered as central for the study of performance functions, we also deal with more special results which may have some interest in various situations and which are simple enough. In particular, after giving general estimates in section 2, we focus section 3 on the study of the performance function in the case the feasible set is fixed, generalizing a famous theorem by Danskin [27]. We present several examples pertaining to such a simple situation in section 3 and we devote section 6 and 7 to the important cases of linear programming and quadratic programming which have been thoroughly investigated in the finite dimensional case but which deserve a closer attention in the infinite dimensional case. In the last section, we show the interest of sensitivity analysis for computational purposes.

We consider as central the results giving estimates of the generalized derivatives of the value function in terms of the data, in particular, in terms of multipliers. One reason for that is the interpretation of multipliers one gets: they appear as far from being an artificial "deus ex machina". On the contrary, they are naturally linked with the problem at hand. Here we do not consider only simple perturbations as in [78], [77] ; we deal with general perturbations for which the Lagrangian function cannot be avoided (sections 4 and 5). We also focus our attention on the role of qualification conditions. In our views, this role is twofold : they provide a linearization property on one hand, and a dualization property on the other hand. We dissociate these two roles, so that the strength of the different conditions one may use is put in clear light. We also show that the role of sufficient optimality conditions can be replaced by stability assumptions on the behavior of the approximate solution multifunction.

In the present paper we limit our study of the generalized derivative of the performance function to the first order, although a second order analysis is useful to deal with the critical case in which the stability assumption of the approximate solution multimapping is of Hölderian type. Also, the behavior of optimal solution multimapping

$$S(w) := \{x \in F(w) : f(w,x) = p(w)\}$$

(when p is studied, otherwise one replaces p by m in this relation) is not considered here, although it is an important matter which has been the object of numerous and interesting studies. The tools required for such a study, for instance the use of generalized equations in the sense of Robinson, are of a different nature. The list of references we give is far from being complete; however, we hope it gives useful hints for a vast literature (see [11], [14], [30], [54] for instance for other sources).

Most of the results of the present paper have been presented in a talk with similar title in the Third Symposium on Mathematical Programming with Data Perturbations in Washington in 1981 and in subsequent talks in Dijon, Paris, Pau, Twente, Warsaw. The mention of this fact gives us the opportunity to thank A. Fiacco for his personal incentive and for his continued dedication to the promotion of studies in mathematical programming with data perturbations.

2 GENERALIZED DERIVATIVES OF PERFORMANCE FUNCTIONS

In the sequel W and X are Banach spaces unless otherwise stated. The reader may assume for simplicity that both of them are finite dimensional, although among our aims is a treatment of extensions to the infinite dimensional case of results in sensitivity analysis which, in certain cases, may raise difficulties (see in particular section 8). Let us start our study by presenting estimates for the contingent derivative of the performance function p. Some other derivatives will be used in the paper, but we consider this directional derivative as the most important tool for dealing with this topic in view of the reasons mentioned below. Recall that the *contingent derivative* (or lower epi-derivative or lower derivative) of p is given by

$$p'(w, u) := dp(w, u) := \lim_{(t, u') \to (0_+, u)} \inf \, t^{-1}(p(w + tu') - p(w)).$$

This derivative is one of the simplest generalized derivatives avalaible. It is also lower than all usual directional derivatives, so that the associated subdifferential (given by relation (3.1) below) is smaller than the usual corresponding subdifferentials, an advantage in terms of precision. This directional derivative has better stability properties than the *(lower) radial derivative*

$$p'_r(w, u) := \lim_{t \to 0_+} \inf \, t^{-1}(p(w + tu) - p(w));$$

in particular it is lower semicontinuous in its second variable and examples in [77] show that this property reflects in a more realistic way the behavior of p. Moreover, p' has a nice geometric interpretation: the epigraph of $p'(w, \cdot)$ is the tangent cone at $(w, p(w))$ to the *epigraph*

$$E_p := \{(w, r) \in W \times \mathbb{R} : r \geq p(w)\}$$

of p, the *tangent cone* (or contingent cone, or Bouligand tangent cone) to a subset E of some normed vector space Z at some $e \in E$ being given by

$$T(E, e) := \lim_{t \to 0_+} \sup \, t^{-1}(E - e)$$

in the Painlevé-Kuratowski's sense. In the sequel, we identify the multimapping F with its graph and we use the *contingent derivative* $DF(w, x)(u)$ of F at (w, x) in the direction u given by

$$DF(w, x)(u) := \{v \in X : (u, v) \in T(F, (w, x))\},$$

as introduced by Aubin [2].

The following estimate is crucial and simple.

PROPOSITION 2.1 For each $u \in W$ and each x_0 in the solution set $S(w_0)$ of $(\mathcal{P}^F_{w_0})$ such that f is Hadamard-differentiable at (w_0, x_0) one has

$$p'(w_0, u) \leq \inf \left\{ f'(w_0, x_0)(u, v) : v \in DF(w_0, x_0)(u) \right\}.$$

Proof. Given $v \in DF(w_0, x_0)(u)$ we can find sequences $(t_n) \searrow 0$, $((u_n, v_n)) \to (u, v)$ such that

$$x_0 + t_n v_n \in F(w_0 + t_n u_n) \quad \forall n \in \mathbb{N}.$$

Thus $f(w_0 + t_n u_n, x_0 + t_n v_n) \geq p(w_0 + t_n u_n)$ and $f(w_0, x_0) = p(w_0)$, so that

$$f'(w_0, x_0)(u, v) \geq \liminf_n t_n^{-1}(p(w_0 + t_n u_n) - p(w_0)) \geq p'(w_0, u).$$

The result follows by taking the infimum over v.□

Let us derive a simple consequence of this result. It stems from the fact that if F is constant around w_0, with value C, then for each $x_0 \in S(w_0)$ and each $u \in W$ one has $DF(w_0, x_0)(u) = T(C, x_0)$ and $f'(w_0, x_0)v \geq 0$ for each each $v \in T(C, x_0)$.

COROLLARY 2.2 If F is constant in a neighborhood of w_0 then for each $u \in W$ one has

$$p'(w_0, u) \leq \inf \left\{ f'(w_0, x_0)(u, 0) : x_0 \in S(w_0) \right\}.$$

EXAMPLE 2.3. If $W = X$ is a Hilbert space, and if $f(w, x) = \frac{1}{2}\|w - x\|^2$, then, if F is the constant multimapping with value a closed convex subset C of X one has $\frac{1}{2}(d_C^2)'(w_0, u) \leq (w_0 - x_0 \mid u)$, where x_0 is the projection of w_0 on C. In fact, this inequality is an equality as we will see later on (section 5).

We will devote the following section to variations on the last corollary.

Now let us consider the case in which F is almost constant. Here X is supposed to be an arbitrary metric space and $d(x, A) := \inf_{a \in A} d(x, a)$ stands for the distance of a point $x \in X$ to a subset A of X.

Given a function $f : W \times X \to \mathbb{R} \cup \{\infty\}$ which is not necessarily differentiable, by f_x we denote the function on W obtained by fixing $x : f_x(w) = f(w, x)$; thus its contingent derivative should be denoted by f'_x. However, in order to avoid confusions with the partial derivative $D_X f := \frac{\partial f}{\partial x}$ of f, we denote it by df_x or $(f_x)'$. When X is a n.v.s. and f is differentiable at (w_0, x_0) we have $df_{x_0}(w_0, u) = f'(w_0, x_0)(u, 0)$, so that the conclusion of the following statement coincides with the one of Corollary 2.2 when F is constant.

PROPOSITION 2.4 Suppose that for some $u \in W$ and some $x_0 \in S(w_0)$ the following assumptions hold:

(a) for each $\varepsilon > 0$ there exist $\tau > 0$ and a neighborhood U of u such that $d(x_0, F(w_0 + tu')) \leq \varepsilon t$ for each $(t, u') \in (0, \tau) \times U$;

(b) there exist $c > 0$, $\tau_0 > 0$ and neighborhoods V of x_0, U_0 of u, such that $\mid f(w, x) - f(w, x_0) \mid \leq cd(x, x_0)$ for each $w \in w_0 + (0, \tau_0)U_0$ and each $x \in V$.
  Then

$$p'(w_0, u) \leq df_{x_0}(w_0, u).$$

Proof. Given $\varepsilon > 0$ we pick τ, τ_0, c and U, U_0 as in assumptions (a), (b); without loss of generality, we suppose $\tau < \tau_0$ and $U \subset U_0$. Then, for $(t, u') \in (0, \tau) \times U$, we pick $x_{t,u'} \in F(w_0 + tu')$ such that $d(x_0, x_{t,u'}) \leq 2\varepsilon t$. Then, as

$$p(w_0 + tu') \leq f(w_0 + tu', x_{t,u'}) \leq f(w_0 + tu', x_0) + 2c\varepsilon t,$$

we get

$$p'(w_0, u) \leq \lim_{(t,u') \to (0_+, u)} \inf t^{-1}(f(w_0 + tu', x_0) + 2c\varepsilon t - f(w_0, x_0)) = df_{x_0}(w_0, u) + 2\varepsilon c,$$

and the result follows, ε being arbitrarily small.$\square$

The preceding result deals with a rather peculiar case. However its assumptions are fulfilled when the feasible set is defined as in the mathematical programming problem $(\mathcal{P}_w)$, when $g'(w_0, x_0)(u, 0) \in T(C, g(w_0, x_0))$, in particular when $g'(w_0, x_0)(u, 0) = 0$ and when the following directional metric regularity condition (or controllability condition) is satisfied for some $\mu, \theta > 0$ and some neighborhood U of u :

$$d(x_0, (g_w)^{-1}(C)) \leq \mu d(g(w, x_0), C) \quad \text{for each } w \in w_0 + (0, \theta)U.$$

This last condition holds when a tangential regularity condition of Robinson's type is satisfied (see condition (R_0) in section 4 below and [80]).

In order to get lower estimates of the contingent derivative of p, let us introduce the approximate solution multifunction S with domain $\mathbb{R}_+ \times W$ which extends to $\mathbb{P} \times W$, with $\mathbb{P} = (0, \infty)$ the solution multimapping by

$$S(\varepsilon, w) \quad : \quad = \{x \in F(w) : f(w, x) \leq p(w) + \varepsilon\} \quad \text{when } p(w) > -\infty$$
$$S(\varepsilon, w) \quad : \quad = \left\{x \in F(w) : f(w, x) \leq -\varepsilon^{-1}\right\} \quad \text{when } p(w) = -\infty.$$

with $S(\varepsilon, w) = S(w)$ if $\varepsilon = 0$ (observe that $S(\varepsilon, w) = X$ when $p(w) = +\infty$, a case we discard in the sequel).

PROPOSITION 2.5 Suppose X is finite dimensional and that for some $u \in W$ and some $x_0 \in S(w_0)$ the following assumptions hold:

(a) for each $\varepsilon > 0$ there exist $c > 0, \tau > 0$ and a neighborhood U of u such that $d(x_0, S(t\varepsilon, w_0 + tu')) \leq ct$ for each $(t, u') \in (0, \tau) \times U$;

(b) f is differentiable at (w_0, x_0).

Then
$$p'(w_0, u) = \inf \left\{f'(w_0, x_0)(u, v) : v \in DF(w_0, x_0)(u)\right\}.$$

Proof. In view of Proposition 2.1 we may suppose $p'(w_0, u) < \infty$. Let $r > p'(w_0, u)$ and let $\varepsilon > 0$ be such that $r - \varepsilon > p'(w_0, u)$. Let $(t_n) \to 0_+$, $(u_n) \to u$ be such that $p'(w_0, u) = \lim_n t_n^{-1}(p(w_0 + t_n u_n) - p(w_0))$. Assumption (a) yields some $c > 0$ and a sequence (x_n) such that $x_n \in S(t_n \varepsilon, w_0 + t_n u_n)$, $\|x_n - x_0\| \leq ct_n$ for each n. Let us set $v_n := t_n^{-1}(x_n - x_0)$; without loss of generality, we may suppose (v_n) converges to some v in cB_X, X being finite dimensional. Then

$$f'(w_0, x_0)(u, v) \quad = \quad \lim_n t_n^{-1}(f(w_0 + t_n u_n, x_0 + t_n v_n) - f(w_0, x_0))$$
$$\leq \quad \lim_n t_n^{-1}(p(w_0 + t_n u_n) + t_n \varepsilon - p(w_0)) = p'(w_0, u) + \varepsilon < r.$$

Taking into account Proposition 2.1, the result follows, r being arbitrarily close to $p'(w_0, u)$.$\square$

When the problem $(\mathcal{P}_w^F)$ is a mathematical programming problem, condition (a) can be ensured by imposing metric regularity assumptions and optimality conditions (see [73], [77] Prop. 5.2). Moreover the result can be dualized in such a case (see section 4 below).

3 THE CASE OF A CONSTANT FEASIBLE SET. EXAMPLES

Let us give a closer look at the important case in which the constraints are fixed in problem $(\mathcal{P}_w^F)$: for some subset C of X one has $F(w) = C$ for each $w \in W$. Here X is an arbitrary topological space.

LEMMA 3.1 The following estimate holds

$$p'(w_0, u) \leq \inf \{ df_{x_0}(w_0, u) : x_0 \in S(w_0) \}.$$

Proof. This follows immediately from the relations $p \leq f_{x_0}$, $p(w_0) = f_{x_0}(w_0)$.□

Similar results hold when the contingent derivative is replaced by the *upper deriva-tive* $p^\sharp$ given by $p^\sharp(w, u) := -(-p)'(w, u)$ (for which the limit superior is substituted to the limit inferior) or by the *incident derivative* (or upper epi-derivative, for which the upper epiderivative is substituted to the lower epiderivative) p^i given by

$$p^i(w, u) := \sup_{V \in \mathcal{N}(u)} \limsup_{t \searrow 0} \inf_{v \in V} t^{-1}(p(w + tv) - p(w));$$

here $\mathcal{N}(u)$ denotes the filter of neighborhoods of u. Of course, the contingent derivative of f_{x_0} has to be replaced by the corresponding derivative in the inequality of the preceding lemma. The following consequence is obtained by changing f into $-f$.

COROLLARY 3.2 Denoting by $S(w_0)$ the solution set of the maximization problem of f_{w_0} on X one has

$$
\begin{aligned}
m^\sharp(w_0, u) &\geq \sup \left\{ (f_{x_0})^\sharp(w_0, u) : x_0 \in S(w_0) \right\}, \\
m'(w_0, u) &\geq \sup \{ df_{x_0}(w_0, u) : x_0 \in S(w_0) \}.
\end{aligned}
$$

These estimates have immediate consequences for the contingent subdifferential of the marginal function m given by

$$\partial m(w) := \{ w^* \in W^* : \forall u \in W \ \langle w^*, u \rangle \leq m'(w, u) \}. \tag{3.1}$$

Here W^* is the (topological) dual space of W.

COROLLARY 3.3 The following inclusion is valid without any assumption:

$$\partial m(w_0) \supset clco \left\{ \bigcup_{x_0 \in S(w_0)} \partial(f_{x_0})(w_0) \right\},$$

$clco(A)$ denoting the closed convex hull of A. In particular, for any $x_0 \in S(w_0)$ such that f_{x_0} is (Hadamard-) differentiable at w_0 one has $(f_{x_0})'(w_0) \in \partial m(w_0)$.

Let us observe that when X is also a normed vector space and $C = X$ one also has

$$p'(w_0, u) \leq \inf \{f'(w_0, x_0)(u, v) : v \in X, \ x_0 \in S(w_0)\},$$
$$\partial p(w_0) \times \{0\} \subset \partial f(w_0, x_0) \quad \forall x_0 \in S(w_0),$$

even when f is not differentiable. This last estimate for contingent derivatives enables one to recover the estimate of Proposition 2.1 by replacing f by $f + i_F$, where i_F is the indicator function of F given by $i_F(z) = 0$ if $z \in F$, $+\infty$ otherwise. In fact, when f is Hadamard differentiable, one has $(f + i_F)'((w_0, x_0), .) = f'(w_0, x_0)(\cdot) + i_{T(F, (w_0, x_0))}(\cdot)$, as easily seen.

In order to obtain equality in the estimate of Lemma 3.1, we may use the following proposition for which we will need a concept introduced in [70] and a variant of it. Such concepts have been used in the recent literature under various guises (for instance, in [M-S] section 6, the sequential version of this property has been renamed "semicompactness").

DEFINITION 3.4 Given topological spaces V and X, $v_0 \in V$ and a subset X_0 of X, a multimapping $M : V \rightrightarrows X$ is said to be lower semicontinuous (l.s.c.) at (v_0, X_0) if for any net $(v_i)_{i \in I}$ with limit v_0 there exists a subnet $(v_j)_{j \in J}$, an element x_0 of X_0 and a net $(x_j)_{j \in J}$ with limit x_0 such that $x_j \in M(v_j)$ for each $j \in J$.

When V is a topological vector space, the multimapping M is said to be l.s.c. at (v_0, X_0) in the direction $u \in V$ if the multimapping $\tilde{M} : \mathbb{P} \times V \rightrightarrows X$ given by $\tilde{M}(t, v) = M(v_0 + tv)$ is l.s.c. at $((0, u), X_0)$.

Obviously, if M is l.s.c. at (v_0, X_0), it is directionally l.s.c. at (v_0, X_0), i.e. l.s.c. at (v_0, X_0) in the direction u for each u in V. Moreover, if X_0 is a singleton $\{x_0\}$, we recover the classical definition of lower semicontinuity at (v_0, x_0) (see [15], [52] for example) ; in particular, when X is a metric space, M is l.s.c. at (v_0, x_0) iff $d(x_0, M(v)) \to 0$ as $v \to v_0$. On the other hand, this condition can be considered as a compactness condition similar to the one studied in [72] (in fact a weaker one). In particular, if M takes its values in a compact subset X_0 of X then it is l.s.c. at (v_0, X_0). The main cases of interest are when X_0 is a subset of $S(w_0)$; then one may say as in [26] that S is semicontinuous at w_0. A closedness assumption on the multifunction S ensures such a property. Such a closedness assumption is usually easy to satisfy. It is satisfied whenever the function f is continuous and the feasible multifunction F is closed and lower semicontinuous.

PROPOSITION 3.5 Suppose that for some $u \in W$, some subset X_0 of $S(w_0)$ and any $\varepsilon > 0$ the approximate solution multimapping S is l.s.c. at $((0, w_0), X_0)$ in the direction (ε, u). Suppose moreover that for each $x \in X_0$ the function f satisfies the following condition:

$$(H) \quad df_x(w_0, u) = \lim_{(t, u', x') \to (0_+, u, x)} \inf \ t^{-1} \left(f(w_0 + tu', x') - f(w_0, x') \right).$$

Then

$$p'(w_0, u) = \inf_{x \in X_0} df_x(w_0, u).$$

Proof. Let $(t_i)_{i \in I} \to 0_+$ and let $(u_i)_{i \in I} \to u$ be such that

$$p'(w_0, u) = \lim_{i \in I} t_i^{-1}(p(w_0 + t_i u_i) - p(w_0)).$$

Given $\varepsilon > 0$, our assumption on S yields some $x \in X_0$, some subnets $(t_j)_{j \in J}$ and $(u_j)_{j \in J}$ of $(t_i)_{i \in I}$ and $(u_i)_{i \in I}$ respectively and a net $(x_j)_{j \in J} \to x$ such that $x_j \in S(t_j \varepsilon, w_0 + t_j u_j)$. When $p(w_0 + t_j u_j) = -\infty$ for j in a cofinal subset K of J we have

$$t_j^{-1}(f(w_0 + t_j u_j, x_j) - f(w_0, x_j)) \leq t_j^{-1}(-t_j^{-1}\varepsilon^{-1} - p(w_0)) \to -\infty$$

and in view of assumption (H) the inequality holds. When $p(w_0 + t_j u_j) > -\infty$ for j large enough this is also the case in view of assumption (H) and of the inequalities

$$t_j^{-1}(p(w_0 + t_j u_j) - p(w_0)) \geq t_j^{-1}(f(w_0 + t_j u_j, x_j) - t_j \varepsilon - f(w_0, x_j)),$$

ε being arbitrary. Thus we get the inequality

$$p'(w_0, u) \geq \min_{x \in X_0} df_x(w_0, u).$$

When X_0 is a subset of $S(w_0)$ the reverse inequality holds by Lemma 3.1. $\square$

When f has a derivative with respect to w which is continuous in both variables w and x at each point (w_0, x_0) with $x_0 \in X_0$, assumption (H) is satisfied. More generally, we have the following sufficient condition.

LEMMA 3.6 Suppose the partial contingent derivative $(u', w, x') \mapsto df_{x'}(w, u')$ is l.s.c. at (u, w_0, x). Then assumption (H) is satisfied at (w_0, x) in the direction u.

Proof. It suffices to prove that for any $r < df_x(w_0, u)$ one has

$$r \leq \lim_{(t, u', x') \to (0_+, u, x)} \inf t^{-1}\left(f(w_0 + tu', x') - f(w_0, x')\right).$$

By assumption, there exist neighborhoods U, W_0, V of u, w_0, x respectively such that

$$r < df_{x'}(w, u')$$

for each $(u', w, x') \in U \times W_0 \times V$. We can choose $\varepsilon > 0$, a smaller neighborhood W_1 of w_0 and shrink U so that $w + tu' \in W_0$ for each $(t, w, u') \in (0, \varepsilon) \times W_1 \times U$. Then, using the fact that the radial lower derivative of $f_{x'}$ is minorized by its contingent derivative and applying the Mean Value Theorem (see [74] for instance), we get

$$f_{x'}(w_0 + tu') - f_{x'}(w_0) \geq rt \quad \text{for } (t, w, u', x) \in (0, \varepsilon) \times W_1 \times U \times V,$$

hence

$$\lim_{(t, u', x') \to (0_+, u, x)} \inf t^{-1}\left(f(w_0 + tu', x') - f(w_0, x')\right)$$
$$\geq \inf_{(t, u', x') \in (0, \varepsilon) \times U \times V} t^{-1}\left(f(w_0 + tu', x') - f(w_0, x')\right) \geq r$$

and the result is proved.$\square$

Supposing f is continuously differentiable and C is compact, we recover the famous theorem of Danskin [27] asserting the directional differentiability of the performance function, a conclusion slightly stronger than the conclusion of the following corollary. The assumptions made here are also weaker. Results of this type and results obtained without existence of solutions at w_0 are to be found in [68], [51]. Let us note a consequence of the preceding proposition with a simpler (but more stringent) assumption.

COROLLARY 3.7 Suppose that for some $u \in W$, some subset X_0 of the solution set $S(w_0)$ the solution multimapping S is l.s.c. at (w_0, X_0) in the direction u. Suppose moreover that for each $x \in X_0$ the function f satisfies condition (H). Then

$$p'(w_0, u) = \inf_{x \in X_0} df_x(w_0, u) = \inf_{x \in S(w_0)} df_x(w_0, u).$$

In the sequel we say that p is *semi-differentiable* at w_0 in the direction u if $p^\sharp(w_0, u) = p'(w_0, u)$, i.e. if the differential quotients $t^{-1}(p(w_0 + tu') - p(w_0))$ converge as $(t, u') \to (0_+, u)$. When the limit $p'(w_0, u)$ is linear in u, it is also linear and continuous, so that one has differentiability of p at w_0 in the sense of Hadamard.

COROLLARY 3.8 Suppose with the assumptions of Corollary 3.4 that for each $x \in X_0$ the function f_x is semi-differentiable at w_0 in the direction u. Then p is semi-differentiable at w_0 in the direction u.

Proof. This follows from the inequalities

$$p'(w_0, u) \le p^\sharp(w_0, u) \le \inf_{x \in S(w_0)} (f_x)^\sharp(w_0, u) = \inf_{x \in S(w_0)} df_x(w_0, u) = p'(w_0, u).$$

The preceding proposition has also consequences pertaining to the contingent subdifferential of p.

COROLLARY 3.9 Suppose the assumptions of Corollary 3.4 hold for each $u \in W$. Then
$$\partial p(w_0) = \bigcap_{x \in X_0} \partial f_x(w_0) = \bigcap_{x \in S(w_0)} \partial f_x(w_0).$$

Without any assumption one just has the inclusion

$$\partial p(w_0) \subset \bigcap_{x \in S(w_0)} \partial f_x(w_0)$$

which stems from Lemma 3.1.

Uniqueness results can be deduced from these relations.

Let us close this section by presenting some examples.

EXAMPLE 3.10: **distance to a closed subset.** Suppose C is a closed subset of a n.v.s. X and that for some $w_0 \in W := X$ there exists a best approximation x_0 of w_0 in C such that any sequence (x_n) in C converges to x_0 whenever $(\|x_n - w_0\|) \to$

$d_C(w_0) := \inf_{x \in C} \|x - w_0\|$. This assumption is satisfied if C is closed convex and if X is a Hilbert space, or more generally an uniformly convex Banach space. If moreover the norm is Gâteaux-differentiable at $w_0 - x_0$ then d_C is Hadamard-differentiable at w_0 with derivative $\| \cdot \|'(w_0 - x_0)$. This result slightly improves [69] Proposition 1.7 in which the norm was supposed to be everywhere Gâteaux-differentiable. The fact that assumption (H) is satisfied in this case follows from the relation

$$t^{-1} \left(\|w_0 + tu' - x'\| - \|w_0 - x'\| \right) \geq \langle y^*, u' \rangle$$

for each $y^* \in \partial \| \cdot \|(w_0 - x')$ and the fact such a y^* converges to $\| \cdot \|'(w_0 - x_0)$ in the weak* topology as $(t, u', x') \to (0_+, u, x_0)$ by a well-known upper semicontinuity property of subdifferentials of convex functions.

EXAMPLE 3.11 : support functions. Let W be a n.v.s. and let C be a weak* closed bounded subset of $X := W^*$. Then the support function

$$h(w) := \sup_{x \in C} \langle x, w \rangle$$

of C has a directional derivative at w_0 given by

$$h'(w_0, u) = \sup_{x \in S(w_0)} \langle x, u \rangle$$

where $S(w_0) := \{x \in C : \langle x, w_0 \rangle = h(w_0)\}$ since the function f given by $f(w, x) := -\langle x, w \rangle$ is continuously differentiable on $W \times X$ (observe that $h = -p$). Of course, since h is a continuous sublinear functional, this result can be established by standard methods of convex analysis.

EXAMPLE 3.12 : eigenvalues. Let W be a n.v.s. and let X be a Hilbert space with scalar product denoted by $(\cdot \mid \cdot)$. Given $w_0 \in W$ and an open neighborhood V of w_0, let $T : V \to L(H, H)$ be a continuously differentiable mapping such that $T(w_0)$ is compact, i.e. such that for some compact subset K of X one has $T(w_0)B_X \subset K$, B_X being the closed unit ball of X. Let $C := S_X$, the unit sphere and set

$$m(w) := \sup_{x \in C}(T(w)x \mid x).$$

Let us suppose X is finite dimensional or $m(w) \neq 0$ for w close to w_0. Then, $S(w)$ is nonempty, $m(w)$ is the greatest eigenvalue of $T(w)$ and $S(w) = E(w) \cap S_X$, where $E(w)$ is the eigenspace corresponding to the eigenvalue $m(w)$.

In many concrete problems the derivative of m with respect to the parameter w has a great physical or technological importance (see [42], [43] for instance). Here the existence of this derivative at w_0 is guaranteed by Corollary 3.8. In fact, since f given by $f(w, x) := -(T(w)x \mid x)$ is continuously differentiable, assumption (H) is satisfied. Let us show that the approximate solution multimapping is l.s.c. at $((0, w_0), S(w_0))$. Let $((\varepsilon_i, w_i))_{i \in I} \to (\varepsilon, w)$ and let $x_i \in S(\varepsilon_i, w_i)$. A subnet $(x_j)_{j \in J}$ of $(x_i)_{i \in I}$ converges weakly to some $x_0 \in B_X$ while $(T(w_0)x_j)_{j \in J}$ has a limit y in the closure of $T(w_0)(B_X)$. Since $(x_j)_{j \in J}$ is contained in B_X and since $\|T(w_j) - T(w_0)\| \to 0$, we also have that $(T(w_j)x_j)_{j \in J} \to y$. Then, for $T_j := T(w_j)$, one has $T_j^* \to T(w_0)^*$, and for each $z \in X$

$$(y \mid z) = \lim(T_j x_j \mid z) = \lim(x_j \mid T(w_0)^* z) = (x_0 \mid T(w_0)^* z) = (T(w_0)x_0 \mid z),$$

so that $y = T(w_0)x_0$. Now, since $(x_j)_{j \in J}$ is bounded, for each $x \in C$ we have

$$
\begin{aligned}
(T(w_0)x_0 \mid x_0) &= \lim(T(w_0)x_0 \mid x_j) = \lim(T(w_j)x_j \mid x_j) \\
&\geq \liminf (m(w_j) - \varepsilon_j) \geq \lim(T(w_j)x \mid x) = (T(w_0)x \mid x),
\end{aligned}
$$

so that $x_0 \in S(w_0)$. Therefore we can apply Corollary 3.8 to obtain that m is semi-differentiable at w_0 with semi-derivative given by

$$
m'(w_0, u) = \sup_{x \in S(w_0)} (T'(w_0)(u)x \mid x).
$$

When $E(w_0)$ is one dimensional, $S(w_0) = E(w_0) \cap S_X$ is the pair $\{x_0, -x_0\}$ and m is Hadamard differentiable at w_0, whereas when $m(w_0)$ is a multiple eigenvalue, $m'(w_0, \cdot)$ is just sublinear. For other results in this direction we refer to [55]-[56], [65], [66], [HU-L], [L].

EXAMPLE 3.13 : infimal regularization. Suppose W is a Banach space and X is a topological space. Recall that a function $h : X \to \mathbb{R} \cup \{+\infty\}$ is said to be inf-compact if if for each $c \in \mathbb{R}$ the set $h^{-1}((-\infty, c])$ is compact. Given a function $K : W \times X \to \mathbb{R}_+ \cup \{+\infty\}$ called a kernel, and $r > 0$, the regularized function f_r associated with a function $f : X \to \mathbb{R} \cup \{+\infty\}$ is given by

$$
f_r(w) := \inf_{x \in X} (f(x) + r^{-1}K(w, x)).
$$

We say that a function $f : X \to \mathbb{R} \cup \{+\infty\}$ is K-minorized if there exists some $a > 0$ such that $f + a^{-1}K(w, \cdot)$ is inf-compact. We shall use the following assumption on the kernel K :

(K_1) the function K is lower semicontinuous (l.s.c.) on $W \times X$;

(K_2) for each $x \in X$ the function K_x is semi-differentiable on W and

$$
K'_x(w, u) = \lim_{(t, u', x') \to (0_+, u, x)} t^{-1}[K(w + tu', x') - K(w, x')];
$$

(K_3) for each $w_0 \in W$ there exist $b, c > 0$ and a neighborhood W_0 of w_0 such that $K(w_0, x) \leq c(K(w, x) + b)$ for each $(w, x) \in W_0 \times X$;

(K_4) the space W is reflexive, X is the space W endowed with its weak topology and $K(w, w) = 0$ for each $w \in W$.

In the usual case, assumption (K_4) is satisfied, $K(w, x) = k(w - x)$ where $k : X \to \mathbb{R}_+$ is convex continuous and coercive, for instance $k = h \circ \| \cdot \|$, where $h : \mathbb{R}_+ \to \mathbb{R}_+$ is convex continuous; then assumption (K_1) is satisfied and conditions (K_2) and (K_3) follow from appropriate assumptions on h and the norm.

PROPOSITION 3.14 Suppose $f : X \to \mathbb{R} \cup \{+\infty\}$ is l.s.c., not everywhere $+\infty$ and K-minorized, where K satisfies assumptions $(K_1) - (K_3)$. Then , for $r > 0$ small enough the regularized function f_r given by

$$
f_r(w) := \inf_{x \in X} (f(x) + r^{-1}K(w, x))
$$

is finitely-valued and semi-differentiable. If moreover (K_4) is satisfied, then for $r > s > 0$ one has $f_r \leq f_s \leq f$ and $(f_r) \to f$ pointwise as $r \to 0_+$. If moreover in assumption (K_2) the limit is uniform for u in bounded sets and $K'_x(w, \cdot)$ is linear and continuous, then f_r is Fréchet differentiable.

Proof. If f is finite at some $\overline{x} \in X$ one has $f_r(w) \leq \overline{m} := f(\overline{x}) + r^{-1}K(w, \overline{x})$. Given $w_0 \in W$, let b, c, W_0 be as in condition (K_3). Then, for $r \in (0, ac^{-1})$, $w \in W_0$, $x \in X$, one has

$$f(x) + r^{-1}K(w, x) \geq f(x) + c^{-1}r^{-1}K(w_0, x) - r^{-1}b \geq f(x) + a^{-1}K(w_0, x) - r^{-1}b,$$

so that for each $t \in \mathbb{R}$ one has the inclusion

$$M_r(w, t) := \left\{x \in X : f(x) + r^{-1}K(w, x) \leq t\right\} \subset M_a(w_0, t + r^{-1}b),$$

and as this last set is compact by assumption. As f and $K(w, \cdot)$ are l.s.c. on X, the infimum in the definition of f_r is attained and for each $w \in W_0$ the set $S_r(w)$ of minimizers of $f + r^{-1}K(w, \cdot)$ is contained in $M_a(w_0, \overline{m} + r^{-1}b)$ which is compact and nonempty. Therefore f is finite on W_0 and the semi-differentiability result follows from the assumptions and from Corollary 3.5. The proof of the last assertions is easy; we refer to [9], [19] for instance.□

4 THE ROLE OF QUALIFICATION CONDITIONS

In the case of the mathematical programming problem $(\mathcal{P}_w)$, another expression for the estimate of p' can be given under the assumption that the constraint is *linearizable* at (w_0, x_0) in the sense that the following relation (in which $z_0 = g(w_0, x_0)$) is satisfied:

$$(L) \quad T(F, (w_0, x_0)) = \{(u, v) : g'(w_0, x_0)(u, v) \in T(C, z_0)\}.$$

PROPOSITION 4.1 If condition (L) is satisfied at (w_0, x_0) then

$$p'(w_0, u) \leq \inf \{f'(w_0, x_0)(u, v) : g'(w_0, x_0)(u, v) \in T(C, z_0)\}.$$

This immediate consequence of Proposition 2.1 can be given several variants ; in particular, a similar inequality can be obtained for the incident derivative $p^i(w_0, \cdot)$ when the tangent cone in (L) can be replaced with the incident cone. Condition (L) is a natural assumption which can be obtained under qualification conditions. The one we present below could be relaxed into a directional metric regularity condition as introduced in [73]. We present it in a more usual form, so that its derivation from [71], [84], [100] is easier. Such a condition has its origins in the classical result of Hoffman about linear inequalities.

LEMMA 4.2 The constraint is linearizable whenever the following metric regularity condition is satisfied:

(MR) there exist $c > 0$, $r > 0$ such that

$$d((w, x), g^{-1}(C)) \leq cd(g(w, x), C) \text{ for any } (w, x) \in B((w_0, x_0), r).$$

In turn, condition (MR) can be obtained under various infinitesimal qualification conditions such as the constant rank condition [49] or the following Robinson's condition (R) which generalizes the condition of Mangasarian and Fromovitz.

LEMMA 4.3 ([71], [84], [100]) The following regularity condition ensures condition (MR), hence condition (L) :

$$(R) \quad g'(w_0, x_0)(W \times X) - \mathbb{R}_+(C - z_0) = Z.$$

When the interior of the cone C is nonempty, it is shown in [71] that condition (R) is equivalent to the condition

$$(\overline{R}) \quad g'(w_0, x_0)(W \times X) - T(C, z_0) = Z$$

or to the Slater's type condition

$$(S) \quad \exists (u, v) \in W \times X \ : g'(w_0, x_0)(u, v) + g(w_0, x_0) \in int(C).$$

Note that here both conditions bear on the perturbation and not on the unperturbed problem. Thus condition (R) is more general than the Robinson's regularity condition relative to the unperturbed problem $(\mathcal{P}_{w_0})$ which reads as

$$(R_0) \quad g'(w_0, x_0)(X) - \mathbb{R}_+(C - z_0) = Z.$$

Note that here and in the sequel we identify X with $\{0\} \times X$ and we write $g'(w_0, x_0)(v)$ instead of $g'(w_0, x_0)(0, v)$; we use similar notations with other mappings.

Clearly, condition (R_0) implies the following condition

$$(R_u) \quad g'(w_0, x_0)(\mathbb{R}_+\{u\} \times X) - \mathbb{R}_+(C - z_0) = Z,$$

which in turn implies the following condition

$$(R'_u) \quad Z_u := g'(w_0, x_0)(\mathbb{R}_+\{u\} \times X) - T(C, z_0) \text{ is a closed vector subspace.}$$

The directional condition (R_u) also appears in [8], [12], [16]; in finite dimensions, it is equivalent to the condition of Gollan [40]. Condition (R'_u), which is reminiscent to a condition of Attouch and Brézis, seems to be new in the context of parametrized optimization; when Z is finite dimensional, it amounts to the simple algebraic relation

$$Z_u = -Z_u.$$

Such a relation is easier to check than a relation of the form

$$0 \in int Z_u.$$

Moreover, it suffices for the use of duality in the preceding estimate which will enable us to formulate it in terms of the Lagrangian L of the problem given by

$$L(w, x, y) := f(w, x) + \langle y, g(w, x) \rangle, \quad (w, x, y) \in W \times X \times Y$$

with $Y := Z^*$ and in terms of the set of multipliers at (w_0, x_0)

$$M(w_0, x_0) := \left\{ y \in C^0 : \langle y, g(w_0, x_0) \rangle = 0, \ L'(w_0, x_0, y)(v) = 0 \ \forall v \in X \right\}.$$

Here we denote by Q^0 the polar cone of a set Q, so that $(T(C, z_0))^0$ is the normal cone to C at $z_0 := g(w_0, x_0)$ and we observe that, as is well-known,

$$N(C, z_0) := (T(C, z_0))^0 = \left\{ y \in C^0 : \langle y, z_0 \rangle = 0 \right\}.$$

PROPOSITION 4.4 Suppose that for some $x_0 \in S(w_0)$, $u \in X$ the following assumptions are satisfied:
(a) condition (L) holds at (w_0, x_0);
(b) condition (R'_u) holds at (w_0, x_0).
Then

$$p'(w_0, u) \leq \sup \left\{ L'(w_0, x_0, y)u : y \in M(w_0, x_0) \right\}.$$

and if $p'(w_0, u) > -\infty$ the set $M(w_0, x_0)$ of multipliers at (w_0, x_0) is nonempty and the supremum is attained.

Condition (R_0) (and also condition (R_u)) can be substituted to assumptions (a), (b) in view of the preceding observations. Moreover, in that case one can replace the preceding estimate by a similar (but stronger) estimate bearing on the upper derivative $p^\sharp(w_0, u)$ (see [80]) as in that case one has the following partial metric regularity condition :

$$(PMR) \quad \text{there exist } c \ > \ 0, \ r > 0 \ \text{such that}$$
$$d(x, (g_w)^{-1}(C)) \ \leq \ c d(g_w(x), C) \text{ for any } (w, x) \in B((w_0, x_0), r).$$

This condition is stronger than condition (MR) ; however, the term "partial" is justified by the fact that the condition bears on the partial map g_w. The proof of the proposition relies on the following lemma which is a refined version of a classical duality result in linear programming (see [75]).

LEMMA 4.5 Let $A : X \to Z$ be linear and continuous, let $b \in Z$, let $c \in X^*$ and let Q be a closed convex cone in Z such that $Z_0 := A(X) - \mathbb{R}_+ b - Q$ is a closed linear subspace of Z. If

$$\alpha := \inf \left\{ c(x) : x \in X, \ Ax - b \in Q \right\} > -\infty$$

then there exists $y \in Q^0$ such that $c + y \circ A = 0$ and

$$\alpha = \langle y, -b \rangle = \max \left\{ \langle y, -b \rangle : y \in Q^0, \ c + y \circ A = 0 \right\}.$$

Moreover, the set $M := \{ y \in Q^0 : \ c + y \circ A = 0, \ \langle y, -b \rangle = \alpha \}$ is nonempty and weak* compact.

Proof of Proposition 4.4. In the preceding lemma, let us observe that α is finite and let us take $A = D_X g(w_0, x_0) := g'(w_0, x_0) \mid X$, $b := -g'(w_0, x_0)u$, $c := D_X f(w_0, x_0)$, $Q := T(C, z_0)$, with $z_0 := g(w_0, x_0)$. With these choices, $M(w_0, x_0)$ is exactly the set of $y \in Q^0$ such that $c + y \circ A = 0$ and Proposition 4.1 ensures that $p'(w_0, u) \leq \alpha$. The result follows.$\square$

COROLLARY 4.6 Suppose that for each $u \in W$ the assumptions of the preceding proposition are satisfied. Then

$$\partial p(w_0) \subset cl\, \{D_W L(w_0, x_0, y) : y \in M(w_0, x_0)\}.$$

When $M(w_0, x_0)$ is bounded, or when the map $(D_W g(w_0, x_0))^T$ transposed of $D_W g(w_0, x_0)$ is a closed mapping for the weak* topologies, in particular when $D_W g(w_0, x_0)$ is an isomorphism, then one can dispense with the closure operation in the preceding inclusion.

Proof. Since the sublinear function q given by

$$q(u) := \sup\, \{D_W L(w_0, x_0, y)u : y \in M(w_0, x_0)\}$$

majorizes $p'(w_0, \cdot)$ we have

$$\partial p(w_0) = \partial p'(w_0, \cdot)(0) \subset \partial q(0).$$

Since $\{D_W L(w_0, x_0, y) : y \in M(w_0, x_0)\}$ is convex as the image by the affine map $h : y \to D_W f(w_0, x_0) + y \circ D_W g(w_0, x_0)$ of the (weak*) closed convex set $M(w_0, x_0)$, the first assertion follows from a standard result of convex analysis (see also the second example of section 3). When this map h is closed, or, equivalently, when $(D_W g(w_0, x_0))^T$ is closed, $\{D_W L(w_0, x_0, y) : y \in M(w_0, x_0)\}$ is closed and convex. Moreover, the map h is continuous for the weak* topologies, so that the image of $M(w_0, x_0)$ is closed when $M(w_0, x_0)$ is compact, or equivalently, bounded.$\square$

The assumption that $(D_W g(w_0, x_0))^T$ is closed (and even an isomorphism) is satisfied when one has $g(w, x) = g_0(x) + w$, in particular when one deals with the simply perturbed problem

$$(S_w) \quad \text{minimize } f_0(x) \quad : g_0(x) + w \in C,$$

where $f_0 : X \to \mathbb{R}$, $g_0 : X \to Z$ are given differentiable mappings.

COROLLARY 4.7 ([78]) Suppose the assumptions of the preceding proposition hold for the simply perturbed problem (S_w). Then

$$\partial p(w_0) \subset M(w_0, x_0).$$

On the other hand, the set $M(w_0, x_0)$ of multipliers at $M(w_0, x_0)$ is bounded when condition (R_0) is satisfied, by a well-known result of [39] extended in [71] and [100].

COROLLARY 4.8 When condition (R_0) holds, one has the following inclusion

$$\partial p(w_0) \subset \{D_W L(w_0, x_0, y) : y \in M(w_0, x_0)\}.$$

A uniqueness result can be deduced from Corollary 4.6 (see also [44], [79]).

COROLLARY 4.9 Suppose the assumptions of Corollary 4.6 are in force.
(a) Suppose that for $y, y' \in M(w_0, x_0)$ one has $D_W L(w_0, x_0, y) = D_W L(w_0, x_0, y')$. Then if $\partial p(w_0)$ is nonempty, it is a singleton.
(b) Suppose that for $x_0 \neq x_1 \in S(w_0)$, $y_0 \in M(w_0, x_0)$, $y_1 \in M(w_0, x_1)$ one has $D_W L(w_0, x_0, y_0) \neq D_W L(w_0, x_1, y_1)$ and condition (R_0) is satisfied at any $x_0 \in S(w_0)$. Then, if $\partial p(w_0)$ is nonempty, $S(w_0)$ is at most a singleton.

5 LOWER ESTIMATE OF THE CONTINGENT DERIVATIVE

A lower estimate of the contingent derivative of the performance function p can be obtained under a generalized convexity assumption and an additional compactness assumption. Recall that a differentiable function is *invex* if any of its critical points is a global minimizer. In particular, any pseudoconvex function (thus any convex function) is invex. For a recent use of the notion of invexity in the context of saddle functions (generalizing earlier results of [26]), we refer to [95]. In fact, we can avoid the use of such a notion provided we introduce an appropriate set of multipliers.

Given a solution $x_0 \in S(w_0)$ of the unperturbed problem $(\mathcal{P}_{w_0})$, the set of x_0-*global multipliers* is the set

$$M'(w_0, x_0) := \left\{ y \in (T(C, g(w_0, x_0)))^0 : \forall x \in X \quad L(w_0, x, y) \geq L(w_0, x_0, y) \right\}.$$

Obviously, this set is contained in the set $M(w_0, x_0)$ of multipliers of $(\mathcal{P}_{w_0})$ at x_0. It also contains the set

$$M(w_0) := \left\{ y \in C^0 : \inf_{x \in X} L(w_0, x, y) = p(w_0) \right\}$$

of global multipliers of $(\mathcal{P}_{w_0})$ since for any $x \in X$, $y \in M(w_0)$, $x_0 \in S(w_0)$ we have $L(w_0, x, y) \geq p(w_0)$ and

$$p(w_0) = f(w_0, x_0) \geq L(w_0, x_0, y) \geq p(w_0),$$

so that $\langle y, g(w_0, x_0) \rangle = 0$ and $L(w_0, x_0, y) = p(w_0) \leq L(w_0, x, y)$ for each $x \in X$. The following obvious lemma justifies the use of generalized convexity assumptions.

LEMMA 5.1 If for some $x_0 \in S(w_0)$, $y_0 \in M(w_0, x_0)$ the function $L(w_0, \cdot, y_0)$ is invex or pseudoconvex, then $y_0 \in M'(w_0, x_0)$. If $f(w_0, \cdot)$ and $g(w_0, \cdot)$ are convex, then $M(w_0, x_0) = M'(w_0, x_0) = M(w_0)$.

We are ready to state and prove our estimate. We observe that its assumptions are weaker than the hypothesis of [94]; moreover, we deal here with the contingent derivative and not with the radial derivative of the performance function p.

PROPOSITION 5.2 Suppose that for some subset X_0 of $S(w_0)$ the approximate solution multimapping S is lower semicontinuous at $((0, w_0), X_0)$ in any direction (ε, u) with $\varepsilon > 0$, $u \neq 0$. Then, for each $u \in W \setminus \{0\}$ one has

$$p'(w_0, u) \geq \inf_{x_0 \in X_0} \sup_{y \in M'(w_0, x_0)} L'(w_0, x_0, y) u.$$

If for each $x_0 \in X_0$ and each $y \in M(w_0, x_0)$ the function $L(w_0, \cdot, y)$ is invex, then one can replace $M'(w_0, x_0)$ by the set $M(w_0, x_0)$ of multipliers in this estimate.

Proof. It suffices to prove the first assertion. Given $\varepsilon > 0$, $u \in W \backslash \{0\}$, and arbitrary sequences $(u_n) \to u$, $(t_n) \to 0_+$, taking subsequences if necessary, one can find some $x_0 \in X_0$ and some sequence $(x_n) \to x_0$ such that $x_n \in S(t_n \varepsilon, w_0 + t_n u_n)$ for each n. Then, for each $y \in M'(w_0, x_0)$ one has $\langle y, g(w_0 + t_n u_n, x_n) \rangle \leq 0$ hence

$$p(w_0 + t_n u_n) + t_n \varepsilon \geq f(w_0 + t_n u_n, x_n) \geq L(w_0 + t_n u_n, x_n, y).$$

On the other hand, by definition of $M'(w_0, x_0)$, one has for each n

$$p(w_0) = f(w_0, x_0) = L(w_0, x_0, y) \leq L(w_0, x_n, y)$$

hence

$$\liminf_n \frac{1}{t_n}(p(w_0 + t_n u_n) - p(w_0)) \geq \liminf_n \frac{1}{t_n}(L(w_0 + t_n u_n, x_n, y) - L(w_0, x_n, y)) - \varepsilon$$
$$\geq L'(w_0, x_0, y)u - \varepsilon.$$

We can conclude by taking the supremum over $y \in M'(w_0, x_0)$ and then the infimum over $x_0 \in C$ and over the sequences (t_n) and (u_n). $\square$

The following result extends previous results of [53], [8], [16], [78]; in particular it extends Theorem 3.2 of [78] from the case of simple perturbations to the case of arbitrary perturbations. For the sake of simplicity, we keep the present definition of the approximate solution multimapping, although one could admit the constraints are just approximatively satisfied, as in [78].

PROPOSITION 5.3 Suppose f and g are Fréchet differentiable at (w_0, x_0), where $x_0 \in S(w_0)$, and that, for some $u \in W$, the approximate solution multimapping satisfies the following assumption :
(A_u) for each $\varepsilon > 0$ $\limsup_{(t,w) \to (0_+, u)} t^{-1} d(x_0, S(\varepsilon t, w_0 + tw)) < \infty$.
Then

$$p'(w_0, u) \geq \sup \{L'(w_0, x_0, y)u : y \in M(w_0, x_0)\}.$$

Proof. The result is obvious when $M(w_0, x_0)$ is empty. Let $y \in M(w_0, x_0)$. Given $\varepsilon > 0$, assumption (A_u) yields some $c > 0$, $\rho > 0$, $\tau > 0$ such that $d(x_0, S(\varepsilon t, w_0 + tw)) < ct$ for $w \in u + \rho B_W$, $t \in (0, \tau)$. Let us pick $x_{t,w} \in S(\varepsilon t, w_0 + tw)$ such that

$$\|x_{t,w} - x_0\| \leq ct.$$

Let $\alpha := \varepsilon(\|u\| + 1 + c)^{-1}$. The differentiability of L yields $\delta \in (0, 1)$ such that for $w \in \delta B_W$, $x \in \delta B_X$ one has

$$| L(w_0 + w, x_0 + x, y) - L(w_0, x_0, y) - L'(w_0, x_0, y)w | \leq \alpha \|(w, x)\|.$$

Let $\sigma \in (0, \tau), \gamma \in (0, \delta)$ be such that $\sigma c \leq \tau$, $\sigma(\|u\| + 1) < \delta$, $\gamma \|L'(w_0, x_0, y)\| \leq \varepsilon$. As, for each $w \in u + \gamma B_W$, $t \in (0, \sigma)$ we have $tw \in \delta B_W$, $x_{t,w} - x_0 \in \delta B_X$,

$\langle y, g(w_0 + tw, x_{t,w}) - g(w_0, x_0) \rangle \leq 0$, the definitions and the preceding inequalities yield

$$
\begin{aligned}
p(w_0 + tw) + \varepsilon t - p(w_0) &\geq f(w_0 + tw, x_{t,w}) - f(w_0, x_0) \\
&\geq L(w_0 + tw, x_{t,w}, y) - L(w_0, x_0, y) \\
&\geq L'(w_0, x_0, y)tw - \alpha\|(tw, x_{t,w} - x_0)\| \\
&\geq tL'(w_0, x_0, y)u - \gamma t\|L'(w_0, x_0, y)\| - \alpha t(\|u\| + 1 + c) \\
&\geq tL'(w_0, x_0, y)u - 2\varepsilon t
\end{aligned}
$$

and the result follows, ε being arbitrarily small. $\square$

COROLLARY 5.4 Suppose that f and g are Fréchet differentiable at (w_0, x_0), where $x_0 \in S(w_0)$, and that, for each $u \in W$, condition (A_u) is satisfied. Then

$$
\partial p(w_0) \supset \{D_W L(w_0, x_0, y) : y \in M(w_0, x_0)\}
$$

Gathering the conclusions of Propositions 4.4 and 5.3 one gets the following result.

THEOREM 5.5 Suppose that f and g are Fréchet differentiable at (w_0, x_0). Suppose that for some $u \in W$ and some $x_0 \in S(w_0)$ the conditions (A_u), (L) and (R'_u) are satisfied. Then

$$
p'(w_0, u) = \sup\{L'(w_0, x_0, y)u : y \in M(w_0, x_0)\}.
$$

Condition (A_u) is clearly weaker than the more familiar condition (see [91])

$(A'_u) \qquad \limsup_{(t,w)\to(0_+,u)} t^{-1}d(x_0, S(w_0 + tw)) < \infty.$

However, as observed in [53], [81], [16], dealing with approximate solutions gives more flexibility and more generality. Let us devote the rest of this section and the following one to a study of conditions ensuring assumption (A_u). For this purpose, let us recall that a point x_0 of a metric space X is said to be a (local) *minimizer of order* $k > 0$ of a function j on X if there exist $\gamma > 0$ and a neighborhood V of x_0 in X such that

$$
j(x) \geq j(x_0) + \gamma d(x, x_0)^k \quad \text{for each } x \in V.
$$

Let us also recall that the *excess* of a subset A of x over a subset B of X is

$$
e(A, B) := \sup_{x\in A} d(x, B) := \sup_{x\in A} \inf_{y\in B} d(x, y).
$$

PROPOSITION 5.6 Suppose that for some $x_0 \in S(w_0)$ and some $u \in W$ such that $p^\sharp(w_0, u) < \infty$ the following assumptions hold:
(a) x_0 is a local minimizer of order one of $j := f_{w_0}$ on $F(w_0)$;
(b) the approximate solution multimapping S is l.s.c. at $((0, w_0), x_0)$;
(c) there exists a neighborhood V of x_0 in X such that

$$
\lim_{(t,w)\to(0_+,u)} \sup \; t^{-1}e(F(w_0 + tw) \cap V, F(w_0)) < \infty;
$$

(d) $\lambda := \limsup_{(t,w,x,x') \to (0_+,u,x_0,x_0)} (t + d(x,x'))^{-1} \mid f(w_0 + tw, x) - f(w_0, x') \mid < \infty.$
Then condition (A_u) is satisfied; more precisely, for any $\varepsilon > 0$ there exist a neighborhood U of u and $\tau > 0$, $c > 0$ such that $S(t\varepsilon, w_0 + tw) \cap V \neq \emptyset$ for $(t,w) \in (0,\tau) \times U$ and

$$e(S(t\varepsilon, w_0 + tw) \cap V, x_0) \leq ct.$$

Proof. Shrinking V if necessary, we may assume that there exists $\gamma > 0$ such that $j(x) \geq j(x_0) + \gamma d(x, x_0)$ for each $x \in V$, where $j := f_{w_0}$ and V is given by assumption (c): for some $\kappa > 0, \tau > 0$ and some neighborhood U of u one has $e(F(w_0 + tw) \cap V, F(w_0)) \leq \kappa t$ for each $(t,w) \in (0,\tau) \times U$. Picking $\mu > p^{\sharp}(w_0, u)$ and taking smaller τ and U if necessary, we may assume $p(w_0 + tw) - p(w_0) \leq \mu t$ for each $(t,w) \in (0,\tau) \times U$. Then for such a pair $(t,w) \in (0,\tau) \times U$ and for each $x_{t,w} \in S(t\varepsilon, w_0 + tw) \cap V$ (by assumption (b) such a point exists, provided τ and U are small enough) we can pick $x'_{t,w} \in F(w_0)$ such that $d(x_{t,w}, x'_{t,w}) \leq 2\kappa t$. Then, shrinking again U and τ, we may assume

$$\mid f(w_0 + tw, x_{t,w}) - f(w_0, x'_{t,w}) \mid \leq (\lambda + 1)(d(x_{t,w}, x'_{t,w}) + t),$$

so that

$$\begin{aligned}
f(w_0, x'_{t,w}) & \leq f(w_0 + tw, x_{t,w}) + (\lambda + 1)(d(x_{t,w}, x'_{t,w}) + t) \\
& \leq p(w_0 + tw) + \varepsilon t + (\lambda + 1)(2\kappa + 1)t \\
& \leq p(w_0) + \mu t + \varepsilon t + (\lambda + 1)(2\kappa + 1)t.
\end{aligned}$$

It follows from assumption (a) that

$$d(x_{t,w}, x_0) \leq \gamma^{-1}(\mu + \varepsilon + (\lambda + 1)(2\kappa + 1))t$$

This inequality (together with $d(x_{t,w}, x'_{t,w}) \leq 2\kappa t$) gives the result. $\square$

Although the assumption that x_0 is a minimizer of order one is fairly restrictive (on the contrary of the assumption that it is a local minimizer of order two), it is satisfied in a number of cases, as the following lemma shows. Here we denote the radial tangent cone to a subset C of X at some point x_0 of C by $T^r(C, x_0) := \{v \in X : \exists (t_n) \searrow 0, \ x_0 + t_n v \in C \ \forall n \in \mathbb{N}\}$; when C is a convex subset of X one has $T^r(C, x_0) = \mathbb{R}_+(C - x_0)$ and its closure is $T(C, x_0)$.

LEMMA 5.7 Let X be a finite dimensional n.v.s. and let $j : X \to \mathbb{R}$ be differentiable at some $x_0 \in C \subset X$. Each of the following conditions ensure that x_0 is a local minimizer of j of order one on C :
(a) $j'(x_0)(v) > 0$ for each $v \in T(C, x_0) \backslash \{0\}$;
(b) j is pseudo-concave, x_0 is an isolated minimizer of j on C and $T(C, x_0) = T^r(C, x_0)$;
(c) j is linear, x_0 is an isolated minimizer of j and C is a polyhedron.

Proof. Clearly, assumption (c) entails assumption (b). Let us show that assumption (b) implies assumption (a). Suppose on the contrary that there exists $v \in T(C, x_0) \backslash \{0\}$ such that $j'(x_0)(v) = 0$. Since j is pseudo-concave, we have $j(x_0 + tv) \leq j(x_0)$ for each $t > 0$. Now, as $v \in T(C, x_0) = T^r(C, x_0)$ we can find a sequence $(t_n) \searrow 0$ such that $x_0 + t_n v \in C$ for each $n \in \mathbb{N}$, a contradiction with the fact that x_0 is a local minimizer of j on C. Now it remains to prove that condition (a) implies that x_0 is a local minimizer of j of order one on C (see also [61]). Suppose the contrary : there exist sequences $(\gamma_n) \searrow 0$ and $(x_n) \to x_0$ in C such that $j(x_n) - j(x_0) < \gamma_n \|x_n - x_0\|$. Then $t_n := \|x_n - x_0\| > 0$ and we may assume that $(v_n) := (t_n^{-1}(x_n - x_0)) \to v$ for some unit vector v. Obviously, $v \in T(C, x_0)$ and since $j'(x_0)(v) = \lim_n t_n^{-1}(j(x_n) - j(x_0))$, we get a contradiction. $\square$

6 THE CASE OF LINEAR PROGRAMMING

We devote the present section to the important case of the parametrized linear programming problem

$$(\mathcal{L}_w) \quad \text{minimize } \langle c(w), x \rangle \ : x \in B, \ A(w)(x) - b(w) \in C$$

where $A : W \to L(X, Z)$, $b : W \to Z$, $c : W \to X^*$ are continuously differentiable mappings and $B \subset X$, $C \subset Z$ are closed convex cones. It corresponds to the case $f(w, x) = c(w)(x)$, $g(w, x) = (x, A(w)(x) - b(w))$, Z (resp. C) being changed into $X \times Z$ (resp. $B \times C$). Although the constraint is linear in the decision variable, the linearization property (L) is not automatic. But other features are specific to the linear case (see [46], [60], [82], [85] for instance). The first one concerns the regularity condition (R_0).

PROPOSITION 6.1 The regularity condition (R_0) does not depend on the choice of an optimal solution x_0. In fact it can be written

$$(R_0) \ A(w_0)(B) - C + \mathbb{R}_+ b = Z.$$

Proof. Given an optimal solution x_0, setting $z_0 := A(w_0)(x_0) - b(w_0)$ we observe that condition (R_0) is equivalent to

$$A(w_0)\mathbb{R}_+(B - x_0) - \mathbb{R}_+(C - z_0) = Z.$$

Then we use the fact that for two convex subsets E, E' of Z containing 0 one has the equivalence

$$\mathbb{R}_+(E - E') = Z \Leftrightarrow \mathbb{R}_+ E - \mathbb{R}_+ E' = Z,$$

taking $E := A(w_0)(B - x_0)$, $E' := C - z_0$. $\square$

The following lemma is not much more than a rephrasing of Lemma 2 in [85].

LEMMA 6.2 Suppose either that X and Z are finite dimensional or the following condition is satisfied:

$$(MR_0) \quad \text{there exists } c \ > \ 0 \text{ such that for } x \in B$$
$$d(x, F(w_0)) \ \leq \ cd(A(w_0)(x) - b(w_0), C).$$

Suppose that the feasible set $F(w_0)$ is bounded and nonempty. Then there exist a neighborhood W_0 of w_0 and $\lambda > 0$ such that for $w \in W_0$ one has

$$e(F(w), F(w_0)) \leq \lambda \max(\|A(w) - A(w_0)\|, \|b(w) - b(w_0)\|).$$

In particular, if A and b are Lipschitzian around w_0 there exist a neighborhood W_0 of w_0 and $\mu > 0$ such that for $w \in W_0$ one has

$$e(F(w), F(w_0)) \leq \mu d(w, w_0).$$

In fact, by a well-known result of Hoffman, condition (MR_0) is satisfied when X and Z are finite dimensional; it is also satisfied under assumption (R_0). The conclusion which can be written

$$F(w) \subset F(w_0) + \mu d(w, w_0) B_X$$

can be adapted to the case of the approximate solution multimapping.

PROPOSITION 6.3 Suppose either that X and Z are finite dimensional or the following condition is satisfied:

(MR'_0) there exists $\gamma > 0$ such that for $x \in B$

$$d(x, S(w_0)) \leq cd(A(w)(x) - b(w), C) + \gamma \mid c(w_0)x - p(w_0) \mid .$$

Suppose that the solution set $S(w_0)$ is weakly compact and nonempty. Then, for each $u \in W$ there exists $x_0 \in S(w_0)$ such that

$$p'(w_0, u) \geq \sup \{ L'(w_0, x_0, y)u : y \in M(w_0, x_0) \} .$$

Proof. We may suppose $p'(w_0, u) < \infty$. Given $s > p'(w_0, u)$ we set

$$G(t, w) := \{ x \in F(w_0 + tw) : c(w_0 + tw)(x) \leq p(w_0) + st \} ,$$

so that $G(0, u) = S(w_0)$. The preceding lemma ensures that there exist $\tau > 0$, a neighborhood W_0 of u and $\lambda > 0$ such that for $(t, w) \in (0, \tau) \times W_0$ one has

$$e(G(t, w), G(0, u)) \leq \lambda \delta(t, w)$$

with

$$\delta(t, w) := \max(\| A(w_0 + tw) - A(w_0) \|, \| b(w_0 + tw) - b(w_0) \|, \| c(w_0 + tw) - c(w_0) \|, st).$$

Let $(t_n) \to 0_+$, $(u_n) \to u$ be such that $\lim_n t_n^{-1}(p(w_0 + t_n u_n) - p(w_0)) = p'(w_0, u)$. For n large enough we have $p(w_0 + t_n u_n) < p(w_0) + st_n$, so that there exists $x_n \in G(t_n, u_n)$. Since A, b and c are continuous, we have

$$d(x_n, G(0, u)) \to 0.$$

Thus there exists a sequence (x'_n) in $G(0, u) = S(w_0)$ such that $(d(x'_n, x_n)) \to 0$. Since $S(w_0)$ is weakly compact, taking a subsequence if necessary, we may suppose (x_n) and (x'_n) have a weak limit x_0. Then, as for any $y \in M(w_0, x_0)$ we have

$$c(w_0 + t_n u_n)(x_n) \geq L(w_0 + t_n u_n, x_n, y)$$

and

$$p(w_0) = c(w_0)(x_0) = L(w_0, x_0, y) \leq L(w_0, x_n, y)$$

we get

$$
\begin{aligned}
s & \geq \lim_n t_n^{-1}(L(w_0 + t_n u_n, x_n, y) - L(w_0, x_n, y)) \\
& = L'(w_0, x_0, y)(u).
\end{aligned}
$$

As s is arbitrarily close to $p'(w_0, u)$, the result is proved. $\square$

In the preceding result, as in the following one, we avoid the strong assumption made in [82] that the feasible set $F(w_0)$ is bounded. In the finite dimensional case one can combine the preceding arguments (using exact solutions) with the main result of [85] to get the following proposition.

PROPOSITION 6.4 Suppose X and $Y = Z^*$ are finite dimensional. Suppose the set of optimal solutions of $(\mathcal{L}_w)$ and of its dual are nonempty and bounded. Then there exists a neighborhood W_0 of w_0 in W and a constant $\gamma > 0$ such that the sets $S(w)$ and $S^*(w)$ of optimal solutions of $(\mathcal{L}_w)$ and of its dual $(\mathcal{D}_w)$ are nonempty, bounded for $w \in W_0$ and satisfy the inequalities

$$e(S(w), S(w_0)) \leq \gamma(\|A(w) - A(w_0)\| + \|b(w) - b(w_0)\| + \|c(w) - c(w_0)\|).$$

Moreover, for each $u \in W$ one has

$$p'(w_0, u) = \inf_{x_0 \in S(w_0)} \max \left\{ L'(w_0, x_0, y_0)(u) : y_0 \in S^*(w_0) \right\}.$$

Let us now examine the infinite dimensional case. Again we make use of duality arguments by considering the dual problem

$$(\mathcal{D}_w) \quad \text{maximize } \langle y, -b(w) \rangle \; : y \in C^0, \; -A^T(w)(y) - c(w) \in B^0.$$

The analogue of the regularity condition (R_0) for this problem is the condition

$$(R_0^*) \; A^T(w_0)(C^0) - B^0 = X^*.$$

PROPOSITION 6.5 Suppose conditions (R_0) and (R_0^*) are satisfied and suppose X and $Y = Z^*$ are reflexive. Then $(\mathcal{L}_w)$ and $(\mathcal{D}_w)$ have optimal solutions for w close enough to w_0 which form bounded sets. Moreover, for each $u \in W$ one has

$$p'(w_0, u) = \inf_{x_0 \in S(w_0)} \max \left\{ L'(w_0, x_0, y_0)(u) : y_0 \in S^*(w_0) \right\}.$$

Proof. The first assertion is given in [46] Theorem 2.2. Since (R_0) holds, and since $S^*(w_0)$ coincides with the set of multipliers at (w_0, x_0) for each $x_0 \in S(w_0)$, the inequality

$$p'(w_0, u) \leq \inf_{x_0 \in S(w_0)} \max \left\{ L'(w_0, x_0, y_0)(u) : y_0 \in S^*(w_0) \right\}$$

holds by Proposition 4.4. Moreover, since condition (R_0) holds, for w close enough to w_0, the set $S^*(w)$ coincides with the set of multipliers of $(\mathcal{L}_w)$ and is contained in a fixed bounded set. Since $(\mathcal{L}_w)$ is the dual of $(\mathcal{D}_w)$ and since condition (R_0^*) holds, the same can be said of $S(w)$. Then the argument of Proposition 6.3 gives the reverse inequality we need. $\square$

7 THE CASE OF QUADRATIC PROGRAMMING

Let us show that in the important case of quadratic programming, the assumption of the preceding theorem are satisfied. In fact, in this case, a direct approach to the differentiability properties of p is possible. The assumptions we make on the problem

$$(\mathcal{Q}_w) \quad \text{minimize } \frac{1}{2}\langle Q(x-c), x-c\rangle \; : \; G(x) = w$$

are weaker than the ones made usually (see for instance [4]). Here $G : X \to W$ is a surjective continuous linear map between the Banach spaces X and W, c is a given element of X, $Q : X \to X^*$ is a continuous linear map which induces a positive definite quadratic form q on the kernel N of G (in the sense :there exists $\alpha > 0$ such that $q(v) \geq \alpha\|v\|^2$ for each $v \in N$,not necessarily for $v \in X$). This amounts to suppose $Q_N := P_N \circ Q \circ J_N$ is definite positive, J_N being the canonical injection of N into X and $P_N : X^* \to N^*$ being its transpose J_N^T. Let us note that our assumptions do not entail convexity of p, as does the standard assumption that Q is semi-definite positive.

LEMMA 7.1 Under the preceding assumptions, for each $w \in W$ the problem $(\mathcal{Q}_w)$ has a unique solution $x_w := S(w)$ characterized by the following relations in which y_w is some element of $Y := W^*$,

$$\begin{aligned} Q(x_w - c) + y_w \circ G &= 0 \\ G(x_w) &= w. \end{aligned}$$

Moreover, for each $w_0 \in W$, the solution mapping S satisfies

$$\lim_{\substack{w \to w_0 \\ \neq}} \sup \|w - w_0\|^{-1} d(S(w_0), S(w)) < \infty.$$

Thus p is subdifferentiable everywhere.

Proof. The proof of the first assertions is standard. Existence stems from the reflexivity of N, the coercivity and the weak lower semicontinuity of the restriction of the quadratic form to $a_w + N$, where a_w is an arbitrary element of $G^{-1}(w)$. Necessity of the condition is a special case of the Lagrange theorem; sufficiency follows from a simple replacement, since for $x \in G^{-1}(w)$ the preceding relations imply that

$$\langle Q(x - c), x - c\rangle - \langle Q(x_w - c), x_w - c\rangle = \langle Q(x - x_w), x - x_w\rangle \geq 0.$$

An easy consequence of the Michael's selection theorem (see [6] p. 355, for instance) ensures that G has a positively homogeneous continuous right inverse H_0 satisfying for some $\gamma > 0$ the relation $\|H_0(w)\| \leq \gamma\|w\|$ for each $w \in W$. Then the mapping H given by $H(w) = x_0 + H_0(w - w_0)$ satisfies the relations $H(w_0) = x_0$, $\|H(w) - H(w_0)\| \leq \gamma\|w - w_0\|$ for each $w \in W$. Setting $a_w := H(w)$, $n_w := x_w - a_w$, we have

$$\begin{aligned} Q J_N n_w &= Q(x_w - c) + Q(c - a_w) \\ &= -y_w \circ G + Q(c - a_w), \end{aligned}$$

hence

$$Q_N(n_w) = (P_N \circ Q)(c - a_w)$$

and

$$x_w = (Q_N^{-1} \circ P_N \circ Q)(c - a_w) + a_w.$$

It follows that there exists some $\sigma > 0$ such that $\|S(w_0) - S(w)\| \le \sigma \|w - w_0\|$. Thus the subdifferentiability of p is a consequence of Corollary 5.4. In fact one can obtain here the subdifferentiability of p in the Fréchet sense (see [78]).□

In order to state our differentiability conclusions, which are stronger than the ones obtained in the general case, let us recall that a mapping $S : W \to X$ is said to be *B-differentiable* (or boundedly semi-differentiable, or Bouligand-differentiable) at w_0 if there exists a continuous positively homogeneous map $S'(w_0) : W \to X$ such that

$$\lim_{\|u\| \searrow 0} \|u\|^{-1} \left(S(w_0 + u) - S(w_0) - S'(w_0)(u) \right) = 0.$$

The proof below uses the fact that if $T : X \to Y$ is B-differentiable at $x_0 := S(w_0)$, in particular if T is linear and continuous or bilinear and continuous, then $T \circ S$ is B-differentiable and its derivative is $T'(x_0) \circ S'(w_0)$.

PROPOSITION 7.2 Under the preceding assumptions, the solution map S, the multiplier map M and the performance function p are B-differentiable on W. Moreover, if the kernel N of G has a topological complement, then p, S and M are infinitely Fréchet differentiable on W.

Proof. The formula obtained above shows that S is B-differentiable at w_0 with derivative $S'(w_0) = (I - Q_N^{-1} \circ P_N \circ Q) \circ H_0$. Since

$$y_w = y_w \circ G \circ H_0 = -Q(x_w - c) \circ H_0,$$

the multiplier mapping M is also B-differentiable with derivative given by

$$M'(w_0)(u) = Q(S'(w_0)(u)) \circ H_0.$$

Here we consider W^* as a subspace of the space $C(B_W)$ of bounded continuous functions on the unit ball of W and we use the fact that $M(w) = T(S(w))$, where $T : X \to C(B_W)$ is the linear and continuous mapping from X into $C(B_W)$ given by $T(x) := -Q(x - c) \circ H_0 \mid B_W$. The derivative $M'(w_0)(u)$ is in fact an element of the dual of W as this function on W is linear and continuous: for some $c > 0$, $\delta > 0$ one has

$$M'(w_0)(u)(z) = \lim_{t \searrow 0} \langle t^{-1}(M(w_0 + tu) - M(w_0)), z \rangle \quad \forall z \in W,$$

$$\mid M'(w_0)(u)(z) \mid \le \sup_{0 < t < \delta} t^{-1} \|M(w_0 + tu) - M(w_0)\| \|z\| \le c \|u\| \|z\| \quad \forall z \in W.$$

Since

$$p(w) = \langle -y_w \circ G, x_w - c \rangle = -\langle y_w, w - G(c) \rangle,$$

the performance function p is also B-differentiable.

When N has a topological complement, one can take H to be linear and continuous, so that S and M are affine continuous and p is quadratic continuous.□

The preceding study can be extended to the case Q, b, c, w are replaced by A_w, b_w, c_w, d_w respectively, where w is an arbitrary parameter in some n.v.s.

8 A COMPUTATIONAL APPLICATION OF SENSITIVITY

From a computational point of view, calculating the derivative of the inverse of a matrix depending on a parameter w may be quite awkward or costly. Therefore it may be more fruitful to consider an adjoint problem, as it is done for instance in shape optimization (see [20], [96] for instance) and for inverse problems ([10]) ; this process is extensively used by petroleum companies. Let us abstract such a situation in a simple and general framework.

Let W, X, Y, Z be Banach spaces, Y being the dual space of Z, let W_0 be an open neighborhood of $w_0 \in W$ and let $G : W_0 \to L(X, Z)$ be differentiable at w_0 and such that $G(w_0)$ is an isomorphism. Let $f : X \to \mathbb{R}$ be differentiable at $x_0 \in X$, where x_0 is the unique solution to

$$G(w_0) x = b$$

where b is a given element of Z. Then, taking W_0 small enough, one may suppose $G(w)$ is invertible for each $w \in W_0$. One is often faced to the problem of studying the function h given by

$$h(w) = f\left(G(w)^{-1} b\right).$$

Then h is easily seen to be differentiable at w_0 with

$$
\begin{aligned}
h'(w_0).w &= -f'(x_0) \circ G(w_0)^{-1}\left(G'(w_0).w\left(G(w_0)^{-1} b\right)\right) \\
&= -f'(x_0) \circ G(w_0)^{-1}.G'(w_0).w.x_0.
\end{aligned}
$$

This formula involves the computation of the inverse of $G(w_0)$, so that another way of computing $h'(w_0)$ is advisable. The method we present below relies on the use of an adjoint equation ; it is widely used. Let us show its links with sensitivity analysis. We use mild assumptions which do not require $G(w_0)$ to be invertible ; they are not the most general hypothesis, as we will see, but they suffice for our purposes.

We consider the differentiation of the performance function

$$p(w) := \inf \{f(x) : g(w, x) = 0\}$$

where $g : W_0 \times X \to Z$ is differentiable at $(w_0, x_0) \in g^{-1}(0)$. Let us assume

(H1) $D_W g(w_0, x_0) : X \to Z$ is surjective

(H2)   the solution multifunction $X : W_0 \rightrightarrows X$ given by

$$S(w) = \{x \in X : f(w, x) = p(w)\}$$

is *tangentially compact* at (w_0, x_0) in the direction $u \in W$ in the following sense: for any sequence (t_n, u_n) in $]0, \infty[\times W$ with limit $(0, u)$ there exists $v \in X$, a subsequence $(t_n, u_n)_{n \in K}$ of (t_n, u_n) and a sequence $(v_n)_{n \in K}$ with limit v such that $x_0 + t_n v_n \in S(w_0 + t_n u_n)$ for each $n \in K$.

Assumption (H2) is obviously fullfiled if S is a single-valued mappping which is differentiable in the usual Hadamard (or Fréchet) sense at w_0. It is also satisfied when X is finite dimensional and S satisfies a kind of Lipschitz (or rather Stepanov) condition as in Theorem 5.5; in fact the following result can be considered as an easy variant of that theorem.

In particular, assumption (H2) is satisfied when $g(w, x) = G(w)x - z$, G is differentiable and $G(w_0)$ is an isomorphism.

PROPOSITION 8.1 Under assumptions (H1), (H2) the performance function p is differentiable at w_0 in the direction u and its derivative is given by

$$p'(w_0, u) = \langle y_0, g'(w_0, x_0)(u, 0) \rangle$$

where y_0 is the Lagrange multiplier corresponding to the constrained problem $(\mathcal{P}_{w_0})$.

Proof. Under assumption (H1), the existence of a Lagrange multiplier y is well-known. Let $(t_n, u_n)_{n \geq 0}$ be a sequence of $]0, \infty[\times W$ with limit $(0, u)$. Taking a subsequence and a sequence $(v_n)_{n \in K}$ with limit v as in in assumption (H2) we get

$$g'(w_0, x_0)(u, v) = 0.$$

Now for $n \in K$ we have

$$
\begin{aligned}
t_n^{-1}(p(w_0 + t_n u_n) - p(w_0)) &= t_n^{-1}(f(x_0 + t_n v_n) - f(x_0)) \\
&\to f'(x_0)v = -\langle y, g'(w_0, x_0)(0, v) \rangle \\
&= \langle y, g'(w_0, x_0)(u, 0) \rangle.
\end{aligned}
$$

Therefore p is differentiable in the direction w and its derivative is given by the announced formula. $\square$

Returning to our original problem by setting

$$g(w, x) = G(w)x,$$

under the surjectivity assumption (H'1) $G(w_0)X = Z$ and assumption (H2) we obtain that $h = p$ is differentiable at w_0 in the direction u with

$$h'(w_0, u) = \langle y_0, G'(w_0)u \cdot x_0 \rangle$$

where y_0 is the solution of the adjoint equation

$$f'(x_0) = -G(w_0)^T(y_0).$$

COROLLARY 8.2 When G is as above, the function h is differentiable at w_0 and its derivative is given by

$$h'(w_0)u = \langle y_0, G'(w_0)u.x_0 \rangle$$

where x_0 is the unique solution to $G(w_0)x = b$ and $y_0 \in Y$ is the solution to the adjoint problem

$$f'(x_0) = -y_0 \circ G(w_0).$$

In some instances, as in [10], the adjoint equation has a simple form, closely related to the primal problem, and it is fruitful to bring it in the picture.

From the computational point of view, the study of the second derivative of h or p would be interesting for the conditioning of algorithms dealing with inverse problems.

References

[1] W. Alt, Stability of solutions for a class of nonlinear cone constrained optimization problems,. Part I: Basic theory, Numer. Funct. Anal. and Optim. 10 : 1053-1065 (1989).

[2] J.-P. Aubin, Contingent derivatives of set-valued maps and existence of solutions to nonlinear inclusions and differential inclusions, Advances in Maths. supplementary studies L. Nachbin, ed. : 160-232 (1981).

[3] J.-P. Aubin, Further properties of Lagrange multipliers in nonsmooth optimization, Applied Maths Optim. 6 :79-90, 1980.

[4] J.-P. Aubin, *Méthodes explicites de l'optimisation*, Dunod, Paris (1982).

[5] J.-P. Aubin and F.H. Clarke,Multiplicateurs de Lagrange en optimisation non convexe et applications, C.R. Acad. Sci. Paris 285 : 451-454, 1977.

[6] J.-P. Aubin and H. Frankowska, *Set-valued analysis*, Birkhauser, Basel (1990).

[7] A. Auslender, Stability in mathematical programming with nondifferentiable data, SIAM J. Control and Opt. 22 : 29-42 (1984).

[8] A. Auslender and R. Cominetti, First and second order sensitivity analysis of nonlinear programs under directional constraint qualification condition, Optimization 21 (1990), 351-353.

[9] H. Attouch and D. Azé, Approximation and regularization of arbitrary functions in Hilbert spaces by the Lasry-Lions method, Ann. Inst. H. Poincaré, Anal. Nonlinéaire 10 : 289-312, 1993.

[10] A. Bamberger, G. Chavent and P. Lailly, About the stability of the inverse problem in 1-D wave equations-Application to the interpretation of seismic profiles, Appl. Math. Optim. 5 : 1-47 (1979).

[11] B. Bank, J. Guddat, D. Klatte et al., *Non-linear parametric optimization*, Akademie Verlag, Berlin (1982).

[12] L. Barbet, *Etude de sensibilité différentielle dans un problème d'optimisation paramétré avec contraintes en dimension infinie*, Thesis, Univ. of Poitiers (1992).

[13] L. Barbet and R. Janin, Analyse de sensibilité différentielle pour un problème d'optimisation paramétré en dimension infinie, C.R. Acad. Sci. Paris 318 Série I: 221-226 (1993).

[14] E. Bednarczuk, Sensitivity in mathematical programming: a review, Control and Cybernetics 23 (4) : 589-604 (1994).

[15] C. Berge, *Espaces topologiques et fonctions multivoques*, Dunod, Paris, (1959), (1966).

[16] J. F. Bonnans, A.D. Ioffe and A. Shapiro, Expansion of exact and approximate solutions in nonlinear programming, in "*Advances in Optimization, Proceedings, Lambrecht, FRG, 1991*", Edited by W. Oettli and D. Pallaschke, Lecture Notes in Economics and Mathematical Systems, No 382, Springer Verlag, Berlin : 103-117 (1992).

[17] J.-M. Bonnisseau and C. Le Van, Sensitivity analysis applied to economic problems, preprint, Univ. Paris 1, (May 1993).

[18] J. Borwein, Stability and regular points of inequality systems, J. Optim. Th. Appl. 48 . 9-52 (1986).

[19] M. Bougeard, J.-P. Penot and A. Pommellet, Towards minimal assumptions for the infimal convolution regularization, J. Approximation Th. 64 (3) :245-270 (1991).

[20] D. Chenais, Optimal design of midsurface of shells: differentiability proof and sensitivity computation, Applied Math. Optim. 16 : 93-133 (1997).

[21] F.H. Clarke, *Optimization and nonsmooth analysis*, Wiley-Interscience (1983).

[22] F.H. Clarke and P. Loewen, The value function in optimal control: sensitivity, controllability and time optimality, SIAM J. Control Optim. 24 : 243-263 (1986).

[23] R. Cominetti, Metric regularity conditions, tangent sets and second order optimality conditions, Applied Math. and Optim. 21 : 265-287 (1990).

[24] B. Cornet and J.-P. Vial, Lipschitz solutions of perturbed nonlinear programming problems, SIAM J. Control Opt. 24:1123-1137 (1986).

[25] B. Cornet and G. Laroque, Lipschitz properties of solutions in mathematical programming, J. Opt. Th. Appl. 53 : 407-427 (1987).

[26] R. Correa and A. Seeger, Directional derivative of a min-max function, Nonlinear Anal. Th. Meth., Appl. 9 :13-22 (1985).

[27] J.M. Danskin, *The theory of max-min*, Springer Verlag, Berlin (1967).

[28] V.F. Demyanov and A.B. Pevnyi, First and second marginal values in of mathematical programming problems, Soviet Math. Dokl. 13 : 1502-1506 (1972).

[29] S. Dolecki, Tangency and differentiation: marginal functions, Advances in Applied Math. 11 : 389-411 (1990).

[30] A.V. Fiacco, *Introduction to Sensitivity and Stability Analysis in Nonlinear Programming*, Academic Press, New York (1983).

[31] A.V. Fiacco and W.P. Hutzler, Basic results in the development of sensitivity and stability analysis in nonlinear programming, Comput. and Oper. Research 9 (1) : 1-28 (1982).

[32] A.V. Fiacco and Y. Ishizuka, Sensitivity and Stability Analysis for Nonlinear Programming, Annals Oper. Res. 27 : 215-236 (1990).

[33] A.V. Fiacco and G.P. McCormick, *Nonlinear programming: sequential unconstrained minimization techniques*, Wiley, New York (1968).

[34] R. Fletcher, Semi-definite constraints in optimization, SIAM J. Control and Optim. 23 : 493-513 (1985).

[35] S. Gautier, *Contributions à l'analyse non-linéaire : inclusions différentielles, différentiabilité des multiapplications*, Thesis, Univ. of Pau (1989).

[36] J. Gauvin, *Theory of nonconvex programming*, Les Publications du CRM, Montréal, (1994).

[37] J. Gauvin and F. Dubeau, Differential properties of the marginal function in mathematical programming, Math. Prog. Study 19 : 101-119 (1982).

[38] J. Gauvin and R. Janin, Directional behavior of optimal solutions in nonlinear mathematical programming, Math. Oper. Res. 13 (4) : 629-649 (1989).

[39] J. Gauvin and J.W. Tolle, Differential stability in nonlinear programming, SIAM J. Control and Opt. 15 : 294-311 (1994).

[40] B. Gollan, On the marginal function in nonlinear programming, Math. Oper. Res. 9 : 208-221 (1984).

[41] E.G. Gol'stein, Duality theory in mathematical programming and its applications, (in Russian), Nauka, Moscow (1971).

[42] E.J. Haug and J. Cea, *Optimization of distributed parameter structures*, Sijthoff and Noordhoff, Alphen aan den Rijn, Holland, (1981).

[43] E.J. Haug and B. Rousselet, Design sensitivity analysis in structural mechanics. II Eigenvalue variations, J. Struct. Mech. 8 (2) : (1981).

[44] D. Henri, Nonuniqueness of solutions in the calculus of variations: a geometric approach, SIAM J. Control and Opt. 18 (6) : 627-639 (1980).

[45] J.-B. Hiriart-Urruty, Gradients généralisés de fonctions marginales, SIAM J. Control Opt. 16 : 310-316 (1978).

[46] N.D. Hoa, Stability of linear programming problems in Banach spaces, Control and Cybernetics, (16) : 73-83 (1987).

[47] W.W. Hogan, Directional derivatives for extremal value function with applications to the completely convex case, Oper; Res. 21 (1) : 188-209 (1973).

[48] A.D. Ioffe, On sensitivity analysis of nonlinear programs in Banach spaces: the approach via composite unconstrained optimization, SIAM J. Optimization 4 (1) : 1-43 (1994).

[49] R. Janin, Directional derivative of the marginal function in nonlinear programming, Math. Prog. Study 21 : 110-126 (1984).

[50] R. Janin, J.-C. Mado and J. Narayaninsamy, Second order multipliers and marginal function in non linear programs, Optimization 22 : 163-176 (1991).

[51] A. Jofre and J.-P. Penot, A note on the directional derivative of a marginal function, Revista Mat. Aplicadas 14: 37-54 (1993).

[52] K. Kuratowski, Topologie vol. I, II, Panstowe Wyd Nauk, Warzawa (1959) and Academic Press, New York, (1966).

[53] F. Lempio and H. Maurer, Differential stability in infinite dimensional programming, Applied Mathematics and Optimization, 6 : 139-152 (1980).

[54] E.S. Levitin, *Perturbation Theory in Mathematical Programming and its Applications,* J. Wiley, Chichester (1994).

[55] A. Lewis, Convex analaysis on the Hermitian matrices, SIAM J. on Optimization 6 : 164-177, 1996.

[56] A.S. Lewis, Derivatives of spectral functions, Mathematics of Oper. Research 21 : 576-588 (1996).

[57] J.-C. Mado, *Conditions d'optimalité en programmation mathématique,* thesis, Univ. of Pau (1985).

[58] K. Malanovski, Second-order conditions and constraint qualifications in stability and sensitivity analysis of solutions to optimization problems in Hilbert spaces, Appl. Math. Opt. : 25 51-79 (1992).

[59] K. Malanowski, Regularity of solutions in stability analysis of optimization and optimal control problems, Control and Cybernetics 23 : 61-86 (1994).

[60] O.L. Mangasarian and R.A. Meyer, Nonlinear perturbations of linear programs, SIAM J. Control and Opt. 17 (6) : 745-752 (1979).

[61] H. Maurer and J. Zowe, First and second-order necessary and sufficient optimality conditions for infinite-dimensional programming problems, Math. Programming 16 : 98-110 (1979).

[62] J. Narayaninsamy, *Condition de qualification et dérivée de la fonction valeur en programmation mathématique,* thèse, Univ. of Pau, (1986).

[63] F. Nozicka, J. Guddat et al., *Theory of linear parametric optimization,* Akademie Verlag, Berlin (1974).

[64] J. Outrata, On generalized gradients in optimization problems with set-valued constraints, Math. Oper. Res. 15 (4) : 626-639 (1980).

[65] M.L. Overton, On minimizing the maximum eigenvalue of a symmetric matrix, SIAM J. on Matrix Anal. Appl. 9 : 256-268 (1988).

[66] M.L. Overton and R.S. Womersley, Optimality conditions and duality theory for minimizing sums of the largest eigenvalues of symmetric matrices, Math. Programming Series B, 62 : 321-357 (1983).

[67] R. Pallu de la Barrière, *Cours d'automatique théorique*, Dunod, Paris, 1966.

[68] J.-P. Penot, The use of generalized subdifferential calculus in optimization theory, Oper. Res. Verf., Methods of Oper. Res. 31, III Symposium on Oper. Res., Mannheim Sept. 1978 : 495-511, Athenaum (1979).

[69] J.-P. Penot, A characterization of tangential regularity, Nonlinear Anal., Th., Methods, Appl., 25 (6) : 625-643 (1981).

[70] J.-P. Penot, Continuity properties of performance functions, in "Optimization, theory and algorithms", J.-B. Hiriart-Urruty, W. Oettli and J. Stoer, eds, Lecture Notes in pure and applied Maths 86, Dekker, New York : 77-90 (1983).

[71] J.-P. Penot, On regularity conditions in mathematical programming, Math. Programming Study 19 : 167-199 (1982).

[72] J.-P. Penot, Compact nets, filters and relations, J. Math. Anal. Appl. 93 : 400-417 (1983).

[73] J.-P. Penot, Differentiability of relations and differential stability of perturbed optimization problems, SIAM J. Control and Optim. 22 (4) : 529-551 (1984).

[74] J.-P. Penot, On the mean value theorem, Optimization 19 : 147-156, 1988.

[75] J.-P. Penot, Optimality conditions in mathematical programming and composite optimization, Math. Progr 67 : 225-245 (1994),.

[76] J.-P. Penot, Sequential derivatives and composite optimization, Revue Roumaine de Math. Pures et Appl. 40 (5-6) : 501-519 (1995).

[77] J.-P. Penot, Generalized derivatives of a performance function and multipliers in mathematical programming, Proceedings Intern. Conf. on Parametric Optimization IV, Enschede (NL), June 6–9, 1995, J. Guddat, H. Th. Jongen, F. NoZicka, G. Still, F. Twilt (eds.) P. Lang, Berlin : 281-298, 1996.

[78] J.-P. Penot, Multipliers and generalized derivatives of performance functions, to appear, J. Optim. Th. Appl. 93 (3), (1997).

[79] J.-P. Penot, Proximal mappings, preprint, Univ. of Pau (March 1996).

[80] J.-P. Penot, A multimapping approach to sensitivity analysis, in preparation.

[81] J.-Ch. Pomerol, The Lagrange multiplier set and the generalized gradient set of the marginal function of a differentiable program in a Banach space, J. Opt. Th. Appl. 38 : 307-317 (1982).

[82] S.M. Robinson, Stability theory for systems of inequalities. Part I :linear systems, SIAM J. Numer. Anal. 12 (5) : 754-769 (1975).

[83] S.M. Robinson, Stability theory for systems of inequalities. Part II : differentiable nonlinear systems, SIAM J. Numer. Anal. 13 : 754-769 (1976).

[84] S.M Robinson, Regularity and stability for convex multivalued functions, Math. Oper. Res. 1 : 130-143 (1976).

[85] S.M. Robinson, A characterization of stability in linear programming, Oper. Research 25 (3) :435-447 (1977).

[86] S.M Robinson, Generalized equations and their solutions, Part I: basic theory, Math. Programming Study 10 : 128-141 (1979).

[87] S.M Robinson, Strongly regular generalized equations, Math. Oper. Res. 5 (1): 43-62 (1980).

[88] S.M Robinson, Local structure of feasible sets in nonlinear programming, part III: stability and sensitivity, Math. Prog. Study 30 : 45-66 (1987).

[89] R.T. Rockafellar, Lagrange multipliers and subderivatives of optimal value functions in nonlinear programming, Math. Prog. Study 17 : 28-66 (1982).

[90] R.T. Rockafellar, Directional differentiability of the optimal value function in a nonlinear programming problem, Math. Prog. Study 21 : 213-226 (1984).

[91] A. Seeger, Second order directionl derivatives in parametric optimization problems, Math. Oper. Res. 13 (1) : 124-139 (1988).

[92] A. Shapiro, Perturbation theory of nonlinear programs when the set of optimal solutions is not a singleton, Appl. Math. Optim. 18 : 215-229 (1988).

[93] A. Shapiro and J.F. Bonnans, Sensitivity analysis of parametrized programs under cone constraints, SIAM J. Control Optim. 30 : 1409-1422 (1992),.

[94] A. Shapiro, Directional differentiability of the optimal value function in convex semi-infinite programming, Math. Programming serie A, 70 : 149-157 (1995).

[95] S. Shiraishi, Sensitivity analysis of nonlinear programming problems via minimax functions, preprint, Faculty of Economics, Toyama University, August 1995.

[96] J. Sokolovski and J.-P. Zolesio, Introduction to shape optimization, Shape sensitivity analysis, Springer Verlag, New York (1992).

[97] L. Thibault, On subdifferential of optimal value functions, SIAM J. Control and Opt. 29 : 1019-1036 (1991).

[98] L. Vandenberghe and S. Boyd, Positive definite programming, in " *Mathematical Programmin. State of the art, 1994*", J.R. Birge and K.G. Murty, eds., Univ. Michigan, 276-308 (1994).

[99] A.C. Williams, Marginal values in linear programming, SIAM J. Appl. Math. 11 : 82-94 (1963).

[100] J. Zowe and S. Kurcyusz, Regularity and stability for the mathematical programming problem in Banach spaces, Appl. Math. Optim. 5 : 49-62 (1979).

The following additional references have come to the attention of the author after submission.

[Be] M. Benlarbi, Caractère lipschitzien et dérivabilité directionnelle des solutions de systèmes d'équations et d'inéquations paramétrés, C. R. Acad. Sci. Paris Série I 323: 137-142 (1996).

[B-C 1] J.F. Bonnans and R. Cominetti, Perturbed optimization in Banach spaces I: a general theory based on a weak directional constraint qualification, SIAM J. Control Opt. 34 (4): 1151-1171 (1996).

[B-C 2] J.F. Bonnans and R. Cominetti, Perturbed optimization in Banach spaces II: a theory based on a strong directional constraint constraint qualification, SIAM J. Control Opt. 34 (4): 1172-1189 (1996).

[B-C 3] J.F. Bonnans and R. Cominetti, Perturbed optimization in Banach spaces III: semi-infinite optimization, SIAM J. Control Opt. 34 (5): 1555-1567 (1996).

[B-S] J.F. Bonnans and A. Shapiro, Optimization problems with perturbations, a guided tour, preprint, INRIA and Georgia Institute of Technology, April 1996.

[HU-L] J.-B. Hiriart-Urruty and A.S. Lewis, The Clarke and Michel-Penot sub-differentials of the eigenvalues of a symmetric matrix, preprint Univ. P. Sabatier, Toulouse, February 1997.

[L] A.S. Lewis, Nonsmooth analysis of eigenvalues, Preprint, Univ. P. Sabatier, Toulouse, Sept. 1996.

[M-S] B. S. Mordukhovich and Y. Shao, Nonsmooth sequential analysis in Asplund spaces, Trans. Amer. Math. Soc.348 (4) (1996), 1235-1280.

Topological Stability of Feasible Sets in Semi-infinite Optimization: A Tutorial

JAN-J. RÜCKMANN   University of Erlangen-Nuremberg, Institute for
Applied Mathematics, Martensstrasse 3, D-91058 Erlangen, Germany
Email: rueckman@am.uni-erlangen.de

ABSTRACT

This tutorial paper deals with feasible sets of semi-infinite optimization problems
(SIP) which are defined by finitely many equality and infinitely many inequality
constraints, where all appearing functions are real-valued and defined on $\mathbb{R}^n$. Many
practical tasks lead to problems of the type (SIP), e.g. in robotics, design centering,
Chebyshev approximation, etc.

An overview is given about results concerning the close relationship between the
overall validity of an appropriate constraint qualification of the Mangasarian-Fromovitz type and certain global topological stability properties of the (noncompact)
feasible set of (SIP), where perturbations in the function space with respect to
the strong (or Whitney-) C^1 topology as well as variations of an additional real
parameter are considered. In particular, several basic geometric ideas of the proofs
of the mentioned results are presented.

1 INTRODUCTION

Let $\mathbb{R}^n$, $n \geq 1$ be the n-dimensional Euclidean space with the Euclidean norm $\|\cdot\|$,
the origin 0_n and $\mathbb{R} = \mathbb{R}^1$. If $U \subseteq \mathbb{R}^n$ is an open subset then denote by $C^\kappa(U, \mathbb{R}^{\bar{n}})$,
$\kappa \geq 1$, $\bar{n} \geq 1$ the set of κ-times continuously differentiable mappings from U to $\mathbb{R}^{\bar{n}}$
and by $Df(\bar{x})$ the $(\bar{n}, n)$-Jacobian of $f \in C^\kappa(U, \mathbb{R}^{\bar{n}})$ at $\bar{x} \in U$. ($D_{x^1}f(\bar{x})$ denotes
the partial Jacobian of f with respect to the subvector x^1 of $\bar{x}$.)

We consider a *semi-infinite optimization problem* of the form

$$\text{Minimize } f(x) \text{ subject to } x \in M(H, G)$$

or, shortly,

(SIP) $\min\{f(x) \mid x \in M(H, G)\}$,

where the *feasible set* $M(H, G)$ is defined by finitely many equality constraints and – perhaps – infinitely many inequality constraints as

$$M(H, G) = \{x \in \mathbb{R}^n \mid h_i(x) = 0, \, i \in I, \, G(x, y) \geq 0, \, y \in Y\}.$$

If $I = \emptyset$ resp. $Y = \emptyset$ we write $M(G)$ resp. $M(H)$. Assume that $I = \{1, \ldots, m\}$, $m < n$, $Y \subset \mathbb{R}^r$, $f : \mathbb{R}^n \to \mathbb{R}$, $H \in C^\kappa(\mathbb{R}^n, \mathbb{R}^m)$ with $H = (h_1, \ldots, h_m)$ and $G \in C^\kappa(\mathbb{R}^n \times \mathbb{R}^r, \mathbb{R})$, $\kappa \geq 1$. Unless stated otherwise, let $\kappa = 1$. Obviously, each index $y \in Y$ represents an inequality constraint. The index set of active inequality constraints at a *feasible point* $\bar{x} \in M(H, G)$ is denoted by

$$Y_0(\bar{x}) = \{y \in Y \mid G(\bar{x}, y) = 0\}.$$

Note that $Y_0(\bar{x})$ can be infinite. Furthermore, let the index set Y be described by

$$Y = \{y \in \mathbb{R}^r \mid u_k(y) = 0, \, k \in K, \, v_\ell(y) \leq 0, \, \ell \in L\},$$

where $K = \{1, \ldots, \bar{k}\}$, $\bar{k} < r$, $L = \{1, \ldots, \bar{\ell}\}$ and $u_k, v_\ell \in C^\infty(\mathbb{R}^r, \mathbb{R})$, $k \in K$, $\ell \in L$. Assume that

- Y is compact and
- at each index $\bar{y} \in Y$ the set $\{Du_k(\bar{y}), Dv_\ell(\bar{y}), k \in K, \ell \in L_0(\bar{y})\}$ (1-1)
  is linearly independent, where $L_0(\bar{y}) = \{\ell \in L \mid v_\ell(\bar{y}) = 0\}$.

There are many practical tasks where problems of the type (SIP) can appear, e.g. design centering problems (cf. e.g. [4,11,23]) and Chebyshev approximation problems (cf. e.g. [13,27]) can be written as semi-infinite problems. For an extensive study on (SIP) models in applications and corresponding references we refer to the review papers [8] and [26].

The goal of the present paper is to give a tutorial on several global topological stability properties of the feasible set $M(H, G)$ of (SIP) which have been published in recent years. These topological stability properties refer to the homeomorphy (of certain subsets) of every sufficiently small perturbed feasible set with (corresponding subsets of) the original set $M(H, G)$ where we consider perturbations in the function space with respect to the strong (or Whitney-) C^1-topology as well as perturbations with respect to an additional real parameter. In the final section we present a consequence of this global topological stability for the use of homotopy methods in nonlinear optimization.

We shall not repeat the proofs of the results but we present the basic tools and illustrate the geometric ideas. In particular, the special role of the Mangasarian-Fromovitz constraint qualification and its modification in this context will be emphasized.

The present paper is organized as follows. In Section 2 we explain some basic results which are used later for the description of the stability results. Section 3 contains topological stability properties of $M(H, G)$; we distinguish three cases: the compact case, the noncompact case and the one-parametric case. Finally, we present a conclusion for homotopy methods in Section 4.

2 BASIC RESULTS

The constraint qualification (EMFCQ)

The following constraint qualification will play an essential role in this paper. It is an appropriate extension of the well-known Mangasarian-Fromovitz constraint qualification (MFCQ) (cf. [22]) to the semi-infinite case (cf. [9,17]).

The *extended Mangasarian-Fromovitz constraint qualification* ($EMFCQ$) is said to hold at $\bar{x} \in M(H, G)$ if

(i) the set $\{Dh_i(\bar{x}), i \in I\}$ is linearly independent and

(ii) there exists a vector $\xi \in \mathbb{R}^n$ satisfying

$$\begin{aligned} Dh_i(\bar{x})\xi &= 0, \quad i \in I \\ D_x G(\bar{x}, y)\xi &> 0, \quad y \in Y_0(\bar{x}). \end{aligned} \tag{2-1}$$

A vector $\xi \in \mathbb{R}^n$ satisfying (ii) is called an *EMF-vector* at $\bar{x}$.

The (EMFCQ) is closely related with several sensitivity and regularity results. As an example we refer to the papers [7,18], where the equivalence of (EMFCQ) and metric regularity is considered. This equivalence entails the important fact that the set of Lagrange multipliers is bounded under (EMFCQ) (Gauvin's Theorem in infinite-dimensional spaces, see, e.g. [19,25,31]).

The validity of (EMFCQ) at $\bar{x} \in M(H, G)$ implies some nice topological properties of the feasible set $M(H, G)$ locally around $\bar{x}$. Assume that (EMFCQ) holds at a point $\bar{x} \in M(H, G)$ with $Y_0(\bar{x}) \neq \emptyset$ and let $\xi^{\bar{x}} \in \mathbb{R}^n$ be an EMF-vector at $\bar{x}$. If $I = \emptyset$ then (2-1) implies that $\xi^{\bar{x}}$ "is looking inside" the feasible set $M(G)$ (cf. Figure 2.1(a)) and, by continuity arguments, there is an open neighbourhood V of $\bar{x}$ such that (EMFCQ) holds at every $x \in V \cap M(G)$ and $\xi^{\bar{x}}$ is an EMF-vector at all $x \in V \cap M(G)$ (cf. Figure 2.1(b)). A short calculation shows that in case of a convex set $M(G)$ (i.e. the functions $G(\cdot, y)$, $y \in Y$ are concave) the validity of (EMFCQ) at a point $\bar{x} \in M(G)$ (local property) is equivalent to the fulfillness of the so-called Slater condition, i.e. there is a point $\tilde{x} \in M(G)$ with $G(\tilde{x}, y) > 0$ for all $y \in Y$ (global property).

If $I \neq \emptyset$ the set $H^{-1}(0) = \{x \in \mathbb{R}^n \mid h_i(x) = 0, i \in I\}$ is locally around $\bar{x}$ a C^1-manifold and, thus, for local considerations being invariant under C^1-coordinate transformations one can consider without loss of generality the (transformed) case $I = \emptyset$. (Note that (EMFCQ) is invariant under local C^1-coordinate transformations (cf. [6, Lemma 2.1] and [17, Lemma 3.1])). However, in case $I \neq \emptyset$ we can construct for every point of an open neighbourhood of $\bar{x}$ an EMF-vector by choosing $b^j \in \mathbb{R}^n$, $j = m + 1, \ldots, n$ and an open neighbourhood V of $\bar{x}$ such that for all $x \in V \cap M(H, G)$:

- $\{Dh_i(x), i \in I, b^j, j = m + 1, \ldots, n\}$ is a linearly independent set,

- the Jacobian $D\Phi(x)$ of the mapping

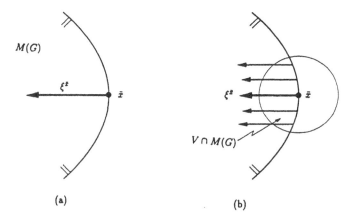

Figure 2.1 The EMF-vector $\xi^{\bar{x}}$ at $x \in V \cap M(G)$.

$$\Phi(x) = \begin{pmatrix} h_1(x) \\ \vdots \\ h_m(x) \\ (b^{m+1})^T(x - \bar{x}) \\ \vdots \\ (b^n)^T(x - \bar{x}) \end{pmatrix}$$

is nonsingular and

- $\xi^{\bar{x}}(x) = D\Phi^{-1}(\Phi(x))D\Phi(\bar{x})\xi^{\bar{x}}$ is an EMF-vector at x.   (2-2)

A topological consequence of (EMFCQ) is given in the subsequent theorem. According to [17, Definition 2.3] a set $\mathcal{M} \subset \mathbb{R}^n$ is called a *Lipschitzian manifold* (with boundary) of dimension d (with $d < n$) if for each $\bar{x} \in \mathcal{M}$ there are neighbourhoods $\mathcal{V}$ of $\bar{x}$ and $\mathcal{U}$ of 0_n as well as a homeomorphism $\psi : \mathcal{V} \to \mathcal{U}$ with the following properties:

- $\psi(\bar{x}) = 0$,

- ψ and $\psi^{-1} : \mathcal{U} \to \mathcal{V}$ are Lipschitz continuous as well as

- either $\psi[\mathcal{V} \cap \mathcal{M}] = \mathcal{U} \cap \{\{0_{n-d}\} \times \mathbb{R}^d\}$ or $\psi[\mathcal{V} \cap \mathcal{M}] = \mathcal{U} \cap \left(\{0_{n-d}\} \times \mathbb{R}^{d-1} \times \{z \in \mathbb{R} \mid z \leq 0\} \right)$, in the latter case $\bar{x}$ is called a boundary point of $\mathcal{M}$.

THEOREM 2.1 (cf. [17, Theorem 2.1]). *Assume that* (*EMFCQ*) *holds at every* $x \in M(H, G)$. *Then,* $M(H, G)$ *is a Lipschitzian manifold of dimension* $n - m$ *with the boundary* $\partial M(H, G) = \{x \in M(H, G) \mid Y_0(x) \neq \emptyset\}$.

The Figure 2.2 illustrates the situation under the assumption of Theorem 2.1 locally around $\bar{x} \in \partial M(H, G)$ with an EMF-vector $\xi^{\bar{x}} \in \mathbb{R}^n$ at $\bar{x} \in M(H, G)$. The corresponding homeomorphism ψ maps the part $\mathcal{V} \cap \partial M(H, G)$ of the boundary of $M(H, G)$ onto $\mathcal{U} \cap \left(\{0_m\} \times \mathbb{R}^{n-m-1} \times \{z \in \mathbb{R} \mid z = 0\} \right)$.

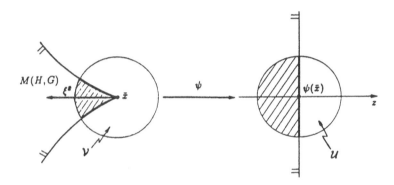

Figure 2.2 The local property of a Lipschitzian manifold.

A straightforward consequence of Theorem 2.1 is given in

COROLLARY 2.2 [15, Lemma 1]. *Assume that* (*EMFCQ*) *holds at* $\bar{x} \in M(H, G)$. *Then there exists an open neighbourhood* V *of* $\bar{x}$ *such that* $V \cap M(H, G)$ *is pathwise connected.*

Furthermore, it follows from Theorem 2.1 and assumption (1-1) that the index set Y is a Lipschitzian manifold of dimension $r - \bar{k}$.

Under the assumption that (EMFCQ) holds at every $x \in M(H, G)$ we construct for our stability investigations in the next section a vector field $\xi(x)$ such that $\xi(\bar{x})$ is an EMF-vector at each $\bar{x} \in M(H, G)$. First we recall the following well-known result from the theory of ordinary differential equations.

LEMMA 2.3 (cf. e.g. [2]). *Let* $U \subseteq \mathbb{R}^n$ *be an open subset and* $F \in C^{\kappa}(U, \mathbb{R}^n)$ *a vector field. Then:*

(i) *For each $\bar{x} \in U$ there exist an open neighbourhood $W(\bar{x})$ of $\bar{x}$, a real $w(\bar{x}) > 0$ and a uniquely determined mapping $\phi \in C^{\kappa}(W(\bar{x}) \times (-w(\bar{x}), w(\bar{x})), U)$ (the flow of F) such that for all $x \in W(\bar{x})$ and all $\tau, \tau_1, \tau_2 \in (-w(\bar{x}), w(\bar{x}))$ with $\overline{|\tau_1 + \tau_2|} < w(\bar{x})$:*

$$D_\tau \phi(x, \tau) = F(\phi(x, \tau)), \quad \phi(x, 0) = x,$$

$$\phi(x, \tau_1 + \tau_2) = \phi(\phi(x, \tau_1), \tau_2).$$

(ii) *If $U = \mathbb{R}^n$ and F is bounded then ϕ is defined on the whole space $\mathbb{R}^n \times \mathbb{R}$ (i.e. completely integrable) and $\phi(\cdot, \tau)$ is a C^{κ}-diffeomorphism for each $\tau \in \mathbb{R}$.*

Now, assume that $h_i \in C^2(\mathbb{R}^n, \mathbb{R})$, $i \in I$ and that (EMFCQ) holds at every point $x \in M(H, G)$. For $\bar{x} \in M(H, G)$ choose an open appropriate neighbourhood $V(\bar{x})$ of $\bar{x}$ and construct a vector field $\xi^{\bar{x}} \in C^1(V(\bar{x}), \mathbb{R}^n)$ as in (2-2) (here, we need $h_i \in C^2$!) such that

- $\|\xi^{\bar{x}}(x)\| \le 1$ for every $x \in V(\bar{x})$ and

- $\xi^{\bar{x}}(x)$ is an EMF-vector at every $x \in V(\bar{x}) \cap M(H, G)$.

For $\bar{x} \in \mathbb{R}^n \setminus M(H, G)$ choose an open neighbourhood $V(\bar{x})$ of $\bar{x}$ such that $V(\bar{x}) \cap M(H, G) = \emptyset$. Then $\{V(\bar{x}), \bar{x} \in \mathbb{R}^n\}$ is an open covering of $\mathbb{R}^n$. By selecting the C^1-vector field $\xi^{\bar{x}}$ on $V(\bar{x})$ for $\bar{x} \in M(H, G)$ and the constant zero vector field 0_n on $V(\bar{x})$ for $\bar{x} \in \mathbb{R}^n \setminus M(H, G)$ as well as by using a partition of unity subordinate to the constructed covering of $\mathbb{R}^n$ we obtain a bounded – and, thus, completely integrable – vector field $\xi \in C^1(\mathbb{R}^n, \mathbb{R}^n)$ such that $\xi(x)$ is an EMF-vector at each $x \in M(H, G)$ (note that $\xi(x)$ is a finite convex combination of EMF-vectors at $x \in M(H, G)$). We summarize this result in

COROLLARY 2.4 *Let $h_i \in C^2(\mathbb{R}^n, \mathbb{R})$, $i \in I$ and assume that (EMFCQ) holds at every $x \in M(H, G)$. Then, there exists a bounded vector field $\xi \in C^1(\mathbb{R}^n, \mathbb{R}^n)$ such that $\xi(\bar{x})$ is an EMF-vector at $\bar{x}$ whenever $\bar{x} \in M(H, G)$.*

The existence of this bounded (so-called EMF-) vector field $\xi(x)$ is basic for the stability investigations of the feasible set in Section 3. Let $\phi(x, \tau)$ denote the flow of ξ. The construction provides for $\bar{x} \in \partial M(H, G)$ that

$$D_\tau \phi(\bar{x}, 0) = \xi(\phi(\bar{x}, 0)) = \xi(\bar{x}) \quad \text{and}$$

$$D_\tau G(\phi(\bar{x}, 0), y) = D_x G(\bar{x}, y)\xi(\bar{x}) > 0 \quad \text{for all} \quad y \in Y_0(\bar{x}),$$

i.e. that the trajectory $\{\phi(\bar{x}, \tau) \mid \tau \in \mathbb{R}\}$ contains exactly one point from the boundary $\partial M(H, G)$:

$$\{\phi(\bar{x}, \tau) \mid \tau \in \mathbb{R}\} \cap \partial M(H, G) = \{\bar{x}\} \quad \text{(cf. Figure 2.3)}.$$

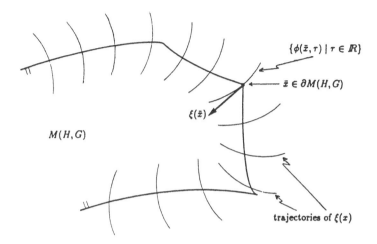

Figure 2.3 The EMF-vector field $\xi(x)$.

The C^1_s-topology

We shall consider perturbations of the appearing functions where we endow the function space $C^1(\mathbb{R}^n, \mathbb{R})$ with the strong (or Whitney-) C^1-topology, which is denoted by C^s_1 (for more details see e.g. [10,14]). A typical C^1_s-base neighbourhood P_ε of the zero function in $C^1(\mathbb{R}^n, \mathbb{R})$ is characterized by a continuous positive function $\varepsilon : x \in \mathbb{R}^n \mapsto \varepsilon(x) > 0$ and defined as

$$P_\varepsilon = \{\bar{g} \in C^1(\mathbb{R}^n, \mathbb{R}) \mid |\bar{g}(x)| + \|D\bar{g}(x)\| < \varepsilon(x) \text{ for all } x \in \mathbb{R}^n \}.$$

A typical C^1_s-base neighbourhood of $g \in C^1(\mathbb{R}^n, \mathbb{R})$ is the set $g+P_\varepsilon$. The C^1_s-topology of the product space $C^1(\mathbb{R}^n, \mathbb{R}^{\bar{n}})$ $(= C^1(\mathbb{R}^n, \mathbb{R}) \times \ldots \times C^1(\mathbb{R}^n, \mathbb{R})$, $\bar{n}$-times) is the induced product-topology. Note that $\inf_{x \in \mathbb{R}^n} \varepsilon(x) = 0$ is possible and, therefore, certain asymptotic effects are taken into the consideration which would not be possible if $\varepsilon(x)$ would be a positive constant for all $x \in \mathbb{R}^n$. This will become more clear in Section 3.

The final lemma in this section summarizes several results from [17, Theorems 2.2 and 2.4], where we define the set

$$\mathcal{F}=\{(H,G)\in C^1(\mathbb{R}^n,\mathbb{R}^m)\times C^1(\mathbb{R}^n\times\mathbb{R}^r,\mathbb{R}) \mid \text{(EMFCQ) holds at every } x\in M(H,G)\}.$$

LEMMA 2.5　(i) $\mathcal{F}$ *is C^1_s-open and C^1_s-dense.*

(ii) *Let $(H,G) \in \mathcal{F}$ and $U \subset \mathbb{R}^n$ be an open subset with $M(H,G) \subset U$. Then, there exists a C^1_s-neighbourhood ϑ of (H,G) in $C^1(\mathbb{R}^n, \mathbb{R}^m) \times C^1(\mathbb{R}^n \times \mathbb{R}^r, \mathbb{R})$ such that $M(\tilde{H}, \tilde{G}) \subset U$ for all $(\tilde{H}, \tilde{G}) \in \vartheta$.*

3   STABILITY OF THE FEASIBLE SET

In this section we investigate global topological stability properties of the feasible set $M(H,G)$, where we consider perturbations of (H,G) with respect to the C_s^1-topology and with respect to the variation of an additional parameter $t \in \mathbb{R}$.

The compact case

The following definition of topological stability of a compact feasible set $M(H,G)$ and the subsequent stability theorem as well as (the technical ideas of) its proof have been firstly presented in [6, Theorem B] for the case that $M(H,G)$ is described by finitely many equality and finitely many inequality constraints. The stability theorem was generalized in [17, Theorem 2.3] for a semi-infinite problem. Both in [6] and [17] it is assumed that $h_i \in C^2(\mathbb{R}^n, \mathbb{R})$, $i \in I$. In [12,28] it is shown that the stability theorem is also valid if $h_i \in C^1(\mathbb{R}^n, \mathbb{R})$, $i \in I$.

According to [6,17] a compact feasible set $M(H,G)$ is called *stable* if there is a C_s^1-neighbourhood ϑ of (H,G) in $C^1(\mathbb{R}^n, \mathbb{R}^m) \times C^1(\mathbb{R}^n \times \mathbb{R}^r, \mathbb{R})$ such that

$$M(H,G) \simeq M(H',G') \tag{3-1}$$

whenever $(H',G') \in \vartheta$ (where '$\simeq$' means 'is homeomorphic with').

The following theorem states the equivalence between

- the *analytical property* that (EMFCQ) holds at every $x \in M(H,G)$ and

- the *topological property* that $M(H,G)$ is stable.

THEOREM 3.1 ([17, Theorem 2.3] for $h_i \in C^2(\mathbb{R}^n, \mathbb{R})$, $i \in I$).
Let $M(H,G)$ be compact. Then, the following two conditions are equivalent:

(i) *(EMFCQ) holds at every $x \in M(H,G)$.*

(ii) *$M(H,G)$ is stable.*

We present the main geometric ideas of the part ((i) $\rightarrow$ (ii)) of the proof of Theorem 3.1. Assume that (EMFCQ) holds at every $x \in M(H,G)$. Then we distinguish two subcases.

Subcase 1: $Y = \emptyset$.

Obviously, $M(H)$ is a C^1-manifold. By continuity and transversality arguments as well as the fact that $C^2(\mathbb{R}^n, \mathbb{R}^m)$ is C_s^1-dense in $C^1(\mathbb{R}^n, \mathbb{R}^m)$ (cf. [10]) we choose a C_s^1-neighbourhood ϑ of H in $C^1(\mathbb{R}^n, \mathbb{R}^m)$ and $\tilde{H} \in C^2(\mathbb{R}^n, \mathbb{R}^m)$ so C_s^1-close to H (cf. Figure 3.1(a) for $I = \{1\}$) that for every $H' \in \vartheta$:

- for every $\bar{x} \in M(H)$ the set $\tilde{N}_{\bar{x}} \cap M(H')$ is a singleton locally in a neighbourhood of $\bar{x}$, where $\tilde{N}_{\bar{x}} = \{\bar{x} - \sum_{i \in I} \lambda_i D\tilde{h}_i(\bar{x})^T \mid \lambda_i \in \mathbb{R}, i \in I\}$ and

- the resulting mapping

$$\psi : \bar{x} \in M(H) \mapsto \psi(\bar{x}) \in \tilde{N}_{\bar{x}} \cap M(H')$$

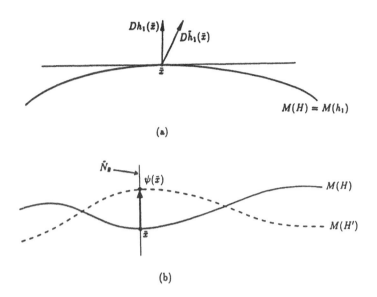

Figure 3.1 The diffeomorphism ψ in Subcase 1: $Y = \emptyset$.

is a (desired) homeomorphism (cf. Figure 3.1(b)) and after an appropriate
extension on the whole space $\mathbb{R}^n$ even a diffeomorphism.

Subcase 2: $I = \emptyset$.

In this case we use the completely integrable EMF-C^1-vector field $\xi(x)$ from Corollary 2.4 with its flow $\phi(x, \tau)$. By continuity arguments and, in particular, Lemma 2.5,
there exist a C_s^1-neighbourhood ϑ of G in $C^1(\mathbb{R}^n \times \mathbb{R}^r, \mathbb{R})$ and a continuous function

$$\alpha : x \in \partial M(G) \mapsto \alpha(x) > 0$$

with the following properties:

- $\mathcal{N}_\varepsilon$ is an open neighbourhood of $\partial M(G)$ for every $\varepsilon \in (0, 3]$, where
  $\mathcal{N}_\varepsilon = \bigcup_{\bar{x} \in \partial M(G)} \mathcal{N}_\varepsilon(\bar{x})$ and

  $$\mathcal{N}_\varepsilon(\bar{x}) = \{\phi(\bar{x}, \tau) \mid \tau \in (-\varepsilon\alpha(\bar{x}), \varepsilon\alpha(\bar{x}))\} \quad \text{for } \bar{x} \in \partial M(G),$$

- for every $G' \in \vartheta$ it is $G' \in \mathcal{F}$ ($\mathcal{F}$ is defined as in Lemma 2.5), $\partial M(G') \subset \mathcal{N}_1$, $\xi(x)$ is an EMF-vector at each $x \in M(G')$ (with respect to $M(G')$), i.e.
  the set $\partial M(G') \cap \mathcal{N}_\varepsilon(\bar{x})$ is a singleton for every $\bar{x} \in \partial M(G)$ and there is a
  Lipschitz continuous function $\hat{T} : \bar{x} \in \partial M(G) \mapsto \hat{T}(\bar{x}) \in (-\alpha(\bar{x}), \alpha(\bar{x}))$ with
  $\phi(\bar{x}, \hat{T}(\bar{x})) \in \partial M(G')$.

By using this function $\hat{T}$ one constructs for every $G' \in \vartheta$ a homeomorphism ψ : $M(G) \to M(G')$ with the following properties:

- $\psi(x) = x$ for every $x \in M(G) \setminus \mathcal{N}_2$,

- $\psi[\mathcal{N}_2(\bar{x}) \cap M(G)] \subset \mathcal{N}_2(\bar{x})$ for every $\bar{x} \in \partial M(G)$,

- $\psi[\partial M(G)] = \partial M(G')$.

Geometrically speaking, one obtains ψ by fixing $\mathbb{R}^n \setminus \mathcal{N}_2$ and stretching $\mathcal{N}_2(\bar{x}) \cap M(G)$ for every $\bar{x} \in \partial M(G)$ along $\mathcal{N}_2(\bar{x})$ such that $\psi(\bar{x}) \in \partial M(G')$ (cf. Figure 3.2).

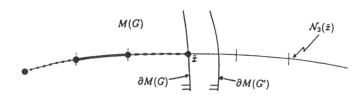

(a):

The trajectory $\mathcal{N}_3(\bar{x}) \cap M(G)$ ...

(b):

... and its image under ψ.

Figure 3.2

In the general case $I \neq \emptyset$, $Y \neq \emptyset$ the (perturbed) equality constraints describe a (perturbed) differentiable manifold in a neighbourhood of $M(H, G)$. Thus, the proof in this general case copies the argumentation of Subcase 2 in local coordinates. In particular, the construction of the corresponding EMF-C^1-vector field (cf. (2-2)) requires that ψ in Subcase 1 is a C^1-diffeomorphism.

Note that the proof of the considered direction ((i) $\to$ (ii)) does not use the compactness of $M(H, G)$. However, the proof of ((ii) $\to$ (i)) uses the compactness of

$M(H,G)$ since the (finite) number of certain topological invariants of higher connectedness of $M(H,G)$ plays an essential role.

The noncompact case

Generally, the set $M(H,G)$ is not compact. Without assuming compactness of $M(H,G)$ the next theorem presents a topological stability property of $M(H,G)$ which is again equivalent to the overall validity of (EMFCQ).

THEOREM 3.2 ([12, Theorem 5]). *The following three conditions are equivalent:*

(i) *(EMFCQ) holds at every $x \in M(H,G)$.*

(ii) *There exists a C_s^1-neighbourhood ϑ of (H,G) in $C^1(\mathbb{R}^n, \mathbb{R}^m) \times C^1(\mathbb{R}^n \times \mathbb{R}^r, \mathbb{R})$ and a family of functions $\varphi^\nu \in C^\infty(\mathbb{R}^n, \mathbb{R})$, $\nu \in \mathbb{N}$, ($\mathbb{N}$ denotes the set of natural numbers) such that for every $\nu \in \mathbb{N}$ the set $L^\nu = \{x \in \mathbb{R}^n \mid \varphi^\nu(x) \geq 0\}$ has the following properties:*

    (a) *L^ν and $\{x \in \mathbb{R}^n \mid \|x\|^2 \leq 1\}$ are diffeomorphic.*

    (b) *L^ν is a proper subset of $L^{\nu+1}$.*

    (c) *$\bigcup_{\nu \in \mathbb{N}} L^\nu = \mathbb{R}^n$.*

    (d) *$M(H',G') \cap L^\nu \simeq M(H,G) \cap L^\nu$ for every $(H',G') \in \vartheta$.*

(iii) *There exists a C_s^1-neighbourhood ϑ of (H,G) and a family of functions as in (ii) having the additional property that (EMFCQ) holds at every $x \in M(H,G,\varphi^\nu)$ (with respect to $M(H,G,\varphi^\nu)$), where*

$$M(H,G,\varphi^\nu) = \{x \in \mathbb{R}^n \mid h_i(x) = 0, i \in I, G(x,y) \geq 0, y \in Y, \varphi^\nu(x) \geq 0\}.$$

The Figure 3.3 illustrates the geometric idea of the condition (ii) of the Theorem 3.2: the sets L^ν, $\nu \in 0, 1, 2$ are concentric disks and for $(H',G') \in \vartheta$ the sets $L^\nu \cap M(H,G)$ (vertically striped for $\nu = 0$) and $L^\nu \cap M(H',G')$ (horizontally striped for $\nu = 0$) are homeomorphic.

Obviously, the sets L^ν, $\nu \in \mathbb{N}$ in Theorem 3.2 can be chosen so sufficiently large that the Theorem 3.1 becomes a corollary of Theorem 3.2 for a compact set $M(H,G)$. In Theorem 3.2 the lack of compactness of $M(H,G)$ is substituted by the condition (ii). The presented local geometric ideas of the proof of Theorem 3.1 are also essentially used in the proof of Theorem 3.2.

We would like to refer to the related paper [30] (excisional stability) where other topological conditions being equivalent with the condition (i) in Theorem 3.2 are discussed.

We see in the following corollary that the choice of a C_s^1-neighbourhood of (H,G) can be more specified under (EMFCQ).

COROLLARY 3.3 ([12, Corollary 1]). *Assume that (EMFCQ) holds at every $x \in M(H,G)$ and let $\delta : x \in \mathbb{R}^n \mapsto \delta(x) > 0$ be a continuous positive function. Then, there exist a C_s^1-neighbourhood ϑ of (H,G) in $C^1(\mathbb{R}^n, \mathbb{R}^m) \times C^1(\mathbb{R}^n \times \mathbb{R}^r, \mathbb{R})$ and, for each $(H',G') \in \vartheta$, a homeomorphism $\psi : \mathbb{R}^n \to \mathbb{R}^n$ satisfying*

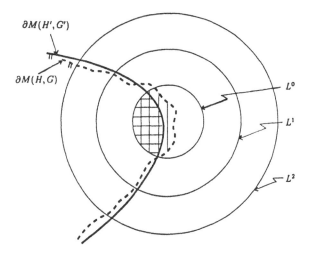

Figure 3.3 The geometric idea of Theorem 3.2(ii).

- $\psi[M(H,G)] = M(H',G')$ and

- $\|x - \psi(x)\| < \delta(x)$ for all $x \in \mathbb{R}^n$.

If $Y = \emptyset$ then ψ is a diffeomorphism and can be chosen in such a way that

$$\max\{\|I^n - D\psi(x)\|, \|x - \psi(x)\|\} < \delta(x) \quad \text{for all } x \in \mathbb{R}^n$$

(where I^n denotes the (n,n)-identity matrix and the chosen matrix norm is the associated $\|\cdot\|_2$ matrix norm).

We learn from the next theorem that under (EMFCQ) the set $M(H,G)$ can be approximated by a set being homeomorphic with $M(H,G)$, where in its description the infinitely many inequality constraints $G(\cdot,y) \geq 0$, $y \in Y$ are substituted by *one* inequality constraint.

THEOREM 3.4 ([12, Theorem 3 and Corollary 2]). *Assume that* $(EMFCQ)$ *holds at every* $x \in M(H, G)$. *Then, there exist a function* $g \in C^{\infty}(\mathbb{R}^n, \mathbb{R})$ *and a* C^1_s-*neighbourhood* ϑ *of* H *in* $C^1(\mathbb{R}^n, \mathbb{R}^m)$ *such that for all* $H' \in \vartheta$:

- $M(H, G) \simeq M(H', g)$,
  where $H' = (h'_1, \ldots, h'_m)$ *and*
  $M(H', g) = \{x \in \mathbb{R}^n \mid h'_i(x) = 0, i \in I, g(x) \geq 0\}$ *as well as*

- $(EMFCQ)$ *holds at every* $x \in M(H', g)$ *(with respect to* $M(H', g)$).

Let us return to the EMF-C^1-vector field $\xi(x)$ described in Corollary 2.4 and assume that the function g in Theorem 3.4 is chosen in such a way that $\xi(x)$ is an EMF-vector at each $x \in M(H, g)$, too. Then, we obtain variations of g by drawing the boundary $\partial M(H, g)$ along the trajectories of $\xi(x)$; we used an analogous construction in the *Subcase 2* $(I = \emptyset)$ of the geometric idea of the proof of Theorem 3.1. This construction is completely described in [12, Remark 2] and illustrated in Figure 3.4 for a special variation $g' \in C^{\infty}(\mathbb{R}^n, \mathbb{R})$: The boundary $\partial M(H, g)$ is drawn along the trajectories of $\xi(x)$ until $\partial M(H, g')$, where g' is chosen in such a way that

$$M(H, G) \subset M(H, g').$$

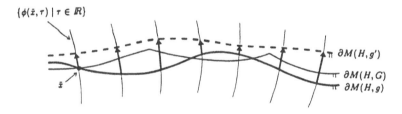

Figure 3.4 Variations of g by drawing $\partial M(H, g)$ along the trajectories.

The one-parametric case

In this case the index set Y is only assumed to be compact and we need not the special structure of Y given in Section 1. Consider the feasible set of a semi-infinite problem which depends on an additional parameter $t \in T$, where $T \subset \mathbb{R}$ is a compact parameter interval. Define

$$M(H, G, T) = \{(x, t) \in \mathbb{R}^n \times T \mid h_i(x, t) = 0, i \in I, G(x, y, t) \geq 0, y \in Y\}$$

and for $\bar{t} \in T$:

$$M(H, G, \bar{t}) = \{x \in \mathbb{R}^n \mid h_i(x, \bar{t}) = 0, i \in I, G(x, y, \bar{t}) \geq 0, y \in Y\},$$

where $H \in C^1(\mathbb{R}^n \times \mathbb{R}, \mathbb{R}^m)$, $G \in C^1(\mathbb{R}^n \times \mathbb{R}^r \times \mathbb{R}, \mathbb{R})$ and the related notations are defined analogously to (SIP). We are interested in a condition which implies

$$M(H, G, t_1) \simeq M(H, G, t_2) \quad \text{for every } t_1, t_2 \in T. \tag{3-2}$$

First we present two examples with finitely many inequality constraints (cf. [28]) where for each $\bar{t} \in T$ (EMFCQ) holds at every $x \in M(H, G, \bar{t})$ (with respect to $M(H, G, \bar{t})$) but (3-2) is not fulfilled.

EXAMPLE 3.5 Let $T = [-1, 1]$ and
$M(H, G, t) = \{(x_1, x_2) \in \mathbb{R}^2 \mid g_1(x_1, x_2, t) = x_2 - t x_1^2 \geq 0, \ g_2(x_1, x_2, t) = x_1 - x_2 \geq 0\}$.
Then,

- $M(H, G, 0)$ is a noncompact connected set (cf. Figure 3.5(a)),

- $M(H, G, t)$, $t > 0$ is a compact connected set (cf. Figure 3.5(b)) and

- the set $M(H, G, t)$, $t < 0$ is not connected (cf. Figure 3.5(c)).

Obviously, (3-2) is not fulfilled.

EXAMPLE 3.6 Let $T = [0, 1]$ and
$M(H, G, t) = \{(x_1, x_2) \in \mathbb{R}^2 \mid g(x_1, x_2, t) = x_1^2 + t x_2^2 - 1 \geq 0\}$. Then,

- $M(H, G, 0) = \{(x_1, x_2) \in \mathbb{R}^2 \mid x_1^2 \geq 1\}$ consists of two disjoint halfspaces (cf. Figure 3.5(d)) and

- $M(H, G, t)$, $t \in (0, 1]$ is the connected complement set of an ellipsoid (cf. Figure 3.5(e)).

Therefore, (3-2) is not fulfilled, too.

We have seen in the compact and noncompact case that the overall validity of (*EMFCQ*) *implies the stability condition* (3-1) which refers to perturbations of the function vector (H, G) in a C_s^1-neighbourhood ϑ of (H, G). Now, we consider perturbations by varying the additional parameter $t \in T$ and the Examples 3.5 and 3.6 show that the overall validity of (*EMFCQ*) *does not imply the stability condition* (3-2). This comes from the fact that a (fixed) C_s^1-neighbourhood ϑ of $(H, G) = (H(\cdot, \bar{t}), G(\cdot, \cdot, \bar{t}))$ in $C^1(\mathbb{R}^n, \mathbb{R}^m) \times C^1(\mathbb{R}^n \times \mathbb{R}^r, \mathbb{R})$ is described by a positive continuous function $\varepsilon(\cdot)$ (cf. Section 2) with, possibly, vanishing $\inf_{x \in \mathbb{R}^n} \varepsilon(x)$. Hence, in general, for a noncompact set $M(H, G, \bar{t})$ there does not exist a neighbourhood $\mathcal{T} \subset T$ of $\bar{t}$ such that $(H(\cdot, \tilde{t}), G(\cdot, \cdot, \tilde{t})) \in \vartheta$ for all $\tilde{t} \in \mathcal{T}$ (cf. Figure 3.6, where ϑ is represented by the shaded area).

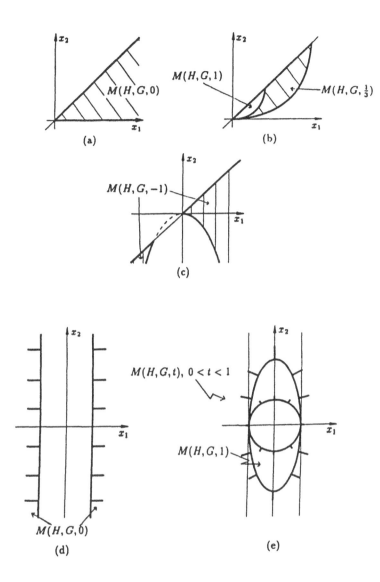

Figure 3.5 (MFCQ) holds but (3-2) is not fulfilled.

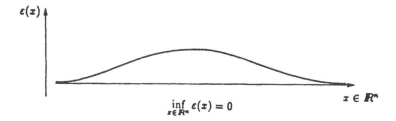

$$\inf_{x \in R^n} \varepsilon(x) = 0$$

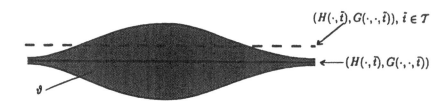

Figure 3.6 The C_s^1-topology allows to consider asymptotic effects.

Therefore, a condition which ensures (3-2) has to take into account an additional information "at infinity".

The *extended Mangasarian-Fromovitz constraint qualification at infinity* (*EMFCQI*) is said to hold at $M(H, G, T)$ if there is a real $\gamma > 0$ such that

$$\inf \left\{ \frac{\left\| \lambda_0 \binom{0_n}{1} - \sum_{i \in I} \lambda_i Dh_i(x,t)^T - \sum_{j=1}^q \mu_j D_{(x,t)} G(x, y^j, t)^T \right\|}{\substack{\lambda_i \in \mathbb{R}, \, i = 0, \ldots, m, \, \mu_j \geq 0, \, y^j \in Y_0(\bar{x}, \bar{t}), \, j = 1, \ldots, q, \, q < \infty, \\ \sum_{i=0}^m |\lambda_i| + \sum_{j=1}^q \mu_j = 1, \, (x,t) \in M(H, G, T)}} \right\} > 0.$$

A short reflexion shows that if (EMFCQI) holds at $M(H, G, T)$ then (EMFCQ) holds at $\bar{x} \in M(H, G, \bar{t})$ (with respect to $M(H, G, \bar{t})$) for every $(\bar{x}, \bar{t}) \in M(H, G, T)$ (but not vice versa!). The condition (EMFCQI) is an appropriate modification of the 'Condition C' in [24]. Before presenting another property under (EMFCQI) we need

DEFINITION 3.7 Let $f \in C^1(\mathbb{R}^n, \mathbb{R})$. Then, a point $\bar{x} \in M(H, G)$ is called a *stationary point* of (SIP) if $\bar{x} - Df(\bar{x}) \in N_{\bar{x}}$, where

$$N_{\bar{x}} = \left\{ \bar{x} - \sum_{i \in I} \lambda_i Dh_i(\bar{x})^T - \sum_{j=1}^{q} \mu_j D_x G(\bar{x}, y^j)^T \;\middle|\; \begin{array}{l} \lambda_i \in \mathbb{R},\, i \in I,\, q < \infty, \\[4pt] \mu_j \geq 0,\, y^j \in Y_0(\bar{x}), \\[2pt] j = 1, \ldots, q \end{array} \right\}.$$

If (EMFCQI) holds at $M(H, G, T)$ then the semi-infinite problem

$$\min\{e(x, t) \mid (x, t) \in M(H, G, T)\}$$

has no stationary point, where $e(x, t) = t$ and, thus, $De(x, t)^T = \binom{0_n}{1}$.
Now, we state the desired result.

THEOREM 3.8 ([16, Theorem]). *Assume that (EMFCQI) holds at $M(H, G, T)$.
Then, it is $M(H, G, t_1) \simeq M(H, G, t_2)$ for every $t_1, t_2 \in T$.*

It is easy to verify that (EMFCQI) does not hold in the Examples 3.5 and 3.6; in
Example 3.5 we obtain for $t \neq 0$, $t \to 0$ and $x_1 = x_2 = \frac{1}{t}$:

$$\lim_{t \to 0} t^2 \cdot Dg_1\left(\frac{1}{t}, \frac{1}{t}, t\right) = \lim_{t \to 0}(-2t^2, t^2, -1) = (0, 0, -1).$$

REMARK 3.9 (cf. [16, Remark 1]). The geometric idea of the proof of Theorem 3.8
is shown in Figure 3.7, where the homeomorphy of $M(H, G, t_1)$ and $M(H, G, t_2)$ for
arbitrarily chosen $t_1, t_2 \in T$ is illustrated. As a simplification we choose $T = [0, 1]$,
$t_1 = 0$, $t_2 = 1$ and put $M(t) = M(H, G, t)$, $t \in T$ as well as $\partial M = \{(x, t) \in \mathbb{R}^n \times T \mid x \in \partial M(t)\}$.

Using (EMFCQI) we construct for every $\bar{x} \in \partial M(0)$ a continuous mapping

$$\mathcal{H}(\bar{x}, \cdot) : t \in T \mapsto \mathcal{H}(\bar{x}, t) \in \partial M(t) \times \{t\}$$

such that:

- $\mathcal{H}(\bar{x}, 0) = (\bar{x}, 0)$ for every $\bar{x} \in \partial M(0)$,

- if $x^1, x^2 \in \partial M(0)$, $x^1 \neq x^2$ then $\mathcal{H}(x^1, t) \neq \mathcal{H}(x^2, t)$ for every $t \in [0, 1]$ and

- $\displaystyle\bigcup_{\bar{x} \in \partial M(0)} \{\mathcal{H}(\bar{x}, t) \mid t \in T\} = \partial M$.

This construction is rather complicated and we do not give more details here. Since
(EMFCQ) holds at every $x \in M(0)$, the notations $\phi(\bar{x}, \tau)$, $\mathcal{N}_\epsilon(\bar{x})$ and $\mathcal{N}_\epsilon$ with
$\tau \in \mathbb{R}$, $\varepsilon \in [0, 3]$ and $\bar{x} \in \partial M(0)$ are defined (for $M(H, G) = M(0)$) as in the proof
of *Subcase* 2 of Theorem 3.1 (cf. Figure 3.7(a)). By fixing $M(0) \setminus \mathcal{N}_2$ and stretching
for every $\bar{x} \in \partial M(0)$ the trajectory $\mathcal{N}_2(\bar{x}) \cap M(0)$ along $\mathcal{N}_2(\bar{x}) \cup \{\mathcal{H}(\bar{x}, t) \mid t \in T\}$
we obtain that

$$M(0) \simeq M(0) \times \{0\} \cup \partial M$$

(cf. Figure 3.7(b) and (c)).

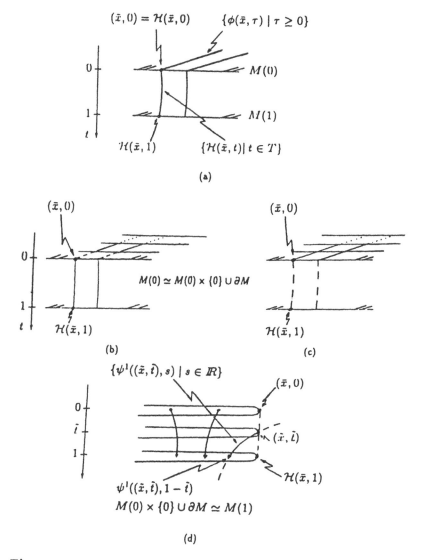

Figure 3.7 The geometric idea of the proof of Theorem 3.8.
(From [16].)

Finally, we construct a bounded (here, we need (EMFCQI) again) vector field $F \in C^2(\mathbb{R}^{n+1}, \mathbb{R}^{n+1})$ with the flow $\psi^1((x,t),s)$ such that for all $(x,t) \in M(H,G,T)$:

$$Dh_i(x,t)F(x,t) = 0, \quad i \in I,$$
$$D_{(x,t)}G(x,y,t)F(x,t) > 0, \quad y \in Y_0(x,t),$$
$$F_t(x,t) = 1, \text{ where } F(x,t) = (F_x(x,t), F_t(x,t)) \in \mathbb{R}^n \times \mathbb{R}.$$

Then, we obtain a homeomorphism

$$(\tilde{x}, \tilde{t}) \in M(0) \times \{0\} \cup \partial M \mapsto \psi^1((\tilde{x}, \tilde{t}), 1 - \tilde{t}) \in M(1) \times \{1\}$$

(cf. Figure 3.7(d)) and, therefore, $M(0) \simeq M(1)$.

The next corollary combines the results on C_s^1-perturbations with the results on parameter perturbations.

COROLLARY 3.10

(i) *Assume that (EMFCQ) holds at $\bar{x} \in M(H,G,\bar{t})$ (with respect to $M(H,G,\bar{t})$) for every $(\bar{x}, \bar{t}) \in M(H,G,T)$. Then, there exists a C_s^1-neighbourhood ϑ' of (H,G) in $C^1(\mathbb{R}^n \times \mathbb{R}, \mathbb{R}^m) \times C^1(\mathbb{R}^n \times \mathbb{R}^r \times \mathbb{R}, \mathbb{R})$ such that for all $(H'', G'') \in \vartheta'$ and all $t \in T$:*

$$M(H,G,t) \simeq M(H'', G'', t).$$

(ii) *Assume that (EMFCQI) holds at $M(H,G,T)$. Then, there exists a C_s^1-neighbourhood ϑ of (H,G) in $C^1(\mathbb{R}^n \times \mathbb{R}, \mathbb{R}^m) \times C^1(\mathbb{R}^n \times \mathbb{R}^r \times \mathbb{R}, \mathbb{R})$ such that for all $(H', G') \in \vartheta$:*

- *$M(H,G,t_1) \simeq M(H',G',t_2)$ for every $t_1, t_2 \in T$ and*
- *(EMFCQI) holds at $M(H',G',T)$.*

The final result in this section shows that the considered constraint qualifications of the Mangasarian-Fromovitz type also imply certain semicontinuity properties of corresponding set-valued mappings, where the definitions of lower and upper semicontinuity are taken from [3].

COROLLARY 3.11

(i) *(cf. [17, Theorem 2.2 for the case $H \in C^2(\mathbb{R}^n, \mathbb{R}^m)$] and [29, Folgerung 5.2]). Assume that (EMFCQ) holds at every $x \in M(H,G)$. Then, there exists a C_s^1-neighbourhood ϑ of (H,G) in $C^1(\mathbb{R}^n, \mathbb{R}^m) \times C^1(\mathbb{R}^n \times \mathbb{R}^r, \mathbb{R})$ such that the set-valued mapping*

$$(\tilde{H}, \tilde{G}) \mapsto M(\tilde{H}, \tilde{G})$$

is both lower semi-continuous and upper semi-continuous at all $(\tilde{H}, \tilde{G}) \in \vartheta$.

(ii) *[16, Remark 5]. Assume that (EMFCQI) holds at $M(H,G,T)$. Then, the set-valued mapping*

$$t \in T \mapsto M(H,G,t)$$

is both lower semi-continuous and upper semi-continuous at all $t \in T$.

4 A CONCLUSION FOR HOMOTOPY METHODS

Finally, we discuss briefly a consequence of the stability result given in Theorem 3.1 for the use of so-called homotopy methods. For a detailed exposé on this subject we refer to [1,5]. A homotopy method can be used for the computation of a stationary point of the finite problem

(P) $\min\{f(x) \mid x \in M\}$,

where $M = \{x \in \mathbb{R}^n \mid h_i(x) = 0, i \in I, g_j(x) \geq 0, j \in J\}$, $J = \{1, \ldots, s\}$ and $(f, h_i, g_j, i \in I, j \in J) \in C^2(\mathbb{R}^n, \mathbb{R}^{1+m+s})$.
The basic idea of a homotopy method is to construct an auxiliary one-parametric problem

$$
\begin{cases}
\tilde{P}(t) \quad \min\{\tilde{f}(x,t) \mid x \in \tilde{M}(t)\}, \text{ where } t \in \mathbb{R}, \\
\tilde{M}(t) = \{x \in \mathbb{R}^n \mid \tilde{h}_i(x,t) = 0, i \in I, \tilde{g}_j(x,t) \geq 0, j \in J\} \text{ and} \\
(\tilde{f}, \tilde{h}_i, \tilde{g}_j, i \in I, j \in J) \in C^2(\mathbb{R}^n \times \mathbb{R}, \mathbb{R}^{1+m+s}),
\end{cases}
$$

with the following properties (cf. Figure 4.1):

(4-1) A stationary point x^0 of $\tilde{P}(0)$ is known.

(4-2) The problems (P) and $\tilde{P}(1)$ are identical.

(4-3) There is a path $\mathcal{C} \subset \mathbb{R}^n \times [0,1]$ connecting $(x^0, 0)$ and $(x^1, 1)$ such that $\tilde{x}$ is a stationary point of $\tilde{P}(\tilde{t})$ for all $(\tilde{x}, \tilde{t}) \in \mathcal{C}$.

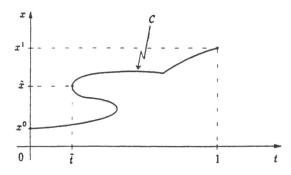

Figure 4.1 The basic idea of homotopy methods.

If (4-1), (4-2), (4-3) and certain conditions for the numerical feasibility of the used algorithm are fulfilled, then a pathfollowing algorithm can approximate the path $\mathcal{C}$ and, in particular, the stationary point x^1 of the original problem (P).

Obviously, (4-1) and (4-2) can be fulfilled by means of an appropriate construction of $\tilde{P}(t)$. However, in general this is not possible for (4-3). If the following four conditions are assumed then (4-3) is fulfilled, too (cf. [5]):

(a) Zero is a regular value of the so-called Kojima-mapping (we do not consider this property and refer to [21]).

(b) There is a compact set $A \subset \mathbb{R}^n$ such that $\tilde{M}(t) \subset A$ for every $t \in [0, 1]$.

(c) There is exactly one stationary point x^0 of $\tilde{P}(0)$ and x^0 is strongly stable in the sense of Kojima [20].

(d) (EMFCQ) holds at every $x \in \tilde{M}(t)$ (with respect to $\tilde{M}(t)$) for all $t \in [0, 1]$.

The restrictive character of the assumptions (a) – (d) becomes obvious by means of the stability property in Theorem 3.1 in the following way.

In order to ensure assumption (c) the auxiliary problem $\tilde{P}(t)$ can be constructed in such a way that $\tilde{f}(\cdot, 0)$ is strictly convex and $\tilde{M}(0)$ is a convex set. If $\tilde{M}(0)$ is convex, the assumptions (b), (d) and the Theorem 3.1 imply that for all $t \in [0, 1]$ the set $\tilde{M}(t)$ – and, in particular, the feasible set $M = \tilde{M}(1)$ of the original problem (P) – *has to be homeomorphic with a convex set*. However, this latter property is only fulfilled for a very restricted family of constraint functions $(h_i, g_j, i \in I, j \in J) \in C^2(\mathbb{R}^n, \mathbb{R}^{m+s})$.

REFERENCES

1. E. Allgower and K. Georg, Introduction to Numerical Continuation Methods, Springer, Berlin, Heidelberg, New York (1990).

2. H. Amann, Gewöhnliche Differentialgleichungen, Walter de Gruyter, Berlin, New York (1983).

3. C. Berge, Espaces topologique, functions multivoques, Dunod, Paris (1966).

4. T.J. Graettinger and B.H. Krogh, The acceleration radius: a global performance measure for robotic manipulators; *IEEE J. Robotics and Automation* 4: 60-69(1988).

5. J. Guddat, F. Guerra and H.Th. Jongen, Parametric Optimization: Singularities, Pathfollowing and Jumps, Wiley, Chichester (1990).

6. J. Guddat, H.Th. Jongen and J.-J. Rückmann, On stability and stationary points in nonlinear optimization, *J. Austral. Math. Soc., Ser. B* 28: 36-56(1986).

7. R. Henrion and D. Klatte, Metric regularity of the feasible set mapping in semi-infinite optimization, *Appl. Math. Opt.* 30: 103-106(1994).

8. R. Hettich and K.O. Kortanek, Semi-infinite programming: theory, methods and applications, *SIAM Rev.* 35: 380-429(1993)3.

9. R. Hettich and P. Zencke, Numerische Methoden der Approximation und semi-infiniten Optimierung, Teubner, Stuttgart (1982).

10. M.W. Hirsch, Differential Topology, Springer, Berlin, Heidelberg, New York (1976).

11. R. Horst and H. Tuy, Global Optimization, Springer, Berlin, Heidelberg, New York (1993).

12. M.A. Jiménez and J.-J. Rückmann, On equivalent stability properties in semi-infinite optimization, *Z. Oper. Res.* 41: 175-190(1995)2.

13. H.Th. Jongen, P. Jonker and F. Twilt, Nonlinear Optimization in R^n, I: Morse Theory, Chebyshev Approximation, Peter Lang, Frankfurt a.M., Bern, New York (1983).

14. H.Th. Jongen, P. Jonker and F. Twilt, Nonlinear Optimization in R^n, II: Transversality, Flows, Parametric Aspects, Peter Lang, Frankfurt a.M., Bern, New York (1986).

15. H.Th. Jongen and J.-J. Rückmann, Nonlinear optimization: on connected components of level sets, *SIAM J. Control Opt.* 31: 86-100(1993)1.

16. H.Th. Jongen, J.-J. Rückmann and G.-W. Weber, One-parametric semi-infinite optimization: on the stability of the feasible set, *SIAM J. Opt.* 4: 637-648(1994)3.

17. H.Th. Jongen, F. Twilt and G.-W. Weber, Semi-infinite optimization: structure and stability of the feasible set, *J. Optim. Theory Appl.* 72: 529-552(1992).

18. D. Klatte, Stable local minimizers in semi-infinite optimization: regularity and second-order conditions, *J. Comp. Appl. Math.* 56: 137-157(1994).

19. D. Klatte, On regularity and stability in semi-infinite optimization, *Set-Valued Analysis* 3: 101-111(1995).

20. M. Kojima, Strongly stable stationary solutions in nonlinear programs, in: S.M. Robinson (ed.), Analysis and Computation of Fixed Points, Academic Press, New York 93-138(1980).

21. M. Kojima and R. Hirabayashi, Continuous deformation of nonlinear programs, *Math. Progr. Study* 21: 150-198(1984).

22. O.L. Mangasarian and S. Fromovitz, The Fritz John necessary optimality conditions in the presence of equality and inequality constraints, *J. Math. Anal. Appl.* 17: 37-47(1967).

23. V.H. Nguyen, J.J. Strodiot and N.V. Thoai, On an optimum shape design problem, Technical Report 85/5, Dept. of Mathematics, Facultes Universitaires de Namur (1985).

24. R.S. Palais and S. Smale, A generalized Morse theory, *Bull. Am. Math. Soc.* 70: 165-172(1964).

25. J.-P. Penot, On regularity conditions in mathematical programming, *Math. Programming Study* 19: 169-199(1982).

26. E. Polak, On the mathematical foundations of nondifferentiable optimization in engineering design, *SIAM Rev.* 29: 21-89(1987).

27. A. Potchinkov and R. Reemtsen, FIR filter design in the complex domain by a semi-infinite programming technique. I. The method, *Int. J. Electronics and Communications* 48: 135-144(1994)3.

28. J.-J. Rückmann, Stability of noncompact feasible sets in nonlinear optimization, in: J. Guddat et al. (eds.), Parametric Optimization and Related Topics III, Peter Lang, Frankfurt a.M., Bern, New York 467-502(1993).

29. J.-J. Rückmann, Stabilitätseigenschaften in der nichtlinearen Optimierung, Habilitationsschrift, Humboldt University Berlin, Dept. of Mathematics (1993).

30. G.W. Weber, Charakterisierung struktureller Stabilität in der nichtlinearen Optimierung, Thesis, RWTH Aachen, Dept. of Mathematics (C) (1992).

31. J. Zowe and S. Kurcyusz, Regularity and stability for the mathematical programming problem in Banach spaces. *Appl. Math. Optim.* 5: 49-62(1979).

Solution Existence for Infinite Qudratic Programming

I. E. SCHOCHETMAN Department of Mathematical Sciences, Oakland University, Rochester, Michigan

R. L. SMITH Department of Industrial and Operations Engineering, The University of Michigan, Ann Arbor, Michigan

S. K. TSUI Department of Mathematical Sciences, Oakland University, Rochester, Michigan

ABSTRACT

We consider an infinite quadratic programming problem with positive semi-definite quadratic costs, equality constraints and unbounded variables. Sufficient conditions are given for there to exist an optimal solution. Specifically, we require that (1) the cost operator be strictly positive definite when restricted to the orthogonal complement of its kernel, and (2) the constraint operator have closed range when restricted to the kernel of the cost operator. Condition (1) is shown to be equivalent to the spectrum of the restricted cost operator being bounded away from zero. Similarly, condition (2) is equivalent to the minimum modulus of the restricted constraint operator being positive. In the presence of separability, we give a sufficient condition for (2) to hold in terms of finite dimensional truncations of the restricted constraint operator. We apply our results to a broad class of infinite horizon optimization problems. In this setting, the finite dimensional truncations can be considered to be finite dimensional approximations to our problem whose limit, in a somewhat formal sense, is our infinite dimensional problem. Each of these approximations has properties (1) and (2) by virtue of their finite-dimensionality, i.e., each admits an optimal solution. However, our infinite dimensional problem may not. Thus, we give sufficient conditions for our problem to also admit an optimal solution. Finally, we illustrate this application in the case of an infinite horizon LQ regulator problem (a production planning problem).

1   INTRODUCTION

We consider the infinite quadratic programming problem ($\mathcal{Q}$) given by:

$$\min \langle x, Qx \rangle$$

subject to $\hspace{9cm}$ ($\mathcal{Q}$)

$$Ax = b,$$

$$x \in H,$$

where H and M are separable, real Hilbert spaces, the constraint operator $A : H \to M$ is a bounded linear operator, $b \in M$, and the cost operator $Q : H \to H$ is a non-zero, (self-adjoint) *positive semi-definite*, bounded linear operator. Recall that Q is positive semi-definite if $\langle x, Qx \rangle \geq 0$, $\forall x \in H$, and that Q is *positive definite* if $\langle x, Qx \rangle > 0$, $\forall x \in H$, $x \neq 0$. In general, the quadratic function $\langle x, Qx \rangle$ evaluated on the linear manifold defined by $Ax = b$ is either unbounded below or Q is positive semi-definite on this manifold. Moreover, in the latter case, if the manifold is also finite-dimensional, then the minimum is attained.

We assume that problem (Q) is non-trivial in that there exists a feasible solution, i.e., b is in the range of A. Our objective in this paper is to give sufficient conditions on the problem data Q, A, b to guarantee the existence of an optimal solution to (Q). (Uniqueness is not an issue here since, in general, multiple optimal solutions will exist in this context.) For convenience, we let $K = ker(Q)$ denote the kernel of Q in H and $L = K^\perp$ its orthogonal complement in H, so that $H = K \oplus L$. Our sufficient conditions require that the operator restriction $Q|L$ of Q to L be *strictly positive definite* (see section 2), i.e., *coercive*, and that the operator restriction $A|K$ of A to K to have *closed range* in M. The strictly positive definite property is equivalent to requiring that the non-zero spectrum of Q be bounded away from zero, and the closed range property is equivalent to requiring that the minimum modulus (Goldberg, 1966) of the restriction $A|K$ be positive.

As a special case, problem (Q) includes *infinite horizon*, discrete-time, linear-quadratic programming. This class of problems includes the important time-varying, discrete-time LQ tracker and regulator problems (Lewis, 1986). Our objective in these cases is to give sufficient conditions, in terms of the time-staged data over increasing finite horizons, for there to exist an optimal solution. Here, decisions, costs and constraints are specified (and vary) over an *infinite*, discrete time-line. By assumption, the operator restrictions referred to above (automatically) have, *in each period*, positive spectra bounded away from zero and positive minimum moduli, respectively. (Note that the positive spectrum of a finite-dimensional operator is the set of positive eigenvalues.) The same is true of the finite direct sums of these restrictions over finite horizon approximations. The problem is that these conditions may not hold for the direct sums of these operator restrictions over the *infinite* horizon. We give sufficient conditions for this to occur, i.e., for $Q|L$ to have positive spectrum bounded away from zero, and for $A|K$ to have positive minimum modulus, in terms of increasing finite-horizon approximations of these operators.

In Schochetman, Smith and Tsui (1995), the authors considered the solution existence question for a problem somewhat similar to (Q). There, the operator analogous to Q above was assumed to be a direct sum $\oplus_{j=1}^{\infty} Q_j$ of positive definite matrices Q_j. The (possibly unbounded) operator Q itself need not be positive definite. We showed that if the (positive) eigenvalues of the Q_j are bounded away from 0, then the optimization problem admits a (unique) solution. Even if Q is a bounded operator which is positive definite, but not *stictly* positive definite (i.e., coercive), then the optimization problem will not admit an optimal solution. See Dontchev and Zolezzi (1992, Theorem 32). For such positive definite Q, we have $K = 0$, so that $H = L$ and $A|K = 0$, which obviously has closed range. Thus, the current paper can be viewed as an extension of Schochetman, Smith and Tsui (1995) to the more general case where Q is only positive semi-definite, i.e., where

$A|K$ is non-trivial.

In section 2, we give sufficient conditions for problem (Q) to have an optimal solution. In particular, we show that if the restriction $A|K$ has closed range in M and the (positive definite) restriction $Q|L$ is strictly positive definite, then there exists an optimal solution for (Q). We then give an equivalent conditon for $A|K$ to have the closed range property. If M is separable, we obtain a sufficient condition for range closure in terms of finite-dimensional truncations of $A|K$, i.e., in terms of finite horizon approximations to $A|K$. (These responses are based on the relevant operator-theoretic results established in the Appendix.) We also give equivalent conditions for $Q|L$ to be strictly positive definite in terms of the finite horizon approximations to $Q|L$. In section 3, we apply our main results to an infinite horizon, discrete-time, non-stationary, linear quadratic programming problem with positive semi-definite cost matrices and lower-staircase constraint structure. We obtain sufficient conditions for such a problem to admit an optimal solution. In section 4, we illustrate these results in a special case of the infinite horizon linear-quadratic regulator problem (a production planning problem).

2   OPTIMAL SOLUTION EXISTENCE

In this section, we establish sufficient conditions for (Q) to admit an optimal solution. Since Q is positive semi-definite, it is not difficult to see that its kernel K is given by

$$K = \{x \in H : \langle x, Qx \rangle = 0\}.$$

Since $H = K \oplus L$, we may let E_K (resp. E_L) denote the orthogonal projection of H onto K (resp. L). Moreover, since Q is self-adjoint, it follows that K and L are invariant under Q. Hence, Q also decomposes into $0 \oplus P$, where 0 is the zero operator on K and $P : L \to L$ is the restriction operator $Q|L$. Note that P is a positive definite, bounded linear operator on L.

Now let F denote the feasible region for (Q), i.e.,

$$F = \{x \in H : Ax = b\}$$
$$= \{\eta + \xi : \eta \in K, \ \xi \in L, \ A(\eta + \xi) = b\},$$

so that (Q) becomes

$$\min_{x \in F} < x, Qx > . \tag{Q}$$

To avoid trivialities, we suppose that $F \neq \emptyset$. Also let

$$F_K = \{\eta \in K : \eta + \xi \in F, \ for \ some \ \xi \in L\} = E_K(F),$$

i.e., F_K is the image of F under E_K. It is non-empty and convex in K, since this is the case for F in H. It is also true that F is closed in H; however, F_K need *not* be closed in K.

Analogously, let

$$F_L = \{\xi \in L : \eta + \xi \in F, \ for \ some \ \eta \in K\} = E_L(F),$$

the image of F under E_L. As with F_K, the set F_L is non-empty and convex, but not necessarily closed in K. Moreover, $F \subseteq F_K \oplus F_L$.

We may now consider the following related problem $(\mathcal{P})$:

$$\min_{\xi \in F_L} \langle \xi, P\xi \rangle \tag{$\mathcal{P}$}$$

where, as we have seen, P is positive definite on L and F_L is a non-empty, convex subset of L. Moreover,

$$\langle \xi, P\xi \rangle = \langle x, Qx \rangle,$$

for all $\xi \in L$, $\eta \in K$ and $x = \eta + \xi$.

Note that solving $(\mathcal{P})$ is equivalent to solving $(\mathcal{Q})$ in the following sense. If ξ^* is an optimal solution to $(\mathcal{P})$, then there exists $\eta^* \in F_K$ such that $x^* = \eta^* + \xi^* \in F$ and x^* is optimal for $(\mathcal{Q})$. Conversely, if x^* is optimal for $(\mathcal{Q})$, then $x^* = \eta^* + \xi^*$, for $\eta^* \in F_K$ and $\xi^* \in F_L$, where ξ^* is optimal for $(\mathcal{P})$.

Given the formulation of problem $(\mathcal{P})$, it is desirable to know when F_L is closed in L. We have the following sufficient condition.

LEMMA 2.1. If $A|K$ has closed range in M, then F_L is closed in L.

PROOF. For the remainder of this paper, it will be convenient to denote $A|K$ by B. Let $\{\xi^n\}$ be a sequence in F_L and $\xi \in L$ such that $\xi^n \to \xi$, as $n \to \infty$. By definition of F_L, for each n, there exists $\eta^n \in K$ such that $\eta^n + \xi^n \in F$, i.e., $A(\eta^n + \xi^n) = b$. But

$$A(\eta^n + \xi^n) = B\eta^n + A\xi^n = b.$$

Hence,

$$B\eta^n = b - A\xi^n,$$

which converges to $b - A\xi$. Consequently, by hypothesis, $b - A\xi$ is necessarily an element of the range $B(K)$ of B. Thus, there exists $\eta \in K$ such that

$$A\eta = B\eta = b - A\xi,$$

i.e.,

$$A\eta + A\xi = b,$$

so that $\eta + \xi \in F$ and $\xi \in F_L$.

We are now in position to prove the main result of this section. We will say that the operator P is *strictly positive definite* if it is coercive, i.e., if there exists $\sigma_P > 0$ satisfying

$$\sigma_P \|\xi\|^2 \le \langle \xi, P\xi \rangle, \quad \forall \xi \in L.$$

(In Lee, Chou and Barr (1972), such an operator P was called positive definite; in Milne (1980) it was called positive-bounded-below.) This condition is known (Dontchev and Zolezzi (1992, p.73)) to be necessary and sufficient for $(\mathcal{P})$ to admit a (unique) optimal solution. Observe that even if F_L is closed in L, $(\mathcal{P})$ may not admit an optimal solution, despite the fact that P is positive definite. See Schochetman, Smith and Tsui (1995) for an example.

THEOREM 2.2. If $B = A|K$ has closed range in M and $P = Q|L$ is strictly positive definite on L, then there exists an optimal solution to $(\mathcal{Q})$.

PROOF. If B has closed range, then F_L is closed in L by Lemma 2.1. Thus, the feasible region in problem $(\mathcal{P})$ is a closed affine subset of L. If, in addition, P is strictly positive definite, then it is well-known that problem $(\mathcal{P})$ admits a (unique) optimal solution (Dontchev and Zolezzi (1992)). (The space L can be equivalently re-normed by $\|\xi\|_P = \sqrt{\langle \xi, P\xi \rangle}$. An optimal solution to $(\mathcal{P})$ is then a best-approximation to the origin in the closed, convex, non-empty subset F_L of the Hilbert space L. Such is well-known to exist (Aubin, 1979).) The proof is then completed by recalling the correspondence between solutions of $(\mathcal{P})$ and $(\mathcal{Q})$.

The previous theorem suggests the following questions. Under what conditions:
(i) is P strictly positive definite on L?
(ii) does B have closed range in M?

To respond to these questions, we require some additional notation. Let $T : X \to Y$ be an arbitrary bounded linear operator from the Hilbert space X to the Hilbert space Y. Define the operator index α_T as follows:

$$\alpha_T = \inf\{\|Tx\| : x \in ker(T)^\perp, \|x\| = 1\}.$$

It is shown in Theorem A.1 of the Appendix that this index is an alternate characterization, in the context of Hilbert space, of the well-known minimum modulus (Goldberg, 1966) of operator T. Thus, in particular,

$$\alpha_B = \inf\{\|B\eta\| : \eta \in ker(B)^{\perp_K}, \|\eta\| = 1\} \geq 0,$$

where $ker(B)^{\perp_K}$ denotes the orthogonal complement of $ker(B)$ in *its* domain K.

Now assume $Y = X$. As usual, let $\sigma(T)$ denote the spectrum of T (Helmberg, 1969). In particular, $\sigma_+(T)$ will denote the *positive* elements of $\sigma(T)$. Recall that if T is self-adjoint and positive semi-definite, then $\sigma(T) \neq \emptyset$ (Helmberg, 1969, p.226) and $\sigma_+(T)$ consists of the non-zero elements of $\sigma(T)$, because $\sigma(T) \subseteq [0, \infty)$. If T is non-zero as well, then $\sigma_+(T) \neq \emptyset$ also. If T is given by a matrix on a Euclidean space with standard basis, then $\sigma(T)$ is simply the set of eigenvalues of the matrix. Finally, if S is any non-empty set of real numbers, then $\inf(S)$ will denote the infimum of the elements of S.

In response to question (1) above, we have the following well-known result.

LEMMA 2.3. The following are equivalent for the self-adjoint, positive definite operator P on L.
 (i) P is strictly positive definite.
 (ii) The quantity $\inf\{\langle \xi, P\xi \rangle / \|\xi\|^2 : \xi \in L, \xi \neq 0\}$ is positive.
 (iii) The operator P is invertible.
 (iv) The quantity $\inf(\sigma(P))$ is positive.

Before proceeding, it is perhaps worthwhile to restate Lemma 2.3 in terms of the original operator Q. Recall that $\sigma_+(Q) \neq \emptyset$ because $Q \neq 0$.

LEMMA 2.3′. The following are equivalent for the self-adjoint, positive semi-definite operator Q on H.
  (i) There exists $\sigma_Q > 0$ such that $\sigma_Q \|x\|^2 \leq \langle x, Qx \rangle$, $\forall x \in H$, $x \perp K$.
  (ii) The quantity $\inf\{\langle x, Qx \rangle / \|x\|^2 : x \in H,\ x \neq 0,\ x \perp K\}$ is positive.
  (iii) The operator $Q|K^\perp$ is invertible.
  (iv) The quantity $\inf(\sigma_+(Q))$ is positive.

In response to question (2) above, we have:

THEOREM 2.4. The operator $B = A|K$ has closed range in M if and only if $\alpha_B > 0$.

PROOF. Apply the Corollary to Theorem A.1 of the Appendix.

Since M is separable, we can obtain a useful lower bound for α_B as follows. Let $\{e_m\}$ be a complete othonormal system for M and $\{m_i\}$ a sequence of strictly increasing positive integers. For each $i = 1, 2, \ldots$, let M_i denote the span of $\{e_1, \ldots, e_{m_i}\}$. Then $\{M_i\}$ is a sequence of finite-dimensional subspaces of M such that $M_i \subseteq M_{i+1}$, $\forall i = 1, 2, \ldots$, and $\cup_{i=1}^\infty M_i$ is dense in M. Let $J_i : M \to M$ be the mapping given by

$$J_i\left(\sum_{m=1}^\infty a_m e_m\right) = \sum_{m=1}^{m_i} a_m e_m, \quad \forall i = 1, 2, \ldots.$$

Note that $J_i(M) = M_i \subseteq M$ and $J_i^* = J_i$, for all i. Define $B_i : K \to M$ by $B_i = J_i B$, $\forall i = 1, 2, \ldots$. Our objective is to investigate the range-closure property for the operator B in terms of its finite-rank truncations B_i, which have (finite-dimensional) closed range.

LEMMA 2.5. Let the notation be as above. Then:
  (i) $\{J_i\}$ converges strongly to the identity operator on M.
  (ii) $\{B_i\}$ converges strongly to B on K.
  (iii) $\{B_i^*\}$ converges strongly to B^* on M.
  (iv) $\|B_i^*\|$, $\|B_i\| \leq \max(\|B^*\|, \|B\|)$, $\forall i = 1, 2, \ldots$.
  (v) $\{B_i B_i^*\}$ converges strongly to BB^* on M, as $i \to \infty$.

PROOF. Items (i) - (iv) are straightforward. To prove (v), apply Remark 2.5.10 of Kadison and Ringrose (1983).

Since the image of B_i is finite-dimensional in M, it follows from Theorem 2.4 that $\alpha_{B_i} > 0$, for all i. However, $\inf_i \alpha_{B_i} \geq 0$, in general, as the next example shows.

EXAMPLE. Let $H = M$ be a separable Hilbert space with $m_i = i$, $\forall i = 1, 2, \ldots$. Define $A : M \to M$ by

$$A\left(\sum_{m=1}^\infty a_m e_m\right) = \sum_{m=1}^\infty \frac{1}{m} a_m e_m.$$

Then A is bounded with norm equal to 1 and $ker(A) = \{0\}$. Moreover, M_i is the span of $\{e_1, \ldots, e_i\}$ and

$$B_i\left(\sum_{m=1}^{\infty} a_m e_m\right) = \sum_{m=1}^{i} \frac{1}{m} a_m e_m, \quad \forall i = 1, 2, \ldots.$$

Hence, $B_i(e_i) = \frac{1}{i} e_i$, so that $\|B_i(e_i)\| = \frac{1}{i}$, i.e., $\alpha_{B_i} \leq \frac{1}{i}$, $\forall i = 1, 2, \ldots$. Consequently, $\inf_i \alpha_{B_i} = 0$. In fact, $\alpha_A = 0$ by Lemma A.3, since the eigenvalues of A are of the form $1/i$, $\forall i = 1, 2, \ldots$.

We next show that the index of the operator B is at least the infimum of the indices of its finite-dimensional truncations B_i.

THEOREM 2.6. Let the notation be as above. Then $\alpha_B \geq \inf_i \alpha_{B_i}$.

PROOF. Let $\epsilon > 0$. By the definition of $\beta_B = \alpha_B$ (Theorem A.1), there exists y_ϵ in M such that $B^* y_\epsilon \neq 0$ and

$$\alpha_B \leq \frac{\|BB^* y_\epsilon\|}{\|B^* y_\epsilon\|} \leq \alpha_B + \frac{\epsilon}{2}.$$

But, by part (2) of Lemma 2.5,

$$\|By_\epsilon\| = \lim_{i \to \infty} \|B_i y_\epsilon\|,$$

so that $B_i y_\epsilon$ is eventually non-zero. Also, by part (5) of Lemma 2.5, we have that $B_i B_i^*$ converges strongly to BB^* on M. Hence,

$$\frac{\|B_i B_i^* y_\epsilon\|}{\|B_i^* y_\epsilon\|} \to \frac{\|BB^* y_\epsilon\|}{\|B^* y_\epsilon\|}, \quad as \; i \to \infty.$$

Therefore, there exists i_ϵ such that

$$\alpha_B - \frac{\epsilon}{2} < \frac{\|B_{i_\epsilon} B_{i_\epsilon}^* y_\epsilon\|}{\|B_{i_\epsilon}^* y_\epsilon\|} < \alpha_B + \epsilon,$$

so that $\alpha_{B_{i_\epsilon}} < \alpha_B + \epsilon$, where ϵ is arbitrary. Consequently, $\alpha_B \geq \inf_i \alpha_{B_i}$.

COROLLARY. Let the B_i be as above. If $\inf_i \alpha_{B_i} > 0$, then B has closed range in M.

We thus have the following numerical version of Theorem 2.2.

THEOREM 2.7. If $\inf_i \alpha_{B_i} > 0$ and $\inf(\sigma_+(Q)) > 0$, then there exists an optimal solution to $(\mathcal{Q})$.

PROOF. Apply Lemma 2.3, as well as Theorems 2.2, 2.4 and 2.6.

REMARK. The previous results from Lemma 2.5 through Theorem 2.7 inclusive are valid more generally for the operator $T : X \to Y$ of the Appendix.

3   AN APPLICATION TO INFINITE HORIZON OPTIMIZATION

As an application of the above, consider the following special case $(\mathcal{S})$ of $(\mathcal{Q})$.

$$\min \sum_{j=1}^{\infty} \langle x_j, Q_j x_j \rangle$$

subject to $(\mathcal{S})$

$$A_{i,i-1} x_{i-1} + A_{ii} x_i = b_i, \quad \forall i = 1, 2, \ldots ,$$

$$x_j \in \mathbf{R}^{n_j}, \quad \forall j = 1, 2, \ldots ,$$

and

$$\sum_{j=1}^{\infty} \|x_j\|_2^2 < \infty,$$

where $A_{10} = 0$, Q_j is a symmetric, positive semi-definite matrix, $\forall j$, and $b_i \in \mathbf{R}^{m_i}$, $\forall i$. Of course, the $A_{i,i-1}$, A_{ii} are matrices of appropriate size. Let H denote the Hilbert sum of the $\mathbf{R}^{n_j}$ and M the Hilbert sum of the $\mathbf{R}^{m_i}$. All Euclidean spaces are assumed to be equipped with their respective standard bases. Note that H and M are separable. In fact, we may let M_i (as in section 2) denote the subspace

$$\mathbf{R}^{m_1} \oplus \ldots \oplus \mathbf{R}^{m_i} \oplus 0 \oplus \ldots$$

of M, $\forall i = 1, 2, \ldots$. In this case, $J_i : M \to M$ is given by $J_i(y) = (y_1, \ldots, y_i, 0, \ldots)$, $\forall y = (y_i) \in M$, $\forall i = 1, 2, \ldots$.

If $\|Q_j\|_2$ denotes the spectral norm (Horn and Johnson, 1988) of Q_j, $\forall j$, we assume that $\sup_j \|Q_j\|_2 < \infty$. This will be the case, for example, if there exist finitely many distinct Q_j. We then obtain a self-adjoint, positive semi-definite operator Q on H given by $Qx = (Q_j x_j)$, for $x \in H$. Moreover,

$$\langle x, Qx \rangle = \sum_{j=1}^{\infty} \langle x_j, Q_j x_j \rangle, \quad x \in H.$$

We also assume that

$$\sup_i \{ \|A_{i,i-1}\|_2, \|A_{ii}\|_2 \} < \infty.$$

As above, this will be the case, for example, if there exist finitely many distinct A_{ij}. We thus obtain a bounded linear operator $A : H \to M$ given by

$$(Ax)_i = A_{i,i-1} x_{i-1} + A_{ii} x_i, \forall i = 1, 2, \ldots, \quad \forall x \in H.$$

The matrix representation of A has the following lower-staircase form:

$$\begin{bmatrix} A_{11} & 0 & 0 & \cdots \\ A_{21} & A_{22} & 0 & \cdots \\ 0 & A_{32} & A_{33} & \cdots \\ \vdots & \vdots & \vdots & \ddots \end{bmatrix}.$$

Finally, assume that $b = (b_i) \in M$, i.e., $\sum_{i=1}^{\infty} \|b_i\|_2^2 < \infty$, and the feasible region

$$F = \{x \in H : Ax = b\}$$

is non-empty.

For each positive integer k, consider the following "finite dimensional" problem $(\mathcal{S}_k)$ which approximates $(\mathcal{S})$:

$$\min \sum_{j=1}^{k} \langle x_j, Q_j x_j \rangle$$

subject to $(\mathcal{S}_k)$

$$A_{i,i-1}x_{i-1} + A_{ii}x_i = b_i, \quad \forall i = 1, \ldots, k,$$

$$x_j \in \mathbf{R}^{n_j}, \quad \forall j = 1, 2, \ldots,$$

and

$$\sum_{j=1}^{\infty} \|x_j\|_2^2 < \infty.$$

Note that $(\mathcal{S}_k)$ is essentially finite dimensional since the objective function depends only on the first k variables and the feasible region consists of those *square-summable extensions* of the first k variables which satisfy the first k constraints. If we let F_k denote the feasible region for $(\mathcal{S}_k)$, then $\{F_k\}$ is a sequence of closed, non-empty subsets of H satisfying $F_{k+1} \subseteq F_k$, $\forall k$, and $F = \cap_{k=1}^{\infty} F_k$, i.e., $\lim_{k\to\infty} F_k = F$, in the sense of Kuratowski (1966). It is then of interest to ask if the problems $(\mathcal{S}_k)$ "converge" to $(\mathcal{S})$ also, in the sense of Fiacco (1974). This would be the case if, in addition to the feasible region convergence, the functions

$$f_k(x) = \sum_{j=1}^{k} \langle x_j, Q_j x_j \rangle, \quad x \in H,$$

converge *uniformly* to the function

$$f(x) = \sum_{j=1}^{\infty} \langle x_j, Q_j x_j \rangle, \quad x \in H,$$

on H as $k \to \infty$. Unfortunately, this is *not* so. However, it is true that the sequence $\{f_k\}$ converges *pointwise* to f on H. Thus, the problems $(\mathcal{S}_k)$ do converge to $(\mathcal{S})$ in this weaker sense.

Since $Q = \oplus_j Q_j$, it follows that the kernel K of Q is given by $K = \oplus_j K_j$, where K_j is the kernel of Q_j in $\mathbf{R}^{n_j}$, $\forall j$. Similarly, $L = \oplus_j L_j$, where L_j is the orthogonal complement of K_j in $\mathbf{R}^{n_j}$, i.e., $\mathbf{R}^{n_j} = K_j \oplus L_j$, $\forall j = 1, 2, \ldots$. If $P_j = Q_j|L_j$, then $Q_j = 0 \oplus P_j$, $\forall j$, and $P = Q|L = \oplus_j P_j$.

LEMMA 3.1. The (real) spectrum of P is equal to the closure of $\cup_{j=1}^{\infty} \sigma(P_j)$ in $\mathbf{R}$, i.e.,

$$\sigma(P) = \overline{\cup_{j=1}^{\infty} \sigma(P_j)}.$$

PROOF. Note that each P_j is positive definite and defined on a subspace of $\mathbf{R}^{n_j}$. Hence, $\sigma(P_j)$ consists of a set of (positive) eigenvalues $\{\lambda_{j1}, \ldots, \lambda_{jk_j}\}$ of P_j, which are also eigenvalues of P. Consequently,

$$\sigma(P_j) \subseteq \sigma(P), \quad \forall j = 1, 2, \ldots,$$

so that

$$\cup_{j=1}^{\infty} \sigma(P_j) \subseteq \sigma(P)$$

and

$$\overline{\cup_{j=1}^{\infty} \sigma(P_j)} \subseteq \sigma(P),$$

since $\sigma(P)$ is closed.

For the other inclusion, let $\lambda \in \mathbf{R}$ and suppose $\lambda \notin \overline{\cup_{j=1}^{\infty} \sigma(P_j)}$. Thus, there exists $r > 0$ such that

$$\{t \in \mathbf{R} : |t - \lambda| < r\} \cap \overline{\cup_{j=1}^{\infty} \sigma(P_j)} = \emptyset,$$

i.e., for each $j = 1, 2, \ldots,$

$$|\lambda_{ji} - \lambda| \geq r, \quad \forall i = i, \ldots, k_j.$$

For each j, consider the bounded linear operator $P_j - \lambda I_j$, where I_j is the identity operator on L_j. Since P_j is self-adjoint and λ is real, we have that each $P_j - \lambda I_j$ is self-adjoint. Moreover, since $\lambda \notin \sigma(P_j)$, it follows that $P_j - \lambda I_j$ is invertible and hence, also self-adjoint. Consequently, if we denote the inverse of $P_j - \lambda I_j$ by V_j, $\forall j = 1, 2, \ldots,$ then by Helmberg (1969, p.227)

$$\sigma(V_j) = \{\rho^{-1} : \rho \in \sigma(P_j - \lambda I_j)\} = \{(\lambda_{ji} - \lambda)^{-1} : i = 1, \ldots, k_j\}$$

and

$$\|V_j\| = \max_{1 \leq i \leq k_j} \{|(\lambda_{ji} - \lambda)|^{-1}\} \leq 1/r,$$

so that the V_j are uniformly bounded. Now let $V = \bigoplus_{j=1}^{\infty} V_j$. Then V is a bounded linear operator since

$$\|V\| = \sup_{1 \leq j < \infty} \|V_j\| \leq 1/r,$$

and

$$V(P - \lambda I) = \bigoplus_{j=1}^{\infty} V_j(P_j - \lambda I_j) = \bigoplus_{j=1}^{\infty} I_j = I,$$

i.e., V is the inverse of $P - \lambda I$ on L. Therefore, $\lambda \notin \sigma(P)$, i.e., we have shown equivalently that

$$\sigma(P) \subseteq \overline{\cup_{j=1}^{\infty} \sigma(P_j)}.$$

REMARKS. The previous lemma can be shown to be true for self-adjoint operators in general. However, it is well-known to be false for arbitrary bounded operators. Moreover, the spectrum of the finite-dimensional operator $\bigoplus_{i=1}^{n} P_i$ consists of its eigenvalues, which are the eigenvalues of the P_i, $i = 1, \ldots, n$, i.e., $\sigma(\bigoplus_{i=1}^{n} P_i) = \cup_{i=1}^{n} \sigma(P_i)$, $n = 1, 2, \ldots$, an increasing sequence of closed sets. On the other hand, $\sigma(\bigoplus_{i=1}^{\infty} P_i) \supseteq \cup_{i=1}^{\infty} \sigma(P_i)$, which may not be closed in general. But

$$\overline{\cup_{i=1}^{\infty} \sigma(P_i)} = \lim_{n \to \infty} \cup_{i=1}^{n} \sigma(P_i)$$

in the sense of Fiacco (1974) (see (Kato, 1980, p.339)), i.e., in the sense of Kuratowski set convergence. Thus, Lemma 3.1 asserts that

$$\sigma\left(\bigoplus_{i=1}^{\infty} P_i\right) = \lim_{n \to \infty} \sigma\left(\bigoplus_{i=1}^{n} P_i\right)$$

in the sense of Kuratowski.

As a consequence of the above discussion, we have that

$$E_K = \begin{bmatrix} E_K^1 & 0 & 0 & \cdots \\ 0 & E_K^2 & 0 & \cdots \\ 0 & 0 & E_K^3 & \cdots \\ \vdots & \vdots & \vdots & \ddots \end{bmatrix},$$

where E_K^j is the orthogonal projection of $\mathbf{R}^{n_j}$ onto K_j, $\forall j = 1, 2, \ldots$. Analogously,

$$E_L = \begin{bmatrix} E_L^1 & 0 & 0 & \cdots \\ 0 & E_L^2 & 0 & \cdots \\ 0 & 0 & E_L^3 & \cdots \\ \vdots & \vdots & \vdots & \ddots \end{bmatrix},$$

where E_L^j is the orthogonal projection of $\mathbf{R}^{n_j}$ onto L_j, $\forall j = 1, 2, \ldots$.

Now let $C_k : \mathbf{R}^{n_1} \oplus \ldots \oplus \mathbf{R}^{n_k} \to \mathbf{R}^{m_1} \oplus \ldots \oplus \mathbf{R}^{m_k}$ denote the matrix operator given by

$$C_k = \begin{bmatrix} A_{11}E_K^1 & 0 & 0 & \cdots & 0 & 0 \\ A_{21}E_K^1 & A_{22}E_K^2 & 0 & \cdots & 0 & 0 \\ 0 & A_{32}E_K^2 & A_{33}E_K^3 & \cdots & 0 & 0 \\ \vdots & \vdots & \vdots & \ddots & \vdots & \vdots \\ 0 & 0 & 0 & \cdots & A_{k,k-1}E_K^{k-1} & A_{kk}E_K^k \end{bmatrix},$$

with minimum modulus given (as in section 2) by

$$\alpha_{C_k} = \inf\{\|C_k(x_1, \ldots, x_k)\|_2 : (x_1, \ldots, x_k) \in ker(C_k)^{\perp}, \|(x_1, \ldots, x_k)\|_2 = 1\},$$

$\forall k = 1, 2, \ldots$. The (doubly-infinite) matrix operator $B_k = J_k(A|K)$ then satisfies

$$B_k E_K = \begin{bmatrix} C_k & 0 & \cdots \\ 0 & 0 & \cdots \\ \vdots & \vdots & \ddots \end{bmatrix}, \quad \forall k = 1, 2, \ldots .$$

The following is the main result of this section.

THEOREM 3.2. Problem $(\mathcal{S})$ admits an optimal solution if $\inf_k \alpha_{C_k} > 0$ and

$$\inf(\cup_{j=1}^\infty \sigma_+(Q_j)) > 0.$$

PROOF. First observe that $ker(B_k E_K) = ker(B_k) \oplus L$, so that $ker(B_k E_K)^\perp = ker(B_k)^{\perp_K} \oplus 0$, $\forall k = 1, 2, \ldots$. From this, it follows that $\alpha_{B_k E_K} = \alpha_{B_k}$, $\forall k$. By a similar argument, we have that $\alpha_{B_k E_K} = \alpha_{C_k}$, i.e., $\alpha_{B_k} = \alpha_{C_k}$, $\forall k$. Thus, by hypothesis, $\alpha_B > 0$ (Theorem A.5).

Now P is strictly positive definite if and only if $\inf(\sigma(P)) > 0$ (Lemma 2.3). But $\sigma(P) = \overline{\cup_{j=1}^\infty \sigma(P_j)}$ by Lemma 3.1. Thus, P is strictly positive definite if and only if $\inf(\cup_{j=1}^\infty \sigma(P_j)) > 0$. However, $\sigma_+(Q_j) = \sigma(P_j)$, $\forall j = 1, 2, \ldots$. Finally, apply Theorems 2.4 and 2.7 to complete the proof.

The previous theorem guarantees an optimal solution for $(\mathcal{S})$ if the two specified quantities are positive. The first quantity is the infimum of the minimum moduli of operators derived from the time-staged constraint matrices. The second quantity is the infimum of the positive eigenvalues of the time-staged cost matrices. Although the second quantity is not difficult to compute in general, determination of the first can lead to significant complications, as the next section shows.

4   AN APPLICATION TO CONTROL THEORY

We consider a special case of the discrete-time, infinite horizon, time-varying LQ regulator problem (Lewis, 1986). Our example may also be viewed as a production planning problem with positive semi-definite quadratic inventory and production costs (Denardo, 1982).

Consider the following problem $(\mathcal{R})$:

$$\min \sum_{j=1}^\infty [r_j y_j^2 + s_j u_j^2]$$

subject to $(\mathcal{R})$

$$y_i = y_{i-1} + u_i - d_i, \quad \forall i = 1, 2, \ldots ,$$

$$y_j \in \mathbf{R}, \quad u_j \in \mathbf{R}, \quad \forall j = 1, 2, \ldots ,$$

where each r_j, $s_j \geq 0$. For the jth period, y_j is the jth state (eg., ending inventory) and u_j is the contributing j^{th} control (eg., production decision). The initial state y_0

is assumed given. (Thus, its cost is constant and may be omitted.) The quantity d_i is assumed to be a known exogenous parameter (eg., demand) in period $i = 1, 2, \ldots$. Since it is customary for control costs to be positive definite (Lewis, 1986), we further assume that $s_j > 0$, for all j. For our purposes, we also assume that $\inf\{r_j : r_j \neq 0\} > 0$ and $\inf\{s_j\} > 0$.

Problem $(\mathcal{R})$ may be rewritten in the form of problem $(\mathcal{S})$ as follows:

$$\min \sum_{j=1}^{\infty} \begin{bmatrix} y_j \\ u_j \end{bmatrix}^t \begin{bmatrix} r_j & 0 \\ 0 & s_j \end{bmatrix} \begin{bmatrix} y_j \\ u_j \end{bmatrix}$$

subject to $\hspace{8cm}$ $(\mathcal{R})$

$$[-I\ I] \begin{bmatrix} y_1 \\ u_1 \end{bmatrix} = -y_0 + d_1, \quad i = 1,$$

$$[I\ 0] \begin{bmatrix} y_{i-1} \\ u_{i-1} \end{bmatrix} + [-I\ I] \begin{bmatrix} y_i \\ u_i \end{bmatrix} = d_i, \quad \forall i = 2, 3, \ldots,$$

$$\begin{bmatrix} y_j \\ u_j \end{bmatrix} \in \mathbf{R}^2, \quad \forall j = 1, 2, \ldots,$$

where

$$x_j = \begin{bmatrix} y_j \\ u_j \end{bmatrix}, \quad \forall j = 1, 2, \ldots,$$

$$Q_j = \begin{bmatrix} r_j & 0 \\ 0 & s_j \end{bmatrix}, \quad \forall j = 1, 2, \ldots,$$

$$A_{i,i-1} = [I\ 0], \quad \forall i = 2, 3, \ldots,$$

$$A_{ii} = [-I\ I], \quad \forall i = 1, 2, \ldots,$$

and

$$b_i = \begin{cases} -y_0 + d_1, & i = 1, \\ d_i, & i = 2, 3, \ldots. \end{cases}$$

Let the remaining notation be as above. In particular, H (resp. M) is the Hilbert sum of countably many copies of $\mathbf{R}^2$ (resp. $\mathbf{R}$). Obviously, $b = (b_i) \in M$ if and only if $\sum_{i=1}^{\infty} \|d_i\|_2^2 < \infty$. To obtain the bounded operator Q, we assume the hypothesis of the next lemma.

LEMMA 4.1. If $\sup_{1 \leq j < \infty}\{r_j, s_j\} < \infty$, then $Q : H \rightarrow H$ is bounded with $\|Q\| \leq \sup_{1 \leq j < \infty}\{r_j, s_j\}$.

PROOF. Left to the reader.

Note that $\sup\{\|A_{i,i-1}\|_2, \|A_{ii}\|_2\} = 1$ in this case, so that the linear operator $A : H \rightarrow M$ is bounded with $\|A\| \leq 1$, where A has the following matrix form:

$$A = \begin{bmatrix} [-I\ I] & 0 & 0 & \cdots \\ [I\ 0] & [-I\ I] & 0 & \cdots \\ 0 & [I\ 0] & [-I\ I] & \cdots \\ \vdots & \vdots & \vdots & \ddots \end{bmatrix}.$$

Recall that K_j is the kernel of Q_j in $\mathbf{R}^2$ and E_K^j is the orthogonal projection of $\mathbf{R}^2$ onto K_j, $j = 1, 2, \ldots$. Consequently, for problem $(\mathcal{R})$,

$$
C_k = \begin{bmatrix}
[-I\ I]E_K^1 & 0 & 0 & \cdots & 0 & 0 \\
[I\ 0]E_K^1 & [-I\ I]E_K^2 & 0 & \cdots & 0 & 0 \\
0 & [I\ 0]E_K^2 & [-I\ I]E_K^3 & \cdots & 0 & 0 \\
\vdots & \vdots & \vdots & \ddots & \vdots & \vdots \\
0 & 0 & 0 & \cdots & [I\ 0]E_K^{k-1} & [-I\ I]E_K^k
\end{bmatrix},
$$

so that $C_k C_k^*$ is given by

$$
\begin{bmatrix}
2E_K^1 & -E_K^1 & 0 & \cdots & 0 & 0 & 0 \\
-E_K^1 & 2E_K^2 + E_K^1 & -E_K^2 & \cdots & 0 & 0 & 0 \\
\vdots & \vdots & \vdots & \ddots & \vdots & \vdots & \vdots \\
0 & 0 & 0 & \cdots & -E_K^{k-2} & 2E_K^{k-1} + E_K^{k-2} & -E_K^{k-1} \\
0 & 0 & 0 & \cdots & 0 & -E_K^{k-1} & 2E_K^k + E_K^{k-1}
\end{bmatrix}, \quad (*)
$$

for $k = 1, 2, \ldots$. Note that our reason for computing $C_k C_k^*$ is contained in Lemma A.2 which gives a lower bound for α_{C_k} in terms of the non-zero eigenvalues of $C_k C_k^*$.

Thus, for each j,

$$
\sigma(Q_j) = \{r_j, s_j\},
$$

and

$$
\sigma_+(Q_j) = \begin{cases} \{r_j, s_j\}, & \text{for } r_j > 0 \\ \{s_j\}, & \text{for } r_j = 0. \end{cases}
$$

Therefore,

$$
\inf(\cup_{j=1}^{\infty} \sigma_+(Q_j)) = \inf\{r_j,\ s_j : r_j > 0\} = \min(\inf\{r_j : r_j > 0\}, \inf\{s_j\}) > 0.
$$

For each $j = 1, 2, \ldots$, define δ_j as follows:

$$
\delta_j = \begin{cases} 0, & \text{for } r_j > 0 \\ 1, & \text{for } r_j = 0. \end{cases}
$$

Then the kernel K_j of Q_j in $\mathbf{R}^2$ is given by $K_j = K_j^1 \oplus 0$, where $K_j^1 = \delta_j \mathbf{R}$, $\forall j = 1, 2, \ldots$. Consequently,

$$
E_K^j = \begin{bmatrix} \delta_j & 0 \\ 0 & 0 \end{bmatrix},
$$

i.e., there are two possible values for each E_K^j, for each j.

Once again, our objective is to apply Theorem 3.2 to problem $(\mathcal{R})$. Since

$$
\inf(\cup_{j=1}^{\infty} \sigma_+(Q_j)) > 0
$$

by our hypotheses, it suffices to show that $\inf_k \alpha_{C_k} > 0$. Observe that

$$
\alpha_{C_k} \geq \frac{1}{\|C_k\|_2} \inf(\sigma_+(C_k C_k^*))
$$

by Lemma A.2. In this case,

$$
C_k = \begin{bmatrix}
[-1\ 1]E_K^1 & 0 & 0 & \cdots & 0 & 0 \\
[1\ 0]E_K^1 & [-1\ 1]E_K^2 & 0 & \cdots & 0 & 0 \\
0 & [1\ 0]E_K^2 & [-1\ 1]E_K^3 & \cdots & 0 & 0 \\
\vdots & \vdots & \vdots & \ddots & \vdots & \vdots \\
0 & 0 & 0 & \cdots & [1\ 0]E_K^{k-1} & [-1\ 1]E_K^k
\end{bmatrix},
$$

and $C_k C_k^*$ is as in (*) above, for $k = 1, 2, \ldots$.

For each k, the spectral norm $\|C_k\|_2$ of C_k is equal to $\sqrt{\|C_k C_k^*\|_2}$, where $\|C_k C_k^*\|_2$ is the largest eigenvalue of $C_k C_k^*$ (Horn and Johnson, 1988). Given the form (*) of $C_k C_k^*$ and the description of the E_K^j, $j = 1, 2, \ldots$, we see from the Gershgorin Circle Theorem that every eigenvalue of $C_k C_k^*$ is at most 5. Consequently, $\|C_k\|_2 \le \sqrt{5}$, so that

$$
\alpha_{C_k} \ge \frac{1}{\sqrt{5}} \inf(\sigma_+(C_k C_k^*)), \quad \forall k = 1, 2, \ldots.
$$

Thus, to show that $\inf_k \alpha_{C_k} > 0$, it suffices to show that the sequence

$$
\{\inf(\sigma_+(C_k C_k^*))\}_{k=1}^\infty
$$

is bounded away from 0.

A careful inspection of the matrix $C_k C_k^*$ reveals that it is of the form

$$
\begin{bmatrix}
D_k^1 & \cdots & 0 \\
\vdots & \ddots & \vdots \\
0 & \cdots & D_k^{p_k}
\end{bmatrix}, \quad k = 1, 2, \ldots,
$$

where D_k^i, for $1 \le i \le p_k$, is necessarily either the 4×4 matrix

$$
\begin{bmatrix}
2 & 0 & -1 & 0 \\
0 & 0 & 0 & 0 \\
-1 & 0 & 3 & 0 \\
0 & 0 & 0 & 0
\end{bmatrix}
$$

with positive eigenvalues $(5 \pm \sqrt{5})/2$, or the 4×4 matrix

$$
\begin{bmatrix}
2 & 0 & -1 & 0 \\
0 & 0 & 0 & 0 \\
-1 & 0 & 1 & 0 \\
0 & 0 & 0 & 0
\end{bmatrix}
$$

with positive eigenvalues $(3 \pm \sqrt{5})/2$, or the $2n \times 2n$ matrix

$$
U_n = \begin{bmatrix}
2 & 0 & -1 & 0 & 0 & 0 & \cdots & 0 & 0 & 0 & 0 & 0 & 0 \\
0 & 0 & 0 & 0 & 0 & 0 & \cdots & 0 & 0 & 0 & 0 & 0 & 0 \\
-1 & 0 & 3 & 0 & -1 & 0 & \cdots & 0 & 0 & 0 & 0 & 0 & 0 \\
0 & 0 & 0 & 0 & 0 & 0 & \cdots & 0 & 0 & 0 & 0 & 0 & 0 \\
\vdots & \vdots & \vdots & \vdots & \vdots & \vdots & \ddots & \vdots & \vdots & \vdots & \vdots & \vdots & \vdots \\
0 & 0 & 0 & 0 & 0 & 0 & \cdots & -1 & 0 & 3 & 0 & -1 & 0 \\
0 & 0 & 0 & 0 & 0 & 0 & \cdots & 0 & 0 & 0 & 0 & 0 & 0 \\
0 & 0 & 0 & 0 & 0 & 0 & \cdots & 0 & 0 & -1 & 0 & 3 & 0 \\
0 & 0 & 0 & 0 & 0 & 0 & \cdots & 0 & 0 & 0 & 0 & 0 & 0
\end{bmatrix}
$$

or the $2n \times 2n$ matrix

$$V_n = \begin{bmatrix} 2 & 0 & -1 & 0 & 0 & 0 & \dots & 0 & 0 & 0 & 0 & 0 & 0 \\ 0 & 0 & 0 & 0 & 0 & 0 & \dots & 0 & 0 & 0 & 0 & 0 & 0 \\ -1 & 0 & 3 & 0 & -1 & 0 & \dots & 0 & 0 & 0 & 0 & 0 & 0 \\ 0 & 0 & 0 & 0 & 0 & 0 & \dots & 0 & 0 & 0 & 0 & 0 & 0 \\ \vdots & \vdots & \vdots & \vdots & \vdots & \vdots & \ddots & \vdots & \vdots & \vdots & \vdots & \vdots & \vdots \\ 0 & 0 & 0 & 0 & 0 & 0 & \dots & -1 & 0 & 3 & 0 & -1 & 0 \\ 0 & 0 & 0 & 0 & 0 & 0 & \dots & 0 & 0 & 0 & 0 & 0 & 0 \\ 0 & 0 & 0 & 0 & 0 & 0 & \dots & 0 & 0 & -1 & 0 & 1 & 0 \\ 0 & 0 & 0 & 0 & 0 & 0 & \dots & 0 & 0 & 0 & 0 & 0 & 0 \end{bmatrix}, \quad n = 3, 4 \dots,$$

or a square zero matrix; each of these is symmetric and positive semi-definite. Since

$$\sigma_+(C_k C_k^*) = \cup_{i=1}^{p_k} \{\sigma_+(D_k^i) : D_k^i \neq 0\},$$

it suffices to determine $\inf(\sigma_+(U_n))$ and $\inf(\sigma_+(V_n))$, $n = 3, 4, \dots$. But

$$\inf(\sigma_+(U_n)) \geq 1, \quad \text{all } n,$$

by the Gershgorin Circle Theorem. Thus, it remains to determine $\inf(\sigma_+(V_n))$, for all n. However,

$$det(V_n - \lambda I) = \pm \lambda^n det(W_n - \lambda I),$$

where W_n is the $n \times n$ symmetric matrix given by

$$W_n = \begin{bmatrix} 2 & -1 & 0 & \dots & 0 & 0 & 0 \\ -1 & 3 & -1 & \dots & 0 & 0 & 0 \\ \vdots & \vdots & \vdots & \ddots & \vdots & \vdots & \vdots \\ 0 & 0 & 0 & \dots & -1 & 3 & -1 \\ 0 & 0 & 0 & \dots & 0 & -1 & 1 \end{bmatrix}, \quad n = 3, 4, \dots .$$

Consequently, $\sigma_+(V_n) = \sigma_+(W_n)$, for all n. Let ϕ_n denote the nth degree characteristic polynomial of W_n, i.e.,

$$\phi_n(\lambda) = det(W_n - \lambda I), \quad n = 3, 4, \dots .$$

It suffices to show that the positive roots of the ϕ_n are bounded away from 0. In particular, we will show that if $\phi_n(\lambda) = 0$, for any $n \geq 3$, then $\lambda \geq .1$.

To this end, define

$$\psi_n(\lambda) = det(Z_n - \lambda I), \quad n = 1, 2, \dots,$$

(where Z_n is as in the first example) so that

$$\psi_n(\lambda) = (3 - \lambda)\psi_{n-1}(\lambda) - \psi_{n-2}(\lambda),$$

$$\phi_n(\lambda) = (1 - \lambda)\psi_{n-1}(\lambda) - \psi_{n-2}(\lambda)$$

and

$$\phi_n(\lambda) = \psi_n(\lambda) - 2\psi_{n-1}(\lambda), \quad n = 3, 4, \ldots.$$

Next define $\rho_1(\lambda) = 3 - \lambda$, $\rho_2(\lambda) = \lambda^2 - 6\lambda + 8$ and

$$\rho_n(\lambda) = \det \begin{bmatrix} 3-\lambda & -1 & 0 & \cdots & 0 & 0 & 0 \\ -1 & 3-\lambda & -1 & \cdots & 0 & 0 & 0 \\ \vdots & \vdots & \vdots & \ddots & \vdots & \vdots & \vdots \\ 0 & 0 & 0 & \cdots & -1 & 3-\lambda & -1 \\ 0 & 0 & 0 & \cdots & 0 & -1 & 3-\lambda \end{bmatrix}, \quad n = 3, 4, \ldots,$$

so that

$$\rho_n(\lambda) = (3-\lambda)\rho_{n-1}(\lambda) - \rho_{n-2}(\lambda),$$

$$\psi_n(\lambda) = (2-\lambda)\rho_{n-1}(\lambda) - \rho_{n-2}(\lambda)$$

and

$$\psi_n(\lambda) = \rho_n(\lambda) - \rho_{n-1}(\lambda), \quad n = 3, 4, \ldots.$$

For convenience, let $\omega = (3 - \lambda)/2$, so that $\lambda = 3 - 2\omega$. Writing $\tau_n(\omega)$ for $\rho_n(3 - 2\omega)$, for all $n \geq 1$, we obtain the recursion

$$\tau_n(\omega) = 2\omega\tau_{n-1}(\omega) - \tau_{n-2}(\omega), \quad n = 3, 4, \ldots,$$

with $\tau_1(\omega) = 2\omega$, $\tau_2(\omega) = 4\omega^2 - 1$, $\omega \in \mathbf{R}$. Restricting ω to the interval $[-1, 1]$ and letting $\theta = \cos^{-1}(\omega)$, $0 \leq \theta \leq \pi$, the solutions to this recursion are well-known (Davis, 1975) to be the Chebychev Polynomials of the Second Kind, i.e.,

$$\tau_n(cos\theta) = \frac{sin(n+1)\theta}{sin\theta}, \quad 0 \leq \theta \leq \pi, \quad n = 3, 4, \ldots.$$

Consequently,

$$\begin{aligned} \psi_n(3 - 2cos\theta) &= \rho_n(3 - 2cos\theta) - \rho_{n-1}(3 - 2cos\theta) \\ &= \tau_n(cos\theta) - \tau_{n-1}(cos\theta) \\ &= \frac{sin(n+1)\theta - sin(n\theta)}{sin\theta} \\ &= \frac{2cos(\frac{2n+1}{2}\theta)sin(\frac{\theta}{2})}{sin\theta} \end{aligned}$$

and

$$\begin{aligned} \phi_n(3 - 2cos\theta) &= \psi_n(3 - 2cos\theta) - 2\psi_{n-1}(3 - 2cos\theta) \\ &= \frac{2cos(\frac{2n+1}{2}\theta)sin(\frac{\theta}{2}) - 4cos(\frac{2n-1}{2}\theta)sin(\frac{\theta}{2})}{sin\theta} \\ &= \frac{2sin(\frac{\theta}{2})[cos(\frac{2n+1}{2}\theta) - 2cos(\frac{2n-1}{2}\theta)]}{sin\theta} \\ &= \frac{cos(\frac{2n+1}{2}\theta) - 2cos(\frac{2n-1}{2}\theta)}{cos(\frac{\theta}{2})}, \quad 0 \leq \theta \leq \pi, \quad n = 3, 4, \ldots. \end{aligned}$$

Hence, for $-1 \leq \omega \leq 1$ and $n = 3, 4, \ldots$, the zeros of ϕ_n correspond to those θ in the interval $[0, \pi]$ for which

$$\cos(\frac{2n+1}{2}\theta) = 2\cos(\frac{2n-1}{2}\theta).$$

Fix $n \geq 3$ and let

$$y_1 = \cos(\frac{2n+1}{2}\theta), \quad 0 \leq \theta \leq \pi,$$

and

$$y_2 = 2\cos(\frac{2n-1}{2}\theta), \quad 0 \leq \theta \leq \pi.$$

Then the zeros of y_1 are given by

$$\theta = \frac{2k+1}{2n+1}\pi, \quad k = 0, 1, \ldots, n,$$

while the zeros of y_2 are given by

$$\theta = \frac{2k+1}{2n-1}\pi, \quad k = 0, 1, \ldots, n-1.$$

Moreover,

$$\frac{2k+1}{2n+1}\pi < \frac{2k+1}{2n-1}\pi \leq \frac{2(k+1)+1}{2n+1}\pi < \frac{2(k+1)+1}{2n-1}\pi, \quad k = 0, 1, \ldots, n-1,$$

where

$$\frac{2k+1}{2n-1}\pi \leq \frac{2k+3}{2n+1}\pi \iff k \leq n-1$$

and

$$\frac{2k+1}{2n-1}\pi = \frac{2k+3}{2n+1}\pi \iff k = n-1.$$

Therefore,

$$\frac{2k+1}{2n-1}\pi < \frac{2k+3}{2n+1}\pi, \quad k = 0, 1, \ldots, n-2,$$

while for $k = n-1$,

$$\frac{2k+1}{2n-1}\pi = \frac{2k+3}{2n+1}\pi = \pi.$$

From the previous discussion, we see that in each of the $n-1$ intervals

$$[\frac{2k+1}{2n-1}\pi, \frac{2k+3}{2n+1}\pi], \quad k = 0, 1, \ldots, n-2,$$

there exists a unique θ for which

$$\cos(\frac{2n+1}{2}\theta) = 2\cos(\frac{2n-1}{2}\theta),$$

i.e., for which $\phi_n(3 - 2\cos\theta) = 0$. For $k = n - 1$, the nth such interval reduces to the point π. In this case, $\phi_n(3 - 2\cos\pi)$ is indeterminate of the form $\frac{0}{0}$. By L'Hopital's Rule, we find that

$$\phi_n(5) = \phi_n(3 - 2\cos\pi) = (-1)^n(6n - 1) \neq 0.$$

Note also that $\phi_n(1) = \phi_n(3 - 2\cos 0) = -1$. Hence, $\phi_n(3 - 2\cos\theta)$ has $n - 1$ roots in the interval $0 \leq \theta \leq \pi$, i.e., $\phi_n(\lambda)$ has $n - 1$ roots in the interval $1 \leq \lambda \leq 5$, $n = 3, 4, \ldots$.

It suffices to show that $\phi_n(.1) > 0$ in order to show that the remaining root of each ϕ_n is strictly between .1 and 1. To this end, define $v_1 = 1.9$, $v_2 = .71$ and $v_n = \phi_n(.1)$, $n = 3, 4, \ldots$. Also define $u_1 = 1.9$, $u_2 = 4.51$ and

$$u_n = \det \begin{bmatrix} 1.9 & -1 & 0 & \ldots & 0 & 0 & 0 \\ -1 & 2.9 & -1 & \ldots & 0 & 0 & 0 \\ \vdots & \vdots & \vdots & \ddots & \vdots & \vdots & \vdots \\ 0 & 0 & 0 & \ldots & -1 & 2.9 & -1 \\ 0 & 0 & 0 & \ldots & 0 & -1 & 2.9 \end{bmatrix}, \quad n = 3, 4, \ldots,$$

so that

$$v_n = .9u_{n-1} - u_{n-2}, \quad n = 3, 4, \ldots.$$

Next define $q_1 = 2.9$, $q_2 = 7.41$ and

$$q_n = \det \begin{bmatrix} 2.9 & -1 & 0 & \ldots & 0 & 0 & 0 \\ -1 & 2.9 & -1 & \ldots & 0 & 0 & 0 \\ \vdots & \vdots & \vdots & \ddots & \vdots & \vdots & \vdots \\ 0 & 0 & 0 & \ldots & -1 & 2.9 & -1 \\ 0 & 0 & 0 & \ldots & 0 & -1 & 2.9 \end{bmatrix}, \quad n = 3, 4, \ldots,$$

so that

$$u_n = 1.9q_{n-1} - q_{n-2}$$

and

$$q_n = 2.9q_{n-1} - q_{n-2}, \quad n = 3, 4, \ldots.$$

Using the techniques of Goldberg (1986), we may solve this recursion with the given initial values to obtain

$$q_n = \frac{2.5^{n+1} - .4^{n+1}}{2.1}, \quad n = 1, 2, \ldots.$$

Consequently, for $n \geq 5$, we have

$$
\begin{aligned}
v_n &= .9u_{n-1} - u_{n-2} \\
&= .9(1.9q_{n-2} - q_{n-3}) - (1.9q_{n-3} - q_{n-4}) \\
&= 1.71q_{n-2} - 2.8q_{n-3} + q_{n-4} \\
&= \frac{1.71}{2.1}[2.5^{n-1} - .4^{n-1}] - \frac{2.8}{2.1}[2.5^{n-2} - .4^{n-2}] + \frac{1}{2.1}[2.5^{n-3} - .4^{n-3}] \\
&= 1.71\sum_{k=0}^{n-2} .4^k 2.5^{n-k-2} - 2.8\sum_{k=0}^{n-3} .4^k 2.5^{n-k-3} + \sum_{k=0}^{n-4} .4^k 2.5^{n-k-4} \\
&= 1.71\sum_{k=0}^{n-3} .4^k 2.5^{n-k-2} + 1.71(.4^{n-2}) - 1.12\sum_{k=0}^{n-3} .4^k 2.5^{n-k-2} + \sum_{k=0}^{n-4} .4^k 2.5^{n-k-4} \\
&= .59\sum_{k=0}^{n-3} .4^k 2.5^{n-k-2} + 1.71(.4^{n-2}) + \sum_{k=0}^{n-4} .4^k 2.5^{n-k-4} \\
&> 0.
\end{aligned}
$$

Since $v_1 = 1.9$, $v_2 = .71$, $v_3 = 2.159$ and $v_4 = 6.0021$, we see that $\phi_n(.1) > 0$, $n = 1, 2, \ldots$. Also, since $\phi_n(1) < 0$, all n, it follows that the remaining n^{th} root of each ϕ_n is in the interval $(.1, 1)$, i.e., the roots of the ϕ_n are bounded away from 0.

We have thus proved that $\inf_k \alpha_{C_k} > 0$. Since we are also assuming that $\inf\{r_j : r_j > 0\} > 0$ and $\inf\{s_j\} > 0$, it follows from Theorem 3.1 that problem $(\mathcal{R})$ admits an optimal solution under our assumptions.

APPENDIX

Let X and Y be Hilbert spaces and $T : X \to Y$ a bounded linear operator with adjoint $T^* : Y \to X$. We consider the question of when the range $T(X)$ of T is closed in Y. In Chapter IV of S. Goldberg (1966), the author showed that this is the case if and only if a certain operator index γ_T is positive. In the context of normed linear spaces, this index is given by

$$
\gamma_T = \inf\left\{\frac{\|Tx\|}{d(x, ker(T))} : x \notin ker(T)\right\},
$$

where, as usual,

$$
d(x, ker(T)) = \inf\{\|x - y\| : y \in ker(T)\}.
$$

One of our objectives here is to give alternate characterizations of γ_T in the Hilbert space context. To this end, we define the following additional non-negative indices for T:

$$\alpha_T = \inf\{\|Tx\| : x \in ker(T)^\perp, \|x\| = 1\}$$

$$= \inf\left\{\frac{\|Tx\|}{\|x\|} : x \in ker(T)^\perp, x \neq 0\right\},$$

$$\beta_T = \inf\left\{\frac{\|TT^*y\|}{\|T^*y\|} : y \in Y, T^*y \neq 0\right\}$$

$$= \inf\{\|TT^*y\| : y \in Y, \|T^*y\| = 1\}.$$

We show that all three indices are the same.

THEOREM A.1. For the bounded operator T, the indices α_T, β_T and γ_T are equal.

PROOF. First we show that $\alpha_T = \beta_T$. Recall (Dunford and Schwartz, 1964) that

$$ker(T)^\perp = \overline{range(T^*)}$$

in X. Hence,

$$\alpha_T \leq \inf\left\{\frac{\|Tx\|}{\|x\|} : x \in range(T^*), x \neq 0\right\} = \beta_T.$$

Conversely, let $x \in ker(T)^\perp$, $x \neq 0$. Then $x \in \overline{range(T^*)}$, so that there exists a sequence $\{y_n\}$ in Y such that $T^*y_n \to x$, i.e., $TT^*y_n \to Tx$, as $n \to \infty$. Consequently, $\|TT^*y_n\| \to \|Tx\|$ and $\|T^*y_n\| \to \|x\| \neq 0$, as $n \to \infty$, so that

$$\lim_{n\to\infty} \frac{\|TT^*y_n\|}{\|T^*y_n\|} = \frac{\|Tx\|}{\|x\|}.$$

Thus, a typical element of the set

$$\left\{\frac{\|Tx\|}{\|x\|} : x \in ker(T)^\perp, x \neq 0\right\}$$

is the limit of a sequence from the set

$$\left\{\frac{\|TT^*y\|}{\|T^*y\|} : y \in Y, T^*y \neq 0\right\}.$$

Now suppose $\alpha_T < \beta_T$. Then, for $\epsilon = \beta_T - \alpha_T > 0$, there exists x in $ker(T)^\perp$, $x \neq 0$, such that

$$\left|\alpha_T - \frac{\|Tx\|}{\|x\|}\right| < \frac{\epsilon}{3}.$$

Also, there exists $y \in Y$ such that $T^*y \neq 0$ and

$$\left|\frac{\|Tx\|}{\|x\|} - \frac{\|TT^*y\|}{\|T^*y\|}\right| < \frac{\epsilon}{3},$$

so that

$$\left| \alpha_T - \frac{\|TT^*y\|}{\|T^*(y)\|} \right| < \frac{2\epsilon}{3} = \frac{2}{3}(\beta_T - \alpha_T)$$

i.e.,

$$\frac{\|TT^*y\|}{\|T^*y\|} < \frac{2}{3}\beta_T + \frac{1}{3}\alpha_T < \beta_T$$

by hypothesis. This is a contradiction, i.e., $\alpha_T = \beta_T$.

We leave the proof of the fact that $\alpha_T = \gamma_T$ to the interested reader. It depends on the fact that X is the direct sum of $ker(T)$ and $ker(T)^\perp$.

COROLLARY. The range $T(X)$ of T is closed in Y if and only if $\alpha_T > 0$.

PROOF. See Goldberg (1966) and Kato (1980).

The following lemma yields a lower bound for α_T.

LEMMA A.2. Let the notation be as above. Suppose $T \neq 0$. Then

$$\alpha_T \geq \frac{1}{\|T\|} \inf\{\lambda : \lambda \neq 0, \ \lambda = eigenvalue \ of \ TT^*\}.$$

PROOF. Let $y \in Y$, $T^*y \neq 0$. Then

$$\frac{\|TT^*y\|}{\|T^*y\|} = \frac{\|TT^*y\|}{\|y\|} \frac{\|y\|}{\|T^*y\|}$$

$$= \frac{\|TT^*y\|}{\|y\|} \Big/ \frac{\|T^*y\|}{\|y\|}$$

$$\geq \inf_{\substack{v \in Y \\ T^*y \neq 0}} \frac{\|TT^*y\|}{\|y\|} \Big/ \sup_{\substack{v \in Y \\ T^*y \neq 0}} \frac{\|T^*y\|}{\|y\|}$$

$$= \frac{1}{\|T^*\|} \inf_{\substack{v \in Y \\ T^*y \neq 0}} \frac{\|TT^*y\|}{\|y\|}$$

$$= \frac{1}{\|T^*\|} \inf\{\lambda : \lambda \neq 0, \ \lambda = eigenvalue \ of \ TT^*\},$$

by the spectral resolution of the identity induced by TT^*. See Theorem 5.2.2 of Kadison and Ringrose (1983). This completes the proof since $\|T^*\| = \|T\|$.

By a similar argument, we obtain the following alternate characterization of α_T in an important special case.

LEMMA A.3. Let T be a non-zero, bounded operator. If T is self-adjoint and positive semi-definite, then $\alpha_T = \inf(\sigma_+(T))$.

ACKNOWLEDGEMENTS

The second author was partially supported by the National Science Foundation under Grant DDM-9214894. The first and third authors were partially supported by Oakland University Fellowships. The authors are grateful to one of the referees for several helpful suggestions and comments.

REFERENCES

1. J.-P. Aubin (1979), *Applied Functional Analysis*, J. Wiley, New York.

2. P. J. Davis (1975), *Interpolation and Approximation*, Dover, New York.

3. E. V. Denardo (1982), *Dynamic Programming. Models and Applications*, Prentice-Hall, Englewood Cliffs.

4. A. L. Dontchev and T. Zolezzi (1992), Well-posed optimization problems, *Lecture Notes in Mathematics*, No. 1543, Springer-Verlag, New York.

5. N. Dunford and J. T. Schwartz (1964), *Linear Operators. Part I*, J. Wiley-Interscience, New York.

6. A. V. Fiacco (1974), Convergence properties of local solutions of sequences of mathematical programming problems in general spaces, *J. of Opt. Theory and Appl.*, **13**: 1-12.

7. S. Goldberg (1966), *Unbounded Linear Operators*, Dover, New York.

8. S. Goldberg (1986), *Introduction to Difference Equations*, Dover, New York.

9. G. Helmberg (1969), *Introduction to Spectral Theory in Hilbert Spaces*, North-Holland, Amsterdam.

10. R. A. Horn and C. R. Johnson (1988), *Matrix Analysis*, Cambridge Univ. Press, Cambridge.

11. R. V. Kadison and J. R. Ringrose (1983), *Fundamentals of the Theory of Operator Algebras. Vol. I. Elementary Theory*, Academic Press, New York.

12. T. Kato (1980), *Perturbation Theory for Linear Operators*, Springer-Verlag, Berlin.

13. C. Kuratowski (1966), *Topologie I*, Academic Press, New York.

14. K. Y. Lee, S. Chow and R. O. Barr (1972), On the control of disrete-time distributed parameter systems, *SIAM J. Control*, **10**: 361-376.

15. F. L. Lewis (1986), *Optimal Control*, J. Wiley, New York.

16. R. D. Milne (1980), *Applied Functional Analysis*, Pitman, Boston.

17. I. E. Schochetman, R. L. Smith and S.-K. Tsui (1995), Solution existence for time-varying infinite horizon control, *J. Mathematical Analysis and Applications*, **195**: 135-147.

Sensitivity Analysis of Nonlinear Programming Problems via Minimax Functions

S. SHIRAISHI   Faculty of Economics,Toyama University,Toyama 930, Japan

Abstract. The marginal function of a parametric nonlinear programming problem can be written as the minimax-function form by the use of Lagrangian. By way of the directional differentiability result of minimax-functions due to Correa and Seeger, we investigate the sensitivity analysis of nonlinear programming problems. We will shed a light on the duality relation of the resulting minimax-function which is a key ingredient of the directional differentiability of minimax-functions.

1 INTRODUCTION

We consider the following parametric nonlinear programming problem:

$$(P_x) \qquad \begin{array}{ll} \text{minimize}_y & f(x,y) \\ \text{subject to} & g_i(x,y) \leq 0, \quad \text{for } i = 1, \ldots, m \end{array}$$

where $f, g_1, \ldots, g_m : \mathbf{R}^k \times \mathbf{R}^n \to \mathbf{R}$ are continuous functions. The first variable $x \in \mathbf{R}^k$ is viewed as a parameter. We shall study the *sensitivity analysis* of knowing the directional derivative $h'(x_0; d)$ of the marginal function:

$$h(x) = \inf_y \{f(x, y) \mid g_i(x, y) \leq 0, \ i = 1, \ldots, m\}.$$

If we introduce the Lagrangian $L(x, y, \lambda) := f(x, y) + \sum_{i=1}^m \lambda_i g_i(x, y)$, then the marginal function $h(x)$ can be rewritten as the minimax-function form:

$$h(x) = \inf_{y \in \mathbf{R}^n} \sup_{\lambda \geq 0} L(x, y, \lambda).$$

Correa and Seeger (1985) showed the directional differentiability of general minimax-functions and applied it to the convex minimization problem to obtain sensitivity results. In this paper we shall also use their directional differentiability result of minimax–functions and give sensitivity results for (P_x) under the generalized convexity of Lagrangian, say, convexlikeness and invexity.

In Section 2, we recall the directional differentiability result of Correa and Seeger. When we apply their result to the minimax-function formed by the Lagrangian, we will recognize the importance of the Lagrangian duality. So in Section 3, we will investigate the duality relation under the convexlike condition of the Lagrangian. As a special case of (P_x), we treat RHS(right hand sided) perturbation in Section 4. If we assume the invexity of the functions involved, we can also prove the duality relation for RHS. As a consequence, in Section 5, we can summarize the sensitivity results obtained from the results of the previous sections.

2   PRELIMINARIES

First of all we recall the notion of directional derivative and generalized directional derivative in the sense of Clarke (1976) which we will use in the sequel. Let X be a Banach space, X^* its topological dual, $x_0, d \in X$, and $F : X \to \mathbf{R}$. F is said to admit a directional derivative at the point x_0 in the direction d if the following limit exists:

$$F'(x_0; d) = \lim_{t \to 0^+} [F(x_0 + td) - F(x_0)]/t.$$

Suppose F is a locally Lipschitz function, i.e., for each $x \in X$ there exist a neighborhood N of x and $K > 0$ such that $|F(x_1) - F(x_2)| \leq K\|x_1 - x_2\|$ whenever $x_1, x_2 \in N$. Clarke's generalized directional derivative $F^\circ(x_0; d)$ and generalized subdifferential $\partial F(x_0)$ are defined by

$$F^\circ(x_0; d) = \limsup_{(x,t) \to (x_0, 0^+)} [F(x + td) - F(x)]/t,$$

$$\partial F(x_0) = \{\zeta \in X^* \mid F^\circ(x_0; d) \geq \langle \zeta, d \rangle, \forall d\}.$$

F is said to be Clarke regular at x_0 if the directional derivative of F exists and $F'(x_0; d) = F^\circ(x_0; d)$ holds for all d.

Now we recall the directional differentiability result of minimax-functions due to Correa and Seeger (1985). We should note that they worked on an infinite dimensional space. However we cite their results in the finite dimensional Euclidean space settings so that we can avoid some technical matter. Let Y_0 and Λ_0 be nonempty closed subsets of the Euclidean spaces $\mathbf{R}^n$ and $\mathbf{R}^m$, respectively. For a real valued function $L : \mathbf{R}^k \times \mathbf{R}^n \times \mathbf{R}^m \to \mathbf{R}$, let us define the maxmin and minimax-functions by

$$h_1(x) = \sup_{\lambda \in \Lambda_0} \inf_{y \in Y_0} L(x, y, \lambda),$$

$$h_2(x) = \inf_{y \in Y_0} \sup_{\lambda \in \Lambda_0} L(x, y, \lambda),$$

and the corresponding solution sets by

$$\Lambda(x) = \{\lambda \in \Lambda_0 \mid h_1(x) = \inf_{y \in Y_0} L(x, y, \lambda)\},$$

$$Y(x) = \{y \in Y_0 \mid h_2(x) = \sup_{\lambda \in \Lambda_0} L(x, y, \lambda)\}.$$

For a given point $x_0 \in \mathbf{R}^k$ and a nonzero direction $d \in \mathbf{R}^k$, the partial directional derivative of L w.r.t. x is defined as follows:

$$L'_x(x_0, y, \lambda; d) = \lim_{t \to 0^+} [L(x_0 + td, y, \lambda) - L(x_0, y, \lambda)]/t.$$

Roughly speaking, Correa and Seeger showed the directional differentiability of minimax-functions under the hypothesis of:

- the semicontinuity of the directional derivative L'_x

- the semicontinuity of the solution set multifunctions $\Lambda(\cdot)$ and $Y(\cdot)$

- the duality condition $h_1 = h_2$.

DEFINITION 1 *[Penot (1983), Definition 1.3] A multifunction $t \in \mathbf{R}_+ \rightrightarrows \Gamma(t)$ is said to be semicontinuous at 0^+ in Penot's sense provided that if for every sequence $\{t_k\}_{k \in N} \to 0^+$, there exist $y \in \Gamma(0)$, a subsequence $\{t_k\}_{k \in K \subset N}$ and a sequence $\{y_k\}_{k \in K}$ converging to y such that $y_k \in \Gamma(t_k)$ for all $k \in K$.*

It is easily verified that $\Gamma(t)$ is semicontinuous at 0^+ if it is upper semicontinuous(u.s.c.),i.e., Γ is uniformly compact near $t = 0^+$ and closed at $t = 0^+$ [as for the definition, see Hogan (1973)].

THEOREM 1 *[Correa and Seeger (1985), Theorem 2.1] Assume that there exists $\delta > 0$ such that for every $(y, \lambda) \in Y_0 \times \Lambda_0$, the function $t \in \mathbf{R}_+ \to L(x_0 + td, y, \lambda)$ is continuous and the directional derivative $L'_x(x_0 + td, y, \lambda; d)$ exists for all $t \in [0, \delta[$ and that the following properties hold:*

(H1) For all $y_0 \in Y(x_0)$, the function $(t, \lambda) \in \mathbf{R}_+ \times \Lambda_0 \to L'_x(x_0 + td, y_0, \lambda; d)$ is u.s.c. at $\{0\} \times \Lambda(x_0)$. For all $\lambda_0 \in \Lambda(x_0)$, the function $(t, y) \in \mathbf{R}_+ \times Y_0 \to L'_x(x_0 + td, y, \lambda_0; d)$ is l.s.c. at $\{0\} \times Y(x_0)$.

(H2) The multifunctions $t \in \mathbf{R}_+ \rightrightarrows \Lambda(x_0 + td)$ and $t \in \mathbf{R}_+ \rightrightarrows Y(x_0 + td)$ are semicontinuous at 0^+.

(H3) the duality relation $h_1(x_0 + td) = h_2(x_0 + td)$ holds for all $t \in [0, \delta[$; this common value is denoted by $h(x_0 + td)$.

Then the directional derivative $h'(x_0; d)$ exists and is characterized by

$$h'(x_0; d) = \sup_{\lambda \in \Lambda(x_0)} \inf_{y \in Y(x_0)} L'_x(x_0, y, \lambda; d),$$

where the operator sup inf commutes.

We shall apply Theorem 1 to the Lagrangian $L(x, y, \lambda) = f(x, y) + \sum_{i=1}^m \lambda_i g_i(x, y)$ of the problem (P_x). From the definition of the Lagrangian, to check (H1) it is sufficient to ensure:

- (H1-1) $t \in \mathbf{R}_+ \to f'_x(x_0 + td, y; d), (g_i)'_x(x_0 + td, y; d)$: u.s.c.,

- (H1-2) $(t, y) \in \mathbf{R}_+ \times Y_0 \to f'_x(x_0 + td, y; d), (g_i)'_x(x_0 + td, y; d)$: l.s.c.

These properties are satisfied by a broad class of functions. Let $F : X \to \mathbf{R}$ be a locally Lipschitz function and directionally differentiable. Then it is well known that $F^\circ(x_0; d) = \limsup_{x \to x_0} F'(x_0; d)$. See Cominetti and Correa (1990). Hence the conditions (H1-1) and (H1-2) are satisfied by the Clarke's regularity condition:

- $f'_x(x_0, y; d) = f^\circ_x(x_0, y; d), (g_i)'_x(x_0, y; d) = (g_i)^\circ_x(x_0, y; d),$

- $(-f)'(x_0, y; d, 0) = (-f)^\circ(x_0, y; d, 0), (-g_i)'(x_0, y; d, 0) = (-g_i)^\circ(x_0, y; d, 0),$

respectively, where $f^\circ_x(x_0, y; d) = \limsup_{(x,t) \to (x_0, 0^+)}[f(x + td, y) - f(x, y)]/t$ and $f^\circ(x_0, y; d, 0) = \limsup_{(x, y', t) \to (x_0, y, 0^+)}[f(x + td, y') - f(x, y')]/t$.

Hence the crucial condition of Theorem 1 may be considered as the duality relation (H3). As we will see later (Propositions 1 and 4), the semicontinuity property (H2) of the solution sets is also implied by the conditions which ensure (H3). In the next section, we will examine the duality relation (H3).

3 ANALYSIS OF THE DUALITY RELATION

When we analyze the directional derivative, we may consider the behavior of functions only over the half line $\{x_0 + td \mid t \in \mathbf{R}_+\}$. Hence for the sake of simplicity we denote $L(x_0 + td, y, \lambda)$ by $L(t, y, \lambda), \ldots$etc. In order to establish the duality condition (H3), in Correa and Seeger (1985) $L(t, \cdot, \cdot)$ is assumed to be a convex-concave function over $Y_0 \times \Lambda_0$. While we consider L as the Lagrangian function, concavity is automatically satisfied. Indeed it is linear w.r.t. λ. Without the convexity of L, we can establish the duality relation. The following technical lemma is a non-convex version of Lemma 3.1 of Correa and Seeger (1985).

LEMMA 1 *Let T be a normed space, $t_0 \in T$ and $g : T \times Y_0 \to \mathbf{R}$. Assume that g is uniformly continuous on $N \times Y_0$ for some neighborhood N of t_0 and $g(t_0, \cdot)$ is inf-compact over Y_0, i.e. the level set $S_r = \{y \in Y_0 \mid g(t_0, y) \leq r\}$ is compact for all $r \in \mathbf{R}$. Then for every sequence $\{t_k\}_{k \in N}$ converging to t_0 and every unbounded sequence $\{y_k\}_{k \in N}$ in Y_0, the sequence $\{g(t_k, y_k)\}_{k \in N}$ is not bounded from above.*

Proof: Suppose that there exists a constant K such that $g(t_k, y_k) \leq K$ for all $k \in N$. By the uniform continuity of g, for any $\varepsilon > 0$ there exists $\delta > 0$ such that $|g(t', y') - g(t, y)| \leq \varepsilon$, whenever $\|t' - t\| \leq \delta$ and $\|y' - y\| \leq \delta$. If we set $t' = t_k$, $t = t_0$ and $y' = y = y_k$, then for sufficiently large k we have $|g(t_k, y_k) - g(t_0, y_k)| \leq \varepsilon$. Hence we also have $g(t_0, y_k) \leq g(t_k, y_k) + \varepsilon \leq K + \varepsilon$, which contradicts to the inf-compactness of $g(t_0, \cdot)$. $\square$

The following proposition asserts the duality relation. The proof is similar to that of Proposition 3.1 of Correa and Seeger (1985) and Theorem 4.3.1 of Hiriart-Urruty and Lemaréchal (1993).

PROPOSITION 1 *Let Y_0 and Λ_0 be nonempty closed convex sets. Suppose $\delta_0 > 0$ be such that:*

    (a) for some $\bar{\lambda} \in \Lambda_0$, $L(\cdot, \cdot, \bar{\lambda})$ is uniformly continuous and $L(0, \cdot, \bar{\lambda})$ is inf-compact over Y_0.

    (b) for some $\bar{y} \in Y_0$, $L(\cdot, \bar{y}, \cdot)$ is uniformly continuous and $L(0, \bar{y}, \cdot)$ is sup-compact over Λ_0.

    (c) for all $t \in [0, \delta_0[$ and any compact convex subsets $\bar{Y} \subset Y_0$ and $\bar{\Lambda} \subset \Lambda_0$, $L(t, \cdot, \cdot)$ has a saddle point on $\bar{Y} \times \bar{\Lambda}$.

Then there exists $\delta > 0$ such that $h_1(t) = h_2(t)$ and the multifunctions $Y(\cdot)$ and $\Lambda(\cdot)$ are nonempty for all $t \in [0, \delta[$. Moreover $Y(\cdot)$ and $\Lambda(\cdot)$ are semicontinuous at $t = 0^+$.

Proof: Denote a closed ball by $B[\cdot, \cdot]$. If we set $Y_n = B[\bar{y}, n] \cap Y_0$ and $\Lambda_n = B[\bar{\lambda}, n] \cap \Lambda_0$, then by assumption (c), for all $t \in [0, \delta_0[$ there exists a saddle point $(y_n(t), \lambda_n(t))$ of the function $L(t, \cdot, \cdot)$ over $Y_n \times \Lambda_n$.

CLAIM. *There exists $\delta > 0$ such that the sequences $\{y_n(t)\}$ and $\{\lambda_n(t)\}$ are bounded for each $t \in [0, \delta[$.*

Indeed if it were not the case, there exists $\{t_k\} \to 0^+$ such that either $\|y_n(t_k)\| \to +\infty$ or $\|\lambda_n(t_k)\| \to +\infty$ occurs when $n \to +\infty$. Set $y_k = y_k(t_k)$ and $\lambda_k = \lambda_k(t_k)$. We may assume that $\lim_{k \to \infty} \|y_k\| = +\infty$. Since (y_k, λ_k) is a saddle point of $L(t_k, \cdot, \cdot)$, we have for sufficiently large k, $L(t_k, y_k, \bar{\lambda}) \leq L(t_k, \bar{y}, \lambda_k)$. Then by Lemma 1 we have

$$+\infty \leftarrow L(t_k, y_k, \bar{\lambda}) \leq L(t_k, \bar{y}, \lambda_k),$$

$k \to +\infty$. Thus $L(t_k, \bar{y}, \lambda_k) \to +\infty$, which can happen only for an unbounded $\{\lambda_k\}$. Using Lemma 1 again we have

$$+\infty \leftarrow L(t_k, y_k, \bar{\lambda}) \leq L(t_k, \bar{y}, \lambda_k) \to -\infty,$$

which leads to the contradiction.

From the claim above, we can take a cluster point $(x(t), y(t))$ of $\{(y_n(t), \lambda_n(t))\}$ to show the existence of a saddle point of $L(t, \cdot, \cdot)$ over $Y_0 \times \Lambda_0$ near $t = 0^+$. This proves that $Y(t)$ and $\Lambda(t)$ are nonempty and the duality $h_1(t) = h_2(t)$ holds near $t = 0^+$.

With the same manner, we can prove that $Y(t)$ and $\Lambda(t)$ are locally bounded near $t = 0^+$. The closedness of $Y(t)$ and $\Lambda(t)$ at $t = 0^+$ is easily verified so that $Y(t)$ and $\Lambda(t)$ are u.s.c., hence semicontinuous in Penot's sense at $t = 0^+$. □

REMARK 1 If we assume

(a') the objective function f is uniformly continuous and $f(x_0, \cdot)$ is inf-compact over Y_0,

(b') $f(\cdot, \bar{y}), g_i(\cdot, \bar{y}), i = 1, \ldots, m$ is uniformly continuous and the Slater condition holds. i.e., for some $\bar{y} \in Y_0$, $g_i(x_0, \bar{y}) < 0$ for $i = 1, \ldots, m$,

then conditions (a) and (b) of Proposition 1 are verified for $\bar{\lambda} = 0$ and $\bar{y}$.

One may think that the condition (c) seems to be rather artificial. Instead of (c), Correa and Seeger assumed that $L(t, \cdot\cdot)$ is convex-concave for all $t \in [0, \delta_0[$. By the classical minimax theorem due to von Neumann (1928), Kakutani (1941) and Ky Fan (1953), the condition (c) is verified under their condition. By the recent development of minimax theorems, several conditions are available to establish (c) (see e.g. Komiya (1982),Takahashi (1988) and references therein). In the rest of this section we recall the Simons' minimax theorem [Simons (1986)] and apply it to Proposition 1.

DEFINITION 2 *Let* $F : Y \times \Lambda \to \mathbf{R}$ *be a real valued function.* F *is said to be convexlike w.r.t.* y *if for any* $y_1, y_2 \in Y$ *and* $\alpha \in [0, 1]$, *there exists* $y_0 \in Y$ *such that*

$$F(y_0, \lambda) \le \alpha F(y_1, \lambda) + (1 - \alpha)F(y_2, \lambda), \qquad \text{for all } \lambda \in \Lambda.$$

Concavelikeness is defined as the same manner.

The following theorem is due to Simons (1986), Corollary 2.3. See also Takahashi (1988).

THEOREM 2 *Let* $F : Y \times \Lambda \to \mathbf{R}$ *be a real-valued function. Suppose* Y *is a compact convex set and*

(1) F *is l.s.c. and convexlike w.r.t.* y,

(2) F *is concavelike w.r.t.* λ,

then we have

$$\min_{y \in Y} \sup_{\lambda \in \Lambda} F(y, \lambda) = \sup_{\lambda \in \Lambda} \min_{y \in Y} F(y, \lambda).$$

If we apply Theorem 2 to the Lagrangian, we have the following.

PROPOSITION 2 *Suppose for any* $t \in [0, \delta_0[$, $L(t, y, \lambda)$ *is convexlike w.r.t.* y. *Then condition (c) holds.*

REMARK 2 In the context of vector valued optimization, one has another definition of convexlikeness. The vector valued function $\mathbf{f} : \mathbf{R}^n \to \mathbf{R}^m$ is said to be convexlike if for any $y_1, y_2 \in \mathbf{R}^n$ and $\alpha \in [0, 1]$, there exists $y_0 \in \mathbf{R}^n$ such that

$$\mathbf{f}(y_0) \leq \alpha \mathbf{f}(y_1) + (1 - \alpha)\mathbf{f}(y_2).$$

If we define the vector valued function

$$\mathbf{L}(x, y) := \begin{pmatrix} f(x, y) \\ g_1(x, y) \\ \vdots \\ g_m(x, y) \end{pmatrix},$$

then convexlikeness in the sense of vector valued function of $\mathbf{L}(x, \cdot)$ assures (c). As for conditions which ensure convexlikeness of general vector valued functions, see, for example, Tardella (1989).

Convexlikeness implies the invexity of the Lagrangian. In the next section, we treat the RHS (right hand side) perturbation problem and show the duality under the invexity assumption.

4 ANALYSIS OF RHS PERTURBATION UNDER INVEXITY

In this section , we are concerned with the right hand side perturbation problem:

$$(P_x^{RHS}) \qquad \begin{aligned} & \text{minimize}_y \quad f(y) \\ & \text{subject to} \quad g_i(y) \leq x_i, \quad \text{for } i = 1, \ldots, m, \end{aligned}$$

where $f, g_1, \ldots, g_m : \mathbf{R}^n \to \mathbf{R}$ and the variable $x = (x_1, \cdots, x_m) \in \mathbf{R}^m$ is viewed as a parameter. Hence the corresponding marginal function and Lagrangian will be

$$h(x) = \inf_y \{f(y)|g_i(y) \leq x_i, \ i = 1, \ldots, m\},$$

and

$$L(x, y, \lambda) = f(y) + \sum_{i=1}^{m} \lambda_i \{g_i(y) - x_i\}.$$

Without loss of generality, we may take the base point $x_0 = 0$. We start with the definition of invex functions. Invexity of functions was originally defined for smooth functions [see e.g. Hanson (1981) and Ben-Israel and Mond (1986)] and later extended for nonsmooth functions by Tanaka, Fukushima and Ibaraki (1989) and Tanaka (1990). We recall here the definition for nonsmooth functions.

DEFINITION 3 *Let* $F : \mathbf{R}^n \to \mathbf{R}$ *be a locally Lipschitz and Clarke regular function.* F *is said to be invex if there exists a mapping* $\eta : \mathbf{R}^n \times \mathbf{R}^n \to \mathbf{R}^n$ *such that, for each* $y, u \in \mathbf{R}$,

$$F(y) - F(u) \geq F'(u; \eta(y, u)).$$

As for properties of invex functions, see Ben-Israel and Mond (1986), Tanaka, Fukushima and Ibaraki (1989) and Tanaka (1990). The following notion of constraint qualification was introduced by Clarke (1976).

DEFINITION 4 *Problem* (P_x^{RHS}) *is said to be calm at* x *if* $h(x)$ *is finite and if*

$$\liminf_{x' \to x}[h(x') - h(x)]/\|x' - x\| > -\infty.$$

We now demonstrate that the Slater's constraint qualification implies calmness under invexity [c.f. Proposition 13 of Clarke (1976)].

PROPOSITION 3 *Let the functions* $g_1, \ldots, g_m : \mathbf{R}^n \to \mathbf{R}$ *be invex for the same* η. *Suppose that* f *is inf-compact and locally Lipschitz. Then if there exists* $\bar{y} \in \mathbf{R}^n$ *such that* $g_i(\bar{y}) < 0$ *for* $i = 1, \ldots, m$, *the problem* (P_x^{RHS}) *is calm at* $x = 0$.

Proof: The hypotheses imply that $h(x)$ is finite near $x = 0$. We will show that the quantity $[h(x) - h(0)]/\|x\|$ is bounded below as $x \to 0$. It suffices to consider x such that $h(x) < h(0)$, and in fact only such x that satisfy $x \geq 0$. Let $\{x_k\}_{k \in N} \to 0^+$ be arbitrary. Choose a sequence $\{y_k\}_{k \in N}$ satisfying $g_i(y_k) \leq (x_k)_i$ ($i = 1, \ldots, m$) and

$$f(y_k) \leq h(x_k) + \|x_k\|^2.$$

We note that $\{y_k\}_{k \in N}$ is bounded since $f(y_k) \leq h(x_k) + \|x_k\|^2 \leq h(0) + \|x_k\|^2$ and f is inf-compact. Hence we may assume $\lim_{k \to \infty} y_k = y_\infty$. If we set $I(y_k) = \{i \mid g_i(y_k) \geq 0\}$, then by the invexity of g_i, we have for all $i \in I(y_k)$

$$0 > g_i(\bar{y}) - g_i(y_k) \geq (g_i)'(y_k; \eta(\bar{y}, y_k)).$$

Hence there exists $\delta_k > 0$ such that for all $i \in I(y_k)$

$$g_i(y_k + t\eta_k) < 0, \qquad \text{for all } t \in]0, \delta_k[,$$

where $\eta_k = \eta(\bar{y}, y_k)$. From the continuity of g_i, we can choose appropriate $\varepsilon_k \in]0, \delta_k[$ such that for all $i = 1, \cdots, m$,

$$g_i(y_k + t\eta_k) \leq 0, \qquad \text{for all } t \in]0, \varepsilon_k]$$

Set $t_k = \min\{\varepsilon_k, \|x_k\|/\|\eta_k\|\}$ (we use the convention $*/0 = \infty$), then we have $h(0) \leq f(y_k + t_k\eta_k)$ and for the Lipschitz constant K of f near y_∞,

$$f(y_k + t_k\eta_k) - f(y_k) \leq K t_k \|\eta_k\| \leq K\|x_k\|.$$

Thus we have

$$h(x_k) - h(0) \geq f(y_k) - f(y_k + t_k\eta_k) - \|x_k\|^2 \geq -K\|x_k\| - \|x_k\|^2.$$

Since $\{x_k\}$ is arbitrary, the calmness holds. $\square$

The following theorem and its corollary are due to Tanaka, Fukushima and Ibaraki (1989). There will be no confusion if we denote $L(0, y, \lambda)$ by $L(y, \lambda)$.

THEOREM 3 *Let $L(\cdot, \lambda)$ be invex for any $\lambda \geq 0$ and (P_x^{RHS}) is calm at $x = 0$. Then y^* is optimal for (P_0^{RHS}) if and only if there exists $\lambda^* \in \mathbf{R}^m$ such that*

$$0 \in \partial f(y^*) + \sum_{i=1}^m \lambda_i^* \partial g_i(y^*)$$

$$\lambda^* \geq 0, g_i(y^*) \leq 0, \lambda^* g_i(y^*) = 0, \qquad i = 1, \ldots, m.$$

COROLLARY 1 *Let $L(\cdot, \lambda)$ be invex for any $\lambda \geq 0$ and (P_x^{RHS}) is calm at $x = 0$. Then y^* is optimal for (P_0^{RHS}) if and only if there exists $\lambda^* \geq 0$ such that (y^*, λ^*) is a saddle point of $L(\cdot, \cdot)$.*

Now we can state the following parametric duality for (P_x^{RHS}).

PROPOSITION 4 *If we suppose that:*

(a") f *is inf-compact over* $\mathbf{R}^n$.

(b") *for some* $\bar{y} \in Y_0$, *the Slater condition $g_i(\bar{y}) < 0$ for $i = 1, \ldots, m$ holds,*

(c") $f, g_1, \ldots, g_m$ *are invex for the same η,*

then the duality relation $h_1(x) = h_2(x)$ holds near $x_0 = 0$. Moreover the multifunctions $Y(\cdot)$ and $\Lambda(\cdot)$ are nonempty and semicontinuous at $x_0 = 0$.

Proof: For $x \in \mathbf{R}^m$, define $\bar{g}_i(y) = g_i(y) - x_i$ for $i = 1, \cdots, m$. Then $L(x, \cdot, \lambda) = f(\cdot) + \sum_{i=1}^m \lambda_i \bar{g}_i(\cdot)$ is invex for any $\lambda \geq 0$. If $\|x\|$ is sufficiently small, the Slater condition $\bar{g}_i(\bar{y}) < 0$ for $i = 1, \ldots, m$ holds, hence calmness by Proposition 3. By Corollary 1, we obtain the duality relation $h_1(x) = h_2(x)$ and nonemptiness of $Y(\cdot)$ and $\Lambda(\cdot)$. It is analogous to Proposition 1 to show the semicontinuity of $Y(\cdot)$ and $\Lambda(\cdot)$. We note that uniform continuity of f is not necessary in the case of RHS perturbation. $\square$

5 SENSITIVITY RESULTS

In this section, we summarize the sensitivity results obtained from Theorem 1 and the results of Sections 3 and 4. In the sequel, we assume the assumptions (H1-1) and (H1-2). We begin with the following general result.

THEOREM 4 *Under the assumptions (H1-1) and (H1-2), let $\delta_0 > 0$ be such that*

(a') *the objective function f is uniformly continuous and $f(x_0, \cdot)$ is inf-compact over $\mathbf{R}^n$.*

(b') $f(\cdot, \bar{y}), g_i(\cdot, \bar{y}), i = 1, \ldots, m$ *is uniformly continuous and the Slater condition holds. i.e., for some $\bar{y} \in \mathbf{R}^n$, $g_i(x_0, \bar{y}) < 0$ for $i = 1, \ldots, m$.*

(c) for all $t \in [0, \delta_0[$ and any compact convex sets $\bar{Y} \subset \mathbf{R}^n$ and $\bar{\Lambda} \subset \mathbf{R}^m_+$, $L(t, \cdot, \cdot)$ has a saddle point on $\bar{Y} \times \bar{\Lambda}$.

Then h is directionally differentiable at x_0 in d and we have

$$h'(x_0; d) = \inf_{y \in Y(x_0)} \sup_{\lambda \in \Lambda(x_0)} L'_x(x_0, y, \lambda; d),$$

where the operator inf-sup commutes.

If we assume convexlikeness, then we have the following.

THEOREM 5 *Under the assumptions (H1-1) and (H1-2) let $\delta_0 > 0$ be such that (a') and (b') of Theorem 4 holds. Suppose also that for any $t \in [0, \delta_0[$ $L(t, y, \lambda)$ is convexlike w.r.t. y. Then h is directionally differentiable at x_0 in d and we have*

$$h'(x_0; d) = \inf_{y \in Y(x_0)} \sup_{\lambda \in \Lambda(x_0)} L'_x(x_0, y, \lambda; d),$$

where the operator inf-sup commutes.

The last one is the result for (P^{RHS}_x) under invexity. We note that in the RHS case (H1-1) is automatically satisfied.

THEOREM 6 *Under the assumption (H1-2), if we suppose that (a''), (b'') and (c'') of Proposition 4 holds. Then h is directionally differentiable at x_0 in d and we have*

$$h'(0; d) = \inf_{y \in Y(0)} \sup_{\lambda \in \Lambda(0)} L'_x(0, y, \lambda; d),$$

where the operator inf-sup commutes.

Acknowledgment. A preliminary version of this paper has been presented at the Workshop on Nonsmooth Analysis and Its Applications, Pau, France, June 27-29, 1995. The author would like to thank Prof. J.-P. Penot of Université de Pau for giving him a chance to present it at the workshop.

REFERENCES

1. Ben-Israel A. and Mond B., (1986). What is invexity?, *Journal of the Australian Mathematical Society Ser. B, 28*: pp.1-9.

2. Clarke F.H., (1976). A new approach to Lagrange multipliers, *Mathematics of Operations Research, 1*: pp.165-174.

3. Cominetti R. and Correa R., (1990). A generalized second-order derivative in nonsmooth optimization, *SIAM Journal on Control and Optimization, 28*: pp.789-809.

4. Correa R. and Seeger A., (1985). Directional derivative of a minimax function, *Nonlinear Analysis, Theory, Methods and Applications, 9*: pp.13-22.

5. Fan K., (1953). Minimax theorems, *Proceedings of the National Academy of Sciences of USA, 39*: pp.42-47

6. Hanson M. A., (1981). On sufficiency of the Kuhn-Tucker conditions, *Journal of Mathematical Analysis and Applications, 80*: pp.545-550

7. Hiriart-Urruty J.-B. and Lemaréchal C., (1993). *Convex analysis and minimization algorithms I*, Springer-Verlag.

8. Hogan W. W., (1973). Point-to-set maps in mathematical programming, *SIAM Review, 15*: pp.591-603.

9. Kakutani S., (1941). A generalization of Browder's fixed point theorem, *Duke Mathematical Journal, 18*: pp.457-459.

10. Komiya H., (1982). Minimax theorems in separation spaces, *RIMS Kokyuroku, Research Institute for Mathematical Sciences, Kyoto University, 789*: pp.1-7.

11. von Neumann J., (1928). Zur Theorie der Gesellschaftsspiele, *Mathematische Annalen, 100*: pp. 295-320.

12. Penot J.-P., (1983). Continuity properties of performance functions, Lecture Notes in Pure and Applied Mathematics, Vol. 86, *Optimization theory and Algorithms*, Edited by Hiriart-Urruty J.-B., Oetly W. and Stoer J., Dekker, pp.77-90.

13. Simons S., (1986). Two-function theorems and variational inequalities for functions on compact and noncompact sets, with some comments on fixed point theorems, *Proceeding of Symposia in Pure Mathematics, American Mathematical Society, 45*: pp.377-392.

14. Takahashi W., (1988). *Nonlinear functional analysis*,(in Japanese) Kindaikagakusha, Tokyo.

15. Tanaka Y., (1990). Note on generalized convex functions, *Journal of Optimization Theory and Applications, 66*: pp.345-349.

16. Tanaka Y., Fukushima M. and Ibaraki T., (1989). On generalized pseudoconvex functions, *Journal of Mathematical Analysis and Applications, 144*: pp.342-355.

17. Tardella F., (1989). On the image of a constrained extremum problem and some applications to the existence of a minimum, *Journal of Optimization Theory and Applications, 60*: pp.93-104.

Parametric Linear Complementarity Problems

KLAUS TAMMER   Humboldt-Universität Berlin, FB Mathematik, Unter den Linden 6, D-10099 Berlin, Germany

Abstract

We study linear complementarity problems depending on parameters in the right-hand side and (or) in the matrix. For the case that all elements of the right-hand side are independent parameters we give a new proof for the equivalence of three different important local properties of the corresponding solution set map (lower semicontinuity, pseudo-Lipschitz continuity and strong regularity) in a neighbourhood of an element of its graph. For one- and multiparametric problems this equivalence does not hold and the corresponding graph may have a rather complicated structure. But we are able to show that for a generic class of linear complementarity problems depending linearly on only one real parameter the situation is much more easier.

1   INTRODUCTION

Linear complementarity problems with parameters in the right-hand side and in the matrix have been extensively studied by many authors (e.g. [1], [3], [5], [7], [10], [13], [22], [23], [26]). Further interesting papers concerning more general problems contain essential consequences also for the special case of parametric linear complementarity problems (cf. [4], [6], [19], [20], [21], [28], [29], [30]).
In our paper we consider parametric linear complementarity problems given by

$$LCP(\lambda): \quad q(\lambda) + K(\lambda)x \geq 0, \; x \geq 0, \; x'(q(\lambda) + K(\lambda)x) = 0,$$

for which generally both the vector $q \in R^n$ and the $(n \times n)$-matrix K depend on a parameter vector $\lambda \in R^d$. Concerning the kind of the parameter dependence we consider two cases, denoted by A_0 and A_1:

Supported by the Deutsche Forschungsgemeinschaft under grant Gu 304/1-4.

A_0 : $q(\cdot) : R^d \to R^n$ and $K(\cdot) : R^d \to R^{n \times n}$ are locally Lipschitz.
A_1 : $q(t) = q^0 + t_1 q^1 + \ldots + t_d q^d$ and K is constant.

In the case that we assume A_1 we denote the parameter vector by t and the corresponding parametric linear complementarity problem by $LCP(t)$.

Only few results will be devoted to the general problem $LCP(\lambda)$ under assumption A_0. Our particular interest concerns its special cases $LCP(q, \lambda)$ given by

$$LCP(q, \lambda) : \qquad q + K(\lambda)x \geq 0, \ x \geq 0, \ x'(q + K(\lambda)x) = 0,$$

where all components of q together with the components of λ are independent parameters, $LCP(q)$ with q as parameter, $LCP(q, K)$ with q and K as parameters and the one-dimensional special case $LCP_1(t)$ of $LCP(t)$ given by

$$LCP_1(t) : \qquad (q^0 + tq^1) + Kx \geq 0, \ x \geq 0, \ x'(q^0 + tq^1 + Kx) = 0.$$

Let us denote the set of all solutions x of $LCP(\lambda)$ $(LCP(t), LCP(q, \lambda), \ldots)$ for the corresponding parameter value by $\Psi(\lambda)$ $(\Psi(t), \Psi(q, \lambda), \ldots)$.

In section 2 of our paper we summarize some essentially known global results on the solution set maps of the considered parametric problems and give certain supplements concerning their polyhedral structure. Motivated by a recent equivalence statement of Dontchev and Rockafellar [4] concerning three different local properties of the solution set map of a more general class of parameter-depending problems (lower semicontinuity, pseudo-Lipschitz continuity and strong regularity) we present in section 3 for problem $LCP(q, \lambda)$ another proof which gives a better insight into the situation. Especially, it will be clear, for which reason lower semicontinuity around a given point of the graph of the solution set map does not hold, if this map is not strongly regular at this point.

Counter examples in the case of problem $LCP_1(t)$ show, that in general lower semicontinuity around a given point of the graph and strong regularity at this point are not equivalent. In section 4 we show that generically the graph of the solution set map of $LCP_1(t)$ has a much more easier structure as in the general case and that only six types of solutions may appear.

Throughout the whole paper we use the symbols "gph" for the graph of a set-valued map, "sgn" for the sign of a real number, "rg" for the rank of a matrix, "dim" for the dimension, "lin" for the linear hull, "int" for the interior, "bd" for the boundary and "ri" for the relative interior of a convex set. The symbol $\mathcal{B}$ stands for the closed unit ball in R^n.

2   SOME GLOBAL RESULTS ON THE SOLUTION SET MAP

In the following theorem we summarize some known properties (cf. [3], [29]) of the solution set map Ψ of $LCP(\lambda)$ and $LCP(t)$, respectively.

THEOREM 1   *1. For the problem $LCP(\lambda)$ the map Ψ is closed (i.e., for each sequence $\{\lambda^\nu\}$, converging to any λ^0, and each sequence $\{x^\nu\}$ with $x^\nu \in \Psi(\lambda^\nu)$, converging to any x^0, we have $x^0 \in \Psi(\lambda^0)$).*

   2. For the problem $LCP(t)$ the map Ψ is polyhedral (i.e., its graph is a union of a finite number of convex polyhedra).

3. *For the problem $LCP(t)$ the map Ψ is locally upper Lipschitz with a uniform modulus (i.e., there is a constant $c > 0$ and for each $t^0 \in R^d$ there exists a neighbourhood V of t^0 such that $\Psi(t) \subseteq \Psi(t^0) + c \parallel t - t^0 \parallel \mathcal{B}$ for all $t \in V$.*

Obviously, Statement 3 of Theorem 1 has the following consequences.

COROLLARY 1 *For the problem $LCP(t)$ the sets $Q_b = \{t : \Psi(t) \text{ is bounded}\}$ as well as $Q_e = \{t : \Psi(t) = \emptyset\}$ are open.*

The following statements on the cardinality $\mid \Psi(q) \mid$ of the set of solutions of problem $LCP(q)$ have been proven in [24] and [30].

THEOREM 2 *For the problem $LCP(q)$ it holds:*

1. *$\mid \Psi(q) \mid \geq 1 \; \forall q \in R^n \Longleftrightarrow K$ is a Q-matrix (for the definition cf. [24]).*

2. *$\mid \Psi(q) \mid < \infty \; \forall q \in R^n \Longleftrightarrow K$ is a nondegenerate matrix (i.e., all its principal subminors are nonzero).*

3. *$\mid \Psi(q) \mid = 1 \; \forall q \in R^n \Longleftrightarrow K$ is a P-matrix (i.e., all its principal subminors are positive).*

REMARK 1 If the matrix K belongs to the class of Q-matrices (nondegenerate matrices, P-matrices, resp.) then for problem $LCP(t)$ we have $\mid \Psi(t) \mid \geq 1 \; \forall \; t \in R^d$ ($\mid \Psi(t) \mid < \infty \; \forall \; t \in R^d$, $\mid \Psi(t) \mid = 1 \; \forall t \in R^d$, resp.). But the reverse statements are not true in general.

For any $\lambda \in R^d$ the set $P(\lambda) = \{x \in R^n : K(\lambda)x + q(\lambda) \geq 0, \; x \geq 0\}$ is a convex polyhedron associated with the problem $LCP(\lambda)$. For any pair (I, J) of index sets $I, J \subseteq \{1, ..., n\}$ let be

$$P^{I,J}(\lambda) = \{x \in P(\lambda) : (K(\lambda)x + q(\lambda))_i = 0, \; i \in I, \; x_j = 0, \; j \in J\},$$

$$\tilde{P}^{I,J}(\lambda) = \{x \in P^{I,J}(\lambda) : (K(\lambda)x + q(\lambda))_i > 0, \; i \in \bar{I}, \; x_j > 0, \; j \in \bar{J}\},$$

$$\mathcal{A}^{I,J} = \{\lambda \in R^d : P^{I,J}(\lambda) \neq \emptyset\} \quad \text{and} \quad \tilde{\mathcal{A}}^{I,J} = \{\lambda \in R^d : \tilde{P}^{I,J}(\lambda) \neq \emptyset\},$$

where $\bar{I} = \{1, ..., n\} \setminus I$. Further let be $\mathcal{S} := \{(I, J) : I \cup J = \{1, ..., n\}\}$. For the special case that $J = \bar{I}$ we write shortly $P^I(\lambda), \tilde{P}^I(\lambda), P^I, \mathcal{A}^I, \tilde{\mathcal{A}}^I$ instead of $P^{I,\bar{I}}(\lambda), \tilde{P}^{I,\bar{I}}(\lambda), P^{I,\bar{I}}, \mathcal{A}^{I,\bar{I}}, \tilde{\mathcal{A}}^{I,\bar{I}}$.

Moreover, we introduce the set $\mathcal{A} = \{\lambda \in R^d : \Psi(\lambda) \neq \emptyset\}$, the matrix $V(I, J; \lambda)$ formed by the rows of $-K(\lambda)$ with the indices $i \in I$ and the rows of the $(n \times n)$-unit matrix with the indices $j \in J$ and the vector $p(I, J; \lambda)$ formed correspondingly by the components $q_i(\lambda)$, $i \in I$ and $n- \mid I \mid$ zero components otherwise.

The following two theorems summarize and supplement corresponding results contained in [1], [3], [9], [10] and [30]. The proof of the first one is obvious.

THEOREM 3 *For the problem $LCP(\lambda)$ it holds:*

1. *For $\lambda \in \mathcal{A}^{I,J}$ the set $P^{I,J}(\lambda)$ is a closed facet of the convex polyhedron $P(\lambda)$ and for $\lambda \in \tilde{\mathcal{A}}^{I,J}$ the set $\tilde{P}^{I,J}(\lambda)$ is the corresponding open facet. For $(I', J') \neq (I'', J'')$ and any $\lambda \in R^d$ it holds $\tilde{P}^{I',J'}(\lambda) \cap \tilde{P}^{I'',J''}(\lambda) = \emptyset$ and $gph\tilde{P}^{I',J'} \cap gph\tilde{P}^{I'',J''} = \emptyset$.*

2. *The sets* $\mathcal{A}$, $gph\Psi$ *and* $\Psi(\lambda)$ *for any* $\lambda \in R^d$ *may be decomposed in the form:*
$$\mathcal{A} = \bigcup_{I \subseteq \{1,\ldots,n\}} \mathcal{A}^I = \bigcup_{(I,J) \in \mathcal{S}} \tilde{\mathcal{A}}^{I,J}, \ gph \ \Psi = \bigcup_{I \subseteq \{1,\ldots,n\}} gph P^I = \bigcup_{(I,J) \in \mathcal{S}} gph \tilde{P}^{I,J}$$
and
$$\Psi(\lambda) = \bigcup_{I \subseteq \{1,\ldots,n\}} P^I(\lambda) = \bigcup_{(I,J) \in \mathcal{S}} \tilde{P}^{I,J}(\lambda).$$

3. *If for any* $\lambda \in R^d$ *a point* $x \in R^n$ *is an isolated solution of* $LCP(\lambda)$ *then* x *corresponds to a vertex* $x(I,J;\lambda) := V^{-1}(I,J;\lambda) \, p(I,J;\lambda)$ *of the convex polyhedron* $P(\lambda)$, *i.e., there is a pair (I,J) of index sets* $I, J \subseteq \{1,\ldots,n\}$ *with* $\mid I \mid + \mid J \mid = n$ *such that the system of (in x) linear equations* $V(I,J;\lambda) \, x = p(I,J;\lambda)$ *has a unique solution* $x(I,J;\lambda)$ *and this solution is also a solution of* $LCP(\lambda)$. *Each vertex* $x(I,J;\lambda)$ *of the convex polyhedron* $P(\lambda)$ *depends locally Lipschitz continuous on the parameter* λ.

We note that for the problem $LCP(t)$ the set $gphP$ is a convex polyhedron and the sets $gphP^I$ respectively $gph\tilde{P}^{I,J}$ are closed respectively open facets of $gphP$. Now we use the submatrix K_{IJ} of K formed by the elements k_{ij} of K with $i \in I$ and $j \in J$.

THEOREM 4 *For the problem* $LCP(q)$ *and each pair* $(I,J) \in \mathcal{S}$ *it holds:*

1. *The set* $\mathcal{A}^{I,J}$ *is a nonempty polyhedral cone and* $\tilde{\mathcal{A}}^{I,J} = ri\mathcal{A}^{I,J}$.

2. $dim P^{I,J}(q) = dim \tilde{P}^{I,J}(q) = d(I,J) \ \forall q \in \tilde{\mathcal{A}}^{I,J}$ *with* $d(I,J) = \mid \bar{J} \mid -rg(K_{I\bar{J}})$ *not depending on q.*

3. $dim\mathcal{A}^{I,J} + d(I,J) + \mid I \mid + \mid J \mid = 2n$ *and* $dim\mathcal{A}^{I,J} + d(I,J) + \mid I \cap J \mid = n$.

4. $dim\mathcal{A}^{I,J} = n \Longleftrightarrow J = \bar{I} \wedge d(I,J) = 0 \Longleftrightarrow J = \bar{I} \wedge K_{II}$ *is nonsingular.*

5. $dim\mathcal{A}^{I,J} = n-1 \Longleftrightarrow a) \ (J = \bar{I} \wedge d(I,J) = 1) \vee b) \ (\mid I \cap J \mid = 1 \wedge d(I,J) = 0)$.

Proof: In our proof we apply the ideas already used in the proof of Theorem 5.4.3 of [1]. To prove Statement 1 we write $\mathcal{A}^{I,J}$ and $\tilde{\mathcal{A}}^{I,J}$ in the form $\mathcal{A}^{I,J} = \{q \in R^n : q = -Kx + y, \ x \geq 0, \ y \geq 0, \ x_j = 0, j \in J, y_i = 0, i \in I\}$ and $\tilde{\mathcal{A}}^{I,J} = \{q \in \mathcal{A}^{I,J} : x_j > 0, j \in \bar{J}, y_i > 0, i \in \bar{I}\}$. Thus, the set $\mathcal{A}^{I,J}$ is the image of a closed facet of the polyhedral cone R_+^{2n} under the linear map which is defined by the matrix $(-K, E)$ and $\tilde{\mathcal{A}}^{I,J}$ the image of the corresponding open facet of R_+^{2n}. But this implies Statement 1. Because of $\tilde{P}^{I,J}(q) \neq \emptyset$ for all $q \in \tilde{\mathcal{A}}^{I,J}$ it follows $\tilde{P}^{I,J}(q) = riP^{I,J}(q)$ and $dim\tilde{P}^{I,J}(q) = dim P^{I,J}(q) = dim L^{I,J} = 2n - \mid I \mid - \mid J \mid - rgB(I,J) = 2n - \mid I \mid - \mid J \mid - \mid \bar{I} \mid -rg(K_{I\bar{J}}) = n - \mid J \mid -rg(K_{I\bar{J}}) = \mid \bar{J} \mid -rg(K_{I\bar{J}})$, where $L^{I,J} = \{(x,y) \in R^{2n} : -Kx + y = 0, \ y_i = 0, i \in I, \ x_j = 0, j \in J\}$ and $B(I,J)$ is the $(n \times (\mid \bar{I} \mid + \mid \bar{J} \mid))$-matrix formed by the columns of $-K$ with the numbers $j \in \bar{J}$ and by the columns of the $(n \times n)$ unit matrix with the numbers $i \in \bar{I}$. This implies Statement 2 and Statement 3, if we use the fact that $dim\mathcal{A}^{I,J} = rg \, B(I,J)$ holds. Statements 4 and 5 follow directly from Statement 3.   q.e.d.

As an immediate consequence of Theorem 4 we mention the following fact.

COROLLARY 2 *For the problem $LCP(q)$ and any $q \in R^n$ we have $| \Psi(q) | = \infty \iff q \in \bigcup_{(I,J) \in \mathcal{S}^*} \tilde{\mathcal{A}}^{I,J}$ with $\mathcal{S}^* := \{(I,J) \in \mathcal{S} : d(I,J) \geq 1\}$, where for all $(I,J) \in \mathcal{S}^*$ it holds $\dim \tilde{\mathcal{A}}^{I,J} < n$.*

For the application in the following sections we give now some additional properties of those sets $\mathcal{A}^{I,J}$ corresponding to problem $LCP(q)$ which have the maximal dimension n. Note that the number of these sets is not zero, since for $I = \emptyset$ and $J = \{1,...,n\}$ we have $\mathcal{A}^{I,J} = R_+^n$. All results follow from the Theorems 3 and 4 and from generally known facts on basic solutions in linear optimization.

REMARK 2   For each set $\mathcal{A}^{I,J}$ with the dimension n corresponding to problem $LCP(q)$ it holds $J = \bar{I}$ and the corresponding matrix K_{II} must be nonsingular (where for the index set $I = \emptyset$ this nonsingularity condition is satisfied by definition). The corresponding set $P^{I,\bar{I}}(q) = P^I(q)$ is formed by a single point, which is a vertex $x(I,q) = \begin{pmatrix} x_I(I,q) \\ x_{\bar{I}}(I,q) \end{pmatrix}$ of $P^I(q)$, where $x_I(I,q) = -K_{II}^{-1}q_I$ and $x_{\bar{I}}(I,q) = 0$. Moreover, it holds $\mathcal{A}^I = \{ q : K_{II}^{-1}q_I \leq 0, q_{\bar{I}} - K_{\bar{I}I}K_{II}^{-1}q_I \geq 0\}$ and $\tilde{\mathcal{A}}^I = \{ q : K_{II}^{-1}q_I < 0, q_{\bar{I}} - K_{\bar{I}I}K_{II}^{-1}q_I > 0\}$.

After introducing slack variables y we can write the convex polyhedron $P(q)$ equivalently in the form $P'(q) = \{(x,y) \in R^{2n} : -Kx + y = q, x \geq 0, y \geq 0\}$ and the complementarity slackness condition can be expressed by $x'y = 0$. The y-part of the vertex $\begin{pmatrix} x(I,q) \\ y(I,q) \end{pmatrix}$ of the convex polyhedron $P'(q)$, which corresponds to the vertex $x(I,q)$ of $P(q)$ is given by $y_I = 0$ and $y_{\bar{I}} = q_{\bar{I}} - K_{\bar{I}I}K_{II}^{-1}q_I$. The corresponding simplex table to this vertex of $P'(q)$ with the vectors of basic variables x_I and $y_{\bar{I}}$ and the vectors of non-basic variables y_I and $x_{\bar{I}}$ is given by

$$
\begin{array}{c|c|c|c}
 & y_I & x_{\bar{I}} & \\
\hline
x_I & -K_{II}^{-1} & K_{II}^{-1}K_{I\bar{I}} & -K_{II}^{-1}q_I \\
\hline
y_{\bar{I}} & -K_{\bar{I}I}K_{II}^{-1} & K_{\bar{I}I}K_{II}^{-1}K_{I\bar{I}} - K_{\bar{I}\bar{I}} & q_{\bar{I}} - K_{\bar{I}I}K_{II}^{-1}q_I
\end{array} \tag{1}
$$

Table (1) contains all coefficients which will be obtained if we transform the system of equations
$$-Kx + y = q \text{ or } \begin{pmatrix} -K_{II} & 0 \\ -K_{\bar{I}I} & E \end{pmatrix}\begin{pmatrix} x_I \\ y_{\bar{I}} \end{pmatrix} + \begin{pmatrix} -K_{I\bar{I}} & E \\ -K_{\bar{I}\bar{I}} & 0 \end{pmatrix}\begin{pmatrix} x_{\bar{I}} \\ y_I \end{pmatrix} = \begin{pmatrix} q_I \\ q_{\bar{I}} \end{pmatrix}$$
in the equivalent form
$$\begin{pmatrix} x_I \\ y_{\bar{I}} \end{pmatrix} = -\begin{pmatrix} -K_{II}^{-1} & K_{II}^{-1}K_{I\bar{I}} \\ -K_{\bar{I}I}K_{II}^{-1} & K_{\bar{I}I}K_{II}^{-1}K_{I\bar{I}} - K_{\bar{I}\bar{I}} \end{pmatrix}\begin{pmatrix} y_I \\ x_{\bar{I}} \end{pmatrix} + \begin{pmatrix} -K_{II}^{-1}q_I \\ q_{\bar{I}} - K_{\bar{I}I}K_{II}^{-1}q_I \end{pmatrix}.$$

3   LOCAL PROPERTIES

Besides global properties of the solution set maps of the considered parametric problems studied in the previous section also local properties are of interest. This means properties of the intersection of the corresponding graph with a sufficiently small neighbourhood of one of its elements. The set of solutions must not be connected or even convex since we do not restrict our considerations to the case that the matrix K has only nonnegative principal subminors. Hence, local properties are not entirely determined by the global ones. Of course, if the set of all solutions for a fixed value of the parameter is finite, then necessarily each solution must be isolated, and, on the other hand, if locally the set of solutions is not finite, then this also must hold globally. But these trivial statements are already almost all relations between local and global properties.

As already done in the paper [4] we are interested to apply the following definitions for general set-valued maps $\Gamma : R^m \rightarrow 2^{R^n}$ to the solution set maps of our parametric linear complementarity problems.

DEFINITION 1 *Let Γ be a set-valued map and $(u^0, v^0) \in gph\Gamma$. Then Γ is called*

(*) *lower semicontinuous around (u^0, v^0), if there are neighbourhoods U of u^0 and V of v^0 such that Γ is lower semicontinuous at every point $(u, v) \in (U \times V) \cap gph\Gamma$ (i.e., for every sequence $\{u^\nu\}$ converging to u there is a sequence $\{v^\nu\}$ with $v^\nu \in \Gamma(u^\nu)$ for ν sufficiently large, converging to v).*

(**) *pseudo-Lipschitz at (u^0, v^0) with a constant $L > 0$, if there are neighbourhoods U of u^0 and V of v^0 such that $\Gamma(u^1) \cap V \subseteq \Gamma(u^2) + L \parallel u^1 - u^2 \parallel \mathcal{B} \quad \forall u^1, u^2 \in U$.*

(***) *strongly regular at (u^0, v^0), if there are neighbourhoods U of u^0 and V of v^0 such that the map $u \rightarrow \Gamma(u) \cap V$ is single-valued and Lipschitz-continuous relative to U.*

The following lemma is the main basis to study parametric linear complementarity problems $LCP(\lambda)$ locally. Suppose that (λ^0, x^0) is any element of the graph of the solution set map Ψ of $LCP(\lambda)$ and we denote

$$I_1 = \{i : (q(\lambda^0) + K(\lambda^0)x^0)_i = 0, x_i^0 > 0\}, \quad I_2 = \{i : (q(\lambda^0) + K(\lambda^0)x^0)_i = 0, x_i^0 = 0\}$$

$$\text{and} \quad I_3 = \{i : (q(\lambda^0) + K(\lambda^0)x^0)_i > 0, \ x_i^0 = 0\}.$$

LEMMA 1 *For each sufficiently small neighbourhood W of $(\lambda^0, x^0) \in gph\Psi$ we have*

$$W \cap gph\ \Psi = W \cap \bigcup_{I \in \mathcal{T}(\lambda^0, x^0)} gph\ P^I = W \cap \bigcup_{(I,J) \in \mathcal{S}(\lambda^0, x^0)} gph\ \tilde{P}^{I,J},$$

where $\mathcal{T}(\lambda^0, x^0) := \{I : I_1 \subseteq I \subseteq I_1 \cup I_2\}$ and $\mathcal{S}(\lambda^0, x^0) := \{(I, J) \in \mathcal{S} : I_1 \subseteq I \subseteq I_1 \cup I_2, I_3 \subseteq J \subseteq I_2 \cup I_3\}$.

Proof: Statement 2 of Theorem 3 implies $W \cap gph\Psi \supseteq W \cap \bigcup_{I \in \mathcal{T}(\lambda^0, x^0)} gphP^I$ and $W \cap gph\Psi \supseteq W \cap \bigcup_{(I,J) \in \mathcal{S}(\lambda^0, x^0)} gph\ \tilde{P}^{I,J}$. For each index set $I \notin \mathcal{T}(\lambda^0, x^0)$ and for each pair $(I, J) \notin \mathcal{S}(\lambda^0, x^0)$ it holds $(\lambda^0, x^0) \notin gphP^I$ and $(\lambda^0, x^0) \notin gphP^{I,J}$. Hence, since the set $gphP^I$ respectively $gphP^{I,J}$ is closed, we get $W \cap gphP^I = \emptyset$ respectively $W \cap gph\tilde{P}^{I,J} = \emptyset$, if we choose W sufficiently small (cf. also [28]). q.e.d.

REMARK 3   For each $(\lambda^0, x^0) \in gph\Psi$ there is a minimal subsystem $\mathcal{Z}(\lambda^0, x^0) = \{P_1, ..., P_k\}$ with $k = k(\lambda^0, x^0)$ of the system of convex polyhedra $P^{I,J}(\lambda^0)$ for $(I, J) \in \mathcal{S}(\lambda^0, x^0)$ such that $\bigcup_{i=1}^{k} P_i = \bigcup_{(I,J)\in\mathcal{S}(\lambda^0,x^0)} P^{I,J}(\lambda^0)$ and $riP_{i_1} \cap riP_{i_2} = \emptyset$ for $i_1 \neq i_2$. For $i = 1, ..., k$ let be $\mathcal{S}_i(\lambda^0, x^0) = \{(I, J) \in \mathcal{S}(\lambda^0, x^0) : P^{I,J}(\lambda^0) = P_i\}$. The sets $\mathcal{S}_i(\lambda^0, x^0)$, $i = 1, ..., k$ are pairwise disjoint and for each $(I, J) \in \mathcal{S}(\lambda^0, x^0) \setminus \bigcup_{i=1}^{k} \mathcal{S}_i(\lambda^0, x^0)$ the convex polyhedron $P^{I,J}(\lambda^0)$ is a closed facet of at least one of the convex polyhedra P_i, $i = 1, ..., k$.

In the following proposition we summarize some immediate observations with respect to the application of Definition 1 to parametric linear complementarity problems.

PROPOSITION 1   *1. For each map Γ we have $(***) \Longrightarrow (**) \Longrightarrow (*)$.*

2. For any element (λ^0, x^0) of the graph of the solution set map of problem $LCP(\lambda)$ condition (*) is equivalent with the existence of neighbourhoods U of λ^0 and V of x^0 satisfying:
(+) For each $(\tilde{\lambda}, \tilde{x}) \in (U \times V) \cap gph\Psi$ there is a neighbourhood $\tilde{U}$ of $\tilde{\lambda}$ such that for each $\lambda \in \tilde{U}$ and each convex polyhedron $P_i \in \mathcal{Z}(\tilde{\lambda}, \tilde{x})$ there is at least one pair $(I, J) \in \mathcal{S}_i(\tilde{\lambda}, \tilde{x})$ with $\lambda \in \mathcal{A}^{I,J}$ and $dimP_i \leq dimP^{I,J}(\lambda)$.

3. For the solution set maps of the problems $LCP(q, \lambda)$ as well as $LCP(t)$ the conditions (*) and (**) are equivalent.

4. For any element (λ^0, x^0) of the graph of the solution set map of problem $LCP(\lambda)$ condition (***) is equivalent with the property that there are neighbourhoods U of λ^0 and V of x^0 such that the map $\lambda \to \Psi(\lambda) \cap V$ is single-valued and continuous on U.

Proof: Statement 1 follows directly from Definition 1 (cf. [4]).
To prove the first direction of Statement 2 we assume that there are neighbourhoods U of λ^0 and V of x^0 having the property (+). Now let $(\tilde{\lambda}, \tilde{x})$ be an arbitrary element of $(U \times V) \cap gph\Psi$, $\{\lambda^\nu\}$ any sequence converging to $\tilde{\lambda}$ and $P_i \in \mathcal{Z}(\tilde{\lambda}, \tilde{x})$. Obviously, it holds $\tilde{x} \in P_i$. According to (+) for each ν sufficiently hight there exists a pair $(I, J) \in \mathcal{S}_i(\tilde{\lambda}, \tilde{x})$ (depending on ν) with $\lambda^\nu \in \mathcal{A}^{I,J}$ and $dimP_i \leq dimP^{I,J}(\lambda^\nu)$. As in the proof of Theorem 3.2.2 in [1] the sequence $\{x^\nu\}$, where x^ν minimizes the Euclidean distance between $\tilde{x}$ and $P^{I,J}(\lambda^\nu)$, converges to $\tilde{x}$. But this means that Ψ is lower semicontinuous at $(\tilde{\lambda}, \tilde{x})$ and, hence, condition (*) is fulfilled at (λ^0, x^0). To prove the second direction of Statement 2 let us suppose that there do not exist any neighbourhoods U of λ^0 and V of x^0 with the property (+). This means that for each neighbourhoods U of λ^0 and V of x^0 there are an element $(\tilde{\lambda}, \tilde{x}) \in (U \times V) \cap gph\Psi$, a sequence $\{\lambda^\nu\}$ converging to $\tilde{\lambda}$ and a convex polyhedron $P_i \in \mathcal{Z}(\tilde{\lambda}, \tilde{x})$ (depending on ν) such for all $\nu = 1, 2, ...$ it holds either $\lambda^\nu \notin \mathcal{A}^{I,J} \ \forall \ (I, J) \in \mathcal{S}_i(\tilde{\lambda}, \tilde{x})$ or for all pairs $(I, J) \in \mathcal{S}_i(\tilde{\lambda}, \tilde{x})$ with $\lambda^\nu \in \mathcal{A}^{I,J}$ we have $dimP_i > dimP^{I,J}(\lambda^\nu)$. Hence, there must be an infinite subsequence of the sequence $\{\lambda^\nu\}$ (for simplicity we denote it again by $\{\lambda^\nu\}$) such that one of the following cases holds true. The first case is that there exists a convex polyhedron $P_i \in \mathcal{Z}(\tilde{\lambda}, \tilde{x})$ such that for all pairs $(I, J) \in \mathcal{S}_i(\tilde{\lambda}, \tilde{x})$ it holds $\lambda^\nu \notin \mathcal{A}^{I,J}$. Lemma 1

and Remark 3 imply that for any element $(\tilde{\lambda}, x^*)$ with $x^* \in riP_i$ sufficiently near to $\tilde{x}$ there can not be any sequence $\{x^\nu\}$ with $x^\nu \in \Psi(\lambda^\nu)$ converging to x^* such that Ψ can not be lower semicontinuous at $(\tilde{\lambda}, x^*)$ and, consequently, (*) is not satisfied at (λ^0, x^0).

The second case is that there exists a convex polyhedron $P_i \in \mathcal{Z}(\tilde{\lambda}, \tilde{x})$ and a nonempty system $\tilde{\mathcal{S}} \subseteq \mathcal{S}_i(\tilde{\lambda}, \tilde{x})$ such that for $\nu = 1, 2, \ldots$ it holds $\tilde{\mathcal{S}} = \{(I, J) \in \mathcal{S}(\tilde{\lambda}, \tilde{x}) : \lambda^\nu \in \mathcal{A}^{I,J}\}$ and $dimP^{I,J}(\lambda^\nu) < dimP^{I,J}(\tilde{\lambda}) \; \forall (I, J) \in \tilde{\mathcal{S}}$. For each pair $(I, J) \in \tilde{\mathcal{S}}$ let be $Q^{I,J} = \{x : \exists\{x^\nu\}, \; x^\nu \in P^{I,J}(\lambda^\nu), \; x^\nu \to x\}$. This set is obviously convex, contained in P_i and can only have a dimension less or equal to the minimal dimension of the sets $P^{I,J}(\lambda^\nu)$. To prove this last condition let us suppose the opposite. Then there must be an infinite subsequence of the sequence $\{\lambda^\nu\}$ (for simplicity we denote it again by $\{\lambda^\nu\}$) such that $d = dimQ^{I,J} > dimP^{I,J}(\lambda^\nu)$ for $\nu = 1, 2, \ldots$. Thus, there must be $d+1$ linearly independent points z^l, $l = 0, 1, \ldots, d$ in $Q^{I,J}$, each of them limit of a sequence $\{x^{\nu l}\}$ with $x^{\nu l} \in P^{I,J}(\lambda^\nu)$. Because of our supposition $dimP^{I,J}(\lambda^\nu) < d$ for each ν there must be a normed vector $c^\nu \in R^d$ satisfying $\sum_{l=1}^{d} c_l^\nu (x^{\nu 0} - x^{\nu l}) = 0$, $\nu = 1, 2, \ldots$. The sequence $\{c^\nu\}$ must have an (again normed) accumulation point c and we obtain (using an infinite subsequence of the sequence $\{c^\nu\}$ converging to c) the relation $\sum_{l=1}^{d} c_l(z^0 - z^l) = 0$ which contradicts our supposition that the points z^l are linearly independent. Hence, it holds $dimQ^{I,J} \leq dimP^{I,J}(\lambda^\nu) < dimP_i$ and, consequently, $Q^{I,J} \subset P_i$ for each pair $(I, J) \in \tilde{\mathcal{S}}$. Using Lemma 1 and Remark 3 this relation implies that in each sufficiently small neighbourhood of $\tilde{x}$ there are elements of the convex polyhedron P_i which may not be a limit of any sequence $\{x^\nu\}$ with $x^\nu \in \Psi(\lambda^\nu)$. But this contradicts (*).

For the problem $LCP(q, \lambda)$ Statement 3 follows from Theorem 1 of [4]. Now let us prove Statement 3 for the problem $LCP(t)$. According to Statement 1 we have only to show $(*) \Rightarrow (**)$. We assume (*) at any element $(t^0, x^0) \in gph\Psi$ and choose polyhedral neighbourhoods $U' \subset U$ of t^0 and $V' \subset V$ of x^0 small enough such that for $W = U' \times V'$ Lemma 1 can be applied. Consider the map Ψ_0 defined for $t \in U'$ by $\Psi_0(t) = V' \cap \bigcup_{(I,J) \in \mathcal{S}(t^0, x^0)} P^{I,J}(t)$, which must be lower semicontinuous on $intU'$. The graph of Ψ_0 is a union of a finite number of convex polyhedra. Consider those edges of these convex polyhedra, which belong to the boundary of the graph but not to the set $bdU' \times V'$. Because of the lower semicontinuity of Ψ_0 all these edges can not be perpendicular to the parameter space R^d. For each such edge we consider its angle to the parameter space R^d. If we now choose L as the maximal absolute value of the tangent of all these angles we find that for arbitrary $t^1, t^2 \in U$ it holds $\Psi_0(t^1) \subseteq \Psi_0(t^2) + L \parallel t^1 - t^2 \parallel \mathcal{B}$ and, hence, condition (**).

The first direction of Statement 4 is trivial, since Lipschitz continuity implies continuity. On the other hand, Statement 3 of Theorem 3 implies that the vector function $x(\cdot)$ is a continuous selection of a finite number of vector functions $x(I, J; \cdot)$, which are locally Lipschitz, and is, thereby, locally Lipschitz itself.   q.e.d.

Unlike the fact that properties (*) and (**) are equivalent for the problem $LCP(t)$ properties (*) and (***) differ generally. The following three examples of the type $LCP_1(t)$ illustrate different possibilities which may appear although (*) holds.

EXAMPLE 1   We define $\Psi_1(t)$ as the set

$$\{x \in R_+^2 : t - x_1 + 2x_2 \geq 0, 3t + 2x_1 + x_2 \geq 0, x_1(t - x_1 + 2x_2) + x_2(3t + 2x_1 + x_2) = 0\}.$$

An easy computation shows $\Psi_1(t) = \begin{cases} \{(0, -3t)', (-t, -t)'\} & \text{for } t \leq 0 \\ \{(0, 0)', (t, 0)'\} & \text{for } t \geq 0 \end{cases}$ such that Ψ_1 satisfies condition (*) but not (***) at the solution $(0, 0)'$ for $t = 0$. Locally (and in this case even globally) this solution for $t = 0$ is unique but in each neighbourhood of $(0, 0)'$ and for each $t \neq 0$ sufficiently near to zero we have more than one element x (namely exactly two) with $(t, x) \in gph\Psi_1$.

EXAMPLE 2   We define $\Psi_2(t)$ as the set

$$\{x \in R_+^3 : -2x_2 + 2x_3 \geq 0, \ 2t - 1 + x_1 + 2x_2 + 2x_3 \geq 0, \ -t + 1 - x_1 - x_2 \geq 0,$$
$$x_1(-2x_2 + 2x_3) + x_2(2t - 1 + x_1 + 2x_2 + 2x_3) + x_3(-t + 1 - x_1 - x_2) = 0\}.$$

Here we get $\Psi_2(t) = \begin{cases} \{(-0.5t + 1, -0.5t, -0.5t)'\} & \text{for } t \leq 0 \\ \{(x_1, 0, 0)' : max(0, 1 - 2t) \leq x_1 \leq 1 - t\} & \text{for } 0 \leq t \leq 1 \end{cases}$ such that Ψ_2 satisfies condition (*) but not (***) at the solution $(1, 0, 0)'$ for $t = 0$. Locally (and again globally) this solution for $t = 0$ is unique but in each neighbourhood of $(1, 0, 0)'$ and for each sufficiently small $t > 0$ we have an infinite number of points x with $(t, x) \in gph\Psi_2$. But globally each connected component of $\Psi_2(t)$ having a nonempty intersection with a sufficiently small neighbourhood of $(1, 0, 0)'$ is bounded.

EXAMPLE 3   We define $\Psi_3(t)$ as the set

$$\{x \in R_+^3 : 2t - 2x_2 + 2x_3 \geq 0, \ -4t + x_1 + 2x_2 + 2x_3 \geq 0, \ -x_1 - x_2 \geq 0,$$
$$x_1(2t - 2x_2 + 2x_3) + x_2(-4t + x_1 + 2x_2 + 2x_3) + x_3(-x_1 - x_2) = 0\}.$$

An easy computation shows $\Psi_3(t) = \begin{cases} \{(0, 0, x_3)' : x_3 \geq -t\} & \text{for } t \leq 0 \\ \{(0, 0, x_3)' : x_3 \geq 2t\} & \text{for } t \geq 0 \end{cases}$ such that also in this case Ψ_3 satisfies condition (*) but not (***) at each solution $(0, 0, x_3)'$ with $x_3 \geq 0$ for $t = 0$. Here we have the situation that the intersection of $gph\Psi_3$ with any neighbourhood of an arbitrary element $(0, 0, 0, x_3)'$ of this graph consists of infinitely many points. But here one component of $\Psi_3(t)$ having a nonempty intersection with a sufficiently small neighbourhood of a solution $(0, 0, x_3)'$ for $t = 0$ is unbounded.

Recently, Dontchev and Rockafellar [4] have shown a general equivalence statement for parametric variational inequalities over polyhedral convex sets, which we formulate here for problem $LCP(q, \lambda)$.

THEOREM 5 *For the problem $LCP(q, \lambda)$ the properties (*), (**) and (***) are equivalent.*

Note that this assertion is valid also for the special cases $LCP(q)$ and $LCP(q, K)$ of $LCP(q, \lambda)$. The essential assumption is only that at least all components of q are independent parameters.

The proof given in [4] is rather abstract and uses a reduction approach, known general properties of projections and normal as well as piecewise linear maps. However, it is not seen immediately, which requirements of (*) would be violated if (***) does not hold. Moreover, it will not intelligible why this proof can not be extended, for instance, to the problem $LCP(t)$. For this reason we will give another proof at the end of this section after some preparations. A recent paper of Kummer [18] is devoted to a corresponding aim, however for the Karush-Kuhn-Tucker conditions for nonlinear and quadratic optimization problems.

According to the decomposition of the whole index set $\{1, ..., n\}$ into the disjoint subsets I_1, I_2 and I_3 we also decompose K in the form $K = \begin{pmatrix} K_{11} & K_{12} & K_{13} \\ K_{21} & K_{22} & K_{23} \\ K_{31} & K_{32} & K_{33} \end{pmatrix}$.

The following necessary and sufficient condition for strong regularity is shown in [28] and [3] (if we use, additionally, Statement 4 of Proposition 1.)

THEOREM 6 *For the problem $LCP(q, \lambda)$ and any element $(q^0, \lambda^0, x^0) \in gph$ Ψ condition (***) is equivalent with*

$$K_{11} \text{ is nonsingular and } N = K_{22} - K_{21}K_{11}^{-1}K_{12} \text{ is a P-matrix.} \qquad (2)$$

In the following for each index set $I \in \mathcal{T}(q^0, \lambda^0, x^0)$ we consider the Jacobian M_I of the linear system, which describes the set $P^I(q)$, namely $M_I = \begin{pmatrix} K_{II} & K_{I\bar{I}} \\ 0 & E \end{pmatrix}$.

Obviously, it holds $det M_I = det K_{II}$. Now we are able to give another equivalent condition for strong stability in problems of the type $LCP(q, \lambda)$, which is already known from [15] for the Karush-Kuhn-Tucker conditions of nonlinear parametric optimization problems. For this case the assertion of the following theorem is shown in [14].

THEOREM 7 *For any $(q^0, \lambda^0, x^0) \in gph\Psi$ condition (***) is equivalent to*

$$sgn \, det M_I = const \neq 0 \qquad \forall I \in \mathcal{T}(q^0, \lambda^0, x^0). \qquad (3)$$

Proof: According to Theorem 6 we only have to show, that (2) and (3) are equivalent. Let (2) be satisfied. Then for $I = I_1$ we have $det M_{I_1} = det K_{I_1 I_1} = det K_{11} \neq 0$.

For any index set I with $I_1 \subset I \subseteq I_1 \cup I_2$ we can write $K_{II} = \begin{pmatrix} K_{11} & K_{I_1 I'} \\ K_{I'I_1} & K_{I'I'} \end{pmatrix}$,

where $I' = I \setminus I_1$. Moreover, using a known determinant rule for Schur complements (cf. [25]), we have $det K_{II} = det K_{11} \cdot det N'$, where $N' = K_{I'I'} - K_{I'I_1}K_{11}^{-1}K_{I_1 I'}$ is a principal submatrix of N having according to (2) a positive determinant. Hence, $sgn \, det M_I = sgn \, det K_{II} = sgn \, det K_{11} = const \neq 0$ as required in (3). The other direction of the proof is similar. We only mention the fact that any principal submatrix of N can be expressed in the form $K_{I'I'} - K_{I'I_1}K_{11}^{-1}K_{I_1 I'}$ with an index set I satisfying $I_1 \subset I \subseteq I_1 \cup I_2$.   q.e.d.

COROLLARY 3 *If the solution set map Ψ of $LCP(q, \lambda)$ does not satisfy condition (***) at any element $(q^0, \lambda^0, x^0) \in gph\Psi$ then one of the following two cases a) or b) holds true.*
Case a) There is an index set $I^* \in \mathcal{T}(q^0, \lambda^0, x^0)$ satisfying $rg K_{I^*I^*} < | I^* |$.
Case b) There are two index sets $I', I'' \in \mathcal{T}(q^0, \lambda^0, x^0)$ and one index $i' \notin I'$ such that $I'' = I' \cup \{i'\}$ and it holds $sgn \, det M_{I'} = -sgn \, det M_{I''} \neq 0$.

Proof: This assertion follows by negation of (3) taking into account that the condition in case a) is only a reformulation of the equation $det M_{I^*} = 0$ and that (if case a) does not come true) the existence of two different index sets $I', I'' \in \mathcal{T}(q^0, \lambda^0, x^0)$ with $sgn\ det M_{I'} = -sgn\ det M_{I''} \neq 0$ implies that there are also two index sets I' and I'' and an index i' with the properties given in case b).   q.e.d.

REMARK 4

1. As shown in [4] even parametric nonlinear complementarity problems satisfying certain differentiability properties can be characterized locally (especially concerning the property of strong regularity) in the same way as it was done here and in former papers for linear problems, namely by analyzing its corresponding linearization. Hence, many results of this section may be used, for instance, to study the Karush-Kuhn-Tucker conditions of nonlinear (and not only quadratic) optimization problems depending on parameters.

2. According to [3] problem $LCP(q, \lambda)$ may be written equivalently as a Lipschitz continuous equation of the form

$$F(z, \lambda) := \qquad K(\lambda)z^- + z^+ = q, \qquad (4)$$

where $z^+ = max(0, z)$ and $z^- = min(0, z)$ componentwise.

The necessary and sufficient conditions (2), (3) (as well as all other equivalent conditions of other papers as [3] and [4]) are equivalent with the nondegeneracy of the projection of the generalized Jacobian $\pi_z \partial F(z, \lambda)$ (in the sense of Clarke [2]) onto the subspace of the z-variables. This follows from a result of [12] and from the fact that the vertices of this projection are closely related to the matrices K_{II} considered in our paper. As we know from [16] this nondegeneracy is in general only a sufficient (but not necessary) condition for the so-called Lipschitz invertibility of systems of the form (4). Only because of a special rank property of the vertices of the mentioned projection (which has been applied already in [12] for the Karush-Kuhn-Tucker condition for nonlinear parametric optimization problems described by C^2-functions) this nondegeneracy condition of Clarke is also necessary for Lipschitz invertibility and, hence, equivalent to a necessary and sufficient condition of Kummer [17] for Lipschitz invertibility of Lipschitz systems and (for the special case of problem $LCP(t)$) to a corresponding necessary and sufficient condition of Scholtes [31] for piecewise linear systems.

Proof of Theorem 5: Because of Statement 1 of Proposition 1 we only have to show $\neg(***) \Rightarrow \neg(*)$. Consider an arbitrary element $(q^0, \lambda^0, x^0) \in gph\Psi$ and suppose, that $(***)$ is not satisfied there. Using Corollary 3 we have to study now more precisely the two cases a) and b) described there.

In the following we delete the dependence on the parameter λ by fixing $\lambda = \lambda^0$ and study the corresponding problem $LCP(q)$. If we can show, that condition $(*)$ is not satisfied at the point (q^0, x^0) of the graph of the solution set map of $LCP(q)$, then, obviously, condition $(*)$ is also not fulfilled at the point (q^0, λ^0, x^0) of the graph of the solution set map of $LCP(q, \lambda)$.

Consider at first case a). According to our assumptions we have $(q^0, x^0) \in gphP^{I^*}$ and thus $q^0 \in \mathcal{A}^{I^*}$. Statements 2 and 3 of Theorem 4 applied to $I = I^*$ and $J = \bar{I}^*$ imply $d(I^*, \bar{I}^*) \geq 1$ and $dim \mathcal{A}^{I^*} < n$. Now we have to distinguish two subcases.

Subcase a_1): If $q^0 \in \tilde{\mathcal{A}}^{I^*}$ then $dim P^{I^*}(q^0) = d(I^*, \bar{I}^*) \geq 1$ and, hence, $| \Psi(q^0) | = \infty$.

But according to Corollary 2 in each neighbourhood U of q^0 there must exist a parameter value q with $\mid \Psi(q) \mid < \infty$ and, hence, $dim P^{I^*}(q) < dim P^{I^*}(q^0)$, where because of $q^0 \in \tilde{\mathcal{A}}^{I^*}$ we have $\mathcal{T}(q^0, x^0) = \{I^*\}$, $\mathcal{Z}(q^0, x^0) = \{P_1\}$, $P_1 = P^{I^*}(q^0)$ and $\mathcal{S}_1 = \{(I^*, \bar{I}^*)\}$. But according to Statement 2 of Proposition 1 this contradicts condition (*).

Subcase a_2): If $q^0 \notin \tilde{\mathcal{A}}^{I^*}$ then q^0 belongs to the relative boundary of $\mathcal{A}^{I^*}$ and for any neighbourhoods U of q^0 and V of x^0 there are points $\tilde{q} \in U \cap \tilde{\mathcal{A}}^{I^*}$ and (because of the fact that according to Corollary 3.4.1.1 in [1] the map P^{I^*} is continuous in the sense of Hausdorff relative to $\mathcal{A}^{I^*}$, shortly H-continuous) $\tilde{x} \in V \cap P^{I^*}(\tilde{q})$, for which our argumentation of subcase a_1) can be repeated.

Now we consider case b). According to our assumptions we have $(q^0, x^0) \in gph P^{I'', \bar{I}'}$ and thus $q^0 \in \mathcal{A}^{I'', \bar{I}'}$. The condition $sgn \, det M_{I'} = -sgn \, det M_{I''} \neq 0$ is equivalent with the condition $sgn \, K_{I'I'} = -sgn \, det K_{I''I''} \neq 0$. Because of $I'' = I' \cup \{i'\}$ with $i' \notin I'$ the matrix $K_{I''I'}$ is formed by $K_{I'I'}$ refilled by one additional row. Hence, it holds $rg(K_{I''I'}) = \mid I' \mid$. Applying Statements 2 and 3 of Theorem 4 we get $d(I'', \bar{I}') = 0$ and $dim \mathcal{A}^{I'', \bar{I}'} = n - 1$. As in the case a) we want to distinguish two subcases.

Subcase b_1): If $q^0 \in \tilde{\mathcal{A}}^{I'', \bar{I}'}$ then $\mathcal{T}(q^0, x^0) = \{I', I''\}$, $\mathcal{Z}(q^0, x^0) = \{P_1\}$, $P_1 = \{x^0\}$ and $\mathcal{S}_1(q^0, x^0) = \{(I', \bar{I}'), (I'', \bar{I}''), (I'', \bar{I}')\}$. Let us consider the simplex tableaus $(1')$ of the vertex $x(I', t)$ as well as $(1'')$ of the vertex $x(I'', t)$ in the form described in (1). Because of $I'' = I' \cup \{i'\}$ tableau $(1'')$ can be generated from tableau $(1')$ by exactly one pivot step with the element $d'_{i'i'}$ of the i'-th row and i'-th column in tableau $(1')$ as pivot element. This pivot element is located in the main diagonal of the submatrix $K_{\bar{I}'I'} K_{I'I'}^{-1} K_{I'\bar{I}'} - K_{\bar{I}'\bar{I}'}$. Let us denote the linear functions of q in the last column in $(1')$ (which is formed according to Remark 2 by the elements of the vectors $-K_{I'I'}^{-1} q_{I'}$ and $(q_{\bar{I}'} + K_{\bar{I}'I'} K_{I'I'}^{-1} q_{I'})$) by $d'_{i0}(q)$ and the corresponding functions in $(1'')$ by $d''_{i0}(q)$. According to the rules of the pivot technique it holds $d'_{i'0}(q) = d'_{i'i'} d''_{i'0}(q)$. Using the already mentioned determinant rule for Schur complements one can show that $det K_{I''I''} = -d'_{i'i'} det K_{I'I'}$ such that because of $sgn \, det K_{I''I''} = -sgn \, det K_{I'I'}$ necessarily $d'_{i'i'} > 0$ follows. According to Remark 2 it holds $\mathcal{A}^{I'} = \{ q : d'_{i0}(q) \geq 0, \ i = 1, ..., n\}$ and $\mathcal{A}^{i''} = \{ q : d''_{i0}(q) \geq 0, \ i = 1, ..., n\}$. Hence, both sets $\mathcal{A}^{I'}$ and $\mathcal{A}^{I''}$ are contained in the same half-space $H_{i'} = \{q : d'_{i'0}(q) \geq 0\}$ and $\mathcal{A}^{I'', \bar{I}'} \subseteq \{q : d'_{i'0}(q) = 0\}$. But due to Statement 2 of Proposition 1 this contradicts (*).

Subcase b_2): If $q^0 \notin \tilde{\mathcal{A}}^{I'', \bar{I}'}$ then q^0 belongs to the relative boundary of $\mathcal{A}^{I'', \bar{I}'}$ and for any neighbourhoods U of q^0 and V of x^0 there are points $\tilde{q} \in U \cap \tilde{\mathcal{A}}^{I'', \bar{I}'}$ and (since the map $P^{I'', \bar{I}'}$ is H-continuous relative to $\mathcal{A}^{I'', \bar{I}'}$) $\tilde{x} \in V \cap P^{I'', \bar{I}'}(\tilde{q})$, for which our argumentation of subcase b_1) can be repeated. q.e.d.

In other words the proof of Theorem 5 says: If at an element $(q^0, \lambda^0, x^0) \in gph \Psi$ condition (***) is violated, then in each neighbourhood of this element there is another element $(\tilde{q}, \lambda^0, \tilde{x})$ of this graph, at which the solution set map is not lower semicontinuous. In (q^0, λ^0, x^0) itself lower semicontinuity may hold or not. The violation of lower semicontinuity at $(\tilde{q}, \lambda^0, \tilde{x})$ may happen for two different reasons. One possibility is that for all sufficiently small neighbourhoods U of $\tilde{q}$ and V of $\tilde{x}$ there exists a value $q \in U$ such that there does not exist any solution x of $LCP(q, \lambda^0)$ in V. The other possibility is that for all sufficiently small neighbourhoods U of $\tilde{q}$ and V of $\tilde{x}$ there exists a value $q \in U$ such that the number of solutions x of

$LCP(q, \lambda^0)$ in V is finite, whereas the number of solutions of $LCP(\tilde{q}, \lambda^0)$ in V is infinite. Corollary 2 shows that this second possibility only leads to a contradiction to condition (*) if all components of q may be perturbed independently of each other. This would be not the case, for instance, in the problem $LCP(t)$ if $d < n$.

4   GENERICITY OF PROBLEMS WITH ONE PARAMETER

Let us study in this section the one-parametric linear complementarity problem $LCP_1(t)$. The examples given in the foregoing section show that even for problems of small size the solution set map of this problem may have a rather complicated structure. As the result of this section we will see that generically the graph of the solution set map has a very easy structure. We can show this with help of the results on the problem $LCP(q)$ given in Theorem 4.

LEMMA 2 *There is an open and dense subset $\tilde{Q} \subseteq R^{2n}$ such that for all $(q^0, q^1) \in \tilde{Q}$ the set $g = \{q \in R^n / q = q^0 + tq^1, \, t \in R\}$ has the following two properties:*

1. For all $(I, J) \in S$ with $\dim \mathcal{A}^{I,J} = n - 1$ it holds $g \not\subseteq \lim \mathcal{A}^{I,J}$.

2. For all $(I, J) \in S$ with $\dim \mathcal{A}^{I,J} \leq n - 2$ it holds $g \cap \mathcal{A}^{I,J} = \emptyset$.

Proof: We show that those values (q^0, q^1), for which 1 or 2 is violated, is contained in the union of a finite number of nondegenerated smooth manifolds of dimension less than or equal to 2n-1.
1. If $g \subseteq \lim \mathcal{A}^{I,J}$ with $\dim \mathcal{A}^{I,J} = n - 1$, then necessarily it follows that (q^0, q^1) belongs to the linear subspace $\lim \mathcal{A}^{I,J} \times \lim \mathcal{A}^{I,J}$ of R^{2n}, which has the dimension 2n-2.
2. If $g \cap \mathcal{A}^{I,J} \neq \emptyset$ with $\dim \mathcal{A}^{I,J} \leq n - 2$, then also $g \cap \lim \mathcal{A}^{I,J} \neq \emptyset$. According to our assumption on the dimension of $\mathcal{A}^{I,J}$ there must are two linear independent vectors $a, b \in R^n$ such that $\lim \mathcal{A}^{I,J} \subseteq L^{n-2}$ with $L^{n-2} = \{q \in R^n : a'q = 0, \, b'q = 0\}$. This means that the two linear equations for one variable t, namely $a'(q^0 + tq^1) = 0$ and $b'(q^0 + tq^1) = 0$ must have a solution. But this implies that either (q^0, q^1) is an element of the linear subspace $L^{n-2} \times L^{n-2}$, which has the dimension 2n-4, or (q^0, q^1) belongs to the set described by $a'q^0 b'q^1 - b'q^0 a'q^1 = 0$, which is outside of $L^{n-2} \times L^{n-2}$ a nondegenerated quadratic manifold of dimension 2n-1.   q.e.d.

The given proof shows that it suffices to perturb slightly only one of the both vectors q^0 or q^1 to reach the set $\tilde{Q}$, if a given pair (q^0, q^1) originally would not belong to $\tilde{Q}$. Only because of the possibility that $q^i \in L^{n-2}$ may come true we can not restrict our perturbations on $q^{i'}$ $(i = 0, 1; \, i' = 1, 0)$.
Using the set $\tilde{Q}$ described in Lemma 2 we are now able to prove in the next theorem an essential generic property (in the sense that this property holds true for all vectors (q^0, q^1) from an open and dense subset $\tilde{Q}$ of R^{2n}) for the graph of the solution set map Ψ of $LCP_1(t)$. For a given element $(t, x) \in \text{gph}\Psi$ we use here the notation $I(t, x) = \{i : (Kx + q^0 + tq^1)_i = 0\}$ and $J(t, x) = \{j : x_j = 0\}$.

THEOREM 8 *For all vectors $(q^0, q^1) \in \tilde{Q}$ we have:*
1. Each connected component of $\text{gph}\Psi$ for $LCP_1(t)$ is a crunode-free edge polygon, which may be either
a) homeomorphic to the real line or

b) homeomorphic to a circle or

c) an isolated point of gphΨ.

2. *Each element $(t, x) \in gph\Psi$ belongs to exactly one of the following six types:*

Type 1: $I(t, x) \cap J(t, x) = \emptyset$ *(strict complementarity)* $\wedge$ $rgK_{II} = \mid I \mid$, *where* $I = I(t, x)$.

Type 2: $I(t, x) \cap J(t, x) = \emptyset$ *(strict complementarity)* $\wedge$ $rgK_{II} = \mid I \mid -1$, *where* $I = I(t, x)$.

Type 3: $I(t, x) \cap J(t, x) = \{i'\}$ $\wedge$ $sgn\ detK_{I'I'} = sgn\ detK_{I''I''} \neq 0$, *where* $I' = I(t, x)$, $J' = J(t, x) \setminus \{i'\}$, $I'' = I(t, x) \setminus \{i'\}$, $J'' = J(t, x)$.

Type 4: $I(t, x) \cap J(t, x) = \{i'\}$ $\wedge$ $sgn\ detK_{I'I'} = -sgn\ detK_{I''I''} \neq 0$, *where* I', J', I'' *and* J'' *are defined as above.*

Type 5: $I(t, x) \cap J(t, x) = \{i'\}$ $\wedge$ $sgn\ detK_{I'I'} \neq 0$ $\wedge$ $sgn\ detK_{I''I''} = 0$ *(or vice versa), where again* I', J', I'' *and* J'' *are defined as above.*

Type 6: $I(t, x) \cap J(t, x) = \{i'\}$ $\wedge$ $sgn\ detK_{I'I'} = sgn\ detK_{I''I''} = 0$, *where again* I', J', I'' *and* J'' *are defined as above.*

3. *For almost all values of t only Type 1 occurs. Almost all elements $(t, x) \in gph\Psi$ are of the types 1 and 2.*

Proof: According to Lemma 2 for $(q^0, q^1) \in \tilde{Q}$ the line g may intersect only those sets $\tilde{A}^{I,J}$ with dimension n or n-1, where the second case may only accur for a finite number of values t. Hence, taking Theorem 4 and Remark 2 into account, the graph of Ψ will be formed by all points (t, x) satisfying either

a) $(q^0 + tq^1) \in \tilde{A}^I$ for any index set $I \in \{1, ..., n\}$ such that $dimA^I = n$, K_{II} is nonsingular and $x = x(I, t) = \begin{pmatrix} x_I(I, t) \\ x_{\bar{I}}(I, t) \end{pmatrix}$ with $x_I(I, t) = -K_{II}^{-1}(q_I^0 + tq_I^1)$ and $x_{\bar{I}}(I, t) = 0$ or

b) $(q^0 + tq^1) \in \tilde{A}^{I,J}$ for any pair $(I, J) \in \mathcal{S}$ such that $dimA^{I,J} = n - 1$ and $x \in P^{I,J}(q^0 + tq^1)$.

According to Theorem 4 and Remark 2 the points $(t, x) \in gph\Psi$ satisfying a) are just those elements of $gph\Psi$ of the type 1 and form together with their boundary points (which also must belong to the closed set $gph\Psi$ and at which b) must be satisfied) a first finite system of edges of the set $gph\Psi$.

For the finite number of values $\bar{t}$, for which there is a point $\bar{x}$ such that $(\bar{t}, \bar{x}) \in gph\Psi$ satisfies b) we can use Statements 2 and 5 of Theorem 4 and Lemma 1. According to Statement 5 of Theorem 4 we must study two subcases of case b).

In the subcase b_1) we have $I \cap J = \emptyset$ and $d(I, J) = 1$. Hence, the points $(t, x) \in gph\Psi$ satisfying b_1) are just those elements of $gph\Psi$ of the type 2 and form together with their boundary points (which must be just those elements of the set $gph\Psi$ of the type 5) a second finite system of edges of the set $gph\Psi$.

In the subcase b_2) we have the situation $I \cap J = \{i'\}$ and $d(I, J) = 0$ such that the set $\{(\bar{t}, x) : x \in P^{I,J}(q^0 + \bar{t}q^1)\}$ is a singleton. Obviously, with the types 3-6 all possibilities for $sgn\ detK_{I'I'}$ and $sgn\ detK_{I''I''}$ under the assumption $I \cap J = \{i'\}$ are exhausted such that the points $(t, x) \in gph\Psi$ satisfying b_2) are just the elements of $gph\Psi$ of the types 3-6. The points of the types 3 and 4 are common boundary points of exactly two adjacent edges of the first system, one of them given by the vertex $x(I', t)$ of the convex polyhedron $P(t)$, the other one by the vertex $x(I'', t)$. This follows by Lemma 1. Again by Lemma 1 we see that the points of the type 5 are the common boundary points of exactly one edge of the first system and exactly one of the second system. Finally, the points of the type 6 are isolated points of $gph\Psi$ since there cannot be any edge of $gph\Psi$ having a point of type 6 as boundary point. Both systems of edges together with the isolated points of type 6 form the whole graph of Ψ. This completes the proof.   q.e.d.

REMARK 5

1. For all elements $(t, x) \in gph\Psi$ of the types 1 and 3-6 the x-part is a vertex of the convex polyhedron $P(t)$. For all elements $(t, x) \in gph\Psi$ of the type 2 the x-part is an inner point of an edge of the convex polyhedron $P(t)$.

2. For all elements $(\bar{t}, \bar{x}) \in gph\Psi$ of the type 6 the corresponding vertices $(\bar{x}, \bar{y})$ of the convex polyhedron $P'(t)$ with $\bar{y} = q^0 + \bar{t}q^1 + K\bar{x}$ are exactly those vertices of $P'(t)$ which actually satisfy the complementarity condition $\bar{x}'\bar{y} = 0$, but for which there does not exist any simplex tableau of the form (1), i.e., there does not exist any basis solution with the property that for each i=1,...,n exactly one of the varibles x_i and y_i is a basic variable and the other one a non-basic variable. Under different assumptions on the matrix K such vertices and, hence, elements $(t, x) \in gph\Psi$ of the type 6 can not exist (cf. [1]).

3. For an arbitrary element $(\bar{t}, \bar{x}) \in gph\Psi$ of one of the types 1 or 3-5 consider a corresponding simplex tableau (1). As in the proof of Theorem 5 let us denote the elements of (1) by d_{ij} and the linear functions of t in the last column of (1) by $d_{i0}(t)$. According to Remark 2 we assume $d_{i0}(\bar{t}) \geq 0$, $i = 1, ..., n$. With help of the data of (1) we can characterize uniquely the type of this point as follows:

a) $(\bar{t}, \bar{x})$ is of the type 1 $\iff d_{i0}(\bar{t}) > 0$, $i = 1, ..., n$.

b) $(\bar{t}, \bar{x})$ is of the type 3 $\iff$ there is exactly one index $i' \in \{1, ..., n\}$ such that $d_{i'0}(\bar{t}) = 0$ and it holds $d_{i'i'} < 0$.

c) $(\bar{t}, \bar{x})$ is of the type 4 $\iff$ there is exactly one index $i' \in \{1, ..., n\}$ such that $d_{i'0}(\bar{t}) = 0$ and it holds $d_{i'i'} > 0$.

d) $(\bar{t}, \bar{x})$ is of the type 5 $\iff$ there is exactly one index $i' \in \{1, ..., n\}$ such that $d_{i'0}(\bar{t}) = 0$ and it holds $d_{i'i'} = 0$.

4. The unique open edge of the second class formed by solutions $(\bar{t}, x) \in gph\Psi$ of the type 2 with an element $(\bar{t}, \bar{x}) \in gph\Psi$ of the type 5 (which satisfies d) of Statement 3) as one boundary point can be constructed with help of the corresponding simplex table (1) to $(\bar{t}, \bar{x})$ as follows: We put $z_{Bi'} = 0$, $z_{Bi} = d_{i0}(\bar{t}) - d_{ii'}s$, $i \neq i'$, $z_{Ni'} = s$, $0 < s < \bar{s}$, $z_{Ni} = 0, i \neq i'$, where z_B stands for the vector of basic variables in (1), z_N for the vector of non-basic variables and $\bar{s} = sup\{s : d_{i0}(\bar{t}) - d_{ii'}s \geq 0, i \neq i'\}$.

5. Concerning the edges of $gph\Psi$ of the second class there are three different cases to distinguish:

Case a) If all elements $d_{ii'}$, $i \neq i'$, are nonpositive then this edge is unbounded ($\bar{s} = \infty$) and is, hence, the first or last edge of the corresponding edge polygon, to which it belongs.

Otherwise the edge is bounded and must have a second boundary point $(\bar{t}', \bar{x}')$. Let be $i'' \in \{i \neq i' : \frac{d_{i0}(\bar{t})}{d_{ii'}} = min_{j:d_{ji'}>0}\frac{d_{j0}(\bar{t})}{d_{ji'}}\}$ and $\bar{s} = \frac{d_{i''0}(\bar{t})}{d_{i''i'}}$. For $(q^0, q^1) \in \tilde{Q}$ the index i" is uniquely determined and it holds $d_{i'i''} \neq 0$. Hence, we can obtain a simplex tableau (1') to $(\bar{t}', \bar{x}')$ from (1) by one (2×2)-pivot step with the (2×2)-matrix

$\begin{pmatrix} d_{i'i'} & d_{i'i''} \\ d_{i''i'} & d_{i''i''} \end{pmatrix}$ as pivot matrix. Because of $d_{i'i'} = 0$, $d_{i''i'} > 0$ and $d_{i'i''} \neq 0$ this matrix is nonsingular, if $(q^0, q^1) \in \tilde{Q}$. The index set I', which corresponds to $(1')$, is formed by I, i' and i'' in the following way. First we put $I^* = I \setminus \{i'\}$ if $i' \in I$ respectively $I^* = I \cup \{i'\}$ otherwise. Analogously, we set $I' = I^* \setminus \{i''\}$ if $i'' \in I$ respectively $I' = I^* \cup \{i''\}$ otherwise. The corresponding matrix $K_{I'I'}$ will always be nonsingular. With respect to sufficiently small neighbourhoods W of $(\bar{t}, \bar{x})$ and W' of $(\bar{t}', \bar{x}')$ the following two possibilities b) and c) may appear.

Case b) If $d_{i'i''} < 0$ then $sgn \, det K_{II} = sgn \, det K_{I'I'}$ and for $t < \bar{t}$ the intersection of W with $gph\Psi$ consists of all points $(t, x(I, t))$ and is empty for $t > \bar{t}$ and the intersection of W' with $gph\Psi$ is empty for $t < \bar{t}$ and consists for $t > \bar{t}$ of all points $(t, x(I', t))$ (or vice versa).

Case c) If $d_{i'i''} > 0$ then $sgn \, det K_{II} = -sgn \, det K_{I'I'}$ and for $t < \bar{t}$ the intersection of W with $gph\Psi$ consists of all points $(t, x(I, t))$ and is empty for $t > \bar{t}$ and the intersection of W' with $gph\Psi$ is empty for $t > \bar{t}$ and consists for $t < \bar{t}$ of all points $(t, x(I', t))$ (or vice versa).

6. At all elements $(t, x) \in gph\Psi$ of the types 1 and 3 condition (***) is satisfied, whereas at all other types 2 and 4-6 even condition (*) is not fulfilled.

7. With respect to an open and dense subset of the $(n \times n)$-dimensional Euclidean space of all elements k_{ij} the matrix K is nondegenerated. Hence, for the problem $LCP_1(t)$ only the types 1, 3 and 4 remains generic, if we permit to perturb beside the vectors q^0 and q^1 also the elements of the matrix K.

8. If all principal minors of the matrix K are nonnegative, then the types 4 and 6 as well as the case c) described in Statement 5 can not appear, the graph of the solution set map of $LCP_1(t)$ consists of exactly one edge polygon and is always homeomorphic to the real line (cf. [1]). If the matrix K is even a P-matrix, then also the types 2 and 5 can not appear, all elements $(t, x) \in gph\Psi$ satisfy (***) and $gph\Psi$ is formed only by edges of the first class.

In the following theorem we characterize the local structure of the graph of the solution set map Ψ of $LCP_1(t)$ for the six different types given above. We use the notations I', I'' and $x(I, t)$ as in Theorem 8.

THEOREM 9 *Let W be a sufficiently small neighbourhood of $(\bar{t}, \bar{x}) \in gph\Psi$.*

1. *If $(\bar{t}, \bar{x})$ is of the type 1 then we have $W \cap gph\Psi = \{(t, x) \in W \, : \, x = x(I, t)\}$, where $I = I(\bar{t}, \bar{x})$.*

2. *If $(\bar{t}, \bar{x})$ is of the type 2 then we have $W \cap gph\Psi = \{(t, x) \in W \, : \, t = \bar{t}, \, x \in P^I(\bar{t})\}$, where $I = I(\bar{t}, \bar{x})$.*

3. *If $(\bar{t}, \bar{x})$ is of the type 3 then we have $W \cap gph\Psi = \{(t, x) \in W \, : \, t \leq \bar{t}, \, x = x(I', t)\} \cup \{(t, x) \in W \, : \, t \geq \bar{t}, \, x = x(I'', t)\}$ or $W \cap gph\Psi = \{(t, x) \in W \, : \, t \geq \bar{t}, \, x = x(I', t)\} \cup \{(t, x) \in W \, : \, t \leq \bar{t}, \, x = x(I'', t)\}$. For $t = \bar{t}$ it holds $x(I', t) = x(I'', t)$.*

4. *If $(\bar{t}, \bar{x})$ is of the type 4 then we have $W \cap gph\Psi = \{(t, x) \in W \, : \, t \leq \bar{t}, \, x = x(I', t)\} \cup \{(t, x) \in W \, : \, t \leq \bar{t}, \, x = x(I'', t)\}$ or $W \cap gph\Psi = \{(t, x) \in W \, : \, t \geq \bar{t}, \, x = x(I', t)\} \cup \{(t, x) \in W \, : \, t \geq \bar{t}, \, x = x(I'', t)\}$. For $t = \bar{t}$ it holds $x(I', t) = x(I'', t)$.*

5. If $(\bar{t}, \bar{x})$ is of the type 5 then we have $W \cap gph\Psi = \{(t, x) \in W : t \leq \bar{t}, x = x(I', t)\} \cup \{(t, x) \in W : t = \bar{t}, x \in P^{I''}(\bar{t})\}$ or $W \cap gph\Psi = \{(t, x) \in W : t \geq \bar{t}, x = x(I', t)\} \cup \{(t, x) \in W : t = \bar{t}, x \in P^{I''}(\bar{t})\}$ or $W \cap gph\Psi = \{(t, x) \in W : t \leq \bar{t}, x = x(I'', t)\} \cup \{(t, x) \in W : t = \bar{t}, x \in P^{I'}(\bar{t})\}$ or $W \cap gph\Psi = \{(t, x) \in W : t \geq \bar{t}, x = x(I'', t)\} \cup \{(t, x) \in W : t = \bar{t}, x \in P^{I'}(\bar{t})\}$.

6. If $(\bar{t}, \bar{x})$ is of the type 6 then we have $W \cap gph\Psi = \{(\bar{t}, \bar{x})\}$.

Proof: The proof follows by Theorem 8 and Remark 5.   q.e.d.

The following figure illustrates the given six types of solutions and the possible situations concerning the structure of the graph of the solution set map. In this example this graph has four connected components, namely two isolated points, one edge polygon which is homeomorphic to the real line and one edge polygon which is homeomorphic to a circle. Moreover, there are three edges of the second class described in Statement 5 of Remark 5.

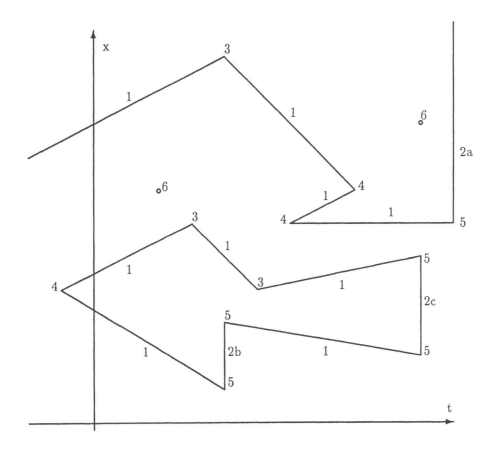

FIGURE 1 The graph of the solution set map

Generic properties of the Karush-Kuhn-Tucker conditions for one-parametric quadratic optimization problems are the common subject of Section 4 of our paper with the papers [11] of Jongen et al and [8] of Henn et al. For the special case of one-parametric linear optimization problems we refer also to the relevant paper [27] of Patewa. The results are partially similar but not identical because of the following essential differences in the assumptions. Firstly, in the papers [11] and [8] the dependence on the parameter t is assumed to be more general (of the type C^3 respectively C^1), whereas we restrict our considerations to the case that only the vector q depends on t and this dependence is linear. This point is connected with the second difference, namely with the fact, that our notion "generic" is based only on small pertubations of the problem in the finite dimensional space R^{2n} of the data q^0 and q^1 with the corresponding topology, whereas in [11] and [8] small perturbations of all underlying functions in the strong C^3-topology respectively strong C^1-topology are allowed. Finally, an exact comparison of the three papers would require, on the one hand, to include the Lagrange multipliers and their dependence on the parameter into the considerations of [11] and [8] and, on the other hand, to include all aspects of [11] which are only relevant for the Karush-Kuhn-Tucker conditions of an optimization problem into our considerations.

Remember that in [11] five types of (generalized) critical points of one-parametric nonlinear optimization problems described by C^3-functions have been identified to be generic, whereas the paper [8] shows that in the corresponding special case of quadratic respectively linear optimization only the types 1, 2 and 5 respectively 1 and 5 from [11] remain generic.

Taking into consideration all differences between the approaches in [11] and in our Section 4 mentioned above, we can see, that the types 1 of both papers are identical and that the two subcases of type 2 of [11] correspond to our types 3 and 4. Type 3 of [11] has some common properties as our type 2 but both are not identical . Finally, type 5 of [11] is related to our types 2 and 5 but is not identical. All other types of both papers differ essentially of each other.

Acknowledgement

The author is indebted to Bernd Kummer and Diethard Klatte for helpful discussions on the subject of this paper as well as to the referees for many valuable hints.

References

[1] Bank, B., Guddat, J., Klatte, D., Kummer, B., and Tammer, K. (1982). Non-linear Parametric Optimization, Akademie-Verlag, Berlin.

[2] Clarke, F. H. (1983). Optimization and Nonsmooth Analysis, Wiley, New York.

[3] Cottle, R. W., Pang, J.-S., and Stone, R. E. (1992). The Linear Complementarity Problem, Academic Press Inc., Boston.

[4] Dontchev, A. L., and Rockafellar, R. T. (1995). Characterization of strong regularity for variational inequalities over polyhedral convex sets, to appear in SIAM J. Optimization.

[5] Gowda, M. S. (1992). On the continuity of the solution set map in linear complementarity problems, SIAM J. Optimization, 2: 619-634.

[6] Gowda, M. S., and Sznajder, R. (1996). On the Lipschitz properties of polyhedral multifunctions, Mathematical Programming, 74: 267-278.

[7] Ha, C. D. (1985). Stability of the linear complementarity problem at a solution point, Mathematical Programming, 31: 327-338.

[8] Henn, M., Jonker, P., and Twilt, F. (1995). On the critical sets of one-parameter quadratic optimization problems, Recent Developments in Optimization (R. Durier, Chr. Michelot, eds.), Springer-Verlag, Berlin, Heidelberg, pp. 183-197.

[9] Jansen, M. J. M. (1983). On the structure of the solution set of a linear complementarity problem, Cahiers du C.E.R.O., 25: 41-48.

[10] Jansen, M. J. M., and Tijs, S. H. (1987). Robustness and nondegenerateness for linear complementarity problems, Mathematical Programming, 37: 293-308.

[11] Jongen, H. Th., Jonker, P., and Twilt, F. (1986). Critical sets in parametric optimization Mathematical Programming, 34 : 333-353.

[12] Jongen, H. Th., Klatte, D., and Tammer, K. (1990). Implicit functions and sensitivity of stationary points Mathematical Programming, 49: 123-138.

[13] Klatte, D. (1985). On the Lipschitz behavior of optimal solutions in parametric problems of quadratic optimization and linear complementarity Optimization, 16: 819-831.

[14] Klatte, D., and Tammer, K. (1990). Strong stability of stationary solutions and Karush-Kuhn-Tucker points in nonlinear optimization Annals of Operations Research, 27: 285-308.

[15] Kojima, M. (1980). Strongly stable stationary solutions in nonlinear programs, Analysis and Computation of Fixed Points (S. M. Robinson, ed.), Academic Press, New York, pp. 93-138.

[16] Kummer,B. (1988). The inverse of a Lipschitz function in R^n: Complete characterization by directional derivatives, Preprint Nr. 195, Humboldt-Universität zu Berlin, Sektion Mathematik.

[17] Kummer, B. (1991). Lipschitzian inverse functions, directional derivatives and applications in $C^{1,1}$-optimization Journal Optimization Theory Appl., 70: 561-582.

[18] Kummer, B. (1995). Lipschitzian and pseudo-Lipschitzian inverse functions and applications to nonlinear optimization, manuscript.

[19] Kyparisis, J. (1988). Perturbed solutions of variational inequality problems over polyhedral sets Journal Optimization Theory Appl., 57: 295-305.

[20] Mangasarian, O. L., Shiau, T.-H. (1987). Lipschitz continuity of solutions of linear inequalities, programs and complementarity problems, SIAM Journal Control Optimization, 25: 582-595.

[21] Meister, H. (1983). Zur Theorie des parametrischen Komplementaritätsproblems, Verlag Anton Hain, Meisenheim am Glan.

[22] Mordukhovich, B. (1994). Stability theory for parametric generalized equations and variational inequalities via nonsmooth analysis, Transactions American Mathematical Soc., 343: 609-657.

[23] Murthy, G. S. R., Parthasarathy, T., and Sabatini, M. (1996). On Lipschitzian Q-matrices, Mathematical Programming, 74: 55-58.

[24] Murty, K. G. (1972). On the number of solutions to the complementarity problem and spanning properties of complementary cones, Linear Algebra Appl., 5: 65-108.

[25] Ouelette, D. V. (1981). Schur complements and statistics, <u>Linear Algebra Appl.</u>, <u>36</u>: 187-295.

[26] Pang, J. S. (1990). Solution differentiability and continuation of Newton's method for variational inequality problems over polyhedral set, <u>Journal Optimization Theory Appl.</u>, <u>66</u>: 121-135.

[27] Patewa, D. D. (1991). On the singularities in linear one-parametric optimization problems, <u>Optimization</u>, <u>22</u>: 193-220.

[28] Robinson, S. M. (1980). Strongly regular generalized equations, <u>Mathematics of Operations Research</u>, <u>5</u>: 43-62.

[29] Robinson, S. M. (1981). Some continuity properties of polyhedral multifunctions, <u>Mathematical Programming Study</u>, <u>14</u>: 206-214.

[30] Samelson, H., Thrall, M., and Wesler, O. (1958). A partition theorem for Euclidean n-space, <u>Proceedings American Mathematical Soc.</u>, <u>9</u>: 805-807.

[31] Scholtes, S. (1994). Introduction to piecewise differentiable equations, Preprint No. 53/1994, Universität Karlsruhe, Institut f. Statistik u. Math. Wirtschaftstheorie.

[32] Walkup, D. W., and Wets, R. J.-B. (1969). A Lipschitzian characterization of convex polyhedra, <u>Proceedings American Mathematical Soc.</u>, <u>20</u>: 167-173.

Sufficient Conditions for Weak Sharp Minima of Order Two and Directional Derivatives of the Value Function

DOUG WARD   Department of Mathematics and Statistics, Miami University, Oxford, Ohio

Abstract: Second-order sufficient optimality conditions play an important role in sensitivity analysis in nonlinear programming. Standard second-order sufficient conditions imply that the point in question is a special type of strict local minimizer, usually termed a strong minimum or strict local minimum of order two.   In this paper, we derive sufficient conditions for weak sharp minima of order two, a larger class of (possibly) non-isolated minima, and show that these sufficient conditions can be used to extend some previous results on first- and second-order directional derivatives of the value function of a family of nonlinear programs with right-hand-side perturbations of the constraints.

1   INTRODUCTION

We consider the family of mathematical programs $P(u)$ given by

$$v(u) := \min\{f(x) \mid g_i(x) \le u_i, \ i \in J, \ h_i(x) = u_i, \ i \in L\},$$

where $J := \{1, \ldots, p\}$, $L := \{p + 1, \ldots, m\}$, $u = (u_1, \ldots, u_m) \in \mathbb{R}^m$, and $f, \ g_i, \ h_i : \mathbb{R}^n \longrightarrow \bar{\mathbb{R}} := [-\infty, +\infty]$.   For $u \in \mathbb{R}^m$, we define

$$R(u) := \{x \in \mathbb{R}^n \mid g_i(x) \le u_i, \ i \in J, \ h_i(x) = u_i, \ i \in L \}.$$

The study of the differential properties of the *value function* v and the continuity properties of the solution set multifunction

$$\Omega(u) := \{x \in R(u) \mid v(u) = f(x) \}$$

is a major research area in mathematical programming (see [2, 7, 14]).   The most complete theories of this subject have been developed with the help of the assumption that for some $\bar{u}$ (say $\bar{u} = 0$) and $\bar{x} \in \Omega(\bar{u})$, a second-order sufficient condition for optimality is satisfied at $\bar{x}$ (e.g. [1, 7-9]). Such sufficient conditions imply that $\bar{x}$ is a special type of isolated local minimizer, usually termed a *strong minimizer* or *strict local minimizer of order two.*

More recently, however, it has been demonstrated that a number of theorems can be extended to the more general situation where $\bar{x}$ belongs to a larger class of possibly non-isolated minima called *weak sharp minima* [12, 13, 17-18].

DEFINITION 1.1. Let $\|\cdot\|$ be the Euclidean norm on $\mathbb{R}^n$. Suppose that $f:\mathbb{R}^n \to \bar{\mathbb{R}}$ is finite and constant on $S \subset \mathbb{R}^n$ and $\bar{x} \in S \cap R(u)$. For $x \in \mathbb{R}^n$, define

$$\text{dist}^2(x,S) := \inf \{ \|y - x\|^2 \mid y \in S \};$$

and for $\varepsilon > 0$, let $B(x,\varepsilon) := \{y \in \mathbb{R}^n \mid \|y - x\| \leq \varepsilon \}$.

We say that $\bar{x}$ is a *weak sharp local minimizer of order two* for problem $P(u)$ if there exist $\beta > 0$, $\varepsilon > 0$ such that

$$f(x) - f(\bar{x}) \geq \beta \, \text{dist}^2(x,S) \quad \forall \, x \in R(u) \cap B(\bar{x},\varepsilon). \qquad (1.1)$$

REMARK 1.2. (a) If $S = \{\bar{x}\}$ in (1.1), then $\bar{x}$ is said to be a *strict local minimizer of order two,* a concept mentioned above. Characterizations of strict local minimizers of order two are discussed in [19, 22].
(b) A simple example of a weak sharp minimizer that is not a strict minimizer can be obtained by letting $n = 2$, $m = 0$, $f(x,y) = x^2$, and $S = \{(0,y) \mid y \in \mathbb{R} \}$. In this example, each point of S is a weak sharp local minimizer of order two, satisfying (1.1) with $\beta = 1$ and any $\varepsilon > 0$.

Several recent papers have addressed the problem of identifying weak sharp minima of order two. Bonnans and Ioffe have established characterizations and sufficient conditions for these minimizers for problems in which the objective and constraint functions are either c^2 convex functions [3] or pointwise maxima of c^2 functions [4]. Necessary conditions not requiring the data to be twice differentiable have been derived by the author in [22], and characterizations and sufficient conditions under minimal smoothness hypotheses on the objective and constraint functions have been investigated by Studniarski and the author in [21].

In the present paper, a sequel to [21], we present additional sufficient conditions for weak sharp minima of order two and explore some of their applications to sensitivity

analysis for problem P(u). We begin in §2 by proving sufficient
conditions for weak sharp minimality of order two in problem
P(0). These conditions differ from those presented in [21] in
two ways. For one thing, they are weak sufficient conditions,
while those in [21] (e.g., [21, Theorem 3.4]) are strong
sufficient conditions. More importantly, they involve a larger
set S than those used in [21], producing conditions that are more
often satisfied (see Example 2.6).

Since many theorems on sensitivity and stability in
nonlinear programming include a sufficient condition for strict
minimality of order two among their hypotheses, Theorem 2.4
suggests an interesting project: find extended versions of these
theorems for problems with weak sharp minima of order two, using
hypotheses like those in Theorem 2.4. In §3, we present a case
study from this overall project, showing that for programs with
C data and right-hand side perturbations of the constraints, the
results of Auslender and Cominetti [1, Theorem 1] on first- and
second-order directional derivatives of v can be extended to
problems with weak sharp minima.

We conclude this section by setting some notation. For a
set $S \subset \mathbb{R}^n$, we denote the *closure of S* by cl S and the *boundary
of S* by bd S. For a sequence $\{x_j\} \subset \mathbb{R}^n$, we will use the notation
$x_j \to_S x$ to indicate that $\{x_j\}$ converges to x with each $x_j \in S$.

Let $\langle \cdot, \cdot \rangle$ denote the Euclidean inner product on $\mathbb{R}^n$. If a
function $f : \mathbb{R}^n \to \bar{\mathbb{R}}$ is finite at $x \in \mathbb{R}^n$, we say that it is
strictly differentiable at x [5] if there exists a linear mapping
$\nabla f(x) : \mathbb{R}^n \to \mathbb{R}$ such that for all $y \in \mathbb{R}^n$,

$$\lim_{(w,v,t) \to (x,y,0^+)} (f(w + tv) - f(w))/t = \langle \nabla f(x), y \rangle.$$

We will say that f is C^2 at x if f is twice differentiable on a
neighborhood of x and the second derivative function $\nabla^2 f$ is
continuous at x.

2   SUFFICIENT CONDITIONS FOR WEAK SHARP MINIMA

In this section, we present sufficient conditions for weak sharp
minima of order two in problem P(0), similar to those derived in
[21]. We will make use of an important geometric concept in
these sufficient conditions: the *normal cone* of Mordukhovich,
also known as the *approximate normal cone* or *limiting proximal
normal cone* [15-16, 11, 6].

DEFINITION 2.1. Let $S \subset \mathbb{R}^n$ be nonempty.

(a) For $x \in \mathbb{R}^n$, define

$$P(S,x) := \{ w \in cl\ S : \|x - w\| = dist(x,S) \}.$$

(b) Let $\bar{x} \in \text{cl } S$. The *normal cone* to S at $\bar{x}$ is defined by

$$N(S,\bar{x}) := \{y \mid \exists \ \{y_j\} \rightarrow y, \ \{x_j\} \rightarrow \bar{x}, \ \{t_j\} \subset (0,+\infty), \ \{s_j\} \subset \mathbb{R}^n$$
$$\text{with} \quad s_j \in P(S,x_j) \text{ and } y_j = (x_j - s_j)/t_j \ \}.$$

The properties of $N(S,\bar{x})$ are discussed in [6, 11, 16] and their references. We mention here in particular that if S is a convex set, then $N(S,\bar{x})$ is the usual normal cone of convex analysis.

Another concept we will use is a special type of directional derivative.

DEFINITION 2.2. Let S be a nonempty closed subset of $\mathbb{R}^n$, and let $f:\mathbb{R}^n \rightarrow \bar{\mathbb{R}}$ be finite-valued on S. For $x \in \text{bd } S$, $y \in \mathbb{R}^n$, we define

$$\underline{d}_S^2 f(x;y) := \lim_{\substack{s \rightarrow_{\text{bd } S} x \\ (t,v) \rightarrow (0^+,y)}} \inf \quad (f(s + tv) - f(s))/t^2,$$

where "$s \rightarrow_{\text{bd } S} x$" means that $s \rightarrow x$ with each $s \in \text{bd } S$, and (x,y) is an allowable choice of (s,v).

The directional derivative $\underline{d}_S^2 f(x;\cdot)$ was introduced in [21] for the purpose of formulating sufficient conditions for weak sharp minimality. When $S = \{x\}$ or x is an isolated point of bd S, $\underline{d}_S^2 f(x;y)$ can be written more simply as

$$\underline{d}^2 f(x;y) := \lim_{(t,v) \rightarrow (0^+,y)} \inf \ (f(x + tv) - f(x))/t^2,$$

a directional derivative used in the study of strict local minima in [20, 22]. If f is sufficiently smooth, $\underline{d}_S^2 f(x;y)$ reduces to an even more familiar object. The following fact is a simple consequence of Taylor's formula.

PROPOSITION 2.3. Let $f:\mathbb{R}^n \rightarrow \mathbb{R}$ be finite and c^2 at $x \in \text{bd } S$. If there exists $\varepsilon > 0$ with $\nabla f(s) = 0$ for all $s \in \text{bd } S \cap B(x,\varepsilon)$, then for all $y \in \mathbb{R}^n$, $\underline{d}_S^2 f(x;y) = \nabla^2 f(x)(y,y)/2$.

In this section, we will assume that the objective and constraint functions in P(0) are strictly differentiable at some $\bar{x} \in \text{bd } \Omega(0)$. For $x \in R(0)$, we single out the set of indices of binding constraints

$$I(x) := \{i \in J \mid g_i(x) = 0\};$$

and for $x \in bd\ \Omega(0)$, the index set

$$I^*(x) := \{i \in I(x) \mid \exists\ \delta > 0 \text{ with } i \in I(y)\ \forall\ y \in B(x,\delta) \cap bd\ \Omega(0)\}.$$

For $\lambda \in \mathbb{R}^m$, we define the *Lagrangian function*

$$L_\lambda(x) := f(x) + \Sigma_{i \in J}\ \lambda_i g_i(x) + \Sigma_{i \in L}\ \lambda_i h_i(x).$$

We will usually assume that λ is an element of the *multiplier set*

$$\Lambda^*(\bar{x}) := \{\lambda \in \mathbb{R}^m \mid \nabla L_\lambda(\bar{x}) = 0,\ \lambda_i \geq 0\ \forall i \in J,\ \lambda_i = 0\ \forall i \in J \setminus I^*(\bar{x})\}.$$

We also define the *cone of critical directions*

$$D^*(\bar{x}) := \{y \mid \langle \nabla f(\bar{x}), y \rangle \leq 0,$$
$$\langle \nabla g_i(\bar{x}), y \rangle \leq 0\ \forall\ i \in I^*(\bar{x}),\ \langle \nabla h_i(\bar{x}), y \rangle = 0\ \forall\ i \in L\ \};$$

and the set

$$S(\bar{x}) := \{x \mid g_i(x) = 0,\ i \in I^*(\bar{x}),\ h_i(x) = 0,\ i \in L,\ v(0) = f(x)\ \},$$

which will serve as the set S in our sufficiency theorem.
    We now give sufficient conditions for weak sharp minimality
of order two in P(0).

THEOREM 2.4. Let f, g_i, $i \in J$, h_i, $i \in L$, be strictly differentiable at $\bar{x} \in bd\ \Omega(0)$, and let $S = S(\bar{x})$. Suppose that for each
$y \in (N(S,\bar{x}) \cap D^*(\bar{x})) \setminus \{0\}$,

$$\sup_{\lambda \in \Lambda^*(\bar{x})}\ d^2_S L_\lambda(\bar{x};y) > 0. \qquad (2.1)$$

Then $\bar{x}$ is a weak sharp local minimizer of order two for P(0).

Proof: We note first that since f, g_i, and h_i are strictly
differentiable at $\bar{x}$, they are continuous at $\bar{x}$, and so in fact $\bar{x} \in$

$\Omega(0)$. If $\bar{x}$ is not a weak sharp local minimizer of order two for P(0), there exists a sequence $\{x_j\} \subset R(0)$ such that $\{x_j\} \longrightarrow \bar{x}$ and

$$f(x_j) - f(\bar{x}) < \text{dist}^2(x_j,S)/j.$$

Since each $x_j \notin S$, there exists $s_j \in \text{bd } S$ with $s_j \in P(S,x_j)$. Let $t_j := \|x_j - s_j\|$, and let $y_j := (x_j - s_j)/t_j$. Taking a subsequence if necessary, we may assume that $\{y_j\} \longrightarrow y$ for some $y \neq 0$. Then $y \in N(S,\bar{x})$. By definition of $S(\bar{x})$,

$$g_i(x_j) - g_i(s_j) \leq 0 \quad \forall \ i \in I^*(\bar{x}),$$

so that

$$\langle \nabla g_i(\bar{x}),y \rangle = \lim_{j \longrightarrow \infty} (g_i(x_j) - g_i(s_j))/t_j \leq 0 \quad \forall \ i \in I^*(\bar{x}).$$

Similarly,

$$\langle \nabla h_i(\bar{x}),y \rangle = 0 \ \forall \ i \in L,$$

and

$$\langle \nabla f(\bar{x}),y \rangle = \lim_{j \longrightarrow \infty} (f(x_j) - f(s_j))/t_j \leq \lim_{j \longrightarrow \infty} t_j/j = 0.$$

Hence $y \in D^*(\bar{x})$.

Now let $\lambda \in \Lambda^*(\bar{x})$. Then for j large enough,

$$(L_\lambda(x_j) - L_\lambda(s_j))/t_j{}^2$$
$$= (f(x_j) - f(s_j) + \Sigma_{i \in I^*(\bar{x})} \lambda_i(g_i(x_j) - g_i(s_j)))/t_j{}^2$$
$$\leq (f(x_j) - f(s_j))/t_j{}^2 \leq 1/j,$$

and so

$$d_S^2 L_\lambda(\bar{x};y) \leq \liminf_{j \longrightarrow \infty} (L_\lambda(x_j) - L_\lambda(s_j))/t_j{}^2 \leq 0.$$

Therefore

$$\sup_{\lambda \in \Lambda^*(\bar{x})} d_S^2 L_\lambda(\bar{x};y) \leq 0,$$

contradicting (2.1).   □

The sufficient conditions in Theorem 2.4 are *weak sufficient conditions*, generalizing those given, for example, in [9, Theorem 2.3]. If the objective and constraint functions are C^2 at $\bar{x}$, then Proposition 2.3 and Theorem 2.4 combine to yield the following result:

COROLLARY 2.5. Let f, g_i, $i \in J$, h_i $i \in L$, be C^2 at $\bar{x} \in$ bd $\Omega(0)$, and let $S = S(\bar{x})$. Assume that there exists $\delta > 0$ such that

$$\Lambda := \bigcap_{s \in \text{ bd } S \cap B(\bar{x},\delta)} \Lambda^*(s) \neq \emptyset;$$

and that for each nonzero $y \in N(S,\bar{x}) \cap D^*(\bar{x})$,

$$\sup_{\lambda \in \Lambda} \nabla^2 L_\lambda(\bar{x})(y,y) > 0. \tag{2.2}$$

Then $\bar{x}$ is a weak sharp local minimizer of order two for $P(0)$.

The assumption that $\Lambda \neq \emptyset$ in Corollary 2.5 holds, in particular, whenever there exists $\delta > 0$ such that $\nabla f(s)$, $\nabla g_i(s)$, $i \in I^*(\bar{x})$, and $\nabla h_i(s)$, $i \in L$, remain constant for all $s \in$ bd $S \cap B(\bar{x},\delta)$. We illustrate with an example.

EXAMPLE 2.6. In problem $P(0)$, let $J = \{1,2\}$, $L = \emptyset$, and define f, g_1, $g_2 : \mathbb{R}^2 \longrightarrow \mathbb{R}$ by $f(x,y) = \sin \pi x$, $g_1(x,y) = x - 1$, and $g_2(x,y) = (x - 1)^2 + y^2 - 1$. Then $\Omega(0) = \{(1,y) \mid -1 \leq y \leq 1\} \cup \{(0,0)\}$. For $\bar{x} = (0,0)$, one can readily see that $I^*(\bar{x}) = \{2\}$, $S(\bar{x}) = S = \{(0,0), (1,\pm 1)\}$, $N(S,\bar{x}) = \mathbb{R}^2$, $D^*(\bar{x}) = \{0\} \times \mathbb{R}$, and $\Lambda^*(\bar{x}) = \{(0,\pi/2)\}$. At this $\bar{x}$, Corollary 2.5 reduces to the second-order sufficient condition of [9, Theorem 2.3]; and since

$$\nabla^2 L_\lambda(\bar{x})((0,y_2),(0,y_2)) = \pi y_2^{\,2} > 0$$

for $\lambda = (0,\pi/2)$ and all $(0,y_2) \in D^*(\bar{x})$, we conclude that $(0,0)$ is a strict local minimizer of order two for $P(0)$.

On the other hand, for any $\bar{x} \in \{(1,y) \mid -1 \leq y \leq 1\}$, we have $I^*(\bar{x}) = \{1\}$, $S(\bar{x}) = S = \{(1,y) \mid y \in \mathbb{R}\}$, $N(S,\bar{x}) = \mathbb{R} \times \{0\}$, $D^*(\bar{x}) = \{0\} \times \mathbb{R}$, and $\Lambda^*(\bar{x}) = \{(\pi,0)\}$, so that (2.2) is satisfied vacuously. Hence all such $\bar{x}$ are weak sharp local minima of order two for $P(0)$.

REMARK 2.7. Observe that $S = S(\bar{x})$ is not necessarily a subset of
$\Omega(0)$, in contrast to the formulation in [21]. In Example 2.6,
our choice of S allows (2.2) to be satisfied at $(1,-1)$ and $(1,1)$.
If instead we had chosen $S = \Omega(0)$, then $N(S,(1,1)) = \mathbb{R} \times [0,\infty)$
and $N(S,(1,-1)) = \mathbb{R} \times (-\infty,0]$, and (2.2) would not hold at these
points.

     We point out that choosing $S \subset \Omega(0)$, as in [21], seems
necessary in certain applications--for example, the exact penalty
results of [21, §4]. For applications requiring (2.2) to hold at
every element of $\Omega(0)$, on the other hand, choosing $S = S(\bar{x})$ seems
to be preferable.

     Of course, Theorem 2.4 also implies *strong sufficiency*
theorems like the following one.

COROLLARY 2.8. Let f, g_i, $i \in J$, h_i, $i \in L$, be strictly differentiable at $\bar{x} \in bd\ \Omega(0)$, and let $S = S(\bar{x})$. Suppose that there
exists $\lambda \in \Lambda^*(\bar{x})$ such that

$$\underline{d}_S^2 L_\lambda(\bar{x};y) > 0 \quad \forall\ y \in (N(S,\bar{x}) \cap D^*(\bar{x}))\backslash\{0\}.$$

Then $\bar{x}$ is a weak sharp local minimizer of order two for P(0).

3   DIRECTIONAL DERIVATIVES OF THE VALUE FUNCTION

In this section, we illustrate the application of sufficient
conditions for weak sharp minima of order two by revisiting one
of the major problems of mathematical programming with data
perturbations: find upper bounds on

$$v'_+(0;u) := \limsup_{t \to 0^+} (v(tu) - v(0))/t,$$

the *upper Dini directional derivative of v* at 0; lower bounds on

$$v'_-(0;u) := \liminf_{t \to 0^+} (v(tu) - v(0))/t,$$

the *lower Dini directional derivative of v* at 0; and conditions
implying that the directional derivative

$$v'(0;u) := \lim_{t \to 0^+} (v(tu) - v(0))/t$$

exists.
     There is an extensive literature on these questions and
their counterparts for second-order directional derivatives of v
(see for example [7, 9, 14] and their references). Some studies

of v′ require only first-order information about f, g_i, and h_i (e.g., [5; 7, §2.3; 10; 23]), while others take advantage of second-order information to obtain stronger conclusions ([1; 7, Chapter 3; 9]). Here we pattern our presentation after [1] and give "weak sharp" analogues of Proposition 2 and Theorem 1 of that paper. As in [1], we will assume throughout this section that f, g_i, and h_i are C^2 functions on $\mathbb{R}^n$.

We first observe that the derivation of upper bounds for $v'_+(0;u)$ is the easiest part of this problem, requiring no second-order sufficient optimality condition. Typical results in this area state that if a Mangasarian-Fromovitz-type constraint qualification holds at some $\bar{x} \in \Omega(0)$, then $v'_+(0;u)$ is bounded above by a support function of the multiplier set

$$\Lambda(\bar{x}) := \{ \lambda \in \mathbb{R}^m |\ \nabla L_\lambda(\bar{x}) = 0,\ \lambda_i \geq 0\ \forall i \in J,\ \lambda_i = 0\ \forall i \in J \backslash I(\bar{x})\ \}.$$

Proofs of these results generally rely on some version of the implicit function theorem. The presentation in [1] fits this basic description, making use of a *directional constraint qualification*.

DEFINITION 3.1. Let $\bar{x} \in \Omega(0)$, $u \in \mathbb{R}^m$. We say that condition [MF_u] holds at $\bar{x}$ if

(a) $\nabla h_i(\bar{x})$, $i \in L$, are independent.

(b) $\exists\ y \in \mathbb{R}^n$ with $\langle \nabla g_i(\bar{x}), y \rangle < u_i\ \forall i \in I(\bar{x})$,
$$\langle \nabla h_i(\bar{x}), y \rangle = u_i\ \forall i \in L.$$

Auslender and Cominetti obtain the following upper bound on $v'_+(0;u)$ in [1, Lemma 2 and Corollary 1]:

PROPOSITION 3.2. [1] If [MF_u] holds at $\bar{x} \in \Omega(0)$, then

$$v'_+(0;u) \leq \sup_{\lambda\ \in\ \Lambda(\bar{x})}\ \langle -\lambda, u \rangle. \qquad (3.1)$$

In particular, $\limsup_{t \to 0^+} v(tu) \leq v(0)$, and so whenever $\{t_j\} \to 0^+$ and $\{x_j\} \to_{\Omega(t_ju)} x_\infty$, it follows that $x_\infty \in \Omega(0)$.

The development of conditions under which $v'_+(0;u) = v'_-(0;u)$ is more complicated. Second-order sufficient optimality conditions are among those typically employed to ensure the existence of $v'(0;u)$. We next state sufficient conditions for a weak sharp minimizer of order two that are suitable for this purpose.

DEFINITION 3.3. Let $\bar{x} \in$ bd $\Omega(0)$, and let $S := S(\bar{x})$. For $u \in \mathbb{R}^m$, define

$$\Lambda_u^*(\bar{x}) := \{\bar{\lambda} \in \Lambda^*(\bar{x}): <-\bar{\lambda},u> = \sup_{\lambda \in \Lambda(\bar{x})} <-\lambda,u>\};$$

and for $\delta > 0$, define

$$\Lambda_\delta^*(\bar{x}) := \Lambda_u^*(\bar{x}) \cap \bigcap_{s \in bd\ S \cap B(\bar{x},\delta)} \Lambda^*(s).$$

We say that $[SOSC_u]$ holds at $\bar{x}$ if there exists $\delta > 0$ such that $\Lambda_\delta^*(\bar{x}) \neq \emptyset$ and for each nonzero $y \in N(S,\bar{x}) \cap D^*(\bar{x})$,

$$\sup_{\lambda \in \Lambda_\delta^*(\bar{x})} \nabla^2 L_\lambda(\bar{x})(y,y) > 0. \qquad (3.2)$$

It follows from Corollary 2.5 that condition $[SOSC_u]$ implies that $\bar{x}$ is a weak sharp local minimizer of order two for $P(0)$. We can use this condition to prove a "weak sharp" version of [1, Proposition 2].

PROPOSITION 3.4. Let $\{t_j\} \longrightarrow 0^+$ and $\{x_j\} \longrightarrow_{\Omega(t_j u)} \bar{x}$ with $\bar{x} \in \Omega(0)$, and let $s_j \in P(S,x_j)$, where $S = S(\bar{x})$. If $[MF_u]$ and $[SOSC_u]$ hold at $\bar{x}$, then

$$\limsup_{j \longrightarrow \infty} \|x_j - s_j\|/t_j < +\infty. \qquad (3.3)$$

Proof: If (3.3) does not hold, there exist $\{t_j\} \longrightarrow 0^+$, $\{x_j\} \in \Omega(t_j u)$ with $\{x_j\} \longrightarrow \bar{x} \in \Omega(0)$, $s_j \in P(S,x_j)$, and $y \in \mathbb{R}^n$ such that

$$\|x_j - s_j\|/t_j \longrightarrow +\infty$$

and

$$y_j := (x_j - s_j)/\|x_j - s_j\| \longrightarrow y \in N(S,\bar{x})\setminus\{0\}.$$

Then for all $i \in I^*(\bar{x})$,

$$<\nabla g_i(\bar{x}),y> = \lim_{j \longrightarrow \infty} (g_i(x_j) - g_i(s_j))/\|x_j - s_j\|$$

$$\leq \lim_{j \longrightarrow \infty} t_j u/\|x_j - s_j\| = 0;$$

and similarly,

$$\langle \nabla h_i(\bar{x}), y \rangle = 0 \quad \forall \ i \in L.$$

Since $x_j \in \Omega(t_j u)$ and $s_j \in S(\bar{x})$, we have $v(t_j u) = f(x_j)$ and $v(0)$ $= f(s_j)$. In addition, (3.1) holds because $[MF_u]$ is satisfied, so there exists $\varepsilon > 0$ such that for j large enough,

$$v(t_j u) - v(0) \leq t_j \varepsilon.$$

It follows that

$$\langle \nabla f(\bar{x}), y \rangle = \lim_{j \to \infty} (f(x_j) - f(s_j))/\|x_j - s_j\|$$
$$\leq \lim_{j \to \infty} t_j \varepsilon / \|x_j - s_j\| = 0,$$

and so $y \in D^*(\bar{x})$.

Now let $\delta > 0$ be such that $\Lambda_\delta^*(\bar{x}) \neq \varnothing$, and let $\lambda \in \Lambda_\delta^*(\bar{x})$ satisfy

$$\nabla^2 L_\lambda(\bar{x})(y, y) > 0.$$

Then

$$L_\lambda(x_j) - L_\lambda(s_j) = f(x_j) - f(s_j) + \sum_{i \in I^*(\bar{x})} \lambda_i (g_i(x_j) - g_i(s_j))$$
$$= v(t_j u) - v(0) + \sum_{i \in I^*(\bar{x})} \lambda_i g_i(x_j)$$
$$\leq v(t_j u) - v(0) + t_j \langle \lambda, u \rangle.$$

Since $\lambda \in \Lambda_u^*(\bar{x})$, Lemma 2 and Proposition 1 of [1] imply that there exists $\alpha \in \mathbb{R}$ such that for all j sufficiently large,

$$v(t_j u) - v(0) \leq -t_j \langle \lambda, u \rangle + \alpha \ t_j^2,$$

and thus

$$L_\lambda(x_j) - L_\lambda(s_j) \leq \alpha \ t_j^2.$$

Hence Proposition 2.3 gives

$$\nabla^2 L_\lambda(\bar{x})(y, y)/2 = \lim_{j \to \infty} (L_\lambda(x_j) - L_\lambda(s_j))/\|x_j - s_j\|^2$$
$$\leq \lim_{j \to \infty} t_j^2 \alpha / \|x_j - s_j\|^2 = 0,$$

a contradiction. $\qquad \square$

By means of Proposition 3.4, we can establish conditions under which v'(0;u) exists. We will make use of a *uniform boundedness* hypothesis, as in [1], in addition to [MF$_u$] and [SOSC$_u$].

DEFINITION 3.5. We will say that hypothesis [H$_u$] holds if there exist $\varepsilon > 0$, $r > 0$ such that $\emptyset \neq \Omega(tu) \subset B(0,r)$ $\forall$ $t \in [0,\varepsilon]$.

We now formulate a "weak sharp" version of [1, Theorem 1(a)]:

THEOREM 3.6. Assume that [H$_u$] holds, and suppose that [MF$_u$], [SOSC$_u$], and $\Lambda(\bar{x}) = \Lambda^*(\bar{x})$ are satisfied for each $\bar{x} \in \Omega(0)$. Then v'(0;u) exists and

$$v'(0;u) = \min_{\bar{x} \in \Omega(0)} \max_{\lambda \in \Lambda(\bar{x})} <-\lambda, u>. \qquad (3.4)$$

Proof: By Proposition 3.2,

$$v'_+(0;u) \leq \inf_{\bar{x} \in \Omega(0)} \sup_{\lambda \in \Lambda(\bar{x})} <-\lambda, u>.$$

By hypothesis [H$_u$], we may choose $\{t_j\} \rightarrow 0^+$ and $r > 0$ such that

$$(v(t_j u) - v(0))/t_j \rightarrow v'_-(0;u)$$

and

$$\emptyset \neq \Omega(t_j u) \subset B(0,r) \quad \forall \ j.$$

Let $x_j \in \Omega(t_j u)$. Since $\{x_j\}$ is bounded, we may assume, taking a subsequence if necessary, that $\{x_j\} \rightarrow x_\infty$. As mentioned in Proposition 3.2, $x_\infty \in \Omega(0)$. Next choose $s_j \in P(S(x_\infty), x_j)$. Then by Proposition 3.4, the sequence $\{(x_j - s_j)/t_j\}$ is bounded.

Now let $\lambda \in \Lambda^*(x_\infty)$. Then $L_\lambda(s_j) = f(s_j) = v(0)$, and

$$L_\lambda(x_j) \leq f(x_j) + t_j <\lambda, u> = v(t_j u) + t_j <\lambda, u>.$$

Putting these facts together, we obtain

$$v(t_j u) - v(0) \geq L_\lambda(x_j) - t_j <\lambda, u> - L_\lambda(s_j)$$
$$= <\nabla L_\lambda(s_j), x_j - s_j> - t_j <\lambda, u> + o(\|x_j - s_j\|);$$

where as usual, $\lim_{j \to \infty} o(\|x_j - s_j\|)/\|x_j - s_j\| = 0$. Thus

$$(v(t_j u) - v(0))/t_j$$
$$\geq \langle \nabla L_\lambda(s_j), (x_j - s_j)/t_j \rangle - \langle \lambda, u \rangle + o(\|x_j - s_j\|)/t_j.$$

Since $\{s_j\} \to x_\infty$, $\lambda \in \Lambda^*(x_\infty)$, and $\{(x_j - s_j)/t_j\}$ is bounded, it follows that

$$\lim_{j \to \infty} \langle \nabla L_\lambda(s_j), (x_j - s_j)/t_j \rangle = 0$$

and

$$\lim_{j \to \infty} o(\|x_j - s_j\|)/t_j = 0.$$

Hence

$$v'_-(0;u) = \lim_{j \to \infty} (v(t_j u) - v(0))/t_j \geq -\langle \lambda, u \rangle.$$

Finally, since $\Lambda(\bar{x}) = \Lambda^*(\bar{x})$ for all $\bar{x} \in \Omega(0)$, we have

$$v'_-(0;u) \geq \inf_{\bar{x} \in \Omega(0)} \sup_{\lambda \in \Lambda^*(\bar{x})} \langle -\lambda, u \rangle$$
$$= \inf_{\bar{x} \in \Omega(0)} \sup_{\lambda \in \Lambda(\bar{x})} \langle -\lambda, u \rangle$$
$$\geq v'_+(0;u).$$

Therefore $v'(0;u)$ exists and (3.4) holds.   □

Let us now apply Theorem 3.6 to the problem in Example 2.6.

EXAMPLE 3.7. In Example 2.6, $[MF_0]$ holds at all $\bar{x} \in \Omega(0)$, so $[MF_u]$ also holds at all $\bar{x} \in \Omega(0)$ for each $u = (u_1, u_2) \in \mathbb{R}^2$. Since $\Lambda^*(\bar{x})$ is a singleton for every $\bar{x} \in \Omega(0)$, $[SOSC_u]$ coincides with (2.2), so $[SOSC_u]$ is satisfied for all $u \in \mathbb{R}^2$. In addition, $H(u)$ holds for each $u \in \mathbb{R}^2$. By Theorem 3.6, $v'(0;u)$ exists and

$$v'(0;(u_1,u_2)) = \min \{-\pi u_2/2, \ -\pi u_1\}.$$

In [1], Auslender and Cominetti also calculate an upper bound for the *upper second-order Dini derivative*

$$v''_+(0;u) := \limsup_{t \to 0^+} 2(v(tu) - v(0) - tv'(0;u))/t^2,$$

and give conditions sufficient to guarantee that the *lower
second-order Dini derivative*

$$v''_-(0;u) := \lim \inf_{t \to 0^+} 2(v(tu) - v(0) - tv'(0;u))/t^2$$

attains this bound, so that the second-order directional derivative

$$v''(0;u) := \lim_{t \to 0^+} 2(v(tu) - v(0) - tv'(0;u))/t^2$$

exists. We next develop a weak sharp analogue of their results.
Again, the proof of an upper bound on $v''_+(0;u)$ is the easiest
part of the process, involving no second-order sufficiency condition. The following upper bound is obtained via Proposition 1
and Lemmas 3 and 4of [1], as shown in the proof of [1, Theorem
1(b)]:

PROPOSITION 3.8. For $\bar{x} \in \Omega(0)$ and $u \in \mathbb{R}^n$, define

$$\Gamma_u^*(\bar{x}) := \{y| <\nabla g_i(\bar{x}),y> \le u_i, \ i \in I(\bar{x}), \ <\nabla h_i(\bar{x}),y> = u_i, \ i \in L,$$

$$<\nabla f(\bar{x}),y> = \sup_{\lambda \in \Lambda(\bar{x})} <-\lambda,u> \}$$

and

$$\Lambda_u(\bar{x}) := \{\bar{\lambda} \in \Lambda(\bar{x}): <-\bar{\lambda},u> = \sup_{\lambda \in \Lambda(\bar{x})} <-\lambda,u>\}.$$

If $[MF_u]$ holds at $\bar{x}$ and $\Lambda(\bar{x}) \ne \varnothing$, then

$$v''_+(0;u) \le \inf_{y \in \Gamma_u^*(\bar{x})} \sup_{\lambda \in \Lambda_u(\bar{x})} \nabla^2 L_\lambda(\bar{x})(y,y) < +\infty. \qquad (3.5)$$

We now formulate and prove an analogue of [1, Theorem 1(b)]
for problems with weak sharp minima of order two.

THEOREM 3.9. Assume that $[H_u]$ holds; that $[MF_u]$, $[SOSC_u]$, and
$\Lambda(\bar{x}) = \Lambda^*(\bar{x})$ are satisfied for each $\bar{x} \in \Omega(0)$; and that for each
$\bar{x} \in \Omega(0)$, there exists $\delta > 0$ such that $\Lambda_u(\bar{x}) = \Lambda_\delta^*(\bar{x})$. Then
$v''(0;u)$ exists and

$$v''(0;u) = \min_{\bar{x} \in \Omega_u(0)} \min_{y \in \Gamma_u^*(\bar{x})} \max_{\lambda \in \Lambda_u(\bar{x})} \nabla^2 L_\lambda(\bar{x})(y,y),$$

where

$$\Omega_u(0) := \{x \in \Omega(0)| \ v'(0;u) = \max_{\lambda \in \Lambda(x)} <-\lambda,u> \}.$$

Proof: As in the proof of Theorem 3.6, there exist sequences $\{t_j\} \to 0^+$ with

$$2(v(t_j u) - v(0) - t_j v'(0;u))/t_j^2 \to v''_-(0;u);$$

$x_j \in \Omega(t_j u)$ with $\{x_j\} \to x_\infty$ for some $x_\infty \in \Omega(0)$; and $s_j \in P(S(x_\infty), x_j)$ such that $\{(x_j - s_j)/t_j\}$ is bounded. Let $y_j := (x_j - s_j)/t_j$. Taking a subsequence if necessary, we may assume that $\{y_j\} \to y$ for some $y \in \mathbb{R}^n$. We will show that $y \in \Gamma_u^*(x_\infty)$.

To that end, we begin by calculating that

$$
\begin{aligned}
\langle \nabla f(x_\infty), y \rangle &= \lim_{j \to \infty} (f(x_j) - f(s_j))/t_j \\
&= \lim_{j \to \infty} (v(t_j u) - v(0))/t_j = v'(0;u);
\end{aligned}
$$

that

$$\langle \nabla h_i(x_\infty), y \rangle = \lim_{j \to \infty} (h_i(x_j) - h_i(s_j))/t_j = u_i \quad \forall\, i \in L;$$

and that

$$\langle \nabla g_i(x_\infty), y \rangle = \lim_{j \to \infty} (g_i(x_j) - g_i(s_j))/t_j \leq u_i \quad \forall\, i \in I^*(x_\infty).$$

Next, we note that by [1, Corollary 1],

$$
\begin{aligned}
v'(0;u) \leq \inf \{\langle \nabla f(x_\infty), w \rangle | \ &\langle \nabla g_i(x_\infty), w \rangle \leq u_i, \ i \in I(x_\infty), \\
&\langle \nabla h_i(x_\infty), w \rangle = u_i, \ i \in L \} \\
= \inf \{\langle \nabla f(x_\infty), w \rangle | \ &\langle \nabla g_i(x_\infty), w \rangle \leq u_i, \ i \in I^*(x_\infty), \\
&\langle \nabla h_i(x_\infty), w \rangle = u_i, \ i \in L \},
\end{aligned}
$$

since $\Lambda(x_\infty) = \Lambda^*(x_\infty)$. We then deduce that $y \in \Gamma_u^*(x_\infty)$. Moreover, we have by linear programming duality [1, Lemma 2] that

$$
\begin{aligned}
\max_{\lambda \in \Lambda(x_\infty)} \langle -\lambda, u \rangle = \inf \{\langle \nabla f(x_\infty), w \rangle | \ &\langle \nabla g_i(x_\infty), w \rangle \leq u_i, \\
&i \in I(x_\infty), \ \langle \nabla h_i(x_\infty), w \rangle = u_i, \ i \in L \},
\end{aligned}
$$

and so $x_\infty \in \Omega_u(0)$.

Now let $\lambda \in \Lambda_u(x_\infty)$. Since $\lambda \in \Lambda_\delta^*(x_\infty)$ for some $\delta > 0$, $\nabla L_\lambda(s_j) = 0$ for all j sufficiently large, and we have

$$v(t_j u) - v(0) - t_j v'(0;u)$$

$$\geq L_\lambda(x_j) - t_j<\lambda,u> - L_\lambda(s_j) + t_j<\lambda,u>$$

$$= <\nabla L_\lambda(s_j),x_j - s_j> + \nabla^2 L_\lambda(s_j)(x_j - s_j,x_j - s_j)/2 + o(\|x_j - s_j\|^2)$$

$$= \nabla^2 L_\lambda(s_j)(x_j - s_j,x_j - s_j)/2 + o(\|x_j - s_j\|^2).$$

Thus

$$v_-''(0;u) = \lim_{j \to \infty} 2(v(t_j u) - v(0) - t_j v'(0;u))/t_j^2$$

$$\geq \nabla^2 L_\lambda(x_\infty)(y,y),$$

and we conclude by (3.5) that

$$v_-''(0;u) \geq \inf_{\bar{x} \in \Omega_u(0)} \inf_{y \in \Gamma_u^*(\bar{x})} \sup_{\lambda \in \Lambda_u(\bar{x})} \nabla^2 L_\lambda(\bar{x})(y,y)$$

$$\geq v_+''(0;u).$$

Therefore $v''(0;u)$ exists and the conclusion follows.   □

    In Theorem 3.9, the assumption that $\Lambda(\bar{x}) = \Lambda^*(\bar{x})$ is satisfied, of course, whenever $I(\bar{x}) = I^*(\bar{x})$. However, this condition can also be satisfied in other situations. For instance, if we know (perhaps on the basis of an appropriate constraint qualification) that $\Lambda(\bar{x})$ is a singleton, and we know in addition that $\Lambda^*(\bar{x}) \neq \emptyset$, it will then follow that $\Lambda(\bar{x}) = \Lambda^*(\bar{x}) = \Lambda_u(\bar{x})$, since in general $\Lambda^*(\bar{x}) \subset \Lambda(\bar{x})$.
    We conclude this section by revisiting Example 2.6.

EXAMPLE 3.10.   From the discussion in Examples 2.6 and 3.7, it is clear that the hypotheses of Theorem 3.9 are satisfied for the problem of Example 2.6.   Let us use Theorem 3.9 to calculate $v''(0;u)$ for some specific choices of u.   For $u = (0,1)$, Example 3.7 implies that $\Omega_u(0) = \{(0,0)\}$.   One can then verify that for $\bar{x} = (0,0)$, we have $\Gamma_u^*(\bar{x}) = \{-.5\} \times \mathbb{R}$, $\Lambda_u(\bar{x}) = \{(0,\pi/2)\}$.   So for $y = (-.5,y_2) \in \Gamma_u^*(\bar{x})$, $\nabla^2 L_\lambda(\bar{x})(y,y) = \pi/4 + \pi y_2^2$.   Therefore $v''((0,0);(0,1)) = \pi/4$ by Theorem 3.9.
    On the other hand, if $u = (1,0)$, we have

$$\Omega_u(0) = \{(1,x_2)| -1 \leq x_2 \leq 1\}.$$

For $\bar{x} \in \Omega_u(0)$, $\Lambda(\bar{x}) = \Lambda^*(\bar{x}) = \Lambda_u(\bar{x}) = \{(\pi, 0)\}$ and

$$\nabla^2 L_\lambda(\bar{x})(y,y) = 0 \quad \forall y \in \mathbb{R}^2,$$

so $v''((0,0);(1,0)) = 0$.

4 CONCLUSION

We have established sufficient conditions for weak sharp minima
of order two in nonlinear programming, generalizing the familiar
sufficient conditions for strict local minima of order two. In
addition, we have shown that such sufficient conditions can be
used to analyze first- and second-order directional derivatives
of the value function of a program with weak sharp minima of
order two. We are hopeful that other aspects of the theory of
sensitivity and stability in nonlinear programming can be
similarly extended.

REFERENCES

1. A. Auslender and R. Cominetti, First and second order
    sensitivity analysis of nonlinear programs under directional
    constraint qualification conditions, *Optimization*, 21:351-
    363 (1990).
2. B. Bank, J. Guddat, D. Klatte, B. Kummer, and K. Tammer,
    Non-Linear Parametric Optimization, Akademie Verlag, Berlin
    (1982).
3. J.F. Bonnans and A.D. Ioffe, Second-order sufficiency and
    quadratic growth for non isolated minima, *Operations
    Research*, 20: 801-817 (1995).
4. J.F. Bonnans and A.D. Ioffe, Quadratic growth and stability
    in convex programming problems with multiple solutions,
    Journal of Convex Analysis, 2: 41-57 (1995).
5. F.H. Clarke, *Optimization and Nonsmooth Analysis*, Wiley, New
    York (1983).
6. F.H. Clarke, *Methods of Dynamic and Nonsmooth Optimization*,
    SIAM Publications, Philadelphia (1989).
7. A.V. Fiacco, *Introduction to Sensitivity and Stability
    Analysis in Nonlinear Programming*, Academic Press, New York
    (1983).
8. A.V. Fiacco and G.P. McCormick, *Nonlinear Programming:
    Sequential Unconstrained Minimization Techniques*, Wiley, New
    York (1968).
9. J. Gauvin, *Theory of Nonconvex Programming*, Les Publications
    CRM, Université de Montréal, Montréal (1994).
10. J. Gauvin and J.M. Tolle, Differential stability in
    nonlinear programming, *SIAM Journal on Control and
    Optimization*, 15: 294-311 (1977).
11. A.D. Ioffe, Approximate subdifferentials and applications I:
    the finite-dimensional theory, *Transactions of the American
    Mathematical Society*, 281: 389-416 (1984).

12. A.D. Ioffe, On sensitivity analysis of nonlinear programs in Banach spaces: the approach via composite unconstrained optimization, *SIAM Journal on Optimization*, 4: 1-43 (1994).
13. D. Klatte, On quantitative stability for non-isolated minima, *Control and Cybernetics*, 23: 183-200 (1994).
14. E.S. Levitin, *Perturbation Theory in Mathematical Programming and its Applications*, Wiley, Chichester, England (1994).
15. B.S. Mordukhovich, Maximum principle in problems of time optimal control with nonsmooth constraints, *Journal of Applied Mathematics and Mechanics*, 40: 960-969 (1976).
16. B.S. Mordukhovich, Generalized differential calculus for nonsmooth and set-valued mappings, *Journal of Mathematical Analysis and Applications*, 183:250-288 (1994).
17. A. Shapiro, Perturbation theory of nonlinear programs when the set of solutions is not a singleton, *Applied Mathematics and Optimization*, 18: 215-229 (1988).
18. A. Shapiro, Perturbation analysis of optimization problems in Banach spaces, *Numerical Functional Analysis and Optimization*, 13: 97-116 (1992).
19. G. Still and M. Streng, Optimality conditions in smooth nonlinear programming, *Journal of Optimization Theory and Applications*, 90: 483-515 (1996).
20. M. Studniarski, Necessary and sufficient conditions for isolated local minima of nonsmooth functions, *SIAM Journal on Control and Optimization*, 24: 1044-1049 (1986).
21. M. Studniarski and D.E. Ward, Weak sharp minima: characterizations and sufficient conditions, preprint, Miami University, Oxford, Ohio (1996).
22. D.E. Ward, Characterizations of strict local minima and necessary conditions for weak sharp minima, *Journal of Optimization Theory and Applications*, 80: 551-571 (1994).
23. D.E. Ward, Dini derivatives of the marginal function of a non-Lipschitzian program, *SIAM Journal on Optimization*, 6: 198-211 (1996).

Index

Printed and bound by CPI Group (UK) Ltd, Croydon, CR0 4YY

21/10/2024

01777097-0017